Electronics in Advanced Research Industries

Electronics in Advanced Research Industries

Industry 4.0 to Industry 5.0 Advances

Alessandro Massaro

WILEY

IEEE PRESS

This edition first published 2022

Registered Offices
John Wiley & Sons, Inc., 111 River Street, Hoboken, NJ 07030, USA
John Wiley & Sons Ltd, The Atrium, Southern Gate, Chichester, West Sussex, PO19 8SQ, UK

Editorial Office
The Atrium, Southern Gate, Chichester, West Sussex, PO19 8SQ, UK

For details of our global editorial offices, customer services, and more information about Wiley products visit us at www.wiley.com.

Wiley also publishes its books in a variety of electronic formats and by print-on-demand. Some content that appears in standard print versions of this book may not be available in other formats.

Library of Congress Cataloging-in-Publication Data

Names: Massaro, Alessandro, 1974– author.
Title: Electronics in advanced research industries : industry 4.0 to industry 5.0 advances / Alessandro Massaro, Dyrecta Lab, Research Institute, Conversano (Ba), Italy.
Description: Hoboken, NJ, USA : Wiley, 2022. | Includes bibliographical references and index.
Identifiers: LCCN 2021028944 (print) | LCCN 2021028945 (ebook) | ISBN 9781119716877 (hardback) | ISBN 9781119716884 (adobe pdf) | ISBN 9781119716891 (epub)
Subjects: LCSH: Industry 4.0. | Automation.
Classification: LCC T59.6 .M37 2022 (print) | LCC T59.6 (ebook) | DDC 658.4/038028563–dc23
LC record available at https://lccn.loc.gov/2021028944
LC ebook record available at https://lccn.loc.gov/2021028945

Cover Design: Wiley
Cover Image: © raigvi/Shutterstock

Set in 9.5/12.5pt STIXTwoText by Straive, Chennai, India
Printed and bound by CPI Group (UK) Ltd, Croydon, CR0 4YY

C9781119716877_280921

To my family: Magda, Andrea, Adriano, and Peggy

Contents

Preface

Modern technologies in production systems open new approaches and concepts of industrial production. The digital Industry 4.0 upgrade provides new elements to control and manage production in all industry sectors. This upgrade allows to improve product quality, and in general the whole supply chain. The new digital technologies include hardware and software tools integrated in infrastructure oriented on the gain of digital knowledge. The fast dynamicity of the markets, the increase of the global competition between companies, and the unpredictable social and health events, imposes the need to think of a new concept of a production system based on full automatisms and self-adaptive processes, predicting production failures and product defects. In this context, the Industry 4.0 facilities can be furthermore upscaled to an intelligent control and actuation system of the production, characterizing the new Industry 5.0 scenario. The new facilities which contribute to Industry 5.0 passage are mainly based on artificial intelligence (AI) implementations in production and information systems, accomplishing predictive maintenance, failure prediction, defect classification, efficient robotic control and actuation, design optimization, testing improvements, and in general technological advances due to the possibility to quickly process data in each production stage. This book analyzes innovative production approaches, and the integration aspects of the AI in different industrial digital technologies, by enhancing specific functionalities. In innovative production systems, AI is fully integrated in information systems and covers cybersecurity, quality processes, business intelligence and intelligent production management. The innovative production is also related to new services associated with the introduction in the market of new technologies such as for the telemedicine sector, and in general for industrial diagnostics, where AI is also adopted for the improvement of inspection services. The main advantage of AI is the self-learning of the algorithms able to learn automatically from the same production data of companies. In an industrial upgrade, the implementation of sensor control and actuation based on intelligent feedback systems is especially important. In this scenario, AI algorithms can accomplish robotic movement, by automatically optimizing the machine parameter setting, by means of image and data processing. The correct use of AI is mainly based on the formulation of the algorithm, and on the dataset adopted to learn the related model. For each application there is an associated AI learning dataset which can be improved by big data systems. In particular, image processing and image segmentation approaches can be improved by AI, enhancing hidden information as defects, or adding new information about the performed production process. Another tool supporting the assembly in supply chains and the coordination of activities is augmented reality, which can be fully integrated in the information company infrastructure. An important step for a new concept of production is the upgrade of the information technology (IT) infrastructure automatically gaining the knowledge. Different IT architectures are proposed for different application fields to enhance technologies more suitable for a self-adaptive production providing decision support systems. A particular interesting topic for the innovative IT is the Internet of Things. The design and the development of an advanced IT infrastructure is the primary action to add for the upgrade in Industry 5.0. The AI concept is extended to the logic condition implementation, acting on signal processing, and on the use of simply electronic circuits representing these logics. The discussed methodologies allow to comprehend how it is possible to move on a competitive production based on the concept of "flexible" production and on new products based on advanced technologies on a micro- and nanoscale. In this scenario, companies working in manufacturing can switch dynamically the production on the new products, thus converting the production in innovative components, machines, materials, sensors, or devices. Following this orientation, this book proposes important approaches to automatize efficiently the new production, by analyzing highly advanced production tools based on nanotechnology. In this direction useful methodologies are analyzed to implement the production of high technology devices, such as reverse engineering and rapid prototyping by showing different examples useful to comprehend the methodologies to apply for an innovative production based on scientific and industrial research development. Particular attention is paid to the procedures to follow to produce a new device, to increase the company capacity to accelerate the industrialization process starting a new innovative prototype fabrication, and to basic approaches for the design modeling and testing. The optimization of the pre-industrialization process to perform is accompanied by a quick check of the basic properties of the new product to fabricate, and by the simultaneous support of the AI application improving analysis. In order to start and to develop a new research activity, also concerning advanced technologies, precise schemes must be followed. The discussed topics facilitate the understanding of the directions of the research for the production upscaling, just to apply the research activity. The last part of the book provides different elements useful for writing an industrial research project, and for the project management. This book deals with multidisciplinary topics including electronics, mechatronics, mechanics, and informatics. All the analyzed topics are useful to know; the key elements are indispensable and useful to move the production from Industry 4.0 to Industry 5.0.

Bari, 28 December 2021 *Alessandro Massaro*

About the Author

Professor Alessandro Massaro (ING/INF/01, FIS/01, FIS/03) carried out scientific research at the Polytechnic University of Marche, at CNR, and at Italian Institute of Technology (IIT) as Team Leader by activating laboratories for nanocomposite sensors for industrial robotics. He was head of the Research and Development section and scientific director of MIUR Research Institute Dyrecta Lab Srl. Actually, he carries out research activities in LUM Enterprise at LUM University -Libera Università Mediterranea- (Casamassima-BA-, Italy), he is in MIUR register as scientific expert in competitive Industrial Research and Social Development, and he is currently Member of the International Scientific Committee of Measurers IMEKO and IEEE Senior Member. He received an award from the National Council of Engineers as Best Engineer of Italy 2018 (Top Young Engineer 2018).

1

State of the Art and Technology Innovation

The chapter is focused on the technological and scientific state of the art about information technology (IT) advances. Starting with Industry 4.0 enabling technologies, the scientific improvements transforming the production lines and machines in intelligent systems following the logic of Industry 5.0 are discussed. The new facilities and the new technologies are oriented on the design of flexible and dynamical production processes, taking into account the market demand which is increasingly unpredictable. Starting with the enabling technologies of Industry 4.0, the specifications of the hardware and software technologies for advances in Industry 5.0 manufacturing industries are introduced. Communication protocols able to improve sensing and actuation in production processes are also discussed. Moreover, the analysis describes the Internet of Things (IoT) protocols, IoT upgrade processes and technological improvements, where of particular interest in monitoring industrial processing is infrared thermography (IRT) for improving thermal measurements in the production environment. The chapter is also focused on the description of different levels of the company information system, where sensors monitoring production constitute the field layer. The discussion is then oriented to provide an overview about sensors communicating with the local network by protocols, and achieving intelligent and efficient sensing and actuation. All the analyzed topics are addressed for integration into an upgraded information infrastructure implementing advanced tools. The analysis is then moved to the production processes in industries by highlighting main interconnections and architectures interfacing different tools. The study also enhances the scientific approaches consolidated in Industry 4.0, by providing limits of the actual technologies and perspectives for future production upscaling. Furthermore, the chapter discusses mainly intelligent information infrastructure suitable for manufacturing industries. The chapter goal is to introduce technological elements such as artificial intelligence (AI), augmented reality (AR) and big data systems, providing knowledge gain (KG). Other important aspects are the horizontal and vertical integrations of the technologies, considering bus-based networks and automatisms in data processing which is significant for the production advances. The chapter provides elements useful to comprehend how technologies can be implemented in flexible information architectures for innovative industrialization processes.

Electronics in Advanced Research Industries: Industry 4.0 to Industry 5.0 Advances, First Edition. Alessandro Massaro.

1.1 State of the Art of Flexible Technologies in Industry

Industry 4.0 introduced digital technologies improving industry productivity and different facilities supporting processes. The main enabling technologies introduced by Industry 4.0 are [1–3]:

- Three-dimensional (3D) printers connected to production software.
- AR oriented on production processes.
- Simulation tools able to optimize production processes by simulating production of different interconnected machines of different production lines.
- Horizontal integration of supply chain elements, such as human resources, supplies, products, transports, logistics, etc., and vertical integration of different production functions including product design, production processes, production quality, and end to end combination of horizontal and vertical functions.
- Cloud computing, cloud data storage, and data management in open data and big data systems.
- Cybersecurity improving security during network operations and in open systems, managing network interconnections.

These main facilities enable smart manufacturing (SM) and computer integrated manufacturing (CIM) industry processes in the fourth industrial revolution. In this scenario of enabling technologies, the information network architecture of companies plays a fundamental rule in production upgrade and in production engineering. The information digitalization is the first step for Industry 4.0 implementation, where the production machines transfer data in the local area network (LAN) and in general in the cloud environment. A particular function in Industry 4.0 improvement is the production monitoring, automated by IoT sensors [4], reading in real time the operation conditions of the whole production lines and allowing intelligent manufacturing. The control performed by sensors is more efficient for *in-line* monitoring procedures, where all sensors are synchronized in order to provide the best production setting of the whole supply chain. All the phases of the supply chain are important to trace. The main parts to trace in the production processes are: (i) warehouse, (ii) production lines, and (iii) logistics. In all these parts, robotics in general improves the processes, by increasing production volumes and by assisting human work. This kind of "joint collaboration" decreases the production errors and consequently the waste materials and related costs. Other technologies such as AR [5, 6] are used for human resources training during production processing, by increasing the worker skills and supporting workers to follow interactively and continuously the production. Augmented reality aided manufacturing (ARAM) is another important topic supporting production quality [5] by means of the programming of machines, robots and production tools, by managing logistics, and by checking assembled products in the whole supply chain. AR is adopted also in manufacturing as a dynamic authoring tool monitoring simultaneously the production activities of several workstations [6], for telerobotics controlling robots from a distance, for waste reduction in production activities, for assembly support, for remote maintenance, and for computer-aided design (CAD) applications [7]. In the Industry 4.0 scenario, AI can furthermore improve the industry production efficiency. AI algorithms are mainly indicated for machine predictive maintenance [8, 9] and for assisted production, where machine working operations are properly and automatically set in order to avoid failures [10], by decreasing or stopping machine in cases of alerting conditions. IoT sensors are very important for control and actuation thus enabling totally automated processes. A broad use of IoT sensing is related to image vision [11, 12] including IRT [13], and temperature and humidity sensors. Moreover, accelerometers provide supplementary information about anomalous vibrations indicating a possible system failure, and other sensors can be applied depending on the manufacturing process to be controlled. IoT signals are processed by AI algorithms to predict the machine status in self learning modality: by analyzing historical data, the AI algorithms create the training models to test for prediction. The AI improvements represent mainly the passage from Industry 4.0 to Industry 5.0 facilities adapting automatically the production with high level efficiency, and optimizing the production processes which are previously simulated. The flexibility of the production is due to the correct choice of the sensor network architecture, of protocols and the possibility to optimize the different layers of the whole communication system of the company. A correct design of the information system allows the disposal of a modular network open to vertical and horizontal integrations introducing innovative tools and algorithms addressing the automatic production control. The layers where it is possible to operate for a flexible production are the input/output (I/O) layer, the user interface layer, the gateway layer, the IoT middleware, the processing layer, and the application layer.

1.1.1 Sensors and Actuators Layer: I/O Layer

The I/O layer is the first layer related to the production field controlled by sensors. The process of machines can be changed by actuation commands provided by the processing layer. The actuation commands must ensure the production synchronization of the whole production lines managing different production steps. In this layer, IoT devices are very important for the accuracy and reliability of the performed measurements. The data sampling is essential for a correct monitoring procedure. When the sensors control different production process steps, it is fundamental to configure and to synchronize all the sensors of the same production line. The IoT technologies are defined for the specific production process to monitor. For example, if the process is fast, it is important to select an image vision technology having a high frame rate, or sensors having a sampling time "following" the production velocity. The technologies for industrial image vision converting light into electrons are charge-coupled device (CCD), complementary metal oxide semiconductor (CMOS), indium antimonide (InSb) infrared (IR) detectors, indium-gallium-arsenide (InGaAs), germanium (Ge), and mercury cadmium telluride (HgCdTe) sensors. Table 1.1 shows the working wavelengths of the IR technology.

Sensor networks are designed after an accurate analysis of the production processes, thus suggesting the correct configurations and connections of possible gateways, routers, and of device combinations. Sensor networks are implemented for point to point, star, extended star, bus, or mesh configuration. In Figure 1.1 are shown the different main network configurations. The design of the network is an important step for the realization of the correct network. The spatial allocation of the production machines and the workflow of the production define the best configuration. The network layout changes with the sensor system: the star or mesh network is typically

Table 1.1 Spectral ranges of infrared technology.

Infrared technology	Spectral range (μm)	References
InSb	0.6–5	[14]
InGaAs	0.9–1.7	[14]
Ge	0.8–1.6	[15]
HgCdTe	1–9.5	[14]

adopted for sensors, besides the bus layout is suitable for production line connections and for the information system. By considering for example a photovoltaic camp with a high number of panels, it is preferable to realize a ring type fiber optic network linking all electrical string panels. The network also assumes a hybrid configuration, especially when a new network is added and linked to an old one. Table 1.2 lists the main advantages and disadvantages of the different network layouts.

The network typology must be compatible with the network information system of the industry. In this way, the hybrid solutions potentially ensure the best network performances and flexibility. Figure 1.2a shows an example of a hybrid extended star network, constructed by merging an extended star with a mesh network, by showing an example of network reconfiguration in cases of connection failures, where the automatic principle of node commutation is managed by an intelligent algorithm detecting and predicting system failures (example of direct interaction between processing layer and production machine layers). The cases of Figure 1.2b–e are related to a possible configuration of the data transmission of a part of the network of Figure 1.2a. This example highlights the importance of adding nodes to avoid the transmission problem. The solution to add nodes to the local network must be "weighted" with the decrease of performance due to the increased complexity of the new hybrid network. The prediction of possible failures of nodes, allows to change anticipatedly a linking configuration, thus avoiding data interruptions, and preserving production control. In the prediction calculation, sensors play an important role because they detect operation conditions of production lines, status machines and product tracking. In Table 1.3 and Table 1.4 are listed the main specifications of traceability sensors able to detect the product in each production stage, and the main characteristics of transmission protocols,

Table 1.2 Advantages and disadvantages of network typologies.

Network type	Advantages	Disadvantages
Star	The star network manages the whole network by a single node behaving as a master node. Each node of the network can be added, removed, and reconfigured by ensuring the network operations. Network simplicity. Easy identification of errors	For a failure of the central node the whole network is out of order. Bandwidth limitation
Bus	Low cost and simple layout. Connection with a simple coaxial or RJ45 cable	For a failure of the bus the whole network is out of order. Additional nodes decrease network velocity. Single direction transmission mode (half duplex)
Ring	Bidirectional transmission mode for dual ring typology (full duplex)	Half duplex modality for basic ring configuration. Transmission security (if a node fails the network stops operating)
Tree	By adding to the tree network a star and a bus layout, it is possible to allow an easy addition of nodes and a network expansion	If the root node is out of order the whole network fails. Network performance decreases for complex hierarchical layouts
Mesh	Reliable and stable network type. Resistance to failure conditions also for complex layouts involving more interconnections	High time for network setting. High computational cost for complex interconnection layouts

respectively. Solutions for the actuation are the plug and play (P&P) solutions and programmable logic controller (PLC) hardware interfaces. For P&P systems the hardware and software components are downloaded and installed at or before run-time. The supervisory control and data acquisition (SCADA) [30] systems are able to read production data and transmit the setpoints to the PLCs. SCADA systems typically are implemented to control system architectures by graphical user interfaces (GUIs), and behaving as a supervisor of peripheral devices such as PLC, and proportional integral derivative (PID) controllers interfacing process plant and production machinery. Typically, SCADA adopts visualization tools and synoptic graphics for real-time data display.

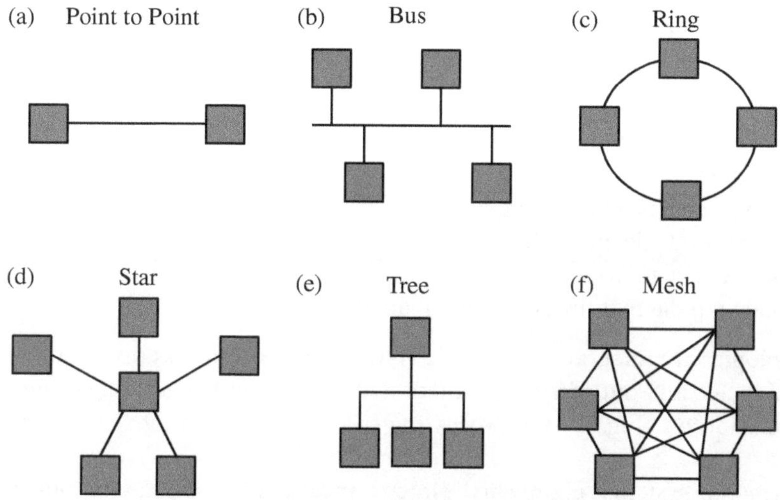

Figure 1.1 Example of network configurations: (a) point to point connection; (b) bus line; (c) ring layout; (d) star connection; (e) tree layout; and (f) node meshing configuration.

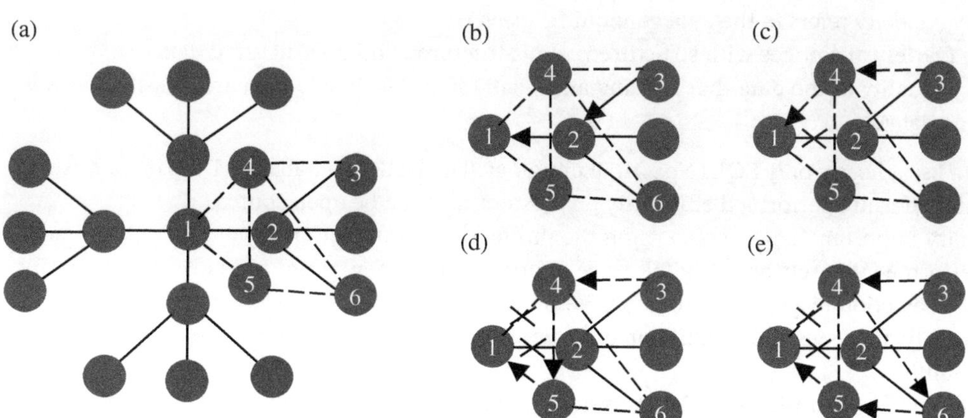

Figure 1.2 Example of hybrid extended star network and failure system reconfiguration for a secure production monitoring: (a) hybrid network structure by extended star and mesh network; (b) normal configuration for data transmission to the manager central node (transmission from node 3 to node 1); (c) example of reconfiguration for an interrupted linking between node 1 (network coordinator) and node 2; (d, e) examples of reconfiguration for interrupted links between node 1 and node 2 and between node 1 and node 4 simultaneously.

The information levels of a dynamic information system, allowing the upgrade from Industry 4.0 to Industry 5.0 production, is interconnected as in Figure 1.3, where the following six main layers are distinguished:

- Sensor and actuator layer.
- Agent, firmware and user interface layer.
- Gateway layer.
- IoT middleware.
- Processing layer.
- Application layer.

Flexible technologies must act in these layers, and are fundamental to automatize all the production processes for:

- *In-line/off-line* production monitoring.
- The elimination of possible failure conditions.
- The decrease of production defects.
- The optimization of human resources and of their work.
- Business intelligence (BI) and strategic marketing.
- The optimization of the warehouse management.
- A dynamic production following the real-time customer requests.

The main flexible technologies are integrated in robotic systems. Robots process information acquired by sensors placed inside and outside the production machines, and generating different outputs suitable for decision making, for the processing coordination, and for the system control.

The flexibility of the information system is mainly in the interconnectivity of all the layers shown in Figure 1.3, representing a standard architecture upgraded by AI and big data system working in the processing layer. Big data systems are characterized by the following features:

- Volume (dataset volumes larger than terabytes [10^{12} byte] and petabytes [10^{15} byte]).
- Velocity (velocity refers to the data generation speed).
- Variety (variety of sources with structured, semi structured and unstructured data).
- Veracity (quality of the data that is being analyzed, the non-valuable data are classified as nose or wrong data).

Big data uses the not only SQL (NoSQL) technology. The NoSQL databases (DBs) do not use the relational model, are performed efficiently on clusters, and can be open source.

A primary important aspect concerning the production optimization is the production traceability, performed by sensors. Digital traceability is fundamental in Industry 4.0 scenarios. Automatic detection by gates installed on the production line at each production stage is able to control quality processes and production in general. Table 1.3 lists the main sensors used for product traceability.

Radio frequency identification (RFID) systems are constituted by a reader and by a TAG or transponder, enabling the electronic identification of the traced product. The active version is equipped with a lithium (Li) battery or is powered by an external source. The passive RFID is not equipped with a battery and is cheaper offering an infinite lifetime. Besides, the active TAGs are more useful when writing operations. Sensors displaced to control production transmit data using specific protocols. Table 1.4 lists some specifications of transmission protocols.

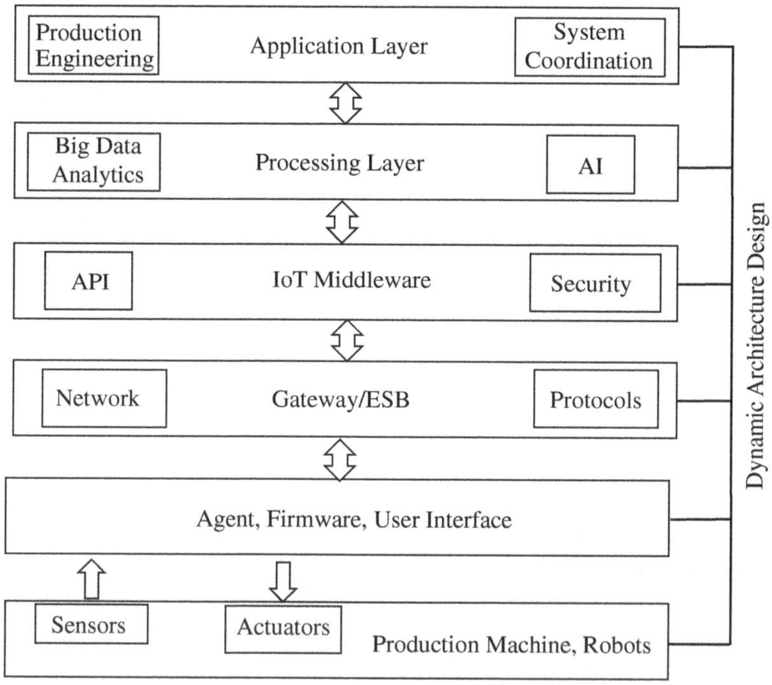

Figure 1.3 Layers of technologies related to an advanced technology.

Table 1.3 Main specifications of sensors used for traceability.

Sensor type	Main specifications	References
Barcode	• Optical laser reading identifying only type of item • Only read • <20 of characters of data capacity	[16, 17]
QRcode	• Optical laser reading identifying only type of item • Only read • Up to 7089 characters of data capacity	[17, 18]
RFID	• Radio frequency detection system • 100–1000 characters of data capacity • Read and write • Standards: 125–134 kHz (LF); 13.56 MHz (HF); 866–915 MHz (UHF); 2.45–5.8 GHz (microwave) • Active tags (100 m reading distance, 64 byte–32 KB of memory capacity) • Passive tags (1 m reading distance, 48 byte–2 KB of memory capacity)	[16, 17, 19–21]
NFC	• Distance of communication: few cm • Technology: RFID based • Frequency: 13.56 MHz • Communication: two way	[21]
iBeacon	• Bluetooth Low-Energy (low power consumption) • 1 m ± 70 m wireless range	[22, 23]

NFC, near field communication; RFID, radio frequency identification.

Table 1.4 Main specifications of sensor transmission protocols.

Standard	Main specifications	References
ZigBee	• Wireless • IEEE802.15.4 standard • WPAN • Frequencies: 868 MHz (Eu); 915 MHz (US); 2.4 GHz • 30–100 m wireless range • Network type: star, mesh, cluster tree, peer to peer • Capacity: 250 Kbit s^{-1}	[24–26]
WiFi	• Wireless • IEEE802.11 standard • Wireless local area network • Frequencies: 2.4–5.4 GHz • 1 km wireless range • Communication type: point to multipoint • Capacity: 54 Mbit s^{-1}	[26]
Bluetooth	• Wireless • IEEE802.15.1 standard • WPAN network • Frequencies: 2.45 GHz • 1–100 m wireless range • Communication type: point to multipoint • Network type: star • Controller: system on chip • Data rate: 1 Mbit s^{-1} • Capacity: 723.1 Kbit s^{-1} (versions 1.1 and 1.2); 3 Mbit s^{-1} (version 2.0) • Bluetooth V4.0	[21, 24]
RS-232	• Wired • LAN/PAN network • Communication type: point to point • Network type: bus • Capacity: 75 bit s^{-1}–115.2 Kbit s^{-1}.	[27]
USB	• Wired • LAN/PAN network • Communication type: point to point • Network type: tree, bus • Capacity: 12 Mbit s^{-1} (version 1.1); 480 Mbit s^{-1} (version 2.0); 4800 Mbit s^{-1} (version 3.0)	[28]
Ethernet	• Wired • LAN network • Communication type: point to point • Network type: star, bus • Capacity: 10 Mbit s^{-1}–10 Gbit s^{-1}	[29]

Table 1.4 (Continued)

Standard	Main specifications	References
ZWAVE	• Radio frequency technology • Sub-GHz communications (900 MHz) • Mesh network • No coordinator node • Master/slave architecture • Data rates: 9.6/40/100 Kbit s^{-1}	[24]
Fifth Generation (5G)	• Cellular network standard • Higher throughput • Lower latency • Artificial intelligence capabilities • Video real-time processing	[24]

WPAN, wireless personal area network.

1.1.2 Agent/Firmware Layer: User Interface Layer

The agent layer interconnects the gateway for device communication with all the other components. It is based on the agent's modular architecture suitable for robotic and production machines controlled by sensors. The IoT agent is linked to the IoT controller, and checks the framework environment during the time, by acquiring data by sensors properly programmed by firmware. Firmware is developed through software and hardware interfaces. Integrated development environment (IDE) platforms are usually adopted for firmware development as user interfaces. Agent-based computing (ABC) tools are able to implement agent functions in modern distributed applications; multi-agent systems (MASs) are ensembles of agents interacting in the same framework [31]. The MASs support edge and cloud computing, interconnecting innovative applications such as machine learning and blockchain.

1.1.3 Gateway and Enterprise Service Bus Layer

The gateway is an important node executing data routing and switching functions of the information network, using different protocol types. The gateway behaves as a provider, and it is interfaced between the IoT middleware and the agent layer. This layer is responsible for publishing and subscribing the services, message routing, and allows the communication between platforms. The enterprise service bus (ESB) enables service-oriented architecture (SOA) applications. Typically, the ESB is adopted to collect the digital data sources of different technologies adopted in the same information network, such as big data systems and AI algorithm datasets, by allowing data transfer between different DB systems [32]. ESB solutions are suitable to integrate old technologies with new ones improving the information system and Industry 4.0 implementations. The gateway system covers the role of collector of the packets coming from different sensors applied in the specific production line field. Communication takes place through IEEE 802.11 (WI-FI/5 GHz) or MODBUS standards, following the most common mechanisms for asynchronous communication including Message Queuing Telemetry Transport (MQTT) protocol [30].

The architecture of the gateway follows Industry 4.0 mechanisms, where the gateway represents the main local component made by:

- A sensor manager, able to locally manage the sensor nodes.
- A local DB manager with regulated accesses.
- An engine optimizer capable of processing advances.
- The application programming interfaces (APIs) integrating the IoT sensor network with the cloud environment by the use of mobile devices.

The IoT gateway is designed to implement the following features:

- IoT network management (local knowledge of connected devices and resources, local management of sensors or network nodes, local data availability).
- System intelligence (ability to perform process optimization locally, integrity of the information received).
- Distributed logic (the data from the nodes or sensors are stored in the local DB and managed for the central system).

The ESB is a software infrastructure carrying out services in complex SOAs and supporting horizontal, vertical and end to end integrations. The main advantage of this infrastructure is to interconnect and to interface heterogeneous technologies including big data systems, ensuring data synchronization, data security, messaging, intelligent routing managed by AI algorithms, and transformations services. The ESB assists the developer in integrating applications, therefore providing the infrastructure necessary to implement routing, translation, and other integration of features.

1.1.4 IoT Middleware

The middleware function [21] is to connect different typologies of programs. This interface is part of the architecture enabling sensor connectivity and application layers. The middleware manages important functionalities, such as collecting and selecting the received data from the IoT devices, by providing access monitoring for applications. The security is mainly performed in the middleware system by [33]:

- User identification.
- Identity management.
- A secure data communication system.
- Secure storage.
- A secure software execution environment.
- Secure contents.
- Security resistance.

The middleware improvement is related to the following specifications:

- Interoperability between devices of different technologies by managing heterogeneous interfaces.
- Managing devices performing load balancing.
- Use of API calls.
- Scalability by supporting the communication between a large number of devices.
- Big data interconnection and big data analytics tools.
- AI algorithm interconnection activating sensor data processing.
- Authentication and implementation of access control improving security and privacy.

- Running algorithms on different cloud services.
- Data extraction and data migration improving innovative DB and big data systems.

Middleware is classified as event based when all the components interact with each other through events, or service oriented when service providers are used for resource management in SOAs, or DB oriented, or application oriented.

1.1.5 Processing Layer

The processing layer is very important for the orientation of the industry network on facilities to Industry 5.0, including big data systems and AI algorithms, able to activate advanced data processing and the setting of all parameters of the architecture of Figure 1.3. The data processing is suitable for predictive maintenance, for machine failure predictions, for the self-adaptive production, and in general for assisted production.

1.1.6 Application Layer

This layer is responsible for delivering various application services. These services are provided through the middleware layer. The application services are suitable for, logistics, BI, rapid prototyping, reverse engineering (RE) and other advanced industry applications.

1.1.7 File Transfer Protocols

File transfer of the company information systems is executed through file transfer protocols. The main protocols adopted by the information network systems are listed in Table 1.5.

Particularly interesting is the LoRaWAN protocol suitable for long range wide area network (WAN) wireless technology tailored for IoT interconnection, and for bidirectional communication systems. The main features of this protocol are the low power consumption, and the possibility to improve scalable wireless networks.

Table 1.5 Other protocols usable in industry.

File transfer protocols	Details	References
File Transfer Protocol	• Client/server • TCP connection • Protocol interpreter; • Data transfer process • Data and command separate connections	[34]
Simple Mail Transfer Protocol	• Connection oriented • Text based • Client/server communication	[35]
TCP	• Network protocol (ISO/OSI transport level) • Transport port (MODBUS TCP): 502, 802 • Encapsulated into Internet Protocol • Application programming interface system call • Security (MODBUS TCP): TLS • Segment with a header and a data section	[30]

(Continued)

Table 1.5 (Continued)

File transfer protocols	Details	References
MODBUS	• Serial MODBUS RS485 (maximum 31 slaves) • Transaction enabled by a master • Cyclic redundancy check (CRC16) algorithm • Ethernet MODBUS TCP/IP	[36]
HTTP	• Application protocol • Hypertext • Infrastructure: Ethernet, WiFi • Model: synchronous • Stateless protocol (communication protocol) • Information web transfer • Mechanism: one to one • Network layer: IPv4 or IPv6 • Transport layer: TCP • Transport port: 80, 443 • Standard: Internet Engineering Taskforce RFC7230 • Encoding: ASCII text • Security: SSL or TLS	[30]
Hypertext Transfer Protocol Secure	• Extension of HTTP • Secure communication • Encryption using encrypted using TLS or SSL • Website authentication • Privacy protection • Digital certificates	[37]
Constrained Application Protocol	• Internet application protocol • Infrastructure: 6LoWPAN • Network layer: IPv6 • Transport layer: UDP • Transport port: 5683 • Mechanism: one to one • Model: asynchronous • Standard: IETF (RFC7252) • Service layer protocol • Wireless sensor network nodes • Devices supporting UDP • Security: Datagram transport layer security • Message length: 4 byte	[30, 38, 39]

Table 1.5 (Continued)

File transfer protocols	Details	References
Message Queuing Telemetry Transport	• Client/server • Infrastructure: Ethernet, WiFi • Messaging transport • Model: asynchronous • Transport port: 502, 802 • Standard: ISO/IEC, OASIS • Publish/subscribe • Machine-to-machine connectivity • Remote connections • Sensor communications • TCP/IP ports: 1883, 8883 • Security: SSL, TLS • UDP transport	[26, 30, 38, 39]
Extensible Messaging and Presence Protocol	• Streaming XML • Multiple communication patterns • Asynchronous messaging • Publish/subscribe • TCP transport • HTTP transport • Request/response	[30, 39]
Advanced Message Queuing Protocol	• Point to point • Publish/subscribe • Queuing • Routing	[39]
LoRa	• Lower physical layer • Radio frequency (433 MHz, 868 MHz, 915 MHz) • Geolocation capabilities • LoRaWAN (managing data device and frequencies transmission) • Long-range connectivity	[25, 39]
KNX	• Network standard • Communications protocol • Twisted pair bus (EHS, BatiBUS, EIB) • Sensing and actuation • Distributed applications • Radio (KNX-RF) • Power-line networking	[40]
Process field net	• Industry technical standard for data communication • XML • PROFIBUS • TCP/IP channel • Real-time channel	[29]

HTTP, Hypertext Transfer Protocol; KNX, Konnex; LoRa, long range; SSL, secure sockets layer; TCP, Transmission Control Protocol; TLS, transport layer security; UDP, User Datagram Protocol; XML, eXtensible Markup Language.

1.2 State of the Art of Scientific Approaches Oriented on Process Control and Automatisms

Technologies and architectures are fundamental for the upgrade of the company production. Different examples are provided to comprehend how innovative tools, including AI, can be applied in a new production scenario.

1.2.1 Architectures Integrating AI

The software and the hardware, in a modern information system oriented on Industry 5.0, must be integrated into a flexible information system infrastructure structured as illustrated in the model of Figure 1.4, where the first layer is represented by the hardware layer constituted by sensors and in general by IoT devices. The firmware layer is programmed in an operating system (OS) framework to provide a defined application including control and actuation processes. The AI interface is the intelligent core of the system, able to update machine setting, by considering prediction of failure conditions, and by automatically adjusting machine and robot working conditions. This automatism is a fundamental aspect of Industry 5.0 systems, where processing functions are managed in auto-adaptive modality. The AI algorithms can be executed directly in the application layer by executing a code (python, java, visual basic, etc.), or by using objects of GUIs. In the architecture model of Figure 1.4, the processor, operates in multitasking and multiuser modality, where the central processing unit (CPU) time is divided for all the programs that work simultaneously. The AI engine and the OS manage the database management system (DBMS) able to collect digital information of the whole supply chain. Furthermore, AI is suitable to exchange data and digital commands to electronic boards, interfacing machines or robots as for intelligent actuators.

Another important element for Industry 4.0 and Industry 5.0 implementations are the PLC systems enabling the actuation. The block scheme of Figure 1.5 represents a variation of the classic Von Neumann microprocessor scheme including the AI capability in the PLC system. The PLC system is constituted by:

- A random access memory (RAM).
- A microprocessor.
- Input and output (I/O) ports.
- An AI engine interfaced with a Programming Unit interface.

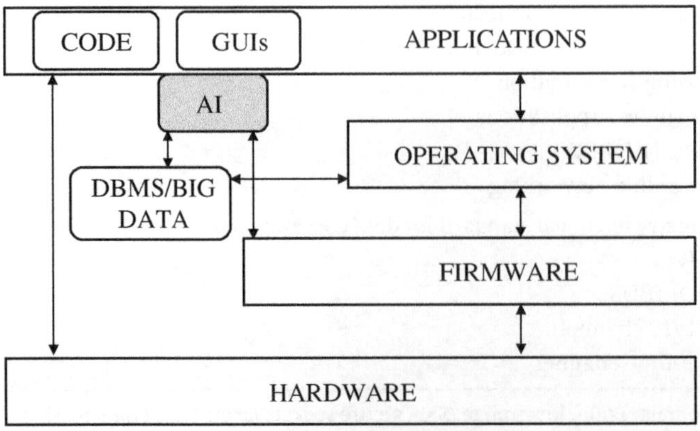

Figure 1.4 Hierarchical scheme of the software in Industry 5.0.

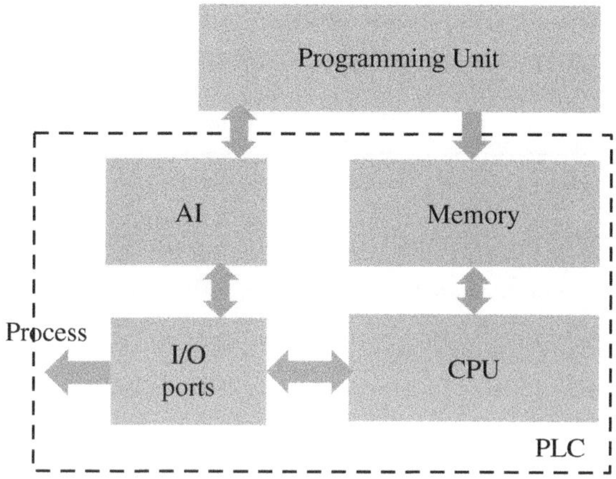

Figure 1.5 Advanced PLC architecture in Industry 5.0: central processing unit (CPU), memory, I/O ports, Program Unit module, and AI upgrading industrial processes.

1.2.2 AI Supervised and Unsupersived Algorithms

Different technologies can be integrated to improve Industry 4.0 production processes and in general digitalization. In industrial cases, supervised and unsupervised AI image vision tools detecting defects can be adopted. In this way, image vision and 3D image reconstruction methods play an important role as for assembling processes in the tire industry [41]. The image processing of raw images requires usually a high computational cost and is applied in post-processing modality. Concerning real-time processing, image segmentation techniques detect most relevant defects but more information about probable defects is hidden and is not easily visible in standard industrial environments. The hidden defects are "extracted" by more accurate inspections and by the application of specific intelligent image processing algorithms enhancing anomalous features. Intelligent algorithms are usually named machine learning (ML). In Table 1.6 the main ML unsupervised and supervised algorithms are classified [42]. The supervised learning algorithm processes a known input dataset and data outputs to learn the regression/classification model. In supervised learning approaches, the training is performed by "labelled" data, selecting specific variables to focus the analysis: some data are already tagged with the requested answer, and the labeled data are adopted for the self-learning of the algorithms predicting outcomes of the labeled variables. Unsupervised learning is the training modality of the algorithm which processes a dataset that is not classified/labeled. In the unsupervised learning approaches the model does not need to be supervised: the models discover information and common features of the variables (attributes) and find all kinds of unknown patterns in the data. The learning phase is structured in the following sequential steps:

- Training dataset construction.
- Features vector extraction.
- Algorithm application setting data processing parameters.
- Training model construction.

Both classes of supervised and unsupervised algorithms are typically applied for data processing applications of image processing for feature classification.

A typical class of unsupervised algorithms are the clustering ones. Clustering methods are able to group objects into homogeneous classes. A cluster is a set of objects that have similarities to each other, but which have dissimilarities with objects in other clusters. The input of a clustering

Table 1.6 Classification of machine learning algorithms.

Machine learning algorithm class	Unsupervised	Supervised
Continuous	• Clustering: - K-means - Mean shift clustering - Density-based spatial clustering of applications with noise - Expectation maximization Clustering using gaussian mixture models - Agglomerative hierarchical clustering • Dimensionality reduction: - Principal component analysis - Singular value decomposition	• Linear regression • Polynomial regression • Artificial neural network • Random forests • Decision trees
Categorical	• Association analysis: - Apriori - FP-growth • Hidden Markov model	Classification: • k-nearest neighbors • Decision trees • Logistic regression • Naïve Bayes • Artificial neural network • Support vector machine

algorithm consists of elements, while the output are clusters where the elements are divided according to a similarity measure. Clustering algorithms also provide a description of the characteristics of each cluster, which is essential for decision-making processes. Concerning the continuous class of unsupervised algorithms, the K-means algorithm is the classical approach estimating the centroid of the clustered data named clusters. The mean shift clustering (MSC) is a sliding-window-based algorithm executed to find dense areas of data points defining the centroids. The density-based spatial clustering of applications with noise (DBSCAN) has the main characteristic to find arbitrarily sized and arbitrarily shaped clusters. The expectation maximization (EM) clustering uses the gaussian mixture model (GMM) approach assuming data points distributed as a Gaussian function and characterized by mean and standard deviation. Agglomerative hierarchical clustering (AHC) group clusters follow a hierarchy represented by a tree or by a dendrogram: the root of the tree is the main cluster grouping all the samples, and the leaves are clusters with only one sample. Principal component analysis (PCA) is able to reduce the dimensionality of a dataset made up of many more or less correlated variables. The single value deposition (SVD) technique is a particular factorization of a matrix based on the use of eigenvalues and eigenvectors. Concerning the categorical class of unsupervised algorithms, the Apriori algorithm is adopted for association rules for frequent item set mining. The algorithm of FP-growth is able to complete a set of frequent patterns by pattern fragment growth, using frequent patterns. The hidden Markov model is a statistical Markov model with hidden states. Concerning the continuous class of supervised algorithms, the regression is a statistical process that tries to establish a relationship between two or more variables. If a regression model is given an x value, this returns the corresponding y value generated by the processing of x. The linear regression differs from the classification, since the latter is limited to discriminating the elements in a given number of classes (label), while for the linear regression approach the input is data and the system gives a real output (unlike the classification method which receives as input data and returns as output a label dataset). Polynomial regression uses the same method as linear regression, but assumes that the function that better describes the data trend is not a straight line, but a polynomial. Artificial neural networks (ANNs) are able to classify and predict data. Specifically, ANNs are made by three types of layer: the input layer, the hidden layers, and the output layer. In the input layer, the neural network receives the data in the form of inputs, activates and processes them according to the classification capacity for which it is trained, and passes the information obtained to the next layer as in neuron propagation. At each step, the starting information takes on an increasingly refined meaning due to the interpretations of the different nodes. Finally, the processed data arrive at the output layer, which collects the results. Concerning ANNs, the network is structured to learn automatically in self-learning modality. Each perceptron has the task of categorizing objects by referring to common characteristics, following a score system calculated on each analyzed element. In AI, the perceptron represents a binary classifier selecting data input, and provides a features vector. The classifier makes its predictions based on a linear predictor function combining a set of weights with the feature vector. In AI machine learning algorithms, the perceptrons are important for their self-learning capacity, thus addressing these tools for auto-adaptive production solutions. A multilayered perceptron (MLP) is a particular class of feedforward ANN characterized by multiple layers of perceptrons having a threshold activation. The MLP consists of at least three layers of nodes: an input layer, a hidden layer, and an output layer. Except for the input nodes, each node is constituted by a neuron implementing an activation function. MLP utilizes a supervised learning technique called back-propagation for model training. In the class of categorical supervised algorithms, the random forest (RFo) represents a type of ensemble model, which uses bagging (the bagging aims to create a set of classifiers of equal importance) as an ensemble method and the decision tree (DT) as an individual model algorithm. Also DTs are a supervised learning tool, mainly solving classification or regression issues, capable of learning nonlinear associations and very easy to interpret and apply. DT algorithms work on both numeric and categorical data, and are categorized with respect to the output variable as categorical DT, and continuous DT. In the categorical class of a supervised algorithm, the k-nearest neighbors (KNN) algorithm is an algorithm used for pattern recognition and for classification based on the characteristics of the objects close to the one considered. The logistic regression algorithm allows to generate a result representing a probability that a given input value belongs to a certain class. In binomial logistic regression problems, the probability that the output belongs to one class is P, while that it belongs to the other class is $1\text{-}P$ (where P is a number between 0 and 1 because it expresses a probability). Naïve Bayes is a supervised learning algorithm suitable for solving binary and multi-class classification problems ant it is based on Bayes' theorem defining the conditional probability: let A and B be two events, and let B be a possible event having a probability of occurrence $P(B) \neq 0$, if $A \cap B$ indicates the intersection of the two events (both occurred events), it is defined the conditional probability $P(A|B)$ (probability of A conditioned by B) as:

$$P(A|B) = \frac{P(A \cap B)}{P(B)} \tag{1.1}$$

The support vector machine (SVM) is a supervised machine learning algorithm that is used for both classification and regression purposes. For a given training dataset, the SVM training algorithm builds a model that assigns new examples to one category or the other, making it a non-probabilistic binary linear classifier.

1.2.3 AI Image Processing

Image segmentation techniques are useful also to detect product anomalies enabling predictive maintenance procedures of production lines [43]: quality monitoring of defect clusters checked on product surface, can be performed by image segmentation and image filtering techniques, predicting defectiveness. Defect prediction in manufacturing processing is achieved by ANNs [44]. The image vision technique [45, 46] also verifies the correct position of each product component, enabling pick and place removal action of defective pieces, according also to *in-line* machine timing processing [45]. Image vision and mechatronic technologies improve the automation of quality check processes according to ISO9001:2015. A typical mechatronic scheme for real-time defect *in-line* control and actuation is illustrated in Figure 1.6 where:

- A product is moved on a conveyor belt.
- The product passes in the image vision-controlled area in a time between t_1 and t_2 (observation window for quasi real-time processing).
- The fixed smart camera locally processes the detected image by estimating the defect tolerance (output O).
- If the measured defect overcomes the tolerance the input command (input i) of the pick and place robotic system is actuated, eliminating the piece.
- The AI engine also predicts defect and machine failures thus improving the quality control of the whole production line.

The real-time image processing in industrial applications permits:

- Execution of a pre-image filtering thus optimizing light and in general noise conditions (camera setting and auto-calibration).
- Extraction of the main features of the product to check contours, shape, color, etc.
- Estimation of defect tolerance.
- Provision of the driven command of mechanic components.
- Classification of defects (defect classification by AI algorithms).

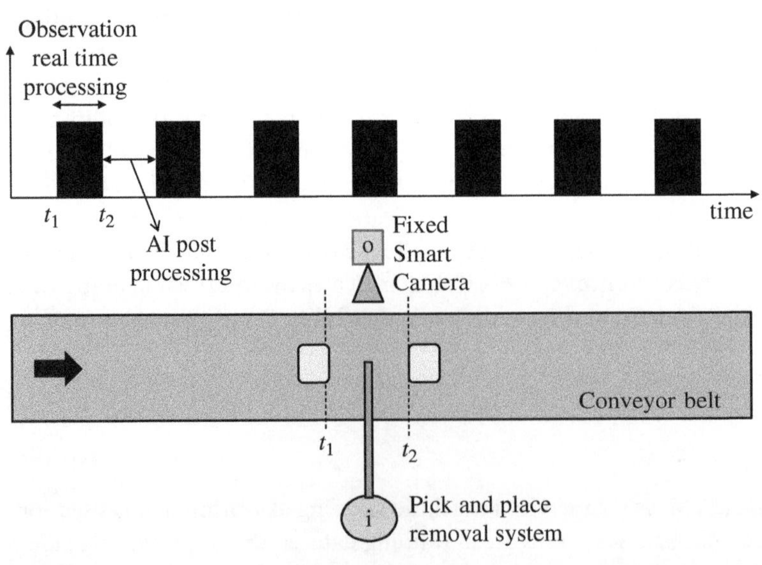

Figure 1.6 Scheme of a pick and place automated system for defect removal, based on image processing.

- Estimation of defect clustering (by clustering algorithms)
- Reconstruction of a 3D image useful to enhance defect features.

The post image processing modality allows prediction of the:

- Number of defects.
- Defect typologies according to the extraction of features (defect classification).
- Machine failures (predictive maintenance application).

Image processing is a part of the whole process monitoring all the production in each stage. The phases of the image processing are shown in Figure 1.7, and are:

- Pre-filtering (image is adjusted by setting brightness, contrast, luminosity, background subtraction, red, green, and blue [RGB] filtering, and other parameters by optimizing noisy images for the post-processing).
- Image segmentation extracting features such as geometries and shapes directly classifying product defects (different techniques can be applied for defect classification such as the thresholding method, region and edge detection-based techniques, clustering-based techniques, watershed-based techniques, partial differential equation-based approaches, morphological segmentation, and others); image segmentation techniques are applied also for radiometric images.
- Application of image post-processing techniques based on AI algorithms classify defects enhancing hidden ones, by means of a training dataset structured by considering the specific defects associated with the production processes.
- Defect classification and defect counting phase enabling defect storage into the DB system.
- Defect detection enabling actuation processes by stopping or decelerating the machine in order to avoid further defective parts (the actuation process acts also to remove wrong pieces).
- Defect prediction estimated by AI algorithms able to process historical production data, collected into the DB and big data system improving the training model.
- Predictive maintenance of production machines analyzing and counting defective parts (the maintenance standard plan is modified by analyzing predictive maintenance results, thus decreasing drastically failure conditions).
- Quality improvement performed by analyzing simultaneously predictive maintenance results and process production mapping.

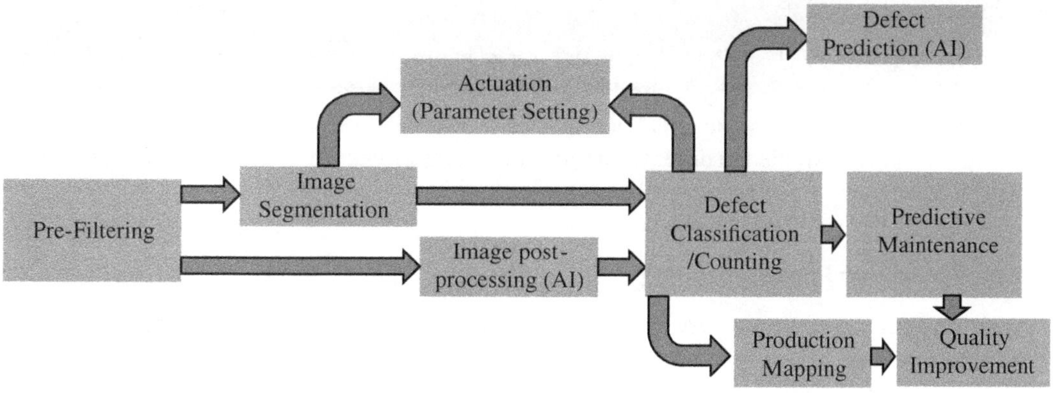

Figure 1.7 Image processing and intelligent system in production processes.

The AI algorithms can be applied directly on the image to process. In this way, the convolutional neural networks (CNNs) [47] represent a good approach for defect classification [48] and laser positioning in industrial applications [49]. The typology of features to extract by the AI algorithms or by traditional image vision are different and are a function of the defect to detect. For example, in electronic welding the defects are classified as [45]:

- Irregular welding geometries.
- Big/small welds.
- Non solder.
- Irregular layout patterns.
- Agglomerates (coupled welds).
- No welds made (detection by welds counting).
- Dirty welds.
- Excess/less deposited welding materials.
- Component missing.
- Component shift.
- Component lift.
- Generic component damage.

Considering tire production industries, the defects are classified as:

- Tread/belt separations.
- Accumulation of contaminants in trapped areas (moisture, grease, oil, etc.).
- Snaked, wavy or off centered steel belts.
- Belt edge deformation (due to the anomalous mechanical stress).
- Interrupted layout patterns (crack identification, tire stretching).
- Swelled layout patters (contaminants accumulation).
- Misaligned patterns.
- Added edges on the basic patterns (contaminants accumulation).
- Different height of the elements constituting the patterns.
- Writing not correctly printed.
- Layout elements unified.
- Different agglomerated contaminant materials (inhomogeneous geometry characterized by different volumes).
- Misalignment between tire and wheel.
- Others.

All these features are classified by particular image features extracted by image processing algorithms. Concerning the welding control of a large metallic surface as for tank production, IRT provides more information about weld homogeneity. Other algorithms, such as the K-means clustering algorithm, are applied in order to enhance for example leakages [13], or in general for radiometric defect detection in photovoltaic systems [48].

1.2.4 Production Process Mapping

Process mapping represents a very important procedure to follow for quality control. All production data provided by sensors, and also results of the image processing tools, are in general stored in DB systems and mapped by diagrams such as fishbone diagrams, p-charts, the Plan-Do-Check-Act (PDCA) cycle, and by XmR charts, thus improving the quality check during the time.

The PDCA cycle, also known as the Deming cycle diagram, is a four-step iterative management method used for the control and continuous improvement of processes and products. Steps in PDCA include:

- The *Plan* concerning the occurring variations.
- The actions to *Do* and to be controlled.
- The *Check* of the executions by validating expected results.
- The *Act* involving all the activities about the process, the standardization and stability, and improving the existing procedure.

The XmR chart, through the upper and lower control limits, provides the limits of an industrial production process mapping at the same time production results. As a further diagram to adopt for process mapping, a fishbone diagram (Ishikawa diagram) is a cause-and-effect approach able to track imperfections, variations, defects, or failures. This diagram is structured as a fish's skeleton with the problem at its head and the causes for the problem following the fish bones. In order to also predict defects or production failures, big data systems facilitate the analysis of the historical data. Big data systems implement NoSQL technology based on the following main categories:

- key-value distributed
- document oriented
- column oriented
- graph oriented

However, some NoSQL DBMSs may have features that belong to more than one of these categories. Some common big data systems are Cassandra, MongoDB, and Hbase.

1.2.5 Technologies of Industry 4.0 and Industry 5.0: Interconnection and Main Limits

The interoperability of technologies is a fundamental element for the upgrade of the information system and in general for the production. Figure 1.8 shows the interoperability of different technologies involved in Industry 4.0 and Industry 5.0 environments. The diagram includes all the main technologies for control and actuation supporting production auto-adaption managed by the AI engine.

The technologies of Figure 1.8 are interconnected in information by exchanging data with different software tools. The components of an innovative information system are divided into two main categories:

- Transactional applications: systems and procedures supporting IT and enterprise resource planning (ERP) systems.
- Decision support system (DSS): systems and procedures supporting strategic decisions and production processes management by data mining.

The ERP systems are developed for finance, logistics and in general for supply chain management. Subsequently ERP systems evolved into material requirements planning (MRP) tools implementing Business Process Re-engineering (BPR) logics. The peculiarities of ERP systems are:

- integrated solutions
- functional modules
- information centralization

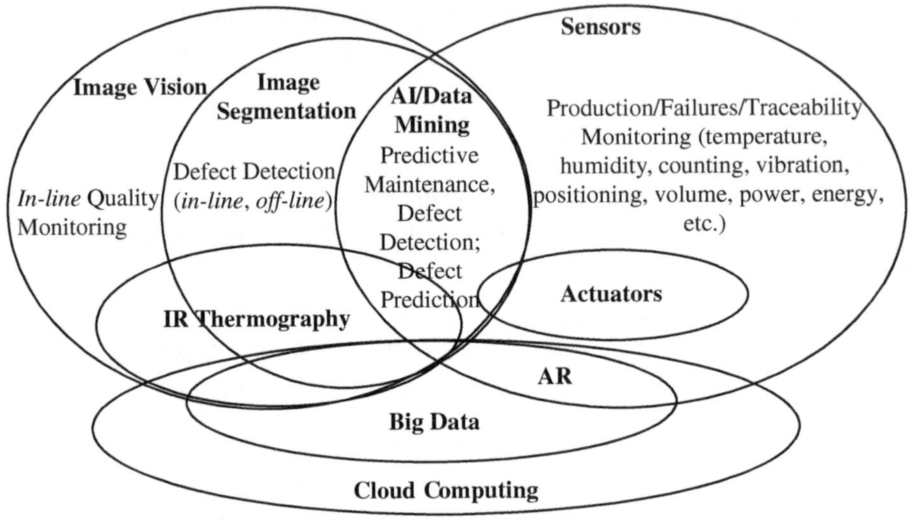

Figure 1.8 Interoperability of different technologies involved in Industry 4.0 and Industry 5.0 environments.

- real-time availability
- fixed functional schemes (standard models)
- customer customizations
- financial compliance

The implementation phases of ERP systems are summarized by the following steps:

- Software selection.
- formulation of the business process modeling (BPM).
- mapping of existing processes ("AS IS" mapping).
- definition of a new implementation scenario concerning future production processes ("TO BE" mapping).
- BPR process redesign.
- Production gap analysis (detailed list of missing production features).
- Formalization of formal documents for the validation of new processes.
- System parameterization.
- Prototype creation.
- Testing of the whole information systems.

The main ERP functions are:

- management control
- management of purchases and receptions
- sales and distribution management
- warehouse and logistics management
- project management
- production, orders, and account processing
- human resources management

The upgrade of the ERP systems involves AI implementations, thus improving DSS dashboards and decision-making processes of the whole supply chain. ERP and other software tools must be interconnected in information infrastructure enabling real-time data processing, sensing, and actuation. According to the architecture of Figure 1.9, information systems interconnect different enabling technologies by means the use of the bus or of the ESB. DBs are synchronized and managed by an advanced ERP having different functions including data migration into the big data system. All the other facilities such as IoT, PLC sensing and actuation, AR and process simulation, are implemented through human–machine interfaces (HMIs) interconnecting the tools with the information system. The information systems are typically managed by a bus or a ESB line.

The bus system is a set of internal connections for the transmission and exchange of signals, supply voltage, and ground potentials. The exchange of signals between the microprocessor and the input and output interfaces therefore takes place via the bus system. The bus lines are classified as:

- Address bus: allowing the access of the addresses of the individual boards.
- Data bus: reading data from the input cards or data transferred to the board outputs.
- Command bus: managing command and control signals within the controller.

Concerning technological limits, the challenge is to improve performance for an innovative Industry 5.0 scenario based on full production automatisms. Table 1.7 lists some possible technological limits characterizing innovative technologies to integrate in the company information system.

An important parameter to check the measurement accuracy of monitoring sensors, is the signal-to-noise ratio (SNR) defined as:

$$\mathrm{SNR}(\mathrm{dB}) = 10\log_{10}\left(\frac{P_{\mathrm{Signal}}}{P_{\mathrm{Noise}}}\right) = 20\log_{10}\left(\frac{A_{\mathrm{Signal}}}{A_{\mathrm{Noise}}}\right) \tag{1.2}$$

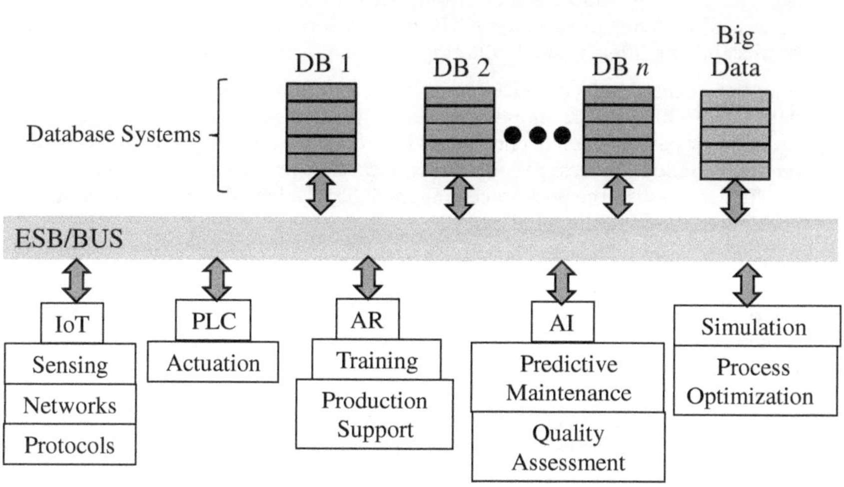

Figure 1.9 Information system interconnecting enabling technologies.

Table 1.7 Possible limits of advanced technologies and technical risks.

Technological limits	Descriptions and mitigations
Production time velocity	Computational cost of sensing and actuation processes must follow the *in-line* production timing. Hardware and data computing can introduce relevant delays for real-time actuation. Cloud data processing or the use of GPUs can be possible solutions
AI training dataset	The training dataset is particularly important in order to define high performance AI models. The training dataset must be constructed with a high quantity of data. In the initial monitoring of the production process, there are few sensor data and the training dataset can provide incorrect outputs. A solution can be the creation of a training dataset based on the creation of artificial data constructed by the initial ones. The dataset must be built with criteria avoiding the creation of redundant data; for example, reproducing in bench test similar data having the features of the production ones
Sensor detection and transmission	The Internet of Things sensor network monitoring production must be designed to detect and to transmit efficiently the signals. The full set of sensors able to monitor a whole production line must be allocated in the precise part of the production line mainly where the production is critical. The sensor sampling frequency is a particularly important parameter to provide enough data to process. In this scenario, a well-structured local area network, based on the efficient solutions as the optical fiber architecture, typically increases the performance of the data transmission system. Another aspect to improve for data transmission is the correct programming of the time windows synchronizing all sensors
Sensor calibration	The sensor calibration is fundamental for the specific application field and machine to control. The calibration must be performed according to the production layout. A correct sensor calibration provides good data to process. In order to calibrate a production layout correctly, it is useful to simulate before all the signals are involved in the whole production line
Noise	Strong and anomalous vibrations, heat sources, inhomogeneous light conditions, acoustic noises, pollution, and electromagnetic induced fields represent possible causes of incorrect measurements. The noise is decreased by means of electronic boards and amplifiers properly filtering and cleaning the sensor signals
Image segmentation techniques	Image segmentation techniques can be inefficient for the specific defect to detect. AI techniques are adopted as alternative approaches for a further quasi real-time or post-processing checking. The processing modality is selected based on the production process to control. If the check is performed at the end of the product processing, the post-processing modality is adopted, thus avoiding the problem of the computational cost
Camera resolution	Camera resolution can be inadequate for image processing. The optical components to combine with the camera are very important for the detection improvement. Angle of view, field of view, and focal lens are the parameters and elements to verify before the choice and installation of the camera system in the production line. The production layout and the characteristics of the production define the optical requirements of the camera system. A correct design of the camera system also takes into account the pollution of the working environment which in general disturbs the image acquisition

Table 1.7 (Continued)

Technological limits	Descriptions and mitigations
IRT	For an IR thermal camera, as well as the optical parameters, the thermal resolution is also important (a good parameter is <20 mK at +30 °C). Heat sources or reflective surfaces near the detection object increase the measurement signal noise and can saturate the camera chip. The correct positioning and orientation of the thermal camera is important to avoid the problem generated by the reflective surfaces
Defect classification	Defect classification is typically performed by AI algorithms. The training model is structured to detect particular typologies of defects according to product features. The correct formulation of the training dataset plays an import role for correct defect detection and classification. For this purpose, it is necessary to know and pre-classify defects, thus optimizing the AI training model to apply
Traceability	The traceability can be inaccurate mainly for indoor applications. The choice of the correct technology is a function of the production layout and production processes. Possible correctly positioned gates reading tags avoid the problem associated with the correct indoor positioning
Computational cost of smart sensors	The computational cost in smart hardware technologies must be adapted to the production velocity. For this reason, local data processing can be performed by high performance processors embedded in smart devices, or by GPUs
Big data system	Big data cannot be connected and synchronized with the ERP database systems. A typical condition is to implement a big data system into a database management system
Big data analytics	The analytics tool can fail if there are empty records or wrong data stored in the big data. The definition of criteria to store and filter data in the big data system ensures the correct creation of the dataset to use for AI modeling
Actuation	An incorrect PLC configuration generates a failure in sensing and actuation processing. Preliminary simulations, checks, and tests of the PLC integrated in the whole production line ensures the best actuation according to the specific feedback systems
AR	AR smart glasses can have a low image resolution and low acceleration sensing according to the application. Standard smart glasses integrate an IR thermal camera having a low thermal resolution. The alternative solution is to use a high-resolution thermal camera wireless interconnected with smartphone cardboard systems avoiding the problem of low thermal resolution
Unstructured data	Many data are not structured and consequently cannot be transferred by the enterprise service bus/bus network or by the wireless system. The association roles, used to extract information from unstructured data, provide more information for the digitized knowledge gain
In-line parameter setting by AI	The AI engine can fail providing wrong outputs for the real-time machine parameter calibration and actuation. The AI must be properly checked before its introduction in the production processes
Automodification of parameter setting	The auto-adaptive system able to change the machine parameter setting can be inaccurate by providing wrong operations and wrong scheduling actions. The automated modification of the parameters can be performed by AI algorithms which must be accurately tested for the specific dataset

AI, artificial intelligence; AR, augmented reality; GPU, graphics processing unit; IR, infrared; IRT, infrared thermography; PLC, programmable logic controller.

where P_{Signal} is the average signal strength, P_{Noise} is the noise strength, A_{Signal} is the root mean square (RMS) of the signal amplitude, and A_{Noise} is the RMS of the signal noise. The SNR parameter can be increased by applying AI as in modern computer numerical control (CNC) machine systems, which are improved by accurate control systems based on ANNs predicting for example surface roughness during tool processing by sound and vibratory signal analyses [50, 51]. The sensor data analyses allow, by means of AI, the prediction of machine failures and malfunctions thus increasing the SNR.

1.2.6 Infrared Thermography in Monitoring Process

The processes controlling the temperature represent many times a fundamental element to guarantee an optimal quality of the worked piece. The temperature check is performed directly on the processed piece or is related to the working environment as for a temperature-controlled room. Excessive rise in temperature may generate thermal damage in the processed materials. Specifically, monitoring the tool's temperature range is of fundamental importance. A function of IRT is to check during the time the correct processing as for cutting, welding, or surface treatments. The basic physical principles of an IR thermal camera in monitoring processes are indicated in Figure 1.10 (a), indicating all the signals involved for a radiometric measurement which are:

$$\mathrm{A} = \varepsilon_{\text{Target}} \cdot \sigma \cdot (T_r)^4 \tag{1.3}$$

$$\mathrm{B} = (1 - \varepsilon_{\text{Target}}) \cdot \sigma \cdot (T_r)^4 \tag{1.4}$$

$$\mathrm{C} = \varepsilon_{\text{Target}} \cdot \tau_a \cdot \sigma \cdot (T_{\text{Target}})^4 \tag{1.5}$$

$$\mathrm{D} = (1 - \varepsilon_{\text{Target}}) \cdot \tau_a \cdot \sigma \cdot (T_{\text{Target}})^4 \tag{1.6}$$

$$\mathrm{E} = (1 - \tau_a) \cdot \sigma \cdot (T_a)^4 \tag{1.7}$$

where C + D + E is the total radiated power, σ is the Stefan–Boltzmann constant, $\varepsilon_{\text{Target}}$ is the emissivity of the object to detect (target), τ_a is the environment transmittance, T_{Target} is the temperature of the object, and T_a is the environment temperature.

The timing of the measurement acquisition and the movement control of the thermal camera are two relevant factors for object temperature checking. Concerning camera systems, the angle of view (AOV) and the field of view (FOV) in degrees are, respectively, expressed in Figure 1.10b as:

$$\mathrm{AOV} = 2 \cdot \mathrm{ArcTan}\left(\frac{\text{Sensor width}}{2 \cdot \text{Focal Length}}\right) \cdot \left(\frac{180}{\pi}\right) \tag{1.8}$$

$$\mathrm{FOV} = 2 \cdot \left(\mathrm{Tan}\left(\frac{\mathrm{AOV}}{2}\right) \cdot \text{distance to object}\right) \tag{1.9}$$

The AOV is a measurement (in degrees) of how much of an object is viewed through the lens, and is measured horizontally, vertically, or diagonally. The FOV is a measurement of object distance and it requires the knowledge of the distance from the optical center of the lens to the object to detect. In a 3D space the FOV is defined horizontally, vertically, and diagonally.

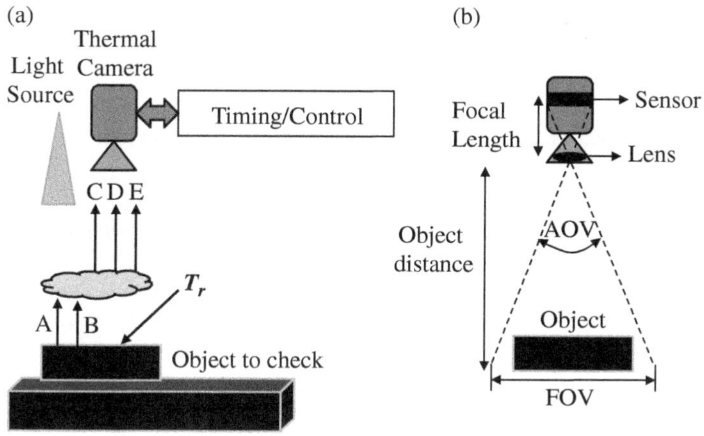

Figure 1.10 (a) Infrared thermal camera signals. (b) AOV and FOV simplified definitions.

1.2.7 Key Parameters in Supply Chain and AI Improving Manufacturing Processes

AI in manufacturing processes [52, 53] follows the production in different forms, including defect classification, defect prediction, sales prediction, assisted production, RE optimization, automated processing, layout and warehouse optimization, DB security predicting network attacks (cybersecurity risk prediction), logistic optimization, and in general supply chain improvements estimating Key Performance Indicators (KPIs). Important KPIs are:

- Count (the amount of product produced in the factory by the last change of machine or the sum of production of the whole shift or week).
- Rejection ratio (production of waste affects profitability goals).
- Production rate (different speed of machines and processes producing goods).
- Production goal (target values for production outputs).
- Cycle time (the amount of time for the completion of a task).
- Overall equipment effectiveness indicating the efficient utilization of available personnel and machinery.
- Idle time (period in which machines are not operating).
- Life cycle of a product.

AI can predict the KPI values thus suggesting decision in a BI key and achieving the main goals indicated in Table 1.8.

In order to improve the KPI, the real issue is to integrate AI in the whole supply chain by following the main scheme of Figure 1.11. The supply chain is characterized by a large number of different variables. A solution to optimize the supply chain is a dynamic monitoring controlling in real time the whole production system. In this way, AI and IoT solutions, by processing digitized data, improve dynamically real-time production processes: if there is a prediction of sales of a particular product, the production line is addressed by the AI directly for the new strategic product to process. AI potentially supports industries to switch production according to needs and the possible adaptions of the available production machines in new requested products. Furthermore, if correctly implemented, AI provides improvements in the production time response and in the machine efficiency of the whole production line, where all the machines of a layout not only work to create a finished product, but also are set for the best efficiency. It is

Table 1.8 Goals achievable in manufacturing industry.

KPI goal	Description
Safety and environment	Number of accidents at work, number of alarms, consumption, waste, recycling of material, etc.
Efficiency	Saved materials and resources, saved energy of production lines, improvement of services, maintenance optimization, less production shutdowns, decrease of production time
Quality	Percentage of finished product that does not meet the quality criteria, percentage of semi-finished and raw products that do not meet the quality criteria, size of production losses and waste, internal and external services, etc.
Production plan tracking	Production time traceability, logistic layout optimization, improvement of warehouse management, programming efficiently picking processes, etc.
Employee's satisfaction	Completed works on time, lost workdays, innovation proposals, product/process innovation proposals, etc.

observed that, in an advanced local communication system, the single machine can communicate the identified setting solution to all the other machines involved in the production. By applying AI, it is possible to achieve the optimum of the economic and operating criteria, by setting the production level of all the individual units that work together. The production setting is managed by means of a supervision and coordination system, where the AI system is suitable to adapt the processing to each reconfiguration of the production programs and of the product mixing to meet customer requirements in the best way.

In manufacturing industries, the AI system is integrated in a more complex multilayer structure, as illustrated in Figure 1.12, with the following layers:

- **Layer 1** (direct control of the single units of the plant). Information about the unit production and about raw materials are executed and transmitted to the highest levels. The first layer is the operational layer enabling production and actuation. The production parameters are optimized by layer 3 related to the AI level.

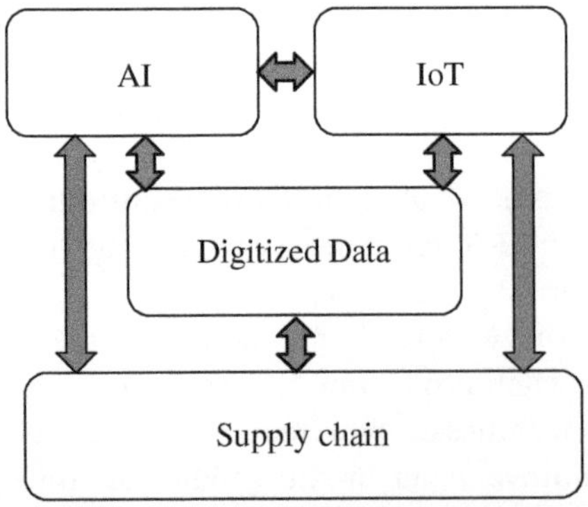

Figure 1.11 Artificial intelligence integrated into the supply chain.

- **Layer 2**. This layer represents the response to any emergency condition arising in its area of competence of the plant, the optimization and the coordination node keeping communications active with the upper and lower levels of the system. The layer is commanded by layer 3 and provides outputs to layer 1.
- **Layer 3** (AI level). This level is suitable for verification and optimization of key production processes in real time. The management of this level is assigned to AI algorithms. Improvement solutions can be implemented in all phases of the production process by automatizing the whole supply chain. This layer is the core of the Industry 5.0 system and includes process mining (process decision making by data mining) [53], process optimization, process prediction and process control. IoT data contribute in this level to enrich the dataset processed by the AI algorithms for better control of processes upgrading the production. Graphical dashboards facilitate the decision-making supported especially by supervised algorithms. The AI allows not only to improve the production process, but also to predict possible failures in a self-adaptive manner.
- **Layer 4** (definition of the activities to coordinate the production resources and to achieve the desired final products). This layer is specific for the production area involving RE facilities, production engineering, quality improvement, data processing, and production switching enabled by the AI level. This layer represents principally the production flexibility, the production quality, and the reliability.

The multilevel model of Figure 1.12 is summarized by the block diagram in Figure 1.13, where AI is the feedback of the whole system, layers 1, 2, and 3 represent the single production station, and layer 4 indicates the whole production line processes performing machine diagnostics.

The proposed methodology is adaptable to different industry sectors and represents a valuable tool for the ISO 9001:2015 check control system.

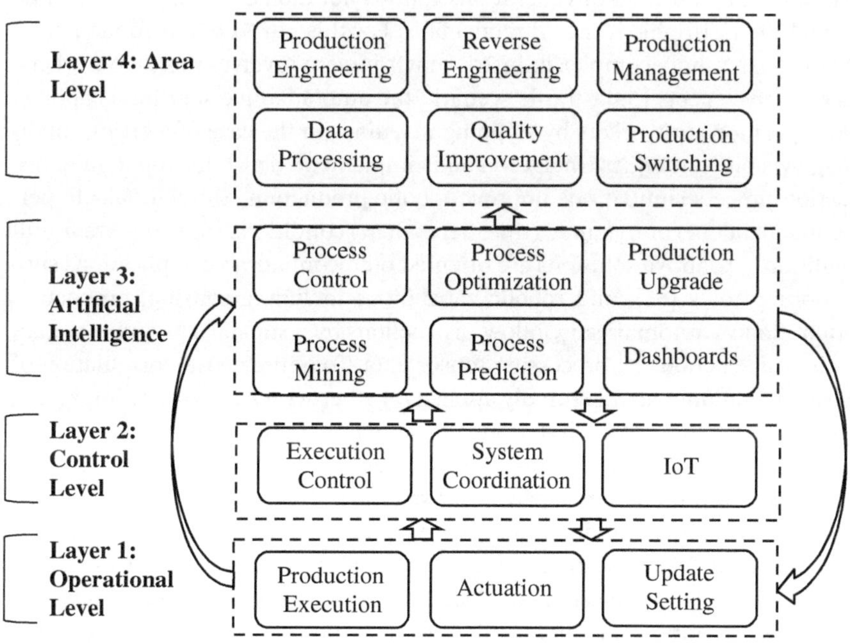

Figure 1.12 Multilevel structure of manufacturing industry processes integrating artificial intelligence tools.

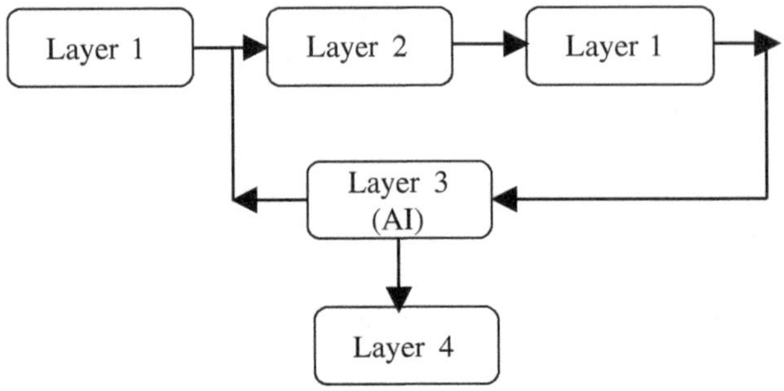

Figure 1.13 Artificial intelligence feedback system in manufacturing processes.

1.3 Intelligent Automatic Systems in Industries

Intelligent systems are implemented by AI algorithms, applied also for the intelligent movements of robotic arms [54]. Motors can control robotic arm movements by commands as coded outputs of the AI algorithms. In this scenario, robotic movements are actuated by image processing and image recognition [55]. In this way, image features classification by AI, and IoT data are fundamental to enable automatic processes. A central workstation [56] can manage the automatic processes by receiving control signals and by activating robotic operations. Robotic manipulation is an important topic in industry automation involving application fields such as self-adaptive robotic fingers optimized for collaborative robots [57]. The AI engine can optimize the production parameters. The parameter configuration can be performed *off-line* [58] or *in-line*. The *in-line* parameter setting by AI, is an issue for the modern adaptive solutions: auto-calibration in real time of machine working parameters can be catastrophic if the AI model fails. For this purpose, it is fundamental to construct robust AI models. A more complex industrial environment involves multiple workers, worker tracking, and security aspects [59]. In this scenario the auto-adaptive solution is applied also to check dangerous positions of workers by stopping machines in the case of alerting conditions, by applying image vision techniques. Image vision is implemented also for robot guidance solutions [11]. Production process simulations improve robotic production. Simulations are performed in static (time independent) or dynamic (time dependent) conditions [60], and are useful to define adaptive conditions. Adaptive solutions are oriented on reconfigurable applications supporting complex positioning tools [61], and robotic capabilities including sensing, production intelligence, and motion. Robots automatically follow instructions by a standard program or generate in real time mechatronic actions by processing sensor data. To optimize the formulation of instructions, image vision techniques are potentially applied together with AR tools, by improving the monitoring of the correct product assembly, by detecting and predicting defects detection, and by adapting actuations. The input data such as historical production data, data sensors and digitized information, are processed for a dynamic parameter setting and for the formulation of new high-performance programs. Historical production data are typically collected into a big data system able to contain massive data. Digitized information is coded into a dataset to be processed by the AI engine. The PLC is typically interfaced to an electronic board reading a standard program (standard parameter set), or actuating command coming from a protocol, by means the decoding of the AI outputs. The production process control is usually performed by image vision techniques and by IoT sensors placed inside the machines (internal IoT) or outside (external IoT). The architecture describing all the functions in advanced manufacturing processes is illustrated in Figure 1.14.

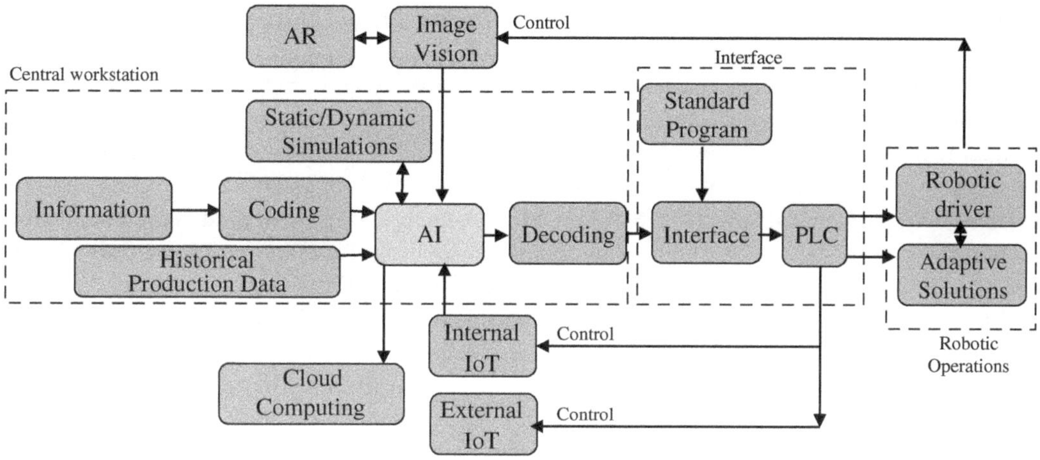

Figure 1.14 Block diagram of adaptive solutions in advanced manufacturing.

1.4 Technological Approaches to Transform the Production in Auto-Adaptive Control and Actuation Systems

System integration plays an important role in the upgrading of production processes. Technological approaches able to transform the production into an auto-adaptive system are horizontal, vertical, and end to end integration [62]. Figure 1.15 defines all the steps involved in the horizontal and vertical integration in Industry 5.0. Specifically, horizontal integration concerns the solution in adaptive robotics, the AI and AR implementations, and the hardware and software tools used for the production upgrade. Besides, vertical integration is mainly oriented on the supply chain including raw material management, production management, manufacturing process applications, assembly processes, logistics and commercial strategies.

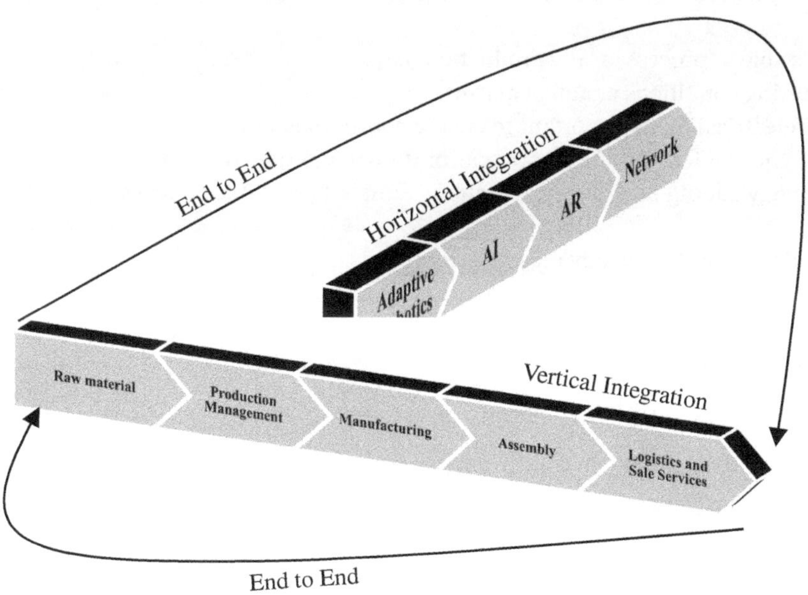

Figure 1.15 Scheme of horizontal, vertical and end to end integration in Industry 5.0.

Among the tools most used by companies to improve production processes are computer-aided manufacturing (CAM) and CAD. The use of these tools makes it possible to significantly speed up the development cycle thus, accelerating the response capacity and product adaptation to preferences and needs of customers. Furthermore, CNC machines can be managed by an AI engine by optimizing the production processing. 3D printing and additive manufacturing [63], complete the full scenario of SM facilities in modern fabrication processes where machine scheduling and process simulation are optimized by AI [64]. Furthermore, ANNs are able to predict the breakage of mechanical components and their degradation, permitting formulation of innovative processes [65]. Concerning CNC processing, the CAD tools are directly interconnected with the machines, and, by means of a feedback system, automatically adapt the working machine parameters, by controlling working tolerances (see the basic feedback system of Figure 1.16). The feedback system is improved by the AI engine able to control production processes, and change production processing according to the optimized CAD design.

Different data mining tools such as Rapid Miner, Weka, R Tool, Konstanz Information Miner (KNIME), and Orange Canvas [66] implement AI algorithms by means of programmable objects using GUIs. The GUIs represent an easy way to integrate the AI tools into the information system. The objects, named blocks or widgets, are programmed to extract data from the DB, and to provide outputs useful for the control and the actuation. Moreover, the GUIs are suitable for parameter setting: each object is a node [67] representing a specific function of data pre-processing, data processing and the output stages interconnected to the PLC. Each node is interconnected to other ones by structuring a data workflow representing the AI algorithm to run. The auto-adaptive solutions are executed by automatically updating the dataset to process, and by planning the automated execution of the AI workflow by cron. The interventions useful to upgrade the transformation of the production optimizing automation are:

- Adding new sensors monitoring critical production point to check.
- Changing some of the old machines of the old production line with new ones, addressing the production for an intelligent actuation.
- Adding new robotic elements able to develop new functions such as the *in line* automatic rejection of defects.
- Using smart local PCs able to process locally production data.
- Setting thresholds alerting conditions enabling alarms.
- Definition of priority electrical loads important to ensure continuous production.
- Interconnection and synchronization of PLC systems of the same production line.
- Upgrading the technology adding possible P&P enabling Simple Object Access Protocol (SOAP) services.
- Upgrading the ERP integrating AI and advanced BI algorithms.
- Implementation of a big data system suitable for data fusion and data collection.

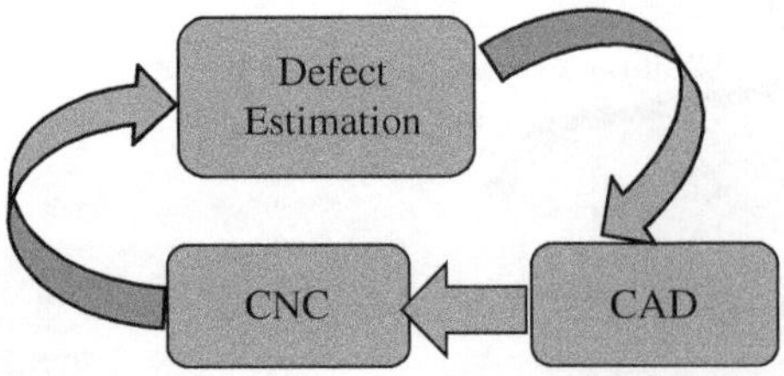

Figure 1.16 CAD and CNC interconnected by a feedback system.

- Introduction of new robotic and mechatronic systems based on a new concept of control.
- Upgrading the CAD design tools evaluating dynamically the working tolerances.
- Advanced RE and rapid prototyping.
- Possibility to launch production process simulations by introducing supervised workflow representing a specific production.
- Adopting AI algorithms integrated in the BPM tool as process mining.
- Introduction of AR technology supporting technology transfer and assembly processes.
- Development of the vertical, horizontal and end to end integrations by means of the new technologies.
- Improvement of the product quality.
- Upgrading the IT infrastructure.
- Implementation of auto-adaptive algorithms supporting production.
- Possibility to switch dynamically the production adapting the marketing in the new market where supply and demand are constantly changing.
- Using new technologies in order to decrease costs.
- Improvement of data security.
- Using CAM tools upgrading RE approaches.

1.5 Basic Concepts of Artificial Intelligence

AI in industrial applications is a discipline to aid data analysis, capable of self-learning from the data, to predict and classify data, and to formulate KPIs. The prediction is referred to a machine failure, or in more extensive cases to the evolution of production malfunctions. AI contributes to main areas in sustainable manufacturing such as:

- supply chain management
- predictive maintenance
- quality control
- energy consumption optimization

Of industrial interest is the self-classification of images classifying and predicting production defects. AI algorithms, as for deep learning (DL), can simultaneously process multiple variables with different "calculation weight." The AI models are specific for the application and therefore are not generic; the training dataset to be analyzed must be chosen accurately. A particular AI algorithm is the ANN. A neural network is composed of:

- **Input layer** (level designed to receive information from outside in order to learn to recognize and process the same information received).
- **Hidden layer** (layer connecting the input level with the output level, and helping the neural network to learn the complex relationships analyzed by the data; often there is more than one hidden level as for DL neural networks).
- **Output layer** (final layer showing the results of what the algorithm has processed).

Each connection between neurons is associated with a weight determining the importance of the input value (input variable x_i). The initial weights are set randomly. Each neuron, on the other hand, is characterized by an activation of a mathematical function. Figure 1.17a shows a basic neural network and Figure 1.17b shows a DL neural network. The ANNs are based on the concept of back-propagation error of the neural network training, consisting of a tuning of the weights

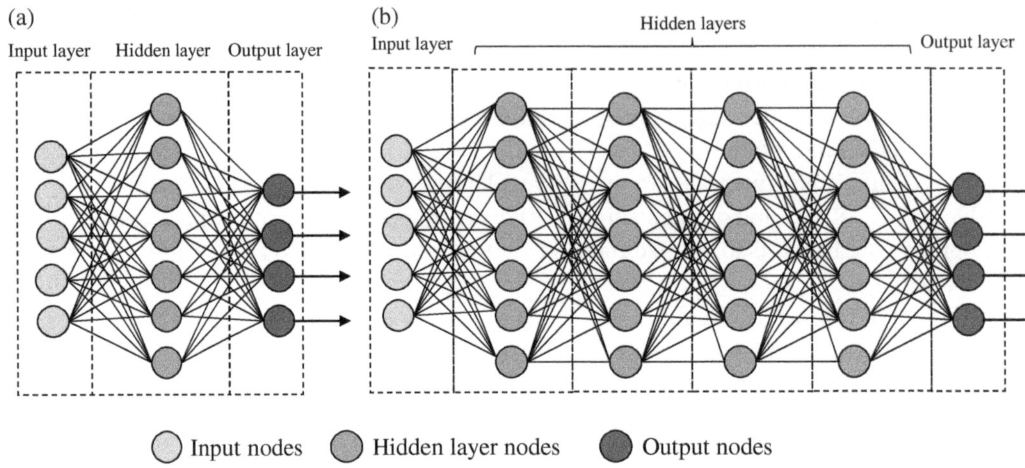

Figure 1.17 (a) Simple ANN. (b) DL neural network.

based on the error rate obtained in the previous iterations. The iteration is named an epoch. Proper refinement of the weights tuning ensures lower error rates, optimizing the model for the specific case of study. In Figure 1.18a is sketched the principle of the back propagation feedback system, enabling self-adjusting weights. Figure 1.18b shows a basic neural network implementing the mathematical function defining node output (unit step functions named activation functions).

The pseudocode of the ANN training process is as follows:

```
1. Train_ANN (fi, wi, oj)
2. For epochs = 1 to N Do
3.       While (j ≤ m) Do
4.             Randomly initialize wi = { w1 , w2 ..., wn};
5.             Input oj = { o1 , o2 ,..., om} in the input layer
forward propagate (fi· wi) though layers until is obtained the pre-
dicted result y;
6.             Compute the error e = y - y²;
7.             Back propagate e from the right to the left of
the ANN network through layers;
8.             Update wi;
9.       End While
10. End For
```

The pseudocode highlights that there are two mechanisms in the ANN network: the forward propagation of the estimation of the predicted output y, and the back propagation of the error function as sketched in Figure 1.18b. The output is estimated by considering the summation of the input contributions and is defined as:

$$y_j = f\left(\sum_{i=1}^{n} x_i w_i\right) \tag{1.10}$$

where f is the activation function. Some examples of activation functions are plotted in Figure 1.19, where the analytical forms are:

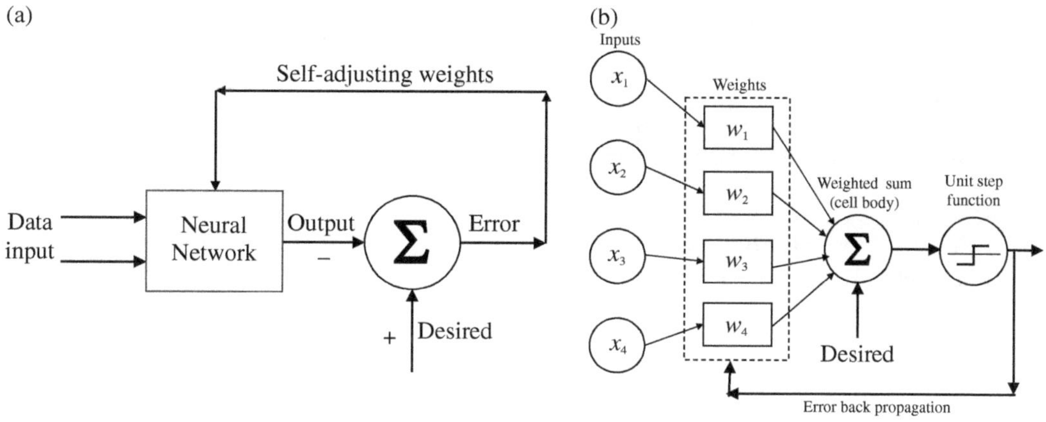

Figure 1.18 (a) Feedback system minimizing calculation error in the training model. (b) Neural network model implementing unit step function.

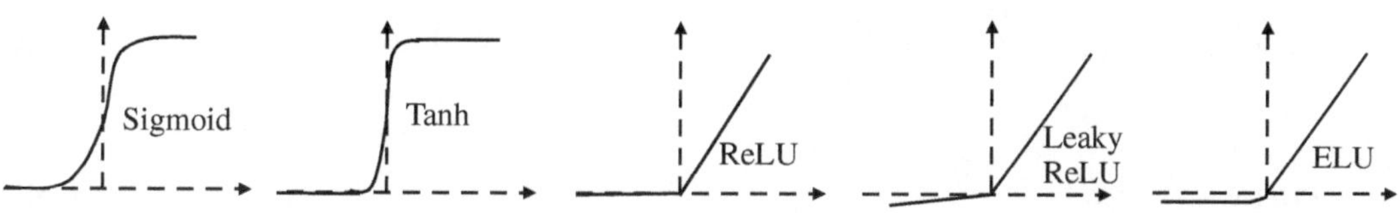

Figure 1.19 Basic mathematical functions defining activation functions.

$$\text{Sigmoid}(x) = \sigma(x) = \frac{1}{1 + \exp(-x)} \tag{1.11}$$

$$\text{Tanh}(x) = \frac{\exp(2x) - 1}{\exp(2x) + 1} \tag{1.12}$$

$$\text{ReLU}(x) = \max(0, x) \tag{1.13}$$

$$\text{Leaky ReLU}(x) = \max(0.1x, x) \tag{1.14}$$

$$\text{ELU}(x) = \begin{cases} x & x \geq 0 \\ \alpha(\exp(x) - 1) & x < 0 \end{cases} \tag{1.15}$$

Other mathematical activation functions are the following [68]:

$$\text{Maxout}(x) = \max\left(w_1^T x + b_1, w_2^T x + b_2\right) \tag{1.16}$$

$$\text{Softsign}(x) = \frac{1}{1 + |x|} \tag{1.17}$$

$$\text{Heaviside}(x) = \begin{cases} 1 & x \geq 0 \\ 0 & x < 0 \end{cases} \tag{1.18}$$

$$\text{HardSigmoid}(x) = \max\left(0, \min\left(1, \frac{(x+1)}{2}\right)\right) \tag{1.19}$$

$$\text{HardTanh}(x) = \begin{cases} -1 & \text{if } x < -1 \\ x & \text{if } 1 = x \leq 1 \\ 1 & \text{if } x > 1 \end{cases} \tag{1.20}$$

$$\text{Soft}\max(x_i) = \frac{\exp(x_i)}{\sum_j \exp(x_j)} \tag{1.21}$$

$$\text{Softplus}(x) = \log(1 + \exp^x) \tag{1.22}$$

$$\text{LReLU}(x) = \alpha x + x = \begin{cases} x & \text{if } x > 0 \\ \alpha x & \text{if } x \leq 0 \end{cases} \tag{1.23}$$

$$\text{PReLU}(x_i) = \begin{cases} x_i & \text{if } x_i > 0 \\ a_i x_i & \text{if } x_i \leq 0 \end{cases} \tag{1.24}$$

$$\text{SReLU}(x_i) = \begin{cases} t_i^r + a_i^r\left(x_i - t_i^r\right) & \text{for } x_i \geq t_i^r \\ x_i & \text{for } t_i^r < x_i < t_i^l \\ t_i^l + a_i^l\left(x_i - t_i^l\right) & \text{for } x_i \leq t_i^l \end{cases} \tag{1.25}$$

$$\text{PELU}(x) = \begin{cases} cx & \text{if } x > 0 \\ \alpha \exp^{\frac{x}{b}} & \text{if } x \leq 0 \end{cases} \tag{1.26}$$

$$\text{RReLU}(x_{ji}) = \begin{cases} x_{ji} & \text{if } x_{ji} \geq 0 \\ a_{\text{ji}} x_{ji} & \text{if } x_{ji} < 0 \end{cases} \tag{1.27}$$

$$\text{SELU}(x) = \begin{cases} x & \text{if } x > 0 \\ \alpha \exp(x) - \alpha & \text{if } x \leq 0 \end{cases} \tag{1.28}$$

$$\text{Swish}(x) = \frac{x}{1 + \exp(-x)} \tag{1.29}$$

$$\text{ELiSH}(x) = \begin{cases} \dfrac{x}{1 + \exp(-x)} & \text{if } x \geq 0 \\ \dfrac{\exp(x) - 1}{1 + \exp(-x)} & \text{if } x < 0 \end{cases} \tag{1.30}$$

$$\text{HardELiSH}(x) = \begin{cases} x \times \max\left(0, \min\left(1, \frac{x+1}{2}\right)\right) & \text{if } x \geq 0 \\ \left(\exp(x) - 1\right) \times \max\left(0, \min\left(1, \frac{x+1}{2}\right)\right) & \text{if } x < 0 \end{cases} \tag{1.31}$$

The activation function represents a basic research element of considerable importance. The correct choice of the activation function defines the best implementation of the logic defining the outputs. The analytical model must therefore be appropriately weighted by the various variables and must be "calibrated" for the specific case study. Another important aspect is the ability of the activation function to self-adapt [69] to the specific case study providing a certain flexibility [70]. Of particular interest is the possibility to consider a combination of activation functions (activation ensemble [71]). The approach to follow is therefore to define a flexible and modular activation function as is the case for the adaptive spline activation function [72].

Concerning the training models, the full dataset of the neural network is divided into a training set, validation set, and test set (Figure 1.20). In particular, the function of the training dataset is to fit the model; the validation set is a small partition of the full dataset able to previously estimate prediction error of the selected model; finally, the test set is used for testing the final model. A correct choice of the three parts depends on the SNR of the full dataset.

The intelligent algorithms which constitute the core of the Industry 5.0 system, are classified in Figure 1.21. The generic algorithms of data analysis, including statistical data processing and classification roles, are classified as computer science and data mining algorithms; other high-level algorithms are included in AI algorithms. The main function of the engine processor enables the managing of big data and of data processing. As previously mentioned, a basic concept of algorithm classification is in the learning supervision: in a supervised learning model, the algorithm learns on a pre-selected dataset with specific labeled attributes filtered by the user; in an unsupervised model all the attributes are unlabeled, and the algorithm tries to extract features and patterns without a guideline. The supervised algorithms mainly support the user to find a solution for a specific problem such as finding a specific defect category or a specific failure system.

A simple way to analyze the data trend is regression analysis, where a linear approach is enough to model a relationship between dependent variables and independent ones (see plot example of Figure 1.22a). Typically, line regression provides information about a linear trend prediction. The classification is based on the concept of data categorization: data are classified by considering a

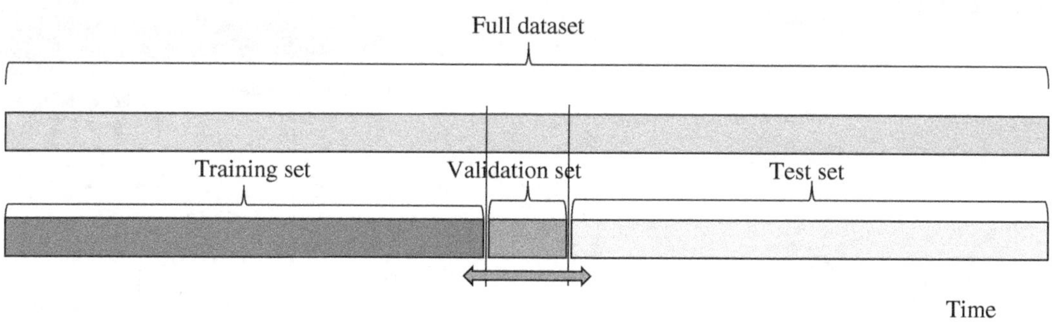

Figure 1.20 Supervised artificial network model: partitioning of the available dataset into training set, validation set, and test set.

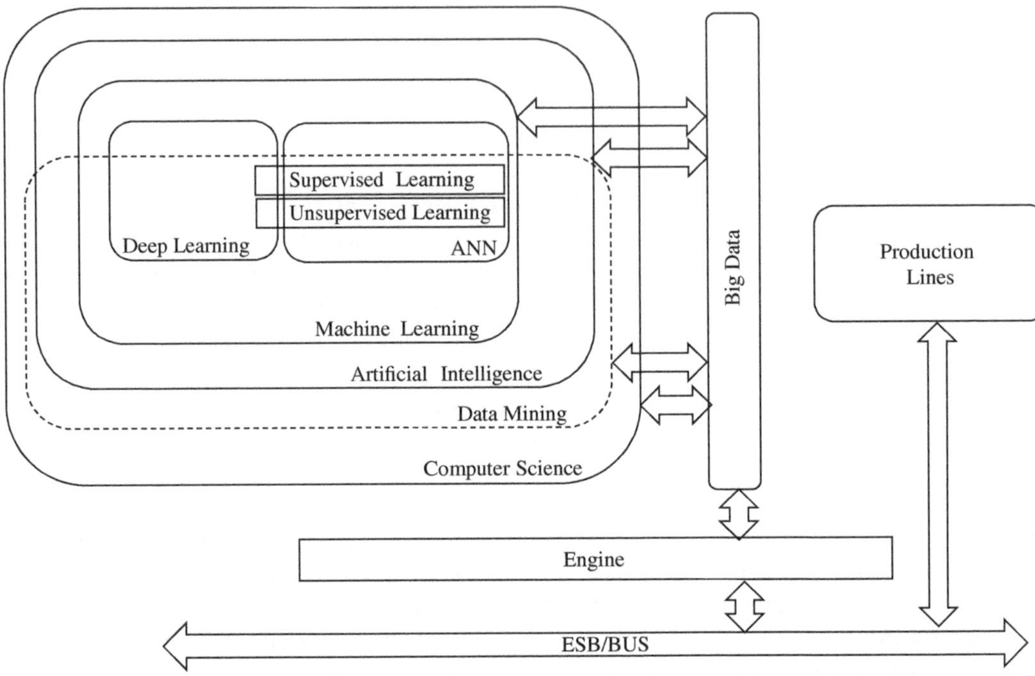

Figure 1.21 Algorithm classification and Industry 5.0 facilities.

generic classification pattern curve defined as shown in Figure 1.22b, where all data above and below the curve appertain to a particular class. Data classification is often used to solve supervised learning models. Finally, data clustering is based on the grouping of datasets forming clusters having similar features. Data clustering is commonly used to solve unsupervised learning models. All analysis can be performed by analyzing a multidimensional domain by taking into account the variable time, which is fundamental for forecasting approaches.

The ensemble approach is an alternative method for data classification. An ensemble is a set of classifiers that learn a target function. By combining different outputs of several classifiers, the risk of selecting a poorly performing classifier is reduced. The typical ensemble procedure is provided by the following pseudocode where T denotes the original training dataset, κ is the number of base classifiers, and B is the test data:

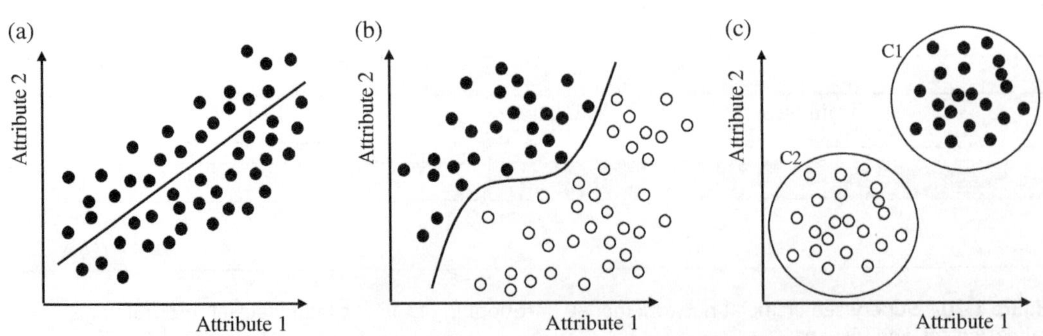

Figure 1.22 (a) Regression analysis, (b) data classification, and (c) data clustering.

```
1. For i=1 to k do
2. Create training set T_i from T.
3. Build a base classifier C_i from T.
4. end For
5. for each test record x∈ B do:
6. C*(x)=Vote (C_1(x),C_2(x),…,C_k(x))
7. end For
```

The individual predictions of classifiers are combined to classify new samples, thus optimizing the classifier performance on a specific domain. Figure 1.23 shows an example of an ensemble approach based on the best classification of a single feature.

The RFo method [73] is a class of ensemble methods specifically designed for DT classifiers. The main property is to combine predictions performed by more DT models: each tree is generated from a part of the training dataset and the values are grouped into a set of random vectors. This algorithm is structured as follows (Figure 1.24):

- *F* input features are randomly selected to split at each node (step 1 of creation of random vectors).
- A linear combination of the input features is created to split at each node (step 2 of using a random vector to build multiple DTs).
- A combination of DTs is created (step 3).

The RFo classification technique is also applied in image processing detecting defect features. The logic of the DT algorithm is reported by the following pseudocode:

```
Decision_Tree Function.
1. Compute Gain values for all attributes and select an attribute
having the highest value creating a node for that attribute.
2. Make a branch from this node for every value of the attribute.
3. Assign all possible values of the attributes to branches.
```

Follow each branch partitioning the dataset to be only instances whereby the value of the branch is present (or for similar values) and then go back to 1.

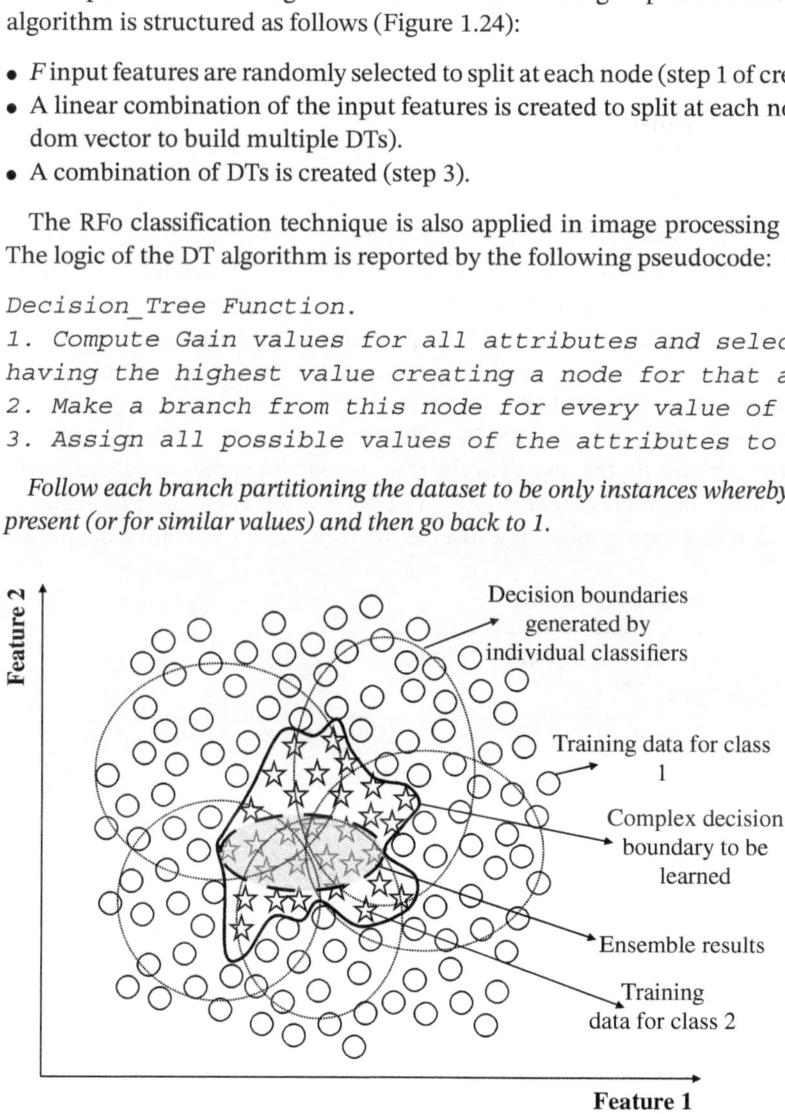

Figure 1.23 Ensemble method and classification.

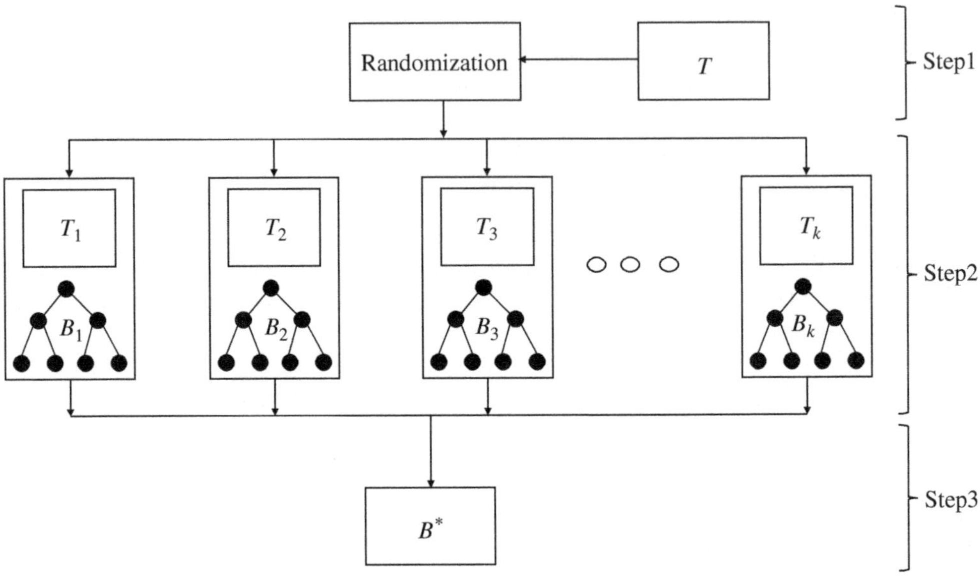

Figure 1.24 Ensemble method and classification.

A particular feature of neural network algorithms are the long short-term memory (LSTM) networks, which are artificial recurrent neural network (RNN) architectures [74] used for DL applications. The basic architecture is shown in Figure 1.25a: the structure is a cell (cell state), an input gate (input gate state activated at the time step *t*), an output gate (output gate state at the time step *t*), and a forget gate (forget gate state at the time step *t*). The gates calculate their activations at time step *t*, by taking into account the activation of the memory cell *C* at time step *t*-1. Figure 1.25b shows the basic model of a network composed of LSTM nodes.

Another important parameter used for the choice of the training dataset is the correlation coefficient, indicating a relationship between two attributes. The correlation coefficient indicates a value between −1 and 1 (−1 when the attributes are inversely correlated, +1 when the attributes

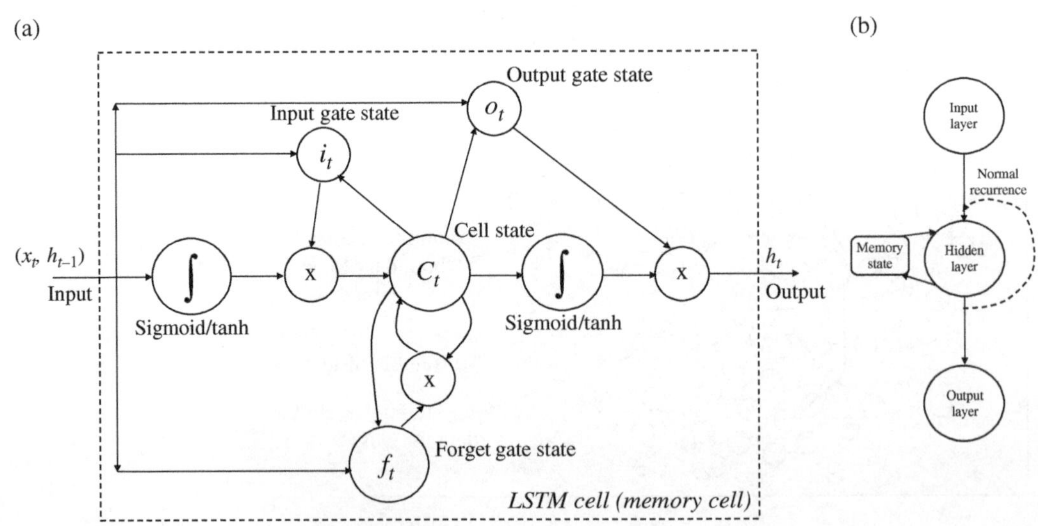

Figure 1.25 (a) LSTM unit cell. (b) LSTM network and its memory.

Table 1.9 Example of a correlation matrix for a model with five attributes (v_1, v_2, v_3, v_4, v_5).

	v_1	v_2	v_3	v_4	v_5
v_1	**1**	0.4	−0.8	0.6	−0.3
v_2	0.4	**1**	0.1	0.8	−0.5
v_3	−0.8	0.1	**1**	0.2	0.1
v_4	0.6	0.8	0.2	**1**	−0.3
v_5	−0.3	−0.5	0.1	−0.3	**1**

are absolutely correlated, and 0 represents no correlations). The correlation coefficient considers the magnitudes of the deviations from the mean value. In particular, the Pearson–Bravais correlation coefficient is estimated as the ratio between the covariance of the two variables and the product of their standard deviations as follows [75]:

$$-1 \leq \rho_{x,y} = \frac{\sigma_{x,y}}{\sigma_x \sigma_y} = \frac{\sum_{i=1}^{n}(x_i - \mu_x)(y_i - \mu_y)}{\sqrt{\sum_{i=1}^{n}(x_i - \mu_x)^2}\sqrt{\sum_{i=1}^{n}(y_i - \mu_y)^2}} \leq +1 \tag{1.32}$$

The correlation coefficients are plotted in a correlation matrix with the structure shown in Table 1.9.

1.6 Knowledge Upgrading in Industries

KG in industry is achieved by upgrading the information system [32] in different stages, as illustrated in Figure 1.26. The knowledge base (KB) is the starting point representing the industry knowhow provided by the digital information. A first increase of the knowledge is obtained by adding new applications concerning software and tools optimizing production such as:

- RE tools.
- Design and simulation tools.
- Rapid prototyping tools.
- Software managing IoT sensor network and monitoring production.
- Building information models implemented in AR systems.
- Digital sensors to be integrated in analogical machines.
- HMI as data entry systems for data digitization according to the Industry 4.0 principle.

All digitized data are typically managed by a unique ERP collecting all data into DB systems. Digitized data travel through the ESB or LAN bus line linking all the software and the information of the industry network [32]. The KG is performed by reading the outputs of advanced analyses related mainly to big data analytics and AI results. The KG is fundamental for industry research. The information system upgrade involves the front-office, the back-office, and the whole network infrastructure, powering contact management, E-commerce, customer relationship management (CRM), BPM, E-learning, and BI algorithms providing association rules and cost predictions [76]. Big data and data mining are important for BI decision-making processes, and are able to [77, 78]:

- Find possible inefficiencies.
- Define human resources KPIs.
- Efficiently manage the scheduling of the activities.
- Classify human resource skills.
- Classify service typologies.
- Define strategic marketing.
- Classify industry locations based on efficiency estimation.

The innovative technology to be implemented in industrial sectors is listed in Table 1.10 (Horizon 2020 topics).

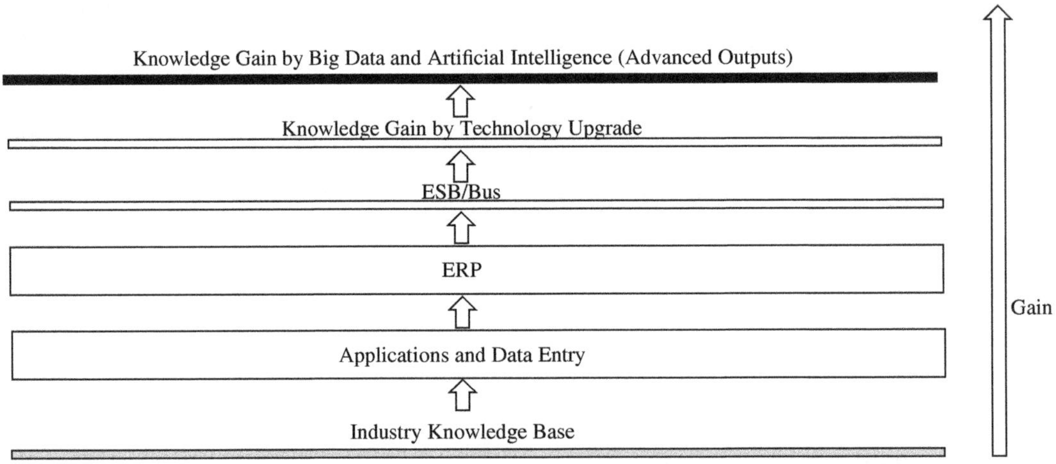

Figure 1.26 Knowledge gain in industry.

Table 1.10 Industrial sectors and advanced application fields.

Industrial sectors	Application fields
• Processes and industrial installations • Manufacturing and advanced transformation	• Methodologies and standards for the machine and system design in complex manufacturing • Tools for design and implementation of high-quality products • Automation and integration of complex manufacturing systems for real-time productions • Information technologies for the integration of supply chain and decision-making process • Control technologies, predictive, maintenance, diagnosis of manufacturing systems improving efficiency and life cycles • Technologies and software solutions for real manufacturing and logistics • Mechatronic components for efficiency production and products • New machines and eco-sustainable systems to optimize energy efficiency and the environmental impact of production systems • Innovative materials for components, machines, and systems • Intelligent adaptive production systems

Table 1.10 (Continued)

Industrial sectors	Application fields
Mobility and transport	• Road/sea/rail transport technologies and solutions to reduce environmental impact, optimize safety, quality, performance, and energy efficiency • Environmentally transport technologies for sustainability • Intelligent transport systems • Intelligent inspection systems • Smart road technologies
Aerospace and aeronautics	• New materials and production processes • Predictive maintenance for aircraft • Eco-friendly engines • Technologies for reduction of environmental impact • Simulation methodologies for integrated design in the aeronautical field • Technologies for autonomous systems • Advanced air traffic management • Integrated technologies for security • Space transport technologies • Space advance operational technologies • Remote sensing and data navigation technologies
Biotechnologies	• Genomics, meta-genomics, proteomics, and metabolomics • Technologies for pharmaceutical chemistry • Biomarkers • Biomolecular production • Technologies and methods for molecular imaging • Minimally invasive technologies • Technologies for advanced and precision therapies • Tissue engineering • Bioinformatics • Technologies related to industrial processes based on biotechnology (chemistry, health, agri-food, etc.)
Chemistry	• Technologies for reducing environmental impact (technologies for recovery of polymeric materials, substitution of high-risk substances, etc.) • Technologies from renewable resources (biorefinery processes, biochemical technologies, raw materials for alternatives energy, etc.) • Nanomaterials for packaging and food • Nanomaterials for chemical processes
Energy and environment	• Transport technologies and energy storage • Energy production technologies and advanced materials • Technologies for the use of biomass and biofuel • Technologies for solar energy enhancement • Technologies for energy efficiency • Smart grid, smart metering, and smart energy • Technologies for the reduction of emissions • Technologies for energy intelligent networks • Renewable energies • Technologies for environmental protection • Energy efficiency and sustainability of buildings • Safety of the built environment

(Continued)

Table 1.10 (Continued)

Industrial sectors	Application fields
Microelectronics and semiconductors	• Integration of electronic systems on silicon • Photovoltaic applications • Alternative materials to silicon • Heterogeneous integrations • Manufacturing techniques and design methods • Sensor technologies integrated on silicon • Silicon photonics
Informatic and telecommunications	• New information technologies (tools and technologies to manage processes, big data storage, human machine interfaces, etc.) • Application platforms of information technologies • New network and service architectures • Mobile networks and services • Technologies for broadband networks • Technologies for information mobility • Technologies for network security • Advanced interfaces and robots • Technologies for systems innovation • Optical and wireless communication • Technologies related to microelectronics, nanoelectronics, and photonics
Advanced materials	• Technologies related to functional, multifunctional, and structural materials (self-repairing materials, self-adaptive materials, biocompatible materials) • Development and transformation of materials (dynamic production) • Techniques and new and innovative systems in assembly, self-assembly, separation, and disassembly • Technologies and materials for low production of carbon emissions • Technologies related to materials for creative industries • Metrology, characterization, standardization, and quality control • Technologies related to optimization of the use of materials
Nanotechnologies	• Nanomaterials, nanodevices, and next generation nanosystems • Scientific tools and platforms for evaluation and risk management throughout the lifecycle of nanomaterials and of nanosystems • New production of nanomaterials in advanced devices as innovative products • Synthesis and fabrication of nanomaterials • Development support technologies for nanomaterial characterization and production • Modeling and design of devices in nanoscale

The enabling technologies of the application fields in the Industry 5.0 scenario are:

- Nanotechnologies.
- Micro- and nanoelectronics.
- Biotechnologies.
- Advanced functionalized materials.
- Photonics.

- Advanced materials.
- Advanced production technologies.
- AI and big data systems.
- Biomaterials.
- Virtual reality and AR.
- Lab-on-chip.
- Advanced electromagnetic sensors and compatibility.
- Advanced high temperature materials.
- Advanced software and hardware production technologies.
- Diagnostic inspection technologies.
- Innovative systems for diagnostics.

References

1 Vaidya, S., Prashant, A., and Bhosle, S. (2018). Industry 4.0 – a glimpse. *Procedia Manufacturing* 20 (1): 233–238.

2 Moon, S.H. (2018). Industry 4.0 for advanced manufacturing and its implementation. *Eurasian Journal of Analytical Chemistry* 13 (6): 491–497.

3 Ruppert, T., Jaskó, S., Holczinger, T., and Abonyi, J. (2018). Enabling technologies for operator 4.0: a survey. *Applied Sciences* 8 (1650): 1–19.

4 Lampropoulos, G., Siekas, K., and Anastasiadis, T. (2019). Internet of Things in the context of industry 4.0: an overview. *International Journal of Entrepreneurial Knowledge* 7 (1): 4–19.

5 Novak-Marcincin, J., Barna, J., Janak, M., and Novakova-Marcincinova, L. (2013). Augmented reality aided manufacturing. *Procedia Computer Science* 25 (1): 23–31.

6 Segovia, D., Mendoza, M., Mendoza, E., and González, E. (2015). Augmented reality as a tool for production and quality monitoring. *Procedia Computer Science* 75 (1): 291–300.

7 Bottani, E. and Vignali, G. (2019). Augmented reality technology in the manufacturing industry: a review of the last decade. *IISE Transactions* 51 (3): 284–310.

8 Cioffi, R., Travaglioni, M., Piscitelli, G. et al. (2020). Artificial intelligence and machine learning applications in smart production: progress, trends, and directions. *Sustainability* 12 (492): 1–26.

9 Lee, W.J., Wu, H., Yun, H. et al. (2019). Predictive maintenance of machine tool systems using artificial intelligence techniques applied to machine condition data. *Procedia CIRP* 80 (1): 506–511.

10 Pandarakone, S.E., Mizuno, Y., and Nakamura, H. (2019). A comparative study between machine learning algorithm and artificial intelligence neural network in detecting minor bearing fault of induction motors. *Energies* 12 (2105): 1–14.

11 Pérez, L., Rodríguez, Í., Rodríguez, N. et al. (2016). Robot guidance using machine vision techniques in industrial environments: a comparative review. *Sensors* 16 (335): 1–26.

12 Massaro, A. and Galiano, A. (2020). Image processing and post-data mining processing for security in industrial applications: security in industry. In: Handbook of Research on Intelligent Data Processing and Information Security Systems (eds. S.M. Bilan and S.I. Al-Zoubi), 117–146. Hershey, PA: IGI Global.

13 Massaro, A. and Galiano, A. (2020). Infrared thermography for intelligent robotic systems in research industry inspections: thermography in industry processes. In: Handbook of Research on Advanced Mechatronic Systems and Intelligent Robotics (ed. M.K. Habib), 98–125. Hershey, PA: IGI Global.

14 Tan, C.L. and Mohseni, H. (2018). Emerging technologies for high performance infrared detectors. *Nanophotonics* 7 (1): 167–197.

15 Kaufmann, R., Isella, G., Sanchez-Amores, A. et al. (2011). Near infrared image sensor with integrated germanium photodiodes. *Journal of Applied Physics* 110 (2): 1–6.

16 Kumar, V., Hallqvist, C., and Ekwall, D. (2017). Developing a framework for traceability implementation in the textile supply chain. *Systems* 5 (33): 1–21.

17 Tzoulis, I. and Andreopoulou, Z. (2013). Emerging traceability technologies as a tool for quality wood trade. *Procedia Technology* 8 (1): 606–611.

18 Agrawal, T.K., Koehl, L., and Campagne, C. (2018). A secured tag for implementation of traceability in textile and clothing supply chain. *The International Journal of Advanced Manufacturing Technology* 99 (1): 2563–2577.

19 Chen, R.-S., Chen, C.-C., Yeh, K.C. et al. (2008). Using RFID technology in food produce traceability. *WSEAS Transactions on Information Science and Applications* 5 (11): 1551–1560.

20 Kelepouris, T., Pramatari, K., and Doukidis, G. (2007). RFID-enabled traceability in the food supply chain. 107 (2): 183–200.

21 Sethi, P., Sarangi, S., and R. (2017). Internet of Things: architectures, protocols, and applications. *Journal of Electrical and Computer Engineering* 2017 (9324035): 1–25.

22 Ke, C.K., Wu, M.Y., Chan, Y.W., and Lu, K.C. (2018). Developing a BLE Beacon-based location system using location fingerprint positioning for smart home power management. *Energies* 11 (3464): 1–18.

23 Lin, Y.-W. and Lin, C.-Y. (2018). An interactive real-time locating system based on Bluetooth low-energy beacon network. *Sensors* 18 (1637): 1–17.

24 Triantafyllou, A., Sarigiannidis, P., and Lagkas, T.D. (2018). Network protocols, schemes, and mechanisms for Internet of Things (IoT): features, open challenges, and trends. *Wireless Communications and Mobile Computing* https://doi.org/10.1155/2018/5349894.

25 Cilfone, A., Davoli, L., Belli, L., and Ferrari, G. (2019). Wireless mesh networking: an IoT-oriented perspective survey on relevant technologies. *Future Internet* 11 (99): 1–35.

26 Froiz-Míguez, I., Fernández-Caramés, T.M., Fraga-Lamas, P., and Castedo, L. (2018). Design, implementation and practical evaluation of an IoT home automation system for fog computing applications based on MQTT and ZigBee-WiFi sensor nodes. *Sensors* 18 (2660): 1–42.

27 Amelia, A., Julham, Sundawa, B.V. et al. (2017). Implementation of the RS232 communication trainer using computers and the ATMEGA microcontroller for interface engineering courses. *Journal of Physics: Conference Series* 890 (012095): 1–6.

28 Dey, M. (2012). Comparision of data transfer protocols over USB. *International Journal of Engineering Research & Technology* 1 (9): 1–6.

29 Riberio, F.M., Costa, T.S., Baratella, A. et al. (2014). Comparative analysis of industrial network profinet, ethernet/IP, and HSE. *International Journal of Innovative Computing, Information and Control* 10 (5): 1931–1945.

30 Jaloudi, S. (2019). Communication protocols of an industrial Internet of Things environment: a comparative study. *Future Internet* 11 (66): 1–18.

31 Savaglio, C., Ganzha, M., Paprzycki, M. et al. (2019). Agent-based Internet of Things: state-of-the-art and research challenges. *Future Generation Computer Systems* 102 (1): 1038–1053.

32 Massaro, A., Calicchio, A., Maritati, V. et al. (2018). A case study of innovation of an information communication system and upgrade of the knowledge base in industry by ESB, artificial intelligence, and big data system integration. *International Journal of Artificial Intelligence and Applications (IJAIA)* 9 (5): 27–43.

33 Burhan, M., Asif Rehman, R., Khan, B., and Kim, B.-S. (2018). IoT elements, layered architectures and security issues: a comprehensive survey. *Sensors* 18 (2796): 1–37.

34 Arsan, T., Günay, F., and Kaya, E. (2014). Implementation of application for huge data file transfer. *International Journal of Wireless & Mobile Networks (IJWMN)* 6 (4): 27–46.

35 Riabov, V.V. and SMTP (Simple Mail Transfer Protocol) (2007). The Handbook of Computer Networks, Volume 2, LANs, MANs, WANs, the Internet, and Global, Cellular, and Wireless Networks (ed. H. Bidgoli), 388–406. Hoboken, NJ: Wiley.

36 Nguyen, T.S. and Huynh, T.-H. (eds.) (2015). Design and implementation of Modbus slave based on ARM Platform and FreeRTOS environment. *Proceedings of International Conference on Advanced Technologies for Communications (ATC)*, Ho Chi Minh City, Vietnam (14–16 October 2015). Piscataway, NJ: IEEE.

37 Rajinder, S. and Satish, K. (2016). An overview of world wide web protocol (Hypertext Transfer Protocol and Hypertext Transfer Protocol Secure). *International Journal of Advanced Research in Computer Science and Software Engineering* 6 (5): 396–399.

38 Ansari, D.B., Rehman, A.U., and Mughal, R.A. (2018). Internet of Things (IoT) protocols: a brief exploration of MQTT and CoAP. *International Journal of Computer Applications* 179 (27): 9–14.

39 Tukade, T.M. and Banakar, R.M. (2018). Data transfer protocols in IoT – an overview. *International Journal of Pure and Applied Mathematics* 118 (16): 121–138.

40 Kastner, W., Neugschwandtner, G., and Kogler, M. (eds.) (2006). An open approach to Eib/Knx software development. *Proceedings of 6th IFAC International Conference*, Puebla, Mexico (14–25 November 2005). Elsevier Ltd.

41 Massaro, A., Manfredonia, I., Galiano, A., and Contuzzi, N. (eds.) (2019). Inline image vision Technique for tires industry 4.0: quality and defect monitoring in tires assembly. *Proceedings of 2019 IEEE International Workshop on Metrology for Industry 4.0 and IoT*, Naples, Italy (4–6 June 2019). Piscataway, NJ: IEEE.

42 Sathya, R. and Abraham, A. (2013). Comparison of supervised and unsupervised learning algorithms for pattern classification. *International Journal of Advanced Research in Artificial Intelligence* 2 (2): 34–38.

43 Massaro, A., Manfredonia, I., Galiano, A. et al. (2019). Sensing and quality monitoring facilities designed for pasta industry including traceability, image vision and predictive maintenance. *Proceeding of 2019 IEEE International Workshop on Metrology for Industry 4.0 and IoT*, Naples, Italy (4–6 June 2019). Piscataway, NJ: IEEE.

44 Massaro, A., Manfredonia, I., Galiano, A., and Xhaysa, B. (2019). Advanced process defect monitoring model and prediction improvement by artificial neural network in kitchen manufacturing industry: a case of study. *Proceeding of IEEE International Workshop on Metrology for Industry 4.0 and IoT*, Naples, Italy (4–6 June 2019). Piscataway, NJ: IEEE.

45 Massaro, A., Vitti, V., and Galiano, A. (2018). Automatic image processing engine oriented on quality control of electronic boards. *Signal & Image Processing: An International Journal (SIPIJ)* 9 (2): 1–14.

46 Kiran, B.R., Thomas, D.M., and Parakkal, R. (2018). An overview of deep learning based methods for unsupervised and semi-supervised anomaly detection in videos. *Journal of Imaging* 4 (36): 1–25.

47 Xu, X., Zheng, H., Guo, Z. et al. (2019). SDD-CNN: small data-driven convolution neural networks for subtle roller defect inspection. *Applied Sciences* 9 (1364): 1–16.

48 Perez, H., Tah, J.H.M., and Mosavi, A. (2019). Deep learning for detecting building defects using convolutional neural networks. *Sensors* 19 (3556): 1–22.

49 Lin, C.-S., Huang, Y.-C., Chen, S.-H. et al. (2018). The application of deep learning and image processing technology in laser positioning. *Applied Sciences* 8 (1542): 1–13.

50 Iskra, P. and Hernàndez, R.E. (2012). Toward a process monitoring of CNC wood router. Sensor selection and surface roughness prediction. *Wood Science and Technology* 46 (1): 115–128.

51 Iskra, P. and Hernàndez, R.E. (2009). The influence of cutting parameters on the surface quality of routed paper birch and surface roughness prediction modeling. *Wood and Fiber Science* 41 (1): 28–37.

52 Contuzzi, N., Massaro, A., Manfredonia, I. et al. (2019). A decision making process model based on a multilevel control platform suitable for Industry 4.0. *Proceeding of 2019 IEEE International Workshop on Metrology for Industry 4.0 and IoT*, Naples, Italy (4–6 June 2019). Piscataway, NJ: IEEE.

53 De Smedt, J., Hasić, F., Vanden Broucke, S.K.L.M., and Vanthienen, J. (2017). Towards a holistic discovery of decisions in process-aware information systems. *Proceedings of the International Conference on Business Process Management*, Barcelona, Spain (10–15 September 2017). Cham: Springer Nature.

54 Tirgul, C.S. and Naik, M.R. (2016). Artificial intelligence and robotics. *International Journal of Advanced Research in Computer Engineering & Technology (IJARCET)* 5 (6): 1787–1793.

55 Kaura, H., Honrao, V.K., Patil, S., and Shetty, P. (2013). Gesture controlled robot using image processing. *International Journal of Advanced Research in Artificial Intelligence* 2 (5): 69–77.

56 Moulianitis, V.C. and Aspragathos, N.A. (2015). IT and mechatronics in industrial robotic workcell design and operation. In: Encyclopedia of Information Science and Technology, 3e (ed. M. Khosrow-Pour), 440–455. Hershey, PA: IGI Global.

57 Birglen, L. (2019). Design of a partially-coupled self-adaptive robotic finger optimized for collaborative robots. *Autonomous Robots* 43 (2): 523–538.

58 Jamshidi, P., Cámara, J., Schmerl, B. et al. (2019). Machine learning meets quantitative planning: enabling self-adaptation in autonomous robots. *Proceedings of the 14th International Symposium on Software Engineering for Adaptive and Self-Managing Systems*, Montreal, Canada (25–26 May 2019). Piscataway, NJ: IEEE.

59 Khalid, A., Kirisci, P., Ghrairi, Z. et al. (2016). A methodology to develop collaborative robotic cyber physical systems for production environments. *Logistics Research* 9 (23): 1–15.

60 Alcácer, V. and Cruz-Machado, V. (2019). Scanning the Industry 4.0: a literature review on technologies for manufacturing systems. *Engineering Science and Technology, an International Journal* 22 (1): 899–919.

61 Vidal, F., Álvarez, M., González, R. et al. (2011). Development of a flexible and adaptive robotic cell for small batch manufacturing. *Contemporary Materials* 2 (1): 1–12.

62 Wang, S., Wan, J., Li, D., and Zhang, C. (2016). Implementing smart factory of industry 4.0: an outlook. *International Journal of Distributed Sensor Networks* 12 (1): 1–10.

63 Mehrpouya, M., Dehghanghadikolaei, A., Fotovvati, B. et al. (2019). The potential of additive manufacturing in the smart factory industrial 4.0: a review. *Applied Sciences* 9 (3865): 1–34.

64 Shahzad, A. and Mebarki, N. (2016). Learning dispatching rules for scheduling: a synergistic view comprising decision trees, tabu search and simulation. *Computers* 5 (3): 1–16.

65 Zhang, X., Xiao, L., and Kan, J. (2015). Degradation prediction model based on a neural network with dynamic windows. *Sensors* 15 (3): 6996–7015.

66 Pynam, V., Spanadna, R.R., and Srikanth, K. (2018). An extensive study of data analysis tools (Rapid Miner, Weka, R Tool, Knime, Orange). *International Journal of Computer Science and Engineering* 5 (9): 4–11.

67 Massaro, A., Maritati, V., Savino, N. et al. (2018). A study of a health resources management platform integrating neural networks and DSS telemedicine for homecare assistance. *Information* 9 (176): 1–20.

68 Nwankpa, C.E., Ijomah, W., Gachagan, A., and Marshall, S. (2018). Activation functions: comparison of trends in practice and research for deep learning. arXiv:1811.03378v1.

69 Agostinelli, F., Hoffman, M., Sadowski, P., and Baldi, P. (2015). Learning activation functions to improve deep neural networks. arXiv:1412.6830v3.

70 Wan, Y., Li, Y., Song, Y., and Rong, X. (2020). The influence of the activation function in a convolution neural network model of facial expression recognition. *Applied Sciences* 10 (5): 1–20.

71 Harmon, M. and Klabjan, D. (2017). Activation ensembles for deep neural networks. arXiv:1702.07790v1.

72 Guarnieri, S., Piazza, F., and Uncini, A. (1999). Multilayer feedforward networks with adaptive spline activation function. *IEEE Transactions on Neural Networks* 10 (3): 672–683.

73 Biau, G. (2012). Analysis of a random forests model. *Journal of Machine Learning Research* 13: 1063–1095.

74 Massaro, A., Maritati, V., Giannone, D. et al. (2019). LSTM DSS automatism and dataset optimization for diabetes prediction. *Applied Sciences* 9 (17): 1–22.

75 Massaro, A., Meuli, G., and Galiano, A. (2018). Intelligent electrical multi outlets controlled and activated by a data mining engine oriented to building electrical management. *International Journal on Soft Computing, Artificial Intelligence and Applications (IJSCAI)* 7 (4): 1–20.

76 Massaro, A., Lisco, P., Lombardi, A. et al. (2019). A case study of research improvements in a service industry upgrading the knowledge base of the information system and the process management: data flow automation, association rules and data mining. *International Journal of Artificial Intelligence and Applications (IJAIA)* 10 (1): 25–46.

77 Massaro, A., Vitti, V., Lisco, P. et al. (2019). A business intelligence platform implemented in a big data system embedding data mining: a case of study. *International Journal of Data Mining & Knowledge Management Process (IJDKP)* 9 (1): 1–20.

78 Massaro, A., Leogrande, A., Lisco, P. et al. (2019). Innovative BI approaches and methodologies implementing a multilevel analytics platform based on data mining and analytical models: a case of study in roadside assistance services. *International Journal on Soft Computing, Artificial Intelligence and Applications (IJSCAI)* 8 (1): 17–36.

2

Information Technology Infrastructures Supporting Industry 5.0 Facilities

This chapter provides important solutions about digitalized information and data monitoring of production lines, merging informatic with electronic aspects. The hardware solutions are combined with the software ones allowing big data analytics, augmented reality (AR) facilities, and predictive maintenance of production machines. Different solutions of sensors trace different scenarios in manufacturing and production processes. The chapter introduces the design approaches to model industry processing and industry 5.0 facilities according to process simulation and modeling. Specifically, tools and approaches to design production processes by integrating sensing and actuation logics are provided. Business process modeling (BPM) and the Unified Modeling Language (UML) are introduced and adopted to describe data flow and data processing by means of different examples. According to the new Industry 5.0 facilities, details are provided about electronic elements useful for the design of advanced information technology (IT) infrastructures, including circuit implementations describing feedback logics to apply improving intelligent production systems. Different workflows and flowcharts indicating process design and data processing are discussed, by highlighting design aspects and industrial applications. The proposed examples include architectures implementing enterprise service bus (ESB) infrastructures able to control and to set machine operations and regulations, and artificial intelligence (AI) algorithms to detect defects for *in-line* and *off-line* production lines, by orienting the research on defect prediction and on defect classification. The production monitoring is assisted by the formulation of maps and tables tracing production steps, defectivity, and providing a framework of tools which can be used for the production upgrade. Different examples are proposed to explain process modeling in advanced systems, such as for AR in a kitchen production process, liquid production, and silo loading in a food company upgraded by AI. AI tools represent the new advances in the Industry 5.0 era, capable of optimizing the production quality or the service quality. In this way, the attention is also focused on the methodology to formulate efficient AI models with high performance by decreasing the calculation error. An overview of electronic logic and its possible implementation in neuron basic models is discussed. Moreover, this chapter describes the implementation of predictive maintenance in industries, by discussing a methodology to perform corrective actions avoiding machine failures and potential break down. Different examples of information platforms integrating AI algorithms are proposed in order to highlight how it is possible to upgrade an information infrastructure. Finally, some E-health architectures are proposed enhancing the possibility to improve care assistance in the cloud environment, by means of the availability of telemedicine decision support systems (DSSs). The goal of this chapter is to provide elements and examples useful for designing and implementing an upgraded information system in industry, allowing the automatic

Electronics in Advanced Research Industries: Industry 4.0 to Industry 5.0 Advances, First Edition. Alessandro Massaro.

control of processes and data flows. Innovative ITs are addressed on the integration of supply chain and decision-making processes for an advanced production.

2.1 Production Process Simulation and Object Design Approaches

A tool suitable to model production processes is the BPM. Standards supporting BPM are classified into three groups:

- Modeling graphic standards.
- Standards for the interchange formats of process models.
- Standards for executing business processes.

In particular, the standard graphic modeling is used for mapping processes and in general for graphical representation of business processes. The graphic standards are useful to represent the information flow, the decision points according to ISO 9001:2015, and to follow the roles of the various business processes by means of diagrams. These standards, compared with other types, are more intuitive and user friendly, as they allow an almost immediate understanding of processes without requiring a particular specific knowledge or training. The most used graphic standard is the Business Process Model and Notation (BPMN), but there are another two widely known "languages", the activity diagram UML (ADU) and the Event-driven process chain (EPC), where:

- The BPMN is a modeling notation for business processes with the aim of eliminating the gap between flowcharts and the process execution language.
- The ADU is an object-oriented visual modeling language, suitable to describe business processes by displaying the sequences of actions to follow, and inserting the description of the data flow control by means of the formulation of activities.
- The EPC is a modeling methodology able to describe processes from different points of view (this methodology includes process managers managing functional flow by a dynamic sequence of activities, the output flows highlighting the relationships between the outputs, and the information describing the used elements).

The standards are often adopted to design production processes by simulating new production phases. Process simulation is an important step for production design and is a "pillar" of Industry 4.0. Tailored processes are mapped by means of workflows indicating process and data flow. The workflows are sketched by symbols. Figure 2.1 lists some of basic symbols useful for workflow design, typical of the BPM tool:

- *Activity* symbol describing the task following the production process.
- *Gateway* merging input flows by defining the output state.
- *Pool* being the container of the workflow to map.
- *Event* symbol being a start event (indicating where a particular process starts) or an intermediate event (occurring between a start event and a terminate event).
- *Connectors* able to link all events and conditions of a mapped process.
- *Error state* setting an error or reacting to a named error.
- *Terminate* symbol defining the end of the mapped process.
- *And* block combining two or more parallel paths into one output path.
- *Or* symbol combining two or more parallel incoming flows in one outgoing flow.

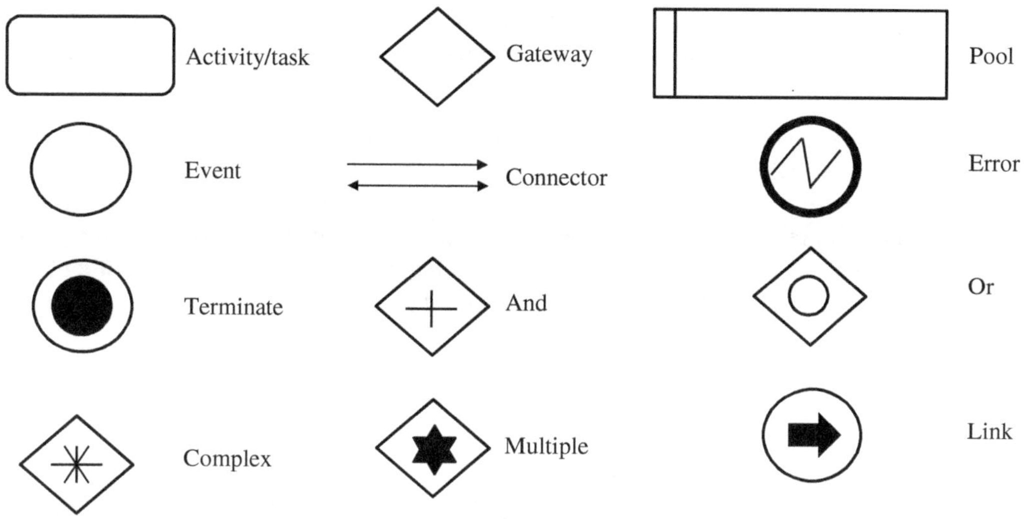

Figure 2.1 Main BPM symbols.

- *Complex* function modeling complex conditions, enabling tasks or activities, and dividing a path into more paths where the activities are developed concurrently, rather than sequentially.
- *Multiple* symbol enabling multiple consequences of ending the process (as in the case of multiple sent messages or emails).
- *Link* symbol representing a mechanism able to connect the final results of one process to the start of another one (process triggering).

The UML model is also appropriate for the design of production processes. This standard can be combined with BPM workflows, thus providing the process implementation matched with the software and firmware design.

- The UML provides multiple advantages about software design such as the:
- Knowledge of the final software output.
- Simplification of the code development, and bug checking.
- Supervision of the whole system function including interconnections with electronic components.

The diagrams to develop within the UML framework are:

- State diagrams (it shows the states that are assumed by the entity or by classes as responses of external events).
- Class diagrams (representing a static structure of classes, by showing attributes, operations, methods, and the relationships among all system objects).
- Object diagram (showing all objects and their relationships).
- Sequence diagram (illustrating the interaction between objects in sequential order).
- Activity diagram (describing the activity flows).
- Component diagram (sketching components, interfaces, ports, and their relationships mainly for service-oriented architectures [SOAs]).
- Deployment diagram (representing a static diagram able to describe a system in terms of hardware resources, nodes, and relationships between them).
- Use case diagram (indicating use cases, actors, and associations).

- Sequence diagram (modeling interactions between objects and actors in a sequential order).
- Collaboration diagram (communication diagram showing relationships and interactions among objects and defining the object roles).

Figure 2.2 shows the main symbols of the UML standard classified in structural elements and basic ones to construct a UML diagram.

The UML diagrams are also used for firmware design, as for the Internet of Energy (IoE) [1], and in general for sensor models [2]. Another generic diagram typology is the generic block diagram, where each block embeds different functions or circuits. Figure 2.3 shows an example of a block diagram representing a basic control model of an advanced production process, where signal and information are converted from analog to digital (A/D) signals, and from digital to analog (D/A)

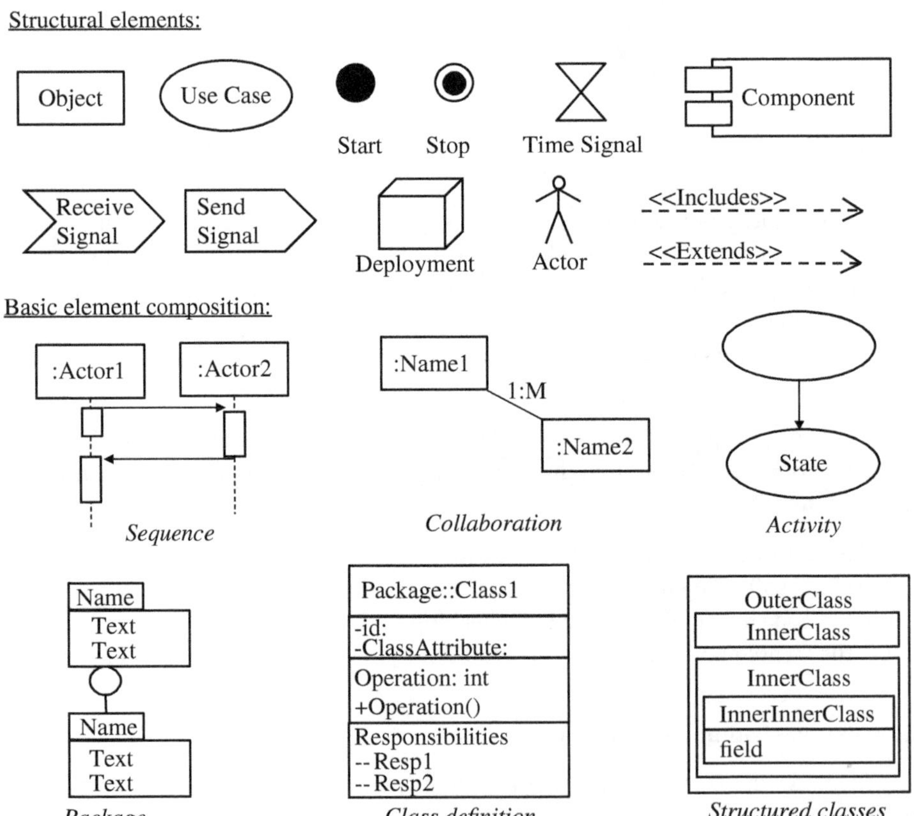

Figure 2.2 Main UML symbols.

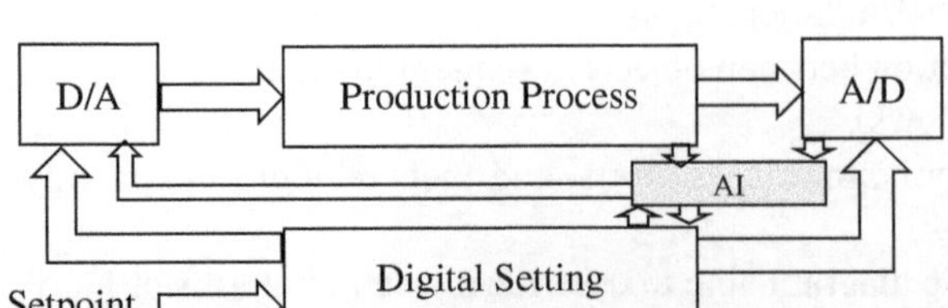

Figure 2.3 Block diagram: digital production process control integrating the AI model.

signals, and an AI block is able to adjust production processes typically as for Industry 5.0 applications. The AI block represents the upgrade of the production process.

2.1.1 Object Design of a Data Mining Algorithm: Block Functions and Parameter Setting

Object oriented software is suitable for a rapid design and development of AI algorithms. The scheme in Figure 2.4 indicates a generic example of a workflow interconnecting objects of a data mining algorithm. The objects are often named blocks or widgets. Each block is considered as an object having some specific function and enabling specific association roles. The design of a production process is structured by assembling different objects into a process workflow. Also, some AI tools require modeling by means of objects linked in workflows. As for Konstanz Information Miner (KNIME), Rapid Miner, and Orange Canvas tools, the objects are configured by means of graphical user interfaces (GUIs) suitable (i) to set parameters for data source extraction, (ii) for data pre-processing, (iii) for data processing, and (iv) for data output formulation, which are the preparatory stages of the whole data processing workflow. Table 2.1 lists some main functions of a data mining block modeling an AI algorithm in each of its phases.

An important aspect in the AI implementation is the estimation of the generated prediction or classification. Table 2.2 lists the main error parameters usable for algorithm performance evaluation. The error function is defined as $ei = (yi - yPi)$, where yi is the measured value at time i and yPi is the predicted value at time i; and N is the forecast horizon. The performance evaluation is fundamental in order to check the AI model accuracy by processing a particular dataset. The performance is optimized by choosing a stable dataset without wrong or ambiguous data, and by considering a sufficient number of historical data to construct the training dataset. A good performance is checked when the training model converges to the testing one.

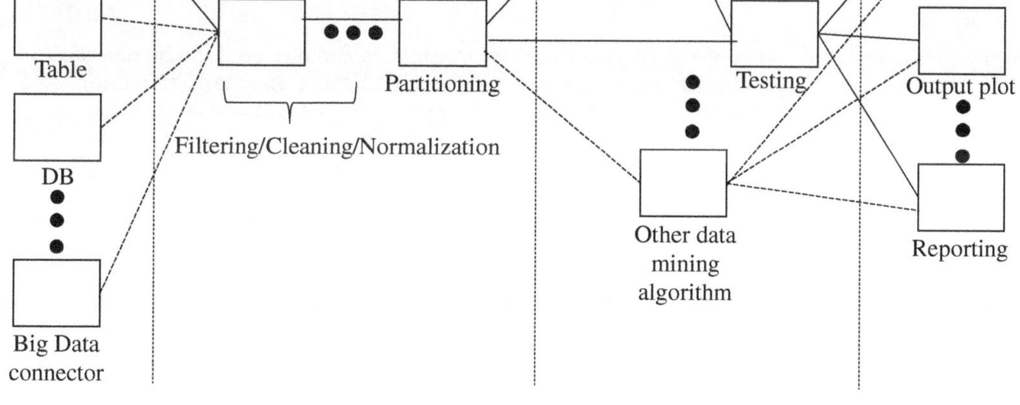

Figure 2.4 Workflow implementing a data mining algorithm with objects (blocks embedding data processing functions).

Table 2.1 Functions of objects constituting a data processing workflow.

Object	Function
	Data input (data source extraction)
Data table	A data source table is generated directly into the local repository of the tool
Data file	Data are imported from a file (*txt*, *xls*, etc.)
Database source	Different connectors are used to extract data directly from a server. For a big data system properly designed connectors are applied as widgets or as Phyton script
Data source workflow	The data to process can be the output of another workflow temporarily stored in a local repository
Data pre-processing	
Data cleaning	Some filtering functions are used to clean spurious data, thus selecting the best dataset to process. As a further function, missing values or wrong data are replaced directly by using the graphical user interfaces, and assigning consistent data. The data cleaning pre-processing stage is particularly important to increase the data processing performance, and to improve the training datasets of the artificial intelligence algorithms.
Data normalization	The calculation error is limited when the data input is properly normalized by taking into account the range scale of the real measured data. The basic normalization process is to set all values in a range between 0 and 1. The data output must be matched with the initial scale range
Data table arrangement	Sometimes the attributes of the dataset to process is non-conformal with the data processing logic or the data mining algorithm to use. For this reason, it is necessary to set the data layout configuration into an executable format
	Data processing
Partitioning	The data source is partitioned in supervised algorithms into a training dataset and a testing dataset
Training model	The training dataset is used to learn the data processing model and constitutes the training model
Testing model	The testing dataset is used to execute the data mining algorithms for the prediction or the classification operations
Direct data processing	Some algorithms are executed directly to the input dataset (no testing and training dataset are required as for unsupervised algorithms)
Data output	
Scoring and algorithm performance	In order to check the performance of data mining algorithms, performance indicators such as the mean square error, root mean square error, mean absolute error, cross-correlation coefficient, standard deviation, recall, precision, accuracy, and receiver operating characteristic curve are estimated
Output visualization and reporting	The algorithm output is visualized by graphical dashboards facilitating the reading of results, and, consequently, the decision-making process

Another methodology to evaluate the prediction performance is the receiver operating characteristic (ROC). Figure 2.5a shows an example of a ROC curve, and Figure 2.5b shows the threshold criterion associated with the ROC curve. The area under the curve (AUC) represents the performance parameter. The ideal case corresponds to an AUC value equal to 1. The AUC is defined as:

$$\mathrm{AUC} = \int \mathrm{TPR}\, d(\mathrm{FPR}) \tag{2.1}$$

Table 2.2 Algorithm performance indicators and related formulas.

Performance indicator	Formula
Mean square error	$= \frac{1}{N}\sum_{i=1}^{N}\left(y_i - y_i^P\right)^2$
Root mean square error	$= \sqrt{\frac{1}{N}\sum_{i=1}^{N}\left(y_i - y_i^P\right)^2}$
Mean absolute error	$= \frac{1}{N}\sum_{i=1}^{N}\frac{\left\|y_i - y_i^P\right\|}{\left\|y_i\right\|}$
Cross-correlation coefficient (R)	$= \frac{\sum_{i=1}^{N}\left(y_i - \overline{y}_i\right)\left(y_i^P - \overline{y_i^P}\right)}{\sqrt{\sum_{i=1}^{N}\left(y_i - \overline{y}_i\right)^2 \sum_{i=1}^{N}\left(y_i^P - \overline{y_i^P}\right)^2}}$
Standard deviation	$= \sum_{i=1}^{N}\frac{\left(y_i - y_i^P\right)^2}{N-1}$
Recall	$= \frac{\text{TP}}{\text{TP}+\text{FN}}$
Precision	$= \frac{\text{TP}}{\text{TP}+\text{FP}}$
Accuracy	$= \frac{\text{TP}+\text{TN}}{\text{TP}+\text{TN}+\text{FP}+\text{FN}}$
Loss	$= \sum_{i=1}^{N}\left(y_i - y_i^P\right)^2$

FN, false negative; FP, false positive; TN, negative; TP, true positive.

where

$$\text{TPR} = \frac{\text{TP}}{\text{TP}+\text{FN}} \tag{2.2}$$

$$\text{FPR} = \frac{\text{FP}}{\text{FP}+\text{TN}} \tag{2.3}$$

Graphically it is possible to estimate the data mining algorithm by comparing the loss function of the training and of the validation model and analyzing underfitting and overfitting conditions [3]. In particular, the validation testing curve must be as much as possible close to the training loss curve versus the iteration number. The error entity is represented by the gap between both loss curves (Figure 2.6). Possible solutions to mitigate overfitting conditions are to:

- Add more training data (historical data extracted from the big data system).
- Create artificial data [1].

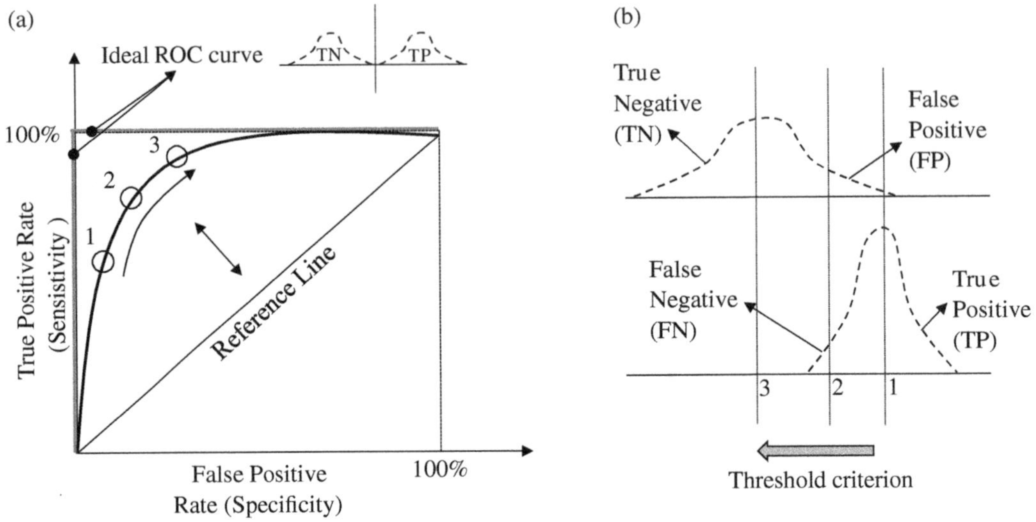

Figure 2.5 (a) ROC curve. (b) Threshold criterion.

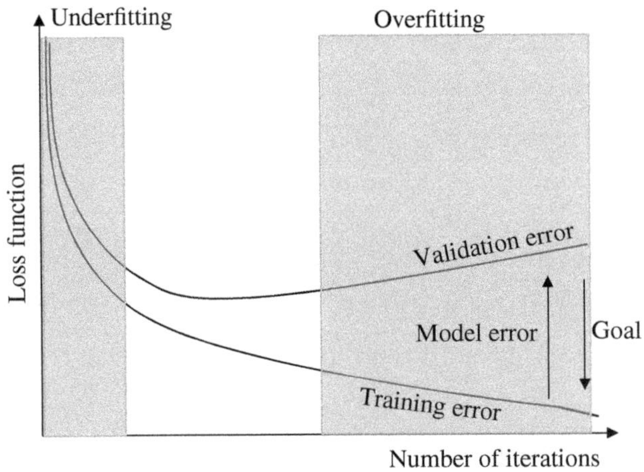

Figure 2.6 Loss function: overfitting and underfitting conditions and performance criterion.

- Decrease redundancies and correlations in choice of the training dataset.
- Properly normalize data.
- Reduce the architecture complexity especially for deep learning networks.

In considering a neural network, an approach to follow for the choice of the best model to adopt is provided by the analysis of different parameters of the network. The main goal is to find the minimum conditions of error by fixing the best network parameters which change with the chosen dataset and with the neural network structure. Figure 2.7a shows a theoretical trend of the mean absolute error (MAE) having an oscillatory trend. The maximum number of iterations oscillates or diverges as for overfitting cases (Figure 2.7b). The minimum MAE condition is also checked for the number of hidden neurons per layer (Figure 2.7c), and for the dataset partitioning methodology for training and testing data processing (Figure 2.7d). The partitioning is performed by taking from the top of the dataset the samples for testing, or using consecutive partitions as test data sets (linear sampling), or random data.

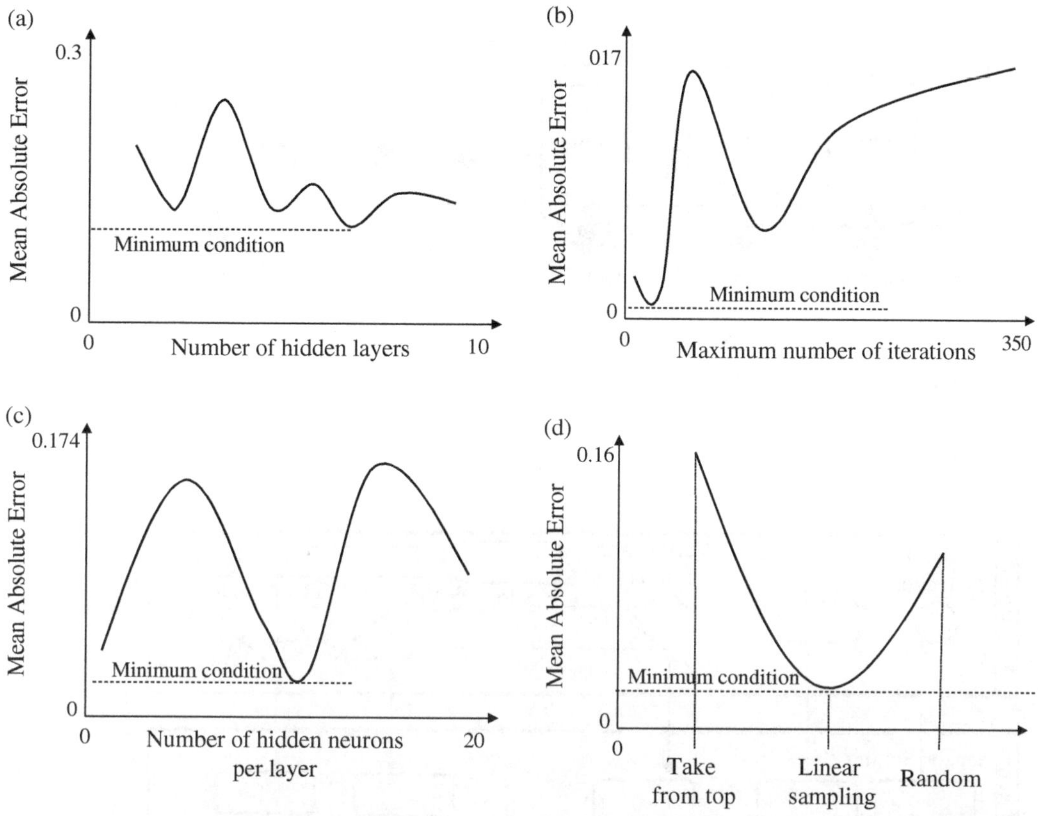

Figure 2.7 Examples of MAE theoretical trends versus neural network parameters: (a) MAE versus the number of hidden layers; (b) MAE versus the maximum number of iterations; (c) MAE versus the number of hidden neurons per layer; and (d) MAE versus different sampling approach for the dataset partitioning.

2.1.2 Example 1: BPM Modeling of Wheat Storage Process for Pasta Production

Process simulation is an important step for production design. Tailored processes are typically mapped by BPM tools using the symbols in Figure 2.1. The BPM approach is applied to simulate processes. Concerning the pasta production process, it is important to detect the silo levels containing wheat [4]. The sensor levels (V1, V2, V3, . . ., Vn) are allocated in order to detect the wheat volumes of the sequential silos (Silo 1, Silo 2, Silo 3, . . ., Silo n) and are of different types such as:

- Capacitive level sensors for solid and liquid level detections (typically used to set different thresholds by positioning different sensors on different silo heights).
- Non-contact level sensors suitable for the environment with high temperature and pressure values.
- Float level sensors (sensors embedded in floats).
- Hydrostatic pressure level sensors.
- Vibrating level sensors changing vibration frequency when the liquid contacts the sensing area.

Process simulation is applied to the wheat storage automated process related to the pasta production. In the simulated process different silos are connected in the cascade layout of Figure 2.8a with sensors detecting the volume level of each one. The simulation of the wheat storage process takes into account the following aspects:

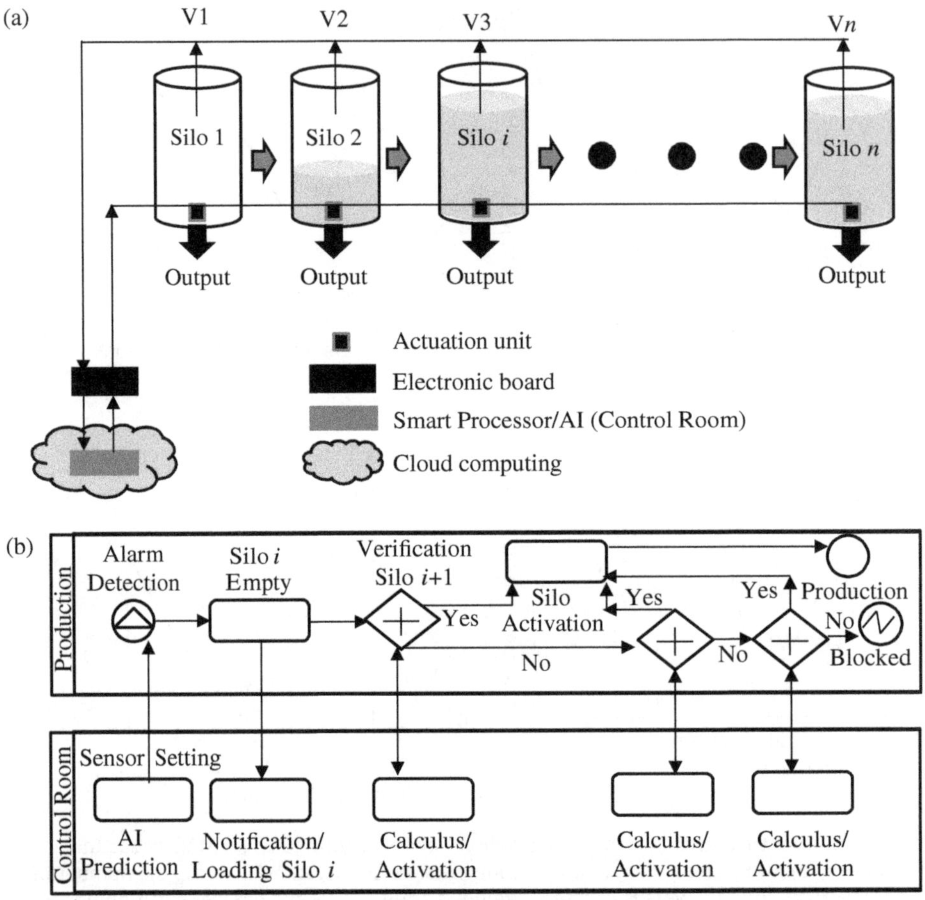

Figure 2.8 (a) Layout controlling silo loading. (b) Implemented workflow of the proposed layout.

- The alarm condition is enabled for volume values under a threshold level.
- When an alarm is detected, automatically the wheat is extracted from the next silo, thus ensuring the production process continuity.

As illustrated in Figure 2.8a, each silo is managed by an actuation unit enabling the silo to be opened (wheat outputs). The signal control is performed by an electronic board having *n* output ports. The electronic board receives the actuation command from a control room accessed in cloud. An external processor controls the level signals and generates the activation signals. The AI engine can provide further information addressing the machine setting, such as:

- Prediction of the pasta needed (useful for material supply management and for the sensor calibration of alert level setting).
- Silo circuits failures by the analysis of log machine signals.

The simulation of the management of *n* silos is modeled by the BPM workflow in Figure 2.8b, where the control room and the local production are distinguished by two main pools. The workflow is structured by the following phases:

- The alarm levels of all the volume sensors are set according to AI prediction about the pasta needed and production velocity.

- The processor verifies if Silo i is under threshold.
- If Silo i is under threshold, it is loaded with the output closed, and the sensor signal of the next Silo $i+1$ is read.
- The iteration continues until a silo is over the threshold (the electronic board enables this silo's output by automatically closing the others which are full or are in the loading phase).
- The electronic board manages more complex output actuations; for example for a production line requiring more silos to be opened during the pasta production processes.

2.1.3 Example 2: Block Diagram Design of a Servo Valve Control and Actuation System

The proposed example shows the circuit block diagram enabling the adjustment of an electrohydraulic servo valve (EHSV) managing the flow rate of an axial piston pump. An EHSV is an electrically operated valve that controls how hydraulic fluid is sent to an actuator. In the proposed example, the pump regulation is adjusted by means of mechanical feedback. A change in electrical current of the servo valve causes regulation of the proportional pump by the actuator. The AI block is adopted to auto-adjust this regulation by reading sensor signals and predicting failures in loading processes requiring different pump regulations. Figure 2.9 shows the feedback system controlling and actuating a load by means of a servo valve.

2.1.4 Example 3: Block Diagram of a Liquid Production System

In an industrial environment, maintaining constant data flow is critical. Today, companies must be fully equipped to receive digital services mainly adopting wireless sensors as for Bluetooth (BL) technology, being highly versatile in a scalable scenario where devices change their roles (a slave device becomes a master and vice versa). Industrial BL is robust enough to withstand critical temperatures, pressures, and vibrations, making it suitable for harsh environments where other forms of digital communication may not work correctly. In the Industry 4.0 era, in which machines must be able to communicate with each other and transmit the data acquired by sensors, it is extremely important to ensure effective, reliable and secure communication between devices. The real-time monitoring is achieved by using ultra-low energy technologies, such as Bluetooth Low Energy (BLE), or by beacons or other small wireless transmitters. In the production environment, the use of BLE sensors is considered an efficient alternative to the installation of wired networks. In fact, wireless technology allows devices to be allocated strategically in industrial areas difficult to reach with cables, as for the

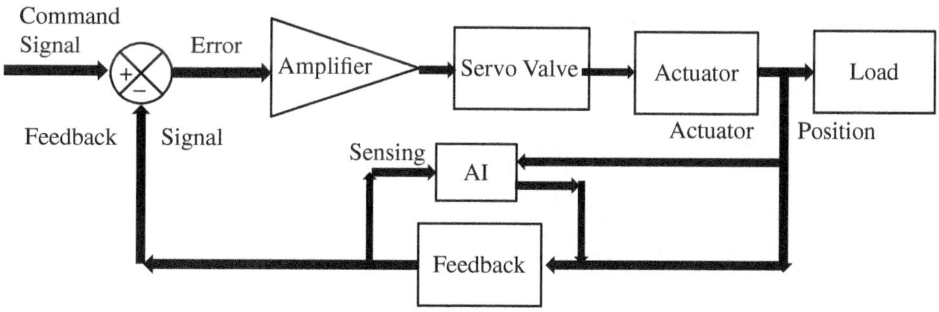

Figure 2.9 Feedback system controlling and actuating a load by means of a servo valve.

case of the proposed example, where liquids and humidity can damage circuits. For complex implementations with a large number of sensors, it is preferable to implement a mesh network layout. In this way, BL mesh is a substantial development for industrial wireless connectivity. This technology adds new features that go beyond the usual "star" and "point-to-point" topologies, enabling connectivity for industries until hundreds of nodes are over a large area through multiple applications of BL mesh networks, behaving as decentralized nodes connected by routers. These communication networks are highly reliable, making them suitable for different industrial environments such as the case in Figure 2.10, representing a design model for the production of a chemical liquid solution controlled by different BLE sensors. The proposed block diagram consists of a container containing the solute, a conveyor belt, and a tank containing the produced solution. The distribution of the solute is carried out by automatically adjusting an electro-hydraulic loading valve and the conveyor belt speed carrying the liquid into the tank. A conductivity sensor measures the concentration value in the tank solution by providing a voltage proportional to the concentration value. The solute is dissolved in the solvent by the mixing action of a mechanical stirrer. The flow rate of the solvent is considered constant according to the chemical composition. The control system can regulate the concentration level according to the sensor signal detecting the concentration as a function of solvent flow rate. A controller is programmed to regulate the valve aperture of the liquid source, the speed velocity of the conveyor belt, and to tune the mixing process speed according to the measured concentration. The reference signal represents the set point of the chemical composition to be improved. The AI engine elaborates historical data of the sensor and predicts the possibility to work in failure conditions generating a liquid having a concentration different to the established value represented by the set point. This prediction allows to adjust all the parameters in order to work in secure conditions by dynamically controlling the tolerance gap of all the parameters.

2.1.5 Example 4: UML Design of a Programmable Logic Controller System

The programmable logic controller (PLC) represents the main control for production machines in Industry 4.0 and for "Smart Factory" applications. The PLC system is considered the central

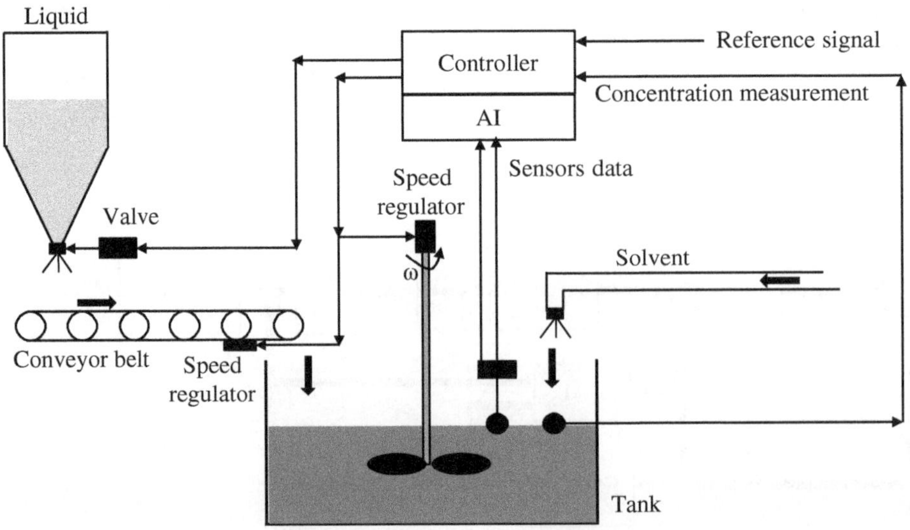

Figure 2.10 Block diagram model of a liquid production system implementing AI.

processor for all real-time decisions of the production process. The first PLCs were designed to support programmability, real-time response, and reliability. Suitable for these original applications, the PLC is now reaching its limits and requires an upgrade. PLCs have been designed as standalone devices that react to local inputs and outputs. The internal control algorithms respond to changes in the actuation processes. The IEC 61131 standard has been oriented to the software implementation and configuration controlling real-time operations. In the Industry 4.0 era, the control exercised by PLCs, and by other control platforms, provides increasingly advanced connection techniques with factory IT systems such as supervisory control and data acquisition (SCADA), the manufacturing execution system (MES), and enterprise resource planning (ERP), and between machines dedicated to the production cycle. This gives rise to the possibility of remote control of the production process from fixed and mobile devices, with the possibility of remote assistance, remote control, remote diagnostics, and continuous monitoring of process parameters through Internet of Things (IoT) technology.

Another important aspect is to make the data of the control systems accessible in real time.

Basically, PLCs remain the pivotal devices for implementing every automation and control solutions. Today, however, with the spread of advanced communication technologies, there is an acceleration of implementations in distributed automations, IoT equipment with delocalized input/output ports, and data collecting information systems. In this way, PLCs are devices capable of connecting machines via the internet with the support of mobile technologies, and of collecting data from the production lines. In addition, PLCs have experienced an increase in computational and functional capabilities due to the use of multicore processors and new software platforms. This permits the decentralization of decision-making process, creating increasingly intelligent stations, and providing information to the central supervisory and control rooms. These characteristics of versatility and adaptability permit users to use different hardware components, to control a single machine, or to connect all the production lines online. The increased flexibility also reduces planning, programming and commissioning times, as well as maintenance interventions. Another fundamental aspect is the integration of the PLC functionality within other industrial devices (gateway, human–machine interface [HMI], inverter, numerical control interface, axis control platform, robot, etc.). The new generation of advanced PLCs is therefore an ideal candidate for high-end applications, including time critical processes, frequency analysis, real-time management of large databases, plant modeling and diagnostics, and multi-variable algorithms.

The main characteristics of PLCs in Industry 4.0 are:

- **High performance control:** intelligent production environments need PLCs to process instructions, service interrupts and integrated high-speed HMI, thus requiring more powerful processors with multiple cores with a cost increase and energy consumption.
- **Connectivity:** machine-to-machine connectivity between disparate machines through the support of multiple industrial Ethernet protocols within a single PLC system.
- **Secure communications:** PLCs are vulnerable to cyberattacks, therefore high-performance security techniques are used; the PLCs must be authenticated before working. Transactions are encrypted and authenticated to prevent interception and data modification.
- **Interoperability between platforms:** by the use of standardized operating systems, different platforms can be interfaced upgrading the whole company information system.

The upgrade of PLC systems in the new era of Industry 5.0 is mainly in the integration of further intelligence provided by AI oriented on auto-adaptive applications.

Figure 2.11 shows a class diagram example related to the design of a packaging auto-adaptive control system integrating PLCs. The packer is coupled to two conveyor belts controlled and

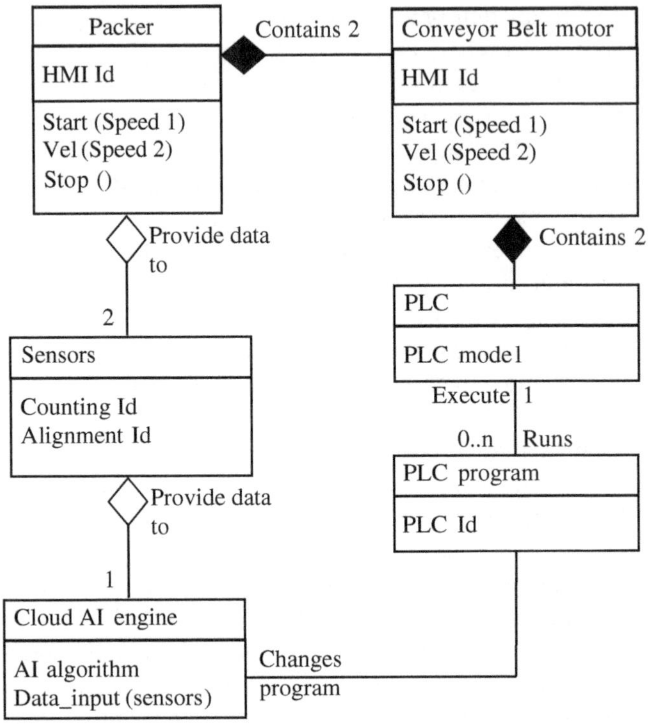

Figure 2.11 UML class diagram implementing AI for a PLC system.

actuated by two PLCs. Sensors counting pieces arriving at the packaging station, and measuring package alignment, are able to control the correct packaging processing. The AI engine can process sensor data in cloud, by predicting possible failure conditions. The possible failure conditions are mainly related to possible package misalignments, and to the mismatching between production velocity and product quality (conveyor belt velocity too fast or too slow).

The self-adaptive program ensures the correct packaging process by analyzing historical data and by dynamically changing the PLC program according to the predicted risks.

2.1.6 Example 5: Electronic Logic Timing Diagram

Electronic logic timing diagrams are important to design electronic circuits, taking into account signal processing and data flow characterization. In this example the scheme of the D flip-flop circuit is described which is typically used as a basic memory storage element. The D flip-flop is constructed by using NAND and NOT ports. The main applications of D flip-flop are to introduce delay in a timing circuit, as a memory buffer, by sampling coded data into time intervals. Figure 2.12a shows the D flip-flop symbol with input and output ports. Figure 2.12b shows a sketch of the equivalent logic circuit constructed with NAND and NOT logic ports. Figure 2.12c shows the related logic timing diagram by analyzing the input and the signals.

2.1.7 Example 6: AR System in Kitchen Production Process

Figure 2.13 shows an example of AR technology adopted in a kitchen production system. The AR is integrated in the following scenario defining the production process:

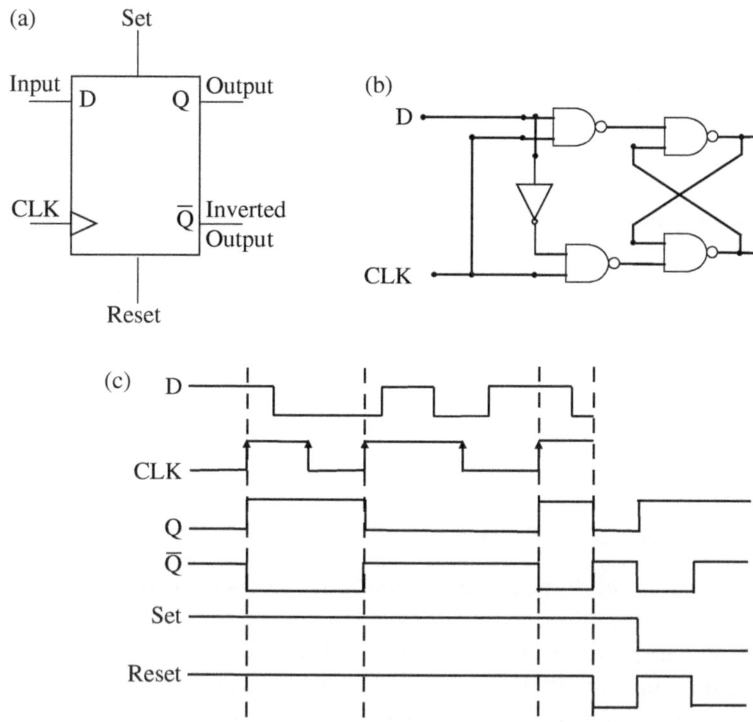

Figure 2.12 (a) D flip-flop symbol. (b) Equivalent circuit adopting logic ports. (c) Electronic logic timing diagram.

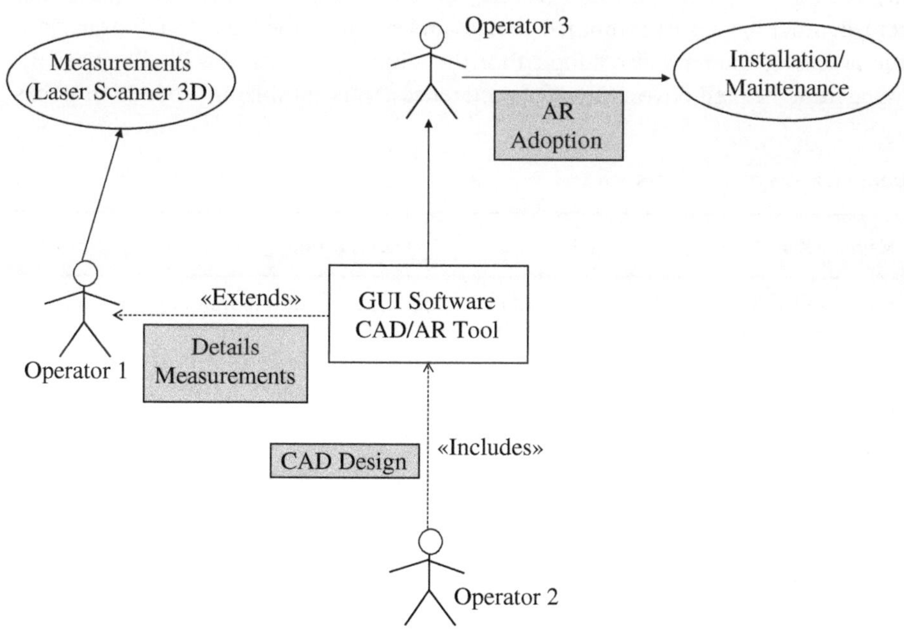

Figure 2.13 UML use case diagram of AR technology adopted in kitchen production activity.

- For the kitchen design, the three-dimensional (3D) kitchen layout is measured by means of a laser scanner.
- The measured data are transmitted wirelessly to a tablet (the wireless command is addressed to control the laser scanner by defining the spatial sampling and by setting all the acquisition parameters).
- Measured data are transferred directly on a remote repository.
- the *.dxf* raw data are processed by means of 3D computer-aided design (CAD) software able to reconstruct the 3D kitchen layout, and to import 3D kitchen blocks from ERP software.
- The 3D full kitchen is uploaded into an AR framework used to apply augmented functions.
- The project visualization occurs through a cardboard viewer or through a visor.

The functions of all the actors involved in the whole production system are summarized in Table 2.3.

It is observed from the listed functions that the operators are interchangeable, i.e. the role of an actor can be played by another actor to optimize the company's human resources. The diagram in Figure 2.13 facilitates the understanding of the actor interoperability.

Other specifications of the technologies used are as follows:

- **AR technology:** the customer from home, by acquiring a photo of the kitchen environment and other basic dimensional data, interacts online in the design of the kitchen, being able to create and view it in a 3D AR (AR rendering) environment by means of a visor or their own smartphone (cardboard); the selection and tests that the user carries out are processed by the AI engine suggesting the modules and the features of the kitchen (colors, textures, models, accessories, etc.), based on the user experience [5] and affective computing (the AI engine acknowledges the environment features and suggests the kitchen blocks to compose by analyzing during the time the choices of other users).
- **AI user experience:** while browsing online possible kitchen configurations, all data are processed by analyzing the customer behavior selecting blocks having specific features (customer user experience), in order to suggest combined choices and speed up the idea for a draft in similar projects; the processed data are also adopted for the planning of business intelligence (BI) strategies deriving from the predictive analysis of preferences, thus optimizing production stocks

Table 2.3 Functions of actors of the AR system.

Actor	Main function	Secondary action
Operator 1	Data acquisition of the 3D environment by using a laser scanner. The laser scanner is used to acquire the most significant points of the environment to detect (corners of the room, pillars, doors, windows, etc.).	Identification of important objects (sockets, vents, windows, etc.) by means of a software interface. This secondary function can also be attributed to Operator 2
Operator 2	Identification of important objects (sockets, vents, widows, etc.) by means of a software interface	Data acquisition of the 3D environment using a laser scanner
Operator 3	Installation or maintenance of kitchens after analysis by means of the AR visor: the operator analyzes the kitchen to assemble or repair before the installation, by means of the AR visor guiding the processing procedure	Functions related to Operator 1 and Operator 2

and defining new products (preference prediction); the AI engine is able to self-learn during the time in order to make the suggestion system automatic; the data of customer choices and attitudes are stored in a database connected to the company information system, and are processed by the AI algorithms

- **Computer numerical control (CNC) system**: the draft project, which is formulated by the customer through the AR application, is optimized by the company's research and development (R&D) department, thus moving production process on to the executive design of the 3D kitchen; the executive design automatically activates the production by means of the CNC system, by speeding up the production and execution times of the project itself.

All the technologies have different effects on the production:

- Increase of the number of kitchens sold compared with previously.
- Decrease of waste materials (implementation of the CNC system, combined with new production processes).
- Speeding up production processes.
- Efficiency of warehouse stocks and their management.
- Efficiency of raw material management.
- Impact of the user experience on the final choices of the customer according to the marketing strategy.

The general advantages of the new upgraded procedures are:

- Reduction of data transfer times for CAD software applications of the technical R&D office (by adopting a defined procedure to detect by means of a laser scanner the most significant points of the kitchen environment, it is possible to decrease the number of points to interpolate through the 3D CAD software).
- Reduction of the design error and therefore of the kitchen assembly imperfections (the precise acquisition by means of laser technology drastically reduces the possibility of measurement errors, and therefore the design office has complete and ready-to-process data).
- The possibility of using the visor allows the installer to have in advance a complete framework, thus reducing installation times and making the installer available for other installations throughout the day (optimization of the delivery phase).
- During the maintenance procedure the technician, who is not where the kitchen is located, interactively communicates remotely with experts in order to receive instructions for repair.
- Reduction of economic losses in general due to errors (an error or missing information can generate a substantial economic loss; the new procedures are oriented on a general reduction of the production error).

Table 2.4 lists the main limits of the AR system specifications.

In order to analyze the impact of AR technology in the production process, it is important to analyze all the variables and facilities involved in the production by means of the new technologies such as laser scanner and AR. Figure 2.14 shows a House of Quality (HoQ) diagram which is a graphical analysis tool to connect production requests to the characteristics of the process that have a direct impact on company satisfaction. The diagram correlates and estimates the performance between the required quality facilities and the quality characteristics. For the specific case of kitchen production, the quality facilities and properties are:

- Installation error reduction (expressed in %).
- Reduction of waste materials.

Table 2.4 Limits of the AR system.

Limits	Description
Performance of the mobile device for cloud computing and data processing (AR contents processing)	The cloud computing of massive data for real-time processing requires high performance technologies. Cloud data processing can be blocked or data can be viewed with interrupts as for the visualization of complex 3D CAD models using a 3G or a 4G network. In addition, the cloud browser system must recognize similar objects in order to intelligently suggest features. This requires an additional computational cost and therefore a rather high-level technology. The 5G network is suitable to overcome this limitation especially for the computational cost of the real-time augmented video processing
Synchronization between cloud computing operations to display AR content and outputs suggested by the intelligent platform	The user interacts with the platform by making specific content choices (choice of blocks, colors, accessories, textures, etc.), and viewing the contents based on the real-time choices. The same choice is processed in the cloud by the AI engine suggesting, in a synchronized way, the further options. Such synchronization can be a limit for cloud data processing
Compatibility of AR technology with different mobile technologies and operating systems	AR contents must be compatible with smartphones and with different technologies for a stable display and for a good virtual immersion interaction
Compatibility of the user's design with the production platform and transition from augmented reality content to a 3D CAD project	The platform needs to efficiently convert AR contents into 3D CAD design contents. This conversion requires an easy remodeling of the layout which is sent automatically to the CNC system of the production line
Production automation by CNC	The CNC production must be synchronized with the production processes to avoid possible conflicts between software and hardware. The production process must be reformulated according to specific priority roles
Reliability of AI algorithms	AI algorithms must work using a well-structured dataset model to guarantee a good reliability of the results. Appropriate parameters such as root mean square error, mean square error, receiver operating characteristic curve, recall, accuracy, or other indicators are used to estimate the data processing error
Error evaluating the user experience	In order to make suggestions for layout content, based on user navigation, it is necessary to consider a large number of data so that the AI model is efficiently trained. A support vector machine algorithm adopting an Extreme Gradient Boosting (XGBoost) approach is suitable to improve computing and user navigation behavior estimation [5]
Decision making and marketing strategies	An incorrect analysis of user behavior in immersive reality can lead to wrong suggestions and incorrect production schedules. Wrong suggestions provided by the intelligent platform can mislead and confuse the user rather than facilitate the formulation of the draft project. In addition, the platform must compare the consumer behavior data with sales data, in order to prove the correct suggestions. This feedback form is also important to better train the intelligence algorithms working in self-learning mode
Cloud computing performance	In order to optimize the computational cost, graphical processing unit technologies are used as replacements for traditional calculation machines, or combined with those of cloud computing

Table 2.4 (Continued)

Limits	Description
Use of appropriate frameworks	The platform must be created using a framework compatible with all the technologies used to integrate the various functionalities (backend, frontend, web service, cloud computing, etc.)

3D, three-dimensional: AI, artificial intelligence; AR, augmented reality; CAD, computer-aided design; CNC, computer numerical control.

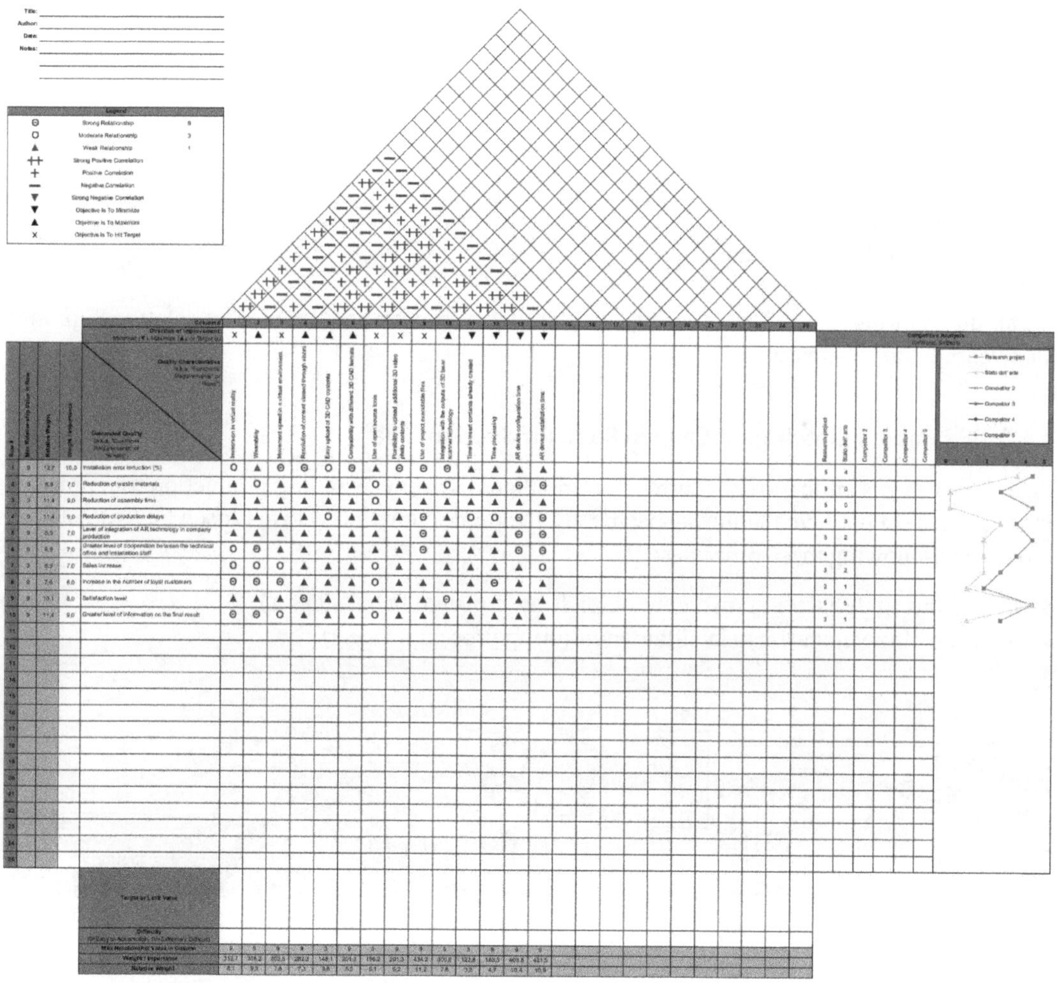

Figure 2.14 HoQ diagram matching a laser scanner with AR technologies.

- Reduction of assembly time.
- Reduction of production delays.
- Level of integration of AR technology in company production.
- Greater level of cooperation between the technical office and installation staff.
- Sales increase.

- Increase in the number of loyal customers.
- Satisfaction level.
- Greater level of information on the final result.

The quality characteristics of the introduced technologies are:

- Immersion in virtual reality.
- Wearability.
- Movement speed in a virtual environment.
- Resolution of content viewed through visors.
- Easy upload of 3D CAD contents.
- Compatibility with different 3D CAD formats.
- Possible to use open-source tools.
- Possible to upload additional 3D video photo contents.
- Possible to use project executable files.
- Integration with the outputs of 3D laser scanner technology.
- Time to insert contents already created.
- Time processing.
- AR device configuration time.
- AR device installation time.

The diagram also provides a scoring about weight, importance, maximum relationship, and about competitor comparison.

2.1.8 Example 7: Intelligent Canned Food Production Line

Figure 2.15 shows an IoT architecture of an intelligent system switching production line processing in the case of predicted failures. Specifically, the production line system is used for canned food production following these processes:

- Load of raw material and movement of loads by conveyor belts (stage 1).
- Vibration of raw material to facilitate monitoring process (stage 2).

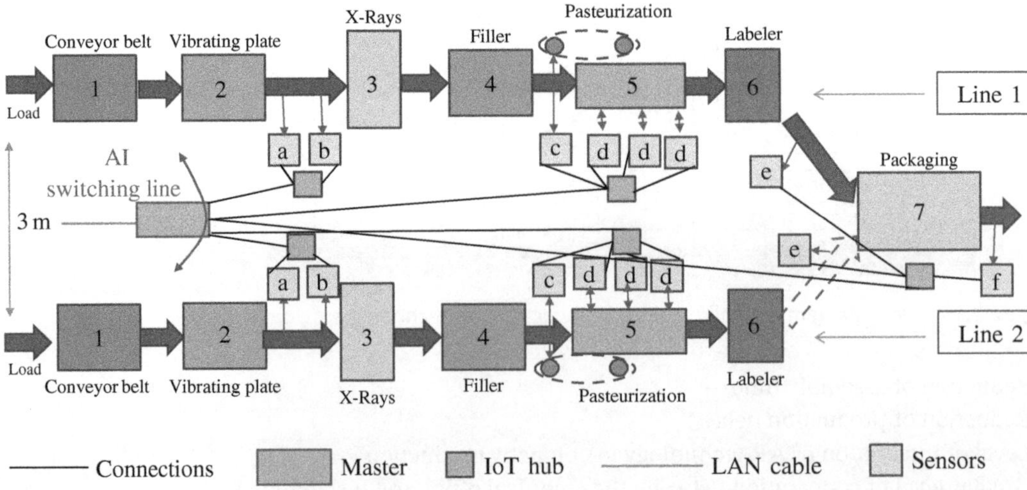

Figure 2.15 AI switching production lines for canned food production. The production lines are 3 m apart.

- X-ray monitoring controlling the intrusions of metallic parts (stage 3).
- Pasteurization process (stage 5).
- Product labeling (stage 6).
- Packaging (stage 7).

The whole production system is controlled by the following different IoT devices, connected by Ethernet local area network (ELAN) line:

- Sensor type a: piece counter sensors.
- Sensor type b: sensors measuring weights.
- Device type c: the inverter controls motor movements at the input of the pasteurizer (the speed must be programmed according to the type of product; there must be a correspondence between speed and regulation frequency, where each frequency corresponds to an operating machine speed).
- Sensor type d: temperature sensors to be inserted in the different areas of the pasteurizer through the three observing windows, or by allocating them inside using external connectors.
- Device type e: video cameras enabling image processing and checking the correspondence of the label design with product inserted (tomato, artichoke, . . .).
- Device type f: barcode sensor tracing production.

The proposed layout combines IoT, local area network and AI facilities and represents an example of a production system upgraded by Industry 5.0 facilities, where the AI algorithms are able to predict the presence of metallic parts by the analysis of historical data, automatizing the switching of the production lines. The possibility to switch the lines permits maintenance of the production machine thus ensuring the production quality in a continuous production. The AI switching can be activated also by the data processing of the other sensors: the measurements of anomalous weights, temperatures or vibrations during the production process are alerting signals predicting serious machine malfunctions. A mechatronic interface enables the automation of the stop condition of the malfunctioning line, the switching of the production to the second line, and the loading of raw material in the selected line. The other electronic boards, improving the control of the production, are the interface between the sensor and the PLC system. The main commands for the PLC are the velocity regulation and the stop machine status condition.

2.2 Electronic Logic Design Oriented on Information Infrastructure of Industry 5.0

In order to show the potential use of the logic ports in electronic implementations, it is important to know their functionalities. Figure 2.16 illustrates the basic logic ports with their symbols, equations, and meaning. The truth tables and graphical sketches of events complete the symbol meaning explanation. The logic ports are implemented to design intelligent logics including association roles, Boolean functions, and process mining algorithms matching with BPM data flows. The logic gates are implemented by N-type metal oxide semiconductor (NMOS), P-type metal oxide semiconductor (PMOS), complementary metal oxide semiconductor (CMOS), resistor–transistor layout, and transistor–transistor layout. The PMOS technology adopts *p*-channel (+) metal-oxide-semiconductor field-effect transistors (MOSFETs). NMOS is made by an *n*-type (−) MOSFET and works by means of an inversion layer in a *p*-type transistor. PMOS transistors behave as an inversion layer (*p*-channel with holes between *p*-type "source" and "drain" ports) in an *n*-type transistor

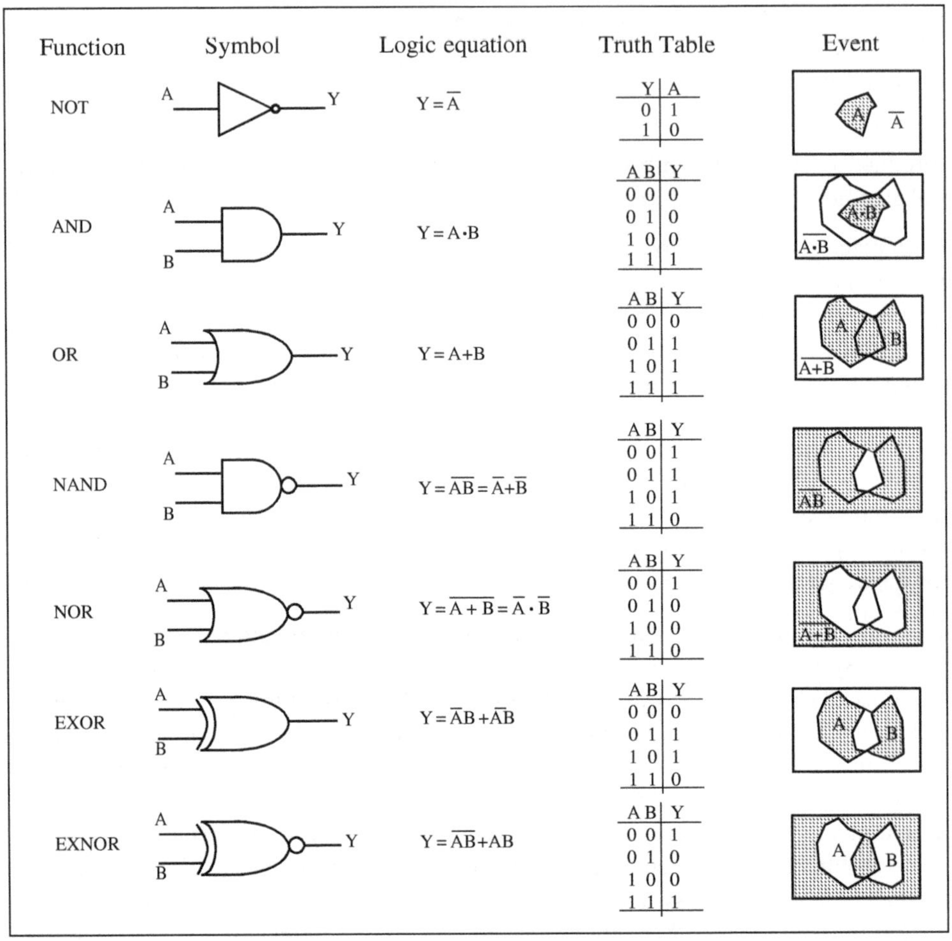

Figure 2.16 Symbols, truth tables, and events characterizing logic ports. A and B indicate the input ports and Y the output port.

structure. CMOS is a typology of MOSFET, adopting complementary and symmetrical pairs of *p*-type and *n*-type MOSFET technologies. Resistor–transistor and transistor–transistor layouts are hybrid configurations involving electrical resistances and transistors.

Logic ports are implemented using diodes or transistors acting as electronic switches. Figure 2.17 shows some examples of electronic implementations of AND, OR, NOT (inverter negation), NAND, NOR, OR/NOR logic ports. Specifically, Figure 2.17a shows an example of implementation of an AND port by means of diode configuration. Figure 2.17b represents an OR port by adopting a diode layout. Figure 2.17c is an example of an inverter transistor configuration. Figure 2.17d illustrates a NOR port made by PMOS transistors. Figure 2.17e is a PMOS NAND port using a negative logic. Figure 2.17f is an OR/NOR port with four input terminals and emitter-coupled logic configuration. The logic ports are adopted also for the implementation of neural networks [6, 7]. The circuital representation by logic ports is an alternative way to implement logic flowcharts based on threshold definitions.

Concerning the McCulloch–Pitts neuron model [8], Figure 2.18a is a model with three binary inputs of the neuron implementing logic conditions. Figure 2.18b represents the related logic

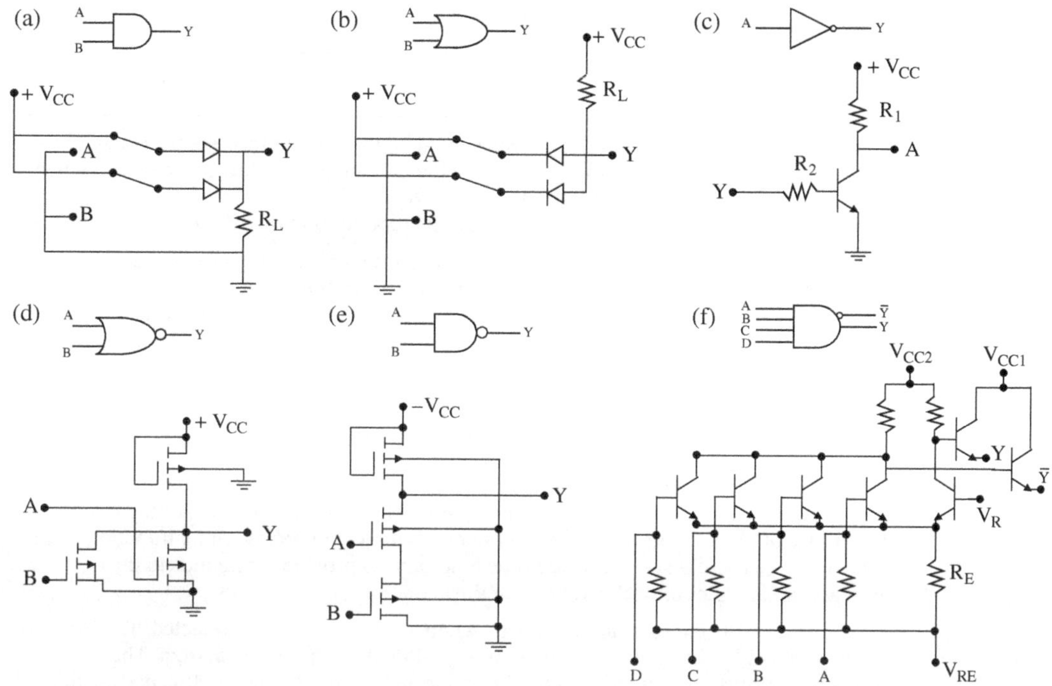

Figure 2.17 Equivalent circuits of (a) AND, (b) OR, (c) NOT, (d) NOR, (e) NAND, and (f) OR/NOR logic ports.

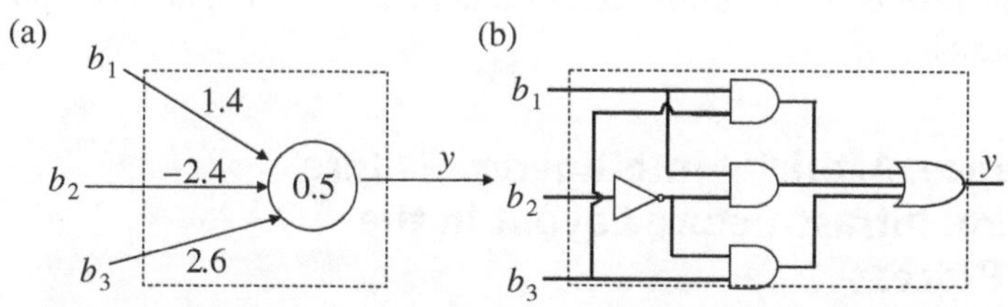

(c)

b_1	b_2	b_3	$\Sigma_{i=1}^{3} b_i \times w_i$	y
0	0	0	0	0
0	0	1	2.6	1
0	1	0	−2.4	0
0	1	1	0.2	0
1	0	0	1.4	1
1	0	1	4	1
1	1	0	−1	0
1	1	1	1.6	1

(d)

b_3 \ b_1b_2	00	01	11	10
0	0	0	0	1
1	1	0	1	1

Figure 2.18 (a) McCulloch–Pitts neuron model with binary input *bi*, weights w_i and a bias value fixed at a value of 0.5. (b) Equivalent logic circuit. (c) Equivalent truth table. (d) Equivalent Karnaugh table.

circuit constituted by AND, OR and NOT ports. The logic conditions are scheduled in Figure 2.18c, indicating the output function y estimated by the weights w_i and by bias value (truth table). Figure 2.18d indicates the equivalent Karnaugh map simplifying the Boolean algebraic expression.

The intelligent logic can be implemented using different technologies such as big data systems, AR, and machine learning. Table 2.5 gives some logics related specifically to big data, AR, and the predictive maintenance of machines.

Table 2.5 Technologies and examples of implementable logics.

Technology	Logic
Big data storage	The logic enables the crons for data storage and data migration processes from a database to a big data system. A good procedure allows the data migration during the period when no data flow occurs in the information infrastructure (this aspect is important to avoid overloading of the information system during the production processing)
Big data analytics	The logic, implemented by informatic tools, enables data processing after a specified query. The queries can be automatized and structured for the specific case (as for predictive maintenance of a machine, data visualization of historical data, etc.). Into the BPM workflow, a dynamic process mining can be implemented activating subprocesses oriented on big data analytics
Predictive maintenance	The failure prediction is estimated when the threshold values are exceeded. The prediction output can switch the production to another line by ensuring the production continuity. By adopting a McCulloch–Pitts neuron model it is possible to set thresholds for machine parameters avoiding the risk of machine failures. A processor can count the frequency of occurred alerts, and can predict the failures, thus selecting priority electric loads. A recurrence of alert conditions represents a potential risk of production malfunctions. The logic ports are useful to set hardware thresholds and to provide commands to the programmable logic controller system for the actuation
Augmented reality	Augmented reality contents are visualized when a specified target is selected. The logic can be enabled by the speech command recognition of the hardware system. The augmented reality backend system can be connected to an artificial intelligence engine able to suggest different options of contents to visualize according to logic association roles

The logic ports can also be implemented for quantum computing involving all optical components to implement the calculus units.

2.3 Predictive Maintenance: Artificial Intelligence Failure Predictions and Information Infrastructure Layout in the Temperature Monitoring Process

Predictive maintenance is an important upgrade of the production quality assessment. A primary requirement for predictive maintenance employment is the measurement of the most important key parameters. One common parameter to analyze is the temperature, important to check during the time the machine status. In order to show the predictive maintenance mechanisms, the monitoring of two lines of milk production is considered [9]. The goal is to control the pasteurization process and, consequently, the temperature is considered as a key parameter. Specifically, in order to check the production, it is important to control the pasteurization temperature within a range of 70–75 °C. An infrastructure suitable to transfer measured data is the ESB line, coupled with multilayered perceptron (MLP) neural networks by a client web-service connection. The MLP network [9] is a class of feedforward artificial neural network (ANN), executing a backpropagation technique for training (supervised technique), integrating the resilient back propagation (Rprop) algorithm as learner [10]. The supervised learning back propagation algorithm is structured into the propagation phase and weight updating phase. These phases are cycled until the performance of the network is checked (root mean square error value under a defined threshold limit). For the back propagation approach, the output is compared in each step with the correct response by estimating the error function. The error function then returns back through the network adjusting the weights of each connection. This refinement is able to decrease the value of the error function. For a sufficient number of training cycles, the network usually converges to the desired results. The MLP network can be implemented by a KNIME workflow enabling data processing and alerting conditions based on predictive analysis. In the proposed example, the temperatures T_1 and T_2 characterize production line 1 and production line 2, respectively. The predictive maintenance of the pasteurization process is performed by comparing the predicted temperature values with real ones of both production lines: in the case of a convergence between the two trends there will be a correct production, otherwise a milk degeneracy will potentially occur. Figure 2.19 shows the architecture of the system integrating the two production lines interconnected to the machine learning engine, and to the big data system.

In the analyzed case, the WSO2 ESB is interfaced with a web service transmitting sensor data to external server machines. The ESB is a real communication bus between the various services. It is a middleware capable of integrating services into a SOA. ESBs are used to connect the different applications and resources, mediating their incompatibilities, orchestrating their interactions, and making them usable for the required purposes. The ESB therefore plays a role of primary importance within the SOA, since it acts both as a register of services and as a centralized management point. In particular, this solution manages problems such as clustering and fault tolerance better. The ESB has the great task of standardizing access to services, and, by means of middleware solutions, makes all applications accessible. Through ESB, all communications between the services are transparent, and it is possible to normalize data for the algorithm execution. The main ESB functionalities are divided into the following four groups:

- **Communication:** the first set of functionalities aims to guarantee transparency, i.e. to ensure that the services are accessible from physical locations; after publishing a service within an ESB,

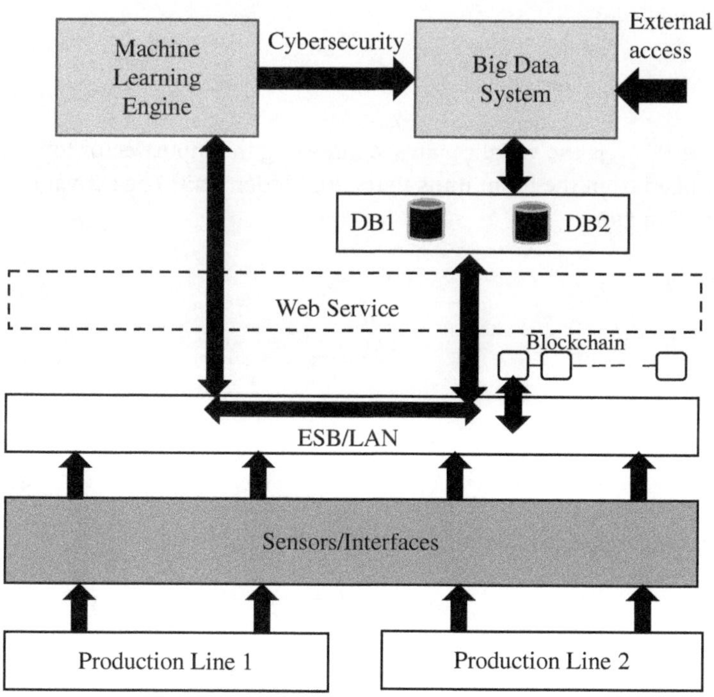

Figure 2.19 Infrastructure of predictive maintenance applied to milk pasteurization production lines.

this is available for all the other services contained therein without knowing the actual location of the service (this requirement is essential to guarantee the maintainability of the services); to ensure correct communication between the services, the ESB correctly routes the exchanged data (it is important to note that the data exchanged are normalized, i.e. they are all based on the same format); routing is not possible without a valid addressing scheme according to the model based on web service connections.

- **Integration**: the integration is necessary to ensure the whole communication systems by embedding various protocols (this requirement is satisfied by creating multiple adapters capable of translating the messages exchanged within the ESB into the relevant access protocols and vice versa); the normalization process is required to correctly route messages within the ESB system.
- **Interaction with services:** an ESB must necessarily guarantee compliance with a SOA interface; to carry out routing, the ESB must know the interface setting and, since an ESB adopts the web service model, it must authorize each service to specify its interface through a Web Service Definition Language (WSDL) file.
- **Management:** for the system administration it is necessary to have an access point allowing analysis of the published services.

The MLP is structured by different layers of connected nodes. Except for the input nodes, each node is a neuron (or processing element) with a nonlinear activation function. From a theoretical point of view, a one-hidden-layer MLP is a function defined as g: $RH \rightarrow RI$, where H is the size of the input data vector x and I is the size of the output vector g(x) defined as:

$$g(x)=G\left(v^{(2)}+F^{(2)}\left(s\left(v^{(1)}+F^{(1)}(x)\right)\right)\right) \tag{2.4}$$

where $v^{(1)}$ and $v^{(2)}$ are the bias functions, $F^{(1)}$ and $F^{(2)}$ are the weight matrices, and G and s are the activation functions. The vector $t(x)$, given by:

$$t(x)=s\left(v^{(1)}+F^{(1)}(x)\right) \tag{2.5}$$

defines the hidden layer, and $F^{(1)} \in R^{HxHh}$ is the weight matrix connecting the input vector to the hidden layer. $Fi^{(1)}$ is the weights linked from the input units to the ith hidden unit. The activation function s behaves as the following function:

$$\tanh(a)=\left(e^{a}-e^{-a}\right)/\left(e^{a}+e^{-a}\right) \tag{2.6}$$

$$\text{sigmoid}(a)=1/\left(1+e^{-a}\right) \tag{2.7}$$

The output vector is calculated as:

$$o(x)=G\left(v^{(2)}+F^{(2)}t(x)\right) \tag{2.8}$$

The MLP training is performed by the following parameter setting:

$$\gamma=\left\{F^{(2)},v^{(2)},F^{(1)},v^{(1)}\right\} \tag{2.9}$$

thus defining the following gradient function of the back propagation algorithm

$$\partial \text{Loss}/\partial \gamma \tag{2.10}$$

The MLP network is modeled by the KNIME workflow of Figure 2.20.

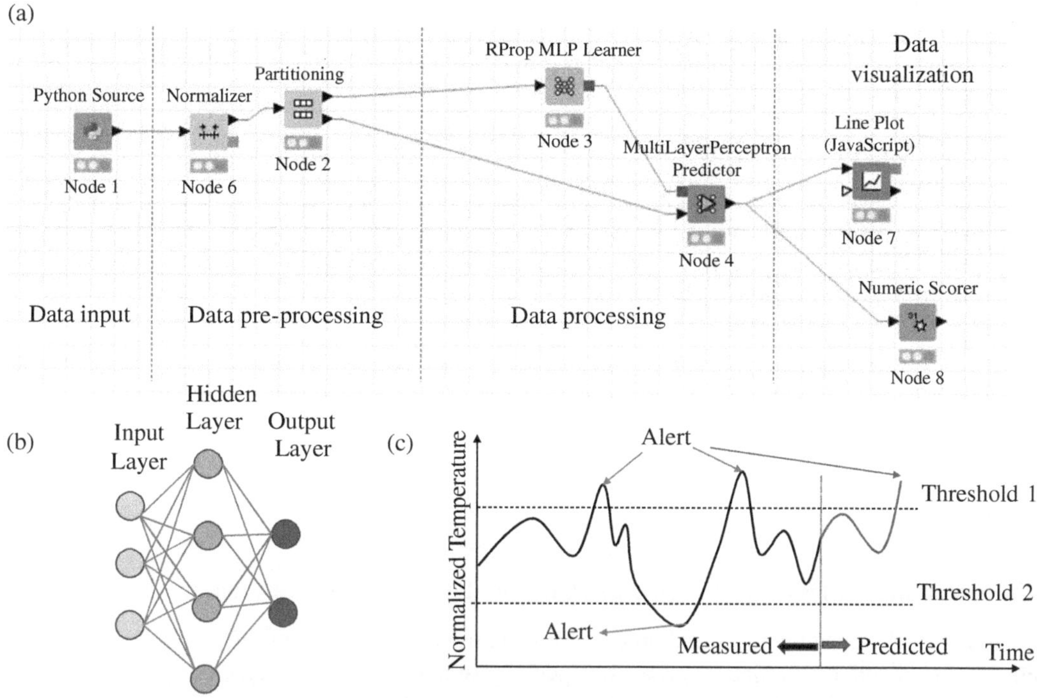

Figure 2.20 (a) KNIME workflow implementing MLP neural network. (b) MLP neural network model. (c) Theoretical plot showing measured and predicted alerting conditions.

The alerting check process is automatized by properly designing the Python script code enabling data extraction and synchronizing data output with actuation commands.

2.4 Defect Estimation and Prediction by Artificial Neural Network

A production process is controlled in different modalities such as: (i) production process control, (ii) defect mapping, and (iii) defect prediction. The first two types are usually enabled by IoT sensors and adopting different standard charts such as Manpower, Material, Methodology, Machine (4M), the Plan-Do-Check-Act (PDCA) chart named the Deming cycle, XmR charts, and *p* control charts. By using, for example, the *p* chart method, a complete approach to check defect production takes into account the following points:

1. By means of *p* charts the defect measurements of samples produced by a prototypal machine are mapped.
2. The measurements are stored in a database, and the historical data trend is mapped to analyze the quality behavior; the historical data are used also for the training of the AI model.
3. All data are processed by the artificial networks predicting defect entities and defective samples.

The *p* charts are defined by the following considerations:

- The *p* chart center line is estimated by the following formula:

$$\overline{p} = \frac{\sum_{i=1}^{k} D_i}{\sum_{i=1}^{k} n_i} \tag{2.11}$$

representing the average value of the defective part, where n is the processed piece number, and D_i is the number of nonconforming units of the ith controlled piece.

- The upper center line (UCL) threshold limit value is the upper limit representing the highest value that the piece can assume; this limit value is calculated as:

$$\text{UCL} = \overline{p} + 3\sqrt{\frac{\overline{p}\left(1-\overline{p}\right)}{n}} \tag{2.12}$$

- The lower center line (LCL) limit value, a lower value that the piece can assume, is calculated as:

$$\text{LCL} = \overline{p} - 3\sqrt{\frac{\overline{p}\left(1-\overline{p}\right)}{n}} \tag{2.13}$$

- The center line(CL) is defined as (UCL + LCL)/2 and indicates the true value.

When a mapped value of a measured defect is outside the range limited by UCL and LCL, it represents an uncontrolled production process, especially when this event is repeatable. Specifically, in [11] this approach has been applied for the monitoring of kitchen top samples processed by a cut machine. The p chart is designed to check the entity of defective portions. The proposed approach is improved by an ANN predicting defect entity values and activating predictive maintenance of the cutting production line. In this case, the ANN is implemented by KNIME workflow [11]. Figure 2.21a shows an example of KNIME workflow implementing an ANN. The workflow is structured as follows:

- The defect measured by means of image vision technique is stored in the big data system or into a local repository as for smart camera systems.
- Measured data are extracted from a database into a file or directly from the big data system, by using a suitable connector (node 1 indicating data input).
- By node 2, the data are partitioned into the training workflow (first partition of node 3), and into a second dataset (second partition indicating the testing dataset) to be processed by the ANN algorithm (node 4).
- The output is visualized by node 5.

For the specific case, training a probabilistic neural network (PNN) is adopted based on the dynamic decay adjustment (DDA) approach. This algorithm generates roles defined by numeric data, and by a Gaussian function characterized by two thresholds. Each Gaussian function is defined by a dispersion center and a standard deviation. Figure 2.21b shows a theoretical trend of a mapped p chart with predicted values useful for predictive maintenance.

The input data are extracted from a big data system by means of ESB directly into the KNIME data repository to be processed. In order to configure the big data connector, it is necessary to program the data input node enabling the server node linking.

The Cassandra big data system is a not only SQL (NoSQL)-based technology. The NoSQL databases are divided into four different types: key-value, document-oriented, graph-oriented,

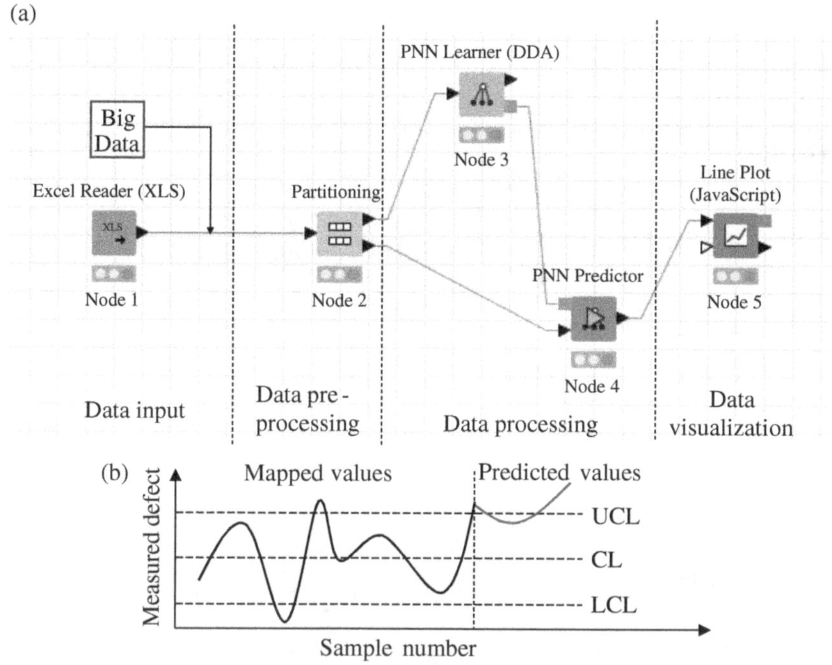

Figure 2.21 (a) KNIME network implementing an artificial neural network. (b) Theoretical example of a *p* chart containing mapped values of measured defects and predicted ones.

and tabular. Key-value databases are the simplest databases in the framework of NoSQL: every single element is saved as an attribute (or key) together with its value. In the document-oriented database, each key is coupled with a complex data structure called a document. Graph-oriented databases are used to store information about networks and relationships. In a tabular database the data are organized into columns of tables. Cassandra NoSQL technology is characterized by a good performance for single node write, delete and read operations if compared with the computational costs of other big data systems such as HBase and MongoDB [12].

2.4.1 Other Methodologies to Map and Read Production Failures and Defects

Control chart are tools useful for production mapping. The 4M chart is a methodology to map production process and supporting quality process. Table 2.6 shows an example of a 4M map suitable for quality check.

In order to reduce the human error to compile the 4M map, the AI engine provides automatically a score by deducting some of the production Key Performance Indicators (KPIs). Figure 2.22 shows a theoretical defect chart indicating unstable points (points over the UCL and the LCL threshold limits), and unstable regions (defects measured at different frequency sampling). The unstable regions represent possible machine failures. An unstable condition is a single point, a set of points, or a trend. Table 2.7 lists the characteristics of other control charts.

Table 2.6 Example of a 4M map.

Production Line 1:				
Worker:				
Production Line Responsible:				
	1 (insufficient)	**2 (enough)**	**3 (good)**	**4 (excellent)**
Machine				
1. Are lubrication and greasing adequate?				
2. Are the inspections adequate?				
3. Is the cycle often interrupted due to mechanical problems?				
4. Is the accuracy level adequate?				
5. Is the noise abnormal?				
6. Is the layout correct?				
7. Are there enough machines/facilities?				
8. Is everything suitable to work?				
9. Does the work-in-process stop?				
Materials				
1. Are the quantities correct?				
2. Are there errors in the specifications?				
3. Is the supplier correct?				
4. Are there any defects or impurities?				
5. Is the inventory level adequate?				
6. Are there any scraps of material?				
7. Is the handling adequate?				
8. Is the quality standard adequate?				
Methods				
1. Are the working standards adequate?				
2. Are the working standards up to date?				
3. Is it a safe method?				
4. Does it guarantee product quality?				
5. Is it efficient?				
6. Is the work sequence adequate?				
7. Are humidity and temperature or other conditions correct?				
8. Are lighting and ventilation sufficient?				
9. The line is correctly connected upstream and downstream				

Table 2.6 (Continued)

Man
1. Does he follow the standards?
2. Is the work efficient?
3. Is he aware of the problem?
4. Is he responsible?
5. Is he qualified?
6. Does he have experience?
7. Is he assigned to the correct position?
8. Does he aspire to improve?
9. Does he maintain good human relationships?
10. Is he correct?
Total

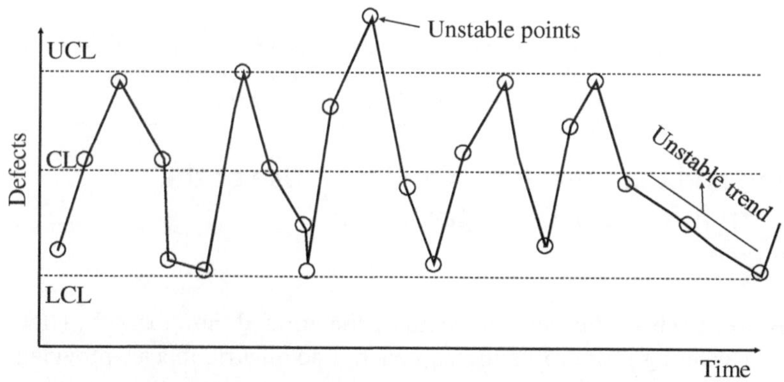

Figure 2.22 Defect chart reading procedure. Unstable regions are provided by unstable points and unstable trends.

Table 2.7 Types of control charts.

Attributes control charts	Description
p Chart	This is suitable for the fraction estimation of defective items in a sample. The fraction defective is calculated as the number of defective items in a sample divided by the total number of items. Typically, the fraction defective estimation is adopted when the sample size varies. For a high percentage of good items of 99%, the fraction defective is too small (0.01). For this analysis, the sample size must also be large for daily production cases
c Chart	This is adopted for estimation of the number of defects in an item. It is based on the Poisson distribution, and requires a constant size of the item, and that each sample has the same defect probability
u Chart	This is for the estimation of the number of defects in a sample and indicates defects per unit. It is adopted for samples having different sizes

(Continued)

Table 2.7 (Continued)

Attributes control charts	Description
np Chart	This is for the estimation of the number of defective items in a sample. It requires that the sample size remains constant. It has two benefits compared with the *p* chart: there is no calculation required of each sample result; it easier to understand and to interpret
Variable charts	
XmR/ImR Chart	This chart is used for only one observation per subgroup and when process variability is not defined. The moving range is used to estimate the variation. It is also applied for destructive sampling and for long production times
XbarR Chart	This indicates the changes happening in the central trend of a production process. The changes might be due to tools or to the adoption of new machine materials/components. R chart values indicate the gain occurring in the dispersion. It measures the central tendency and dispersion
XbarS Chart	This indicates the loss occurring in the dispersion. It measures the central tendency and dispersion
Prediction control chart	
Artificial neural networks	These are adopted to predict possible unstable points and unstable trends by supporting decisions about maintenance. The model uses historical data in order to optimize the prediction of the defect estimation

2.5 Defect Clustering and Classification: Combined Use of the K-Means Algorithm with Infrared Thermography for Predictive Maintenance

An innovative methodology able to check defects, is to combine the infrared thermography (IRT) technique together with the clustering post-processing approach. Radiometric images provide a dataset of pixels each one characterized by a temperature measurement. For its simplicity of implementation, the data mining K-means algorithm is also applied in many application fields, including computer vision. It is an iterative unsupervised algorithm that refines the division of objects at each cycle, and it is based on the concept of centroid definition in a multidimensional space. A centroid is a point in space that basically represents the center of a cluster corresponding to the midpoint of the clust-er space. In the example in Figure 2.23, three clusters with their centroids are

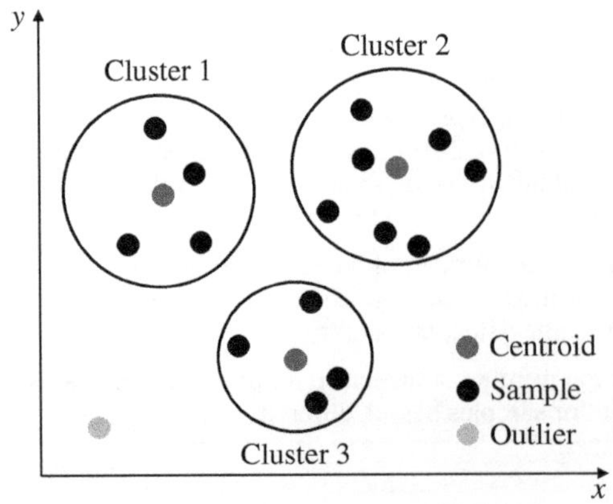

Figure 2.23 Clusters and related centroids.

represented. The number of iterations clustering samples is a function of the adopted methodology. It is possible to fix a number of iterations, or to make sure that the algorithm exits the cycle immediately after an iteration in which no objects are classified. The choice of the initial centroids greatly influences the final result.
The pseudocode of the K-means algorithm explaining its computational logic is as follows:

```
### K-Means Clustering applied on a calculus region ###
1. Choose the number of clusters
### parameter k indicating the clustering region to segment the
detected image; in this phase is set the initial cluster center iden-
tifying different thermal regions made by cluster of pixels (each
pixel is a radiometric measurement) ###
2. Place the centroids o_1, o_2, ..... o_k randomly (cluster initiali-
zation phase);
3. Repeat steps 4 and 5 until obtaining convergence or the error is
close to zero;
### the algorithm converges to zero by an iterative process ###
4. for each data point y_i:
- find the nearest centroid (o_1, o_2 .... o_k) updating centroid
position;
- assign the data point to that cluster;
5. for each cluster j = 1, … k;
- new centroid (all points assigned to that cluster)
6. End.
```

The K-means algorithm is applied to enhance clusters of radiometric data, defining regions of pixels having similar temperatures [13]. The best choice of the K parameter (cluster number) is related to the particular case of study and to the number of measurements to group. Concerning defect clustering, the k-Means algorithm:

- Enhances the defect region having a different temperature (welding defects, geometrical defects, etc.).
- Enhances a different region at low efficiency (photovoltaic panels, electric lines, energy routers, etc.).
- Supports the image analysis to confirm the presence of a defect (other image processing algorithms reinforce the defect analysis).

By considering a subset containing different measurement data, the Euclidean distance is defined as (Figure 2.24):

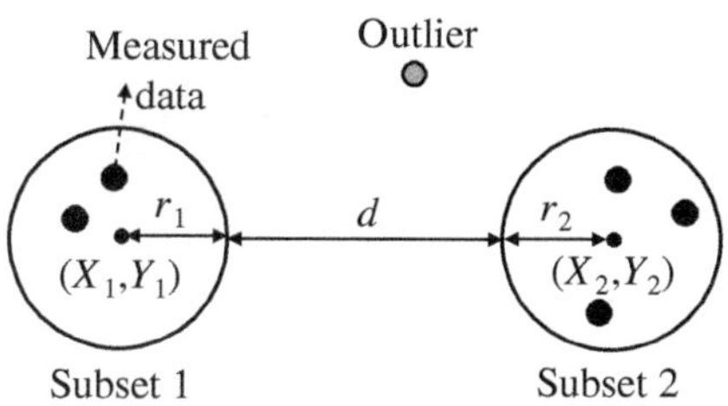

Figure 2.24 Parameters for the estimation of the Euclidean distance.

$$d = \sqrt{(Y_2 - Y_1)^2 + (X_2 - X_1)^2} - (r_1 + r_2) \tag{2.14}$$

where (X_1,Y_1) and (X_2,Y_2) are the centroid coordinates of the two clusters. By considering a dataset having a dimension *n,* data normalization must be applied to the Euclidean distance as:

$$\mathrm{d}(P,Q) = \mathrm{d}(Q,P) = \sqrt{(q_1 - p_1)^2 + \ldots + (q_n - p_n)^2} = \sqrt{\sum_{i=1}^{n}(q_i - p_i)^2} \tag{2.15}$$

where *P* and *Q* are two points of the domain space.

2.6 Facilities of a Prototype Network Implementing Advanced Technology: Example of an Advanced Platform Suitable for Industry 5.0 Integrating Predictive Maintenance

In this section a model of a modern informatic platform addressing Industry 5.0 upgrade (see architecture of Figure 2.25) enabling different advanced functionalities of the communication systems between operators is discussed. A software platform is improved by a virtual "control room" integrating AR tools supporting the management of the automations of the production process, and mechatronics enabling machinery actuation. The ESB network is developed as a new information infrastructure associated with the new production, by allowing sensor data interconnection with the big data system and AI algorithms integrated into the software platform. The platform is also able to manage the production test operations and to track the production stages of treatment using blockchain technology. The intelligent platform is characterized by:

- Connectors enabling interconnection with an innovative ERP platform based on the dynamic implementation of processes, and with the framework of AI algorithms.

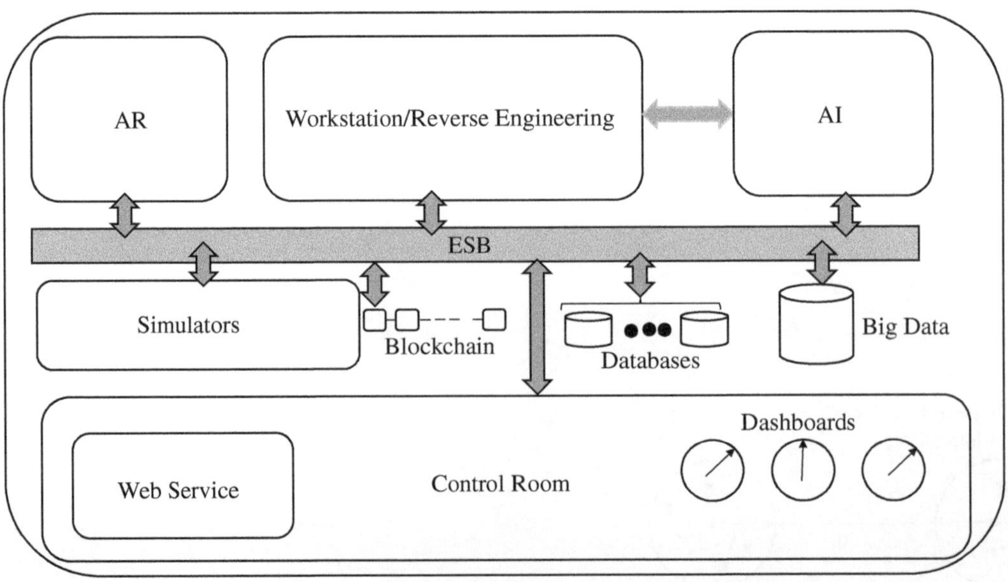

Figure 2.25 Architecture implementing Industry 4.0 and Industry 5.0 facilities.

- Connection to sensors monitoring production and allowing a direct linking of data storage in the connected big data system.
- "Embedded" algorithms integrated as objects in the ERP (deep learning, neural networks and other data mining algorithms).
- ESB bus connectivity to facilitate the interconnection between AI algorithms and the big data system.
- Management and synchronization ports for video and audio streaming communication of different users remotely connectable to the control room (users working in production testing).
- Graphic dashboards facilitating the activity managing and production traceability.
- Advanced BPM, based on the concept of process mining (the AI algorithms are interconnected and synchronized to the BPM which traces all production processes).
- Integration of graphic workflows in the BPM tool (graphic object workflows) capable of simulating production processes and of activating predictive maintenance procedures.
- Connection ports by means of wireless transmission of the testing devices of the control room platform.
- Software management module dedicated to bidirectional communications.
- Supervised backend system properly designed to control the whole production.
- Data transfer and data migration modules synchronized by crons (synchronization and timing of data transfer).
- IoT module (module for managing and transferring the data sensor of the production lines).
- Industry 4.0 module (mainly constituted by PLC systems of production lines and of sublines).
- Industry 5.0 module: adaptation of the Industry 4.0 module improving the adaptive logics of Industry 5.0 (self-control and self-regulation of production machinery, automated parameter optimization, predictive maintenance, etc.).
- AR module: to overcome the considerable difficulties encountered in the advanced production systems and in the use of innovative production machinery, a tool is required capable of supporting production and operational management problems; in this scenario AR facilities permit the operators to communicate during their work by exchanging opinion remotely with the control room by smart glasses enabling a videocall in streaming modality; in this way, the operator of the control room leads the worker from his/her terminal, seeing through the webcam placed on the smart glass; the audio-video communication is important for interaction with the staff, to analyze and modify photos in order to highlight the point-of-interest, and to indicate specific instructions by interactive markers; AR client–server architecture enables the following features:
  - Bidirectional voice call with the control room.
  - Video streaming activation.
  - Display messages and images sent by the operator through a specific chat.
  - Control of operators through a web browser to receive call notifications, voice call, video streaming, and screenshot of the video streaming, to edit video screenshots, to save video screenshots on the big data system, and to send messages and images via chat.

The platform of Figure 2.25 also takes into account the management of the simultaneous functioning of the various communications between operators and the control room, by the use of the latest generation of smart glasses connected to the network through a Wi-Fi electronic communication system of smartphone behaving as hub node. The operator connected to the platform is alerted through his/her device (PC, tablet, and smartphone equipped with the web browser), receiving the call notification. The smart glasses are authenticated by different actors of control room operators and workers. The operator on site, or the operator who is authorized for maintenance or to operate an intervention line, initiates an audio call to the control room with subsequent activation of video streaming: all this is automatized using the keys placed on smart glasses or voice commands. During communication with the control room, chat appears on the display on which all the information received remotely is scrolled (scrolling of the highlighted parts or of messages and screenshots of the point-of-interest). The platform manages all the accesses through a control panel. Before accessing the platform, the operator must authenticate him/herself to be available for the service: authentication is necessary to avoid the intrusion of attackers who can replace the operator with other communicating operations not suitable for the intervention. During video streaming, screenshots are edited, highlighting the areas of interest in real time (real-time interactivity) and stored data on the big data server. In addition, through the chat function, it is possible to interact with operators working on the production line by sending messages or captured images. Table 2.8 summarizes the main specifications of the AR communication system.

2.7 Predictive Maintenance Approaches

Maintenance procedures are fundamental to ensure a good production quality for a long period. Different actions are performed in this direction by acting on corrective, preventive, or predictive maintenance. Table 2.9 describes the three types of maintenance.

Industry 5.0 facilities include AI algorithms for the correct maintenance of production machines, ensuring a high standard of production quality. For each production machine there is a

Table 2.8 Main specifications of the AR communication system.

AR specification	Description
Software installation	The plug and play installation mode is carried out following the setting of the smartphone as hub node. The smartphone is connected to the smart glass via the Wi-Fi network. This operation is performed only once and is easily configurable by any end-user
Variability of the devices to use	Device able to connect with a high-performance network including 5G
System remote management using a web browser	Through an ad-hoc web platform, the user has the possibility of accessing various functions, such as: Video streaming enabled by the smart-glass Voice and chat Sending screenshots with the point-of-interest highlighted
Internet connection	The smartphones behave as "bridge connectors" of the smart-glass, and the server can transmit data by 3G/4G/5G/LTE connection for high resolution video stream according to the specific application
Audio and video quality	High resolution video streaming (1080p)
Data security management	Data security management involves controlled access through different authorization levels and the adoption of specific criteria and procedures, such as: The use of an identifying username and password to access the data The planning of periodic data backups Application of other approaches including attack classification by artificial intelligence
Data storage process	The detected images are stored in a database

Table 2.9 Maintenance definitions.

Maintenance type	Description
Corrective maintenance	This action is performed after the incident has taken place and the failure has already produced its effects. The cost of the failure is incurred by the company which initiates a repair operation and, possibly, maintenance of the equipment connected, or which may have been impacted. This case is the least favorable for the company
Preventive maintenance	This scheme is the most traditional. It consists of planning regular maintenance operations on the environments based on the average time between failures. As this process is purely statistical, it entails actions that are sometimes costly without being really justified. To avoid incidents, there is the likelihood of doing more than necessary and therefore incurring unnecessary additional costs
Predictive maintenance	The ideal model consists of relying on weak signals or environmental elements that make it possible to anticipate a failure and therefore to trigger a targeted maintenance action. Artificial intelligence offers effective tools to address this issue

corresponding maintenance plan provided by the company which produced the machine indicated by the standard maintenance plan line of Figure 2.26, which usually concern a long period. By means of the IoT sensors it is possible to predict machine wear and machine failures, thus applying a predictive maintenance of the production lines. Predictive maintenance is performed by AI algorithms predicting production failures and running on the network of Figure 2.25. The prediction estimation possibly changes the initial plan by anticipating or by delaying the maintenance procedures. The main cases are (Figure 2.26):

- The initial plan, of long period maintenance, remains the same because the analysis predicts failures and defects according to the standard tolerances (plot ABC).
- The initial plan is delayed because there is a low use of the production machine, and there are no particular stress working conditions (plot AD), allowing maintenance in a very long period (in this case it is not necessary analyze sensor data).

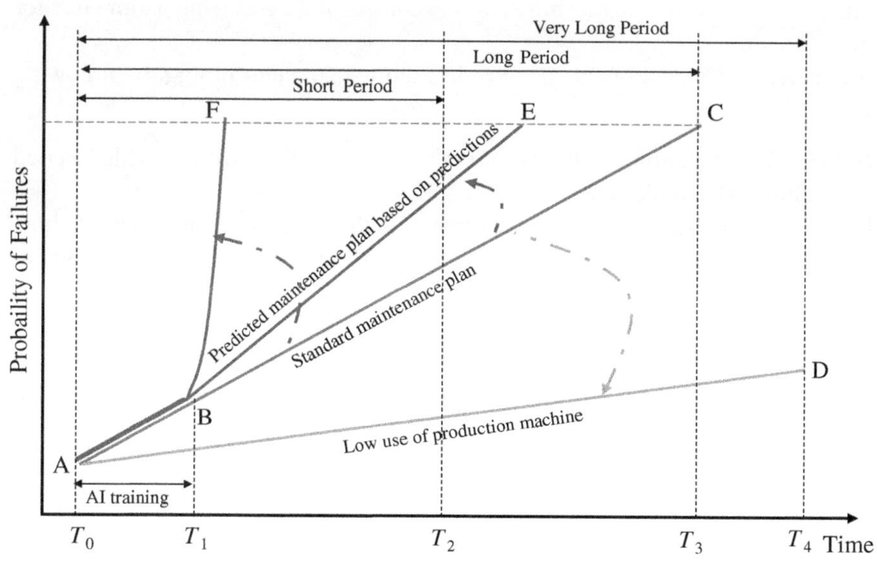

Figure 2.26 Maintenance plan variation based on prediction results.

- The initial maintenance plan is anticipated in the long period because there is a rather low probability of machine failures (plot ABE).
- The initial maintenance is anticipated in the short period, because there is a high failure probability (see the exponential plot ABF).

The observing periods are:

- Initial use of the production machine (period T_0).
- Data acquisition period for the collection of enough data for AI algorithm execution (period T_1).
- Short period anticipating maintenance for high failure probability (period T_2).
- Long period related to the standard maintenance planned time (period T_3).
- Very long period related to the maintenance delay (period T_4).

In order to follow a decision making suggested by AI, it is important to construct a model characterized by a stable dataset containing data of normal behavior of the machine by considering different working conditions (different velocities, different operating parameters, etc.). A big data system supports the analysis by allowing the use of a lot of historical data related to measurements performed on the machine.

The predictive maintenance model is implemented by considering three different danger levels [14, 15] such as:

- Warning level due to only IoT data monitoring (level 1).
- Level of corrective and preventive actions (level 2).
- Level of predictive actions (level 3).

The adjustment of the production parameters during the production period can avoid maintenance anticipations. The predictive maintenance approach is based on the concept to analyze the deviation between the standard maintenance plan and the real maintenance plan [16, 17]. This is possible by the training of the AI algorithms estimating predictions (samples detected during the time $\Delta T = T_1 - T_0$). Each production machine is characterized by a standard maintenance plan plot having a specific slope. The first approach to follow to apply predictive maintenance is to construct this plot, by defining at the same time the machine parts to control and measuring parameters for the prediction.

Figure 2.27 is a more detailed model of the diagram in Figure 2.26, determining the following observation points:

- **Machine starting time:** the machine starts its production (if the initial testing provided a good check, there are no failure risk conditions).
- **Detection of abnormal vibrations:** by the accelerometer analysis, including frequency behavior study of the vibrations, after one or more months, it is possible to check anomalous oscillations as signals of machinery malfunctions.
- **Audible noise:** more accentuated vibrations or component frictions can degenerate indicating a possible generation of a product defectivity; when the noise becomes audible this risk is very high; the audible noises appear usually after one or a few weeks after the anomalous vibration condition.
- **Anomalous heat:** the continuous mechanical component frictions induce a local heating which is significant for possible machine malfunctions or failures; the IRT facilitates this thermal analysis; the anomalous heat appears after some days after the noises.
- **Smoke condition:** the smoke represents a pre-breakdown condition and appears some minutes after the observation of anomalous heating.

- **Breakdown:** the machine or component breakdown condition totally stops production, significantly increasing the production costs.

Different corrective actions must be performed in order to avoid the critical break down condition. The corrective actions refer to (Figure 2.27):

- **Predictive maintenance stage I:** the corrective action 1 recalibrates the mechanical components or machine parameters setting; the vibration analysis is difficult to perform, but AI algorithms facilitate this study.
- **Predictive maintenance stage II:** the corrective action 2 is urgent in order to avoid product defect or in general machine malfunctions; acoustic sensors enable noise analysis by also distinguishing frequencies not easily audible to the worker.
- **Predictive maintenance stage III:** the corrective action 3 is automatized by reading thermograms and applying suitable image processing algorithms; it may be that this action is applied with a delay thus requiring a further and quick corrective action (corrective action 4), avoiding the risk of breakdown (when smoke appears it is too late).

All the corrective actions can be automatized as auto-adaptive actions by applying the update of the maintenance plan, and by changing components or tools. Theoretically the auto-adaptive corrective action allows to totally avoid not only the breakdown of the machine, but also the occurrence of defective conditions. The auto-adaptive process is supported by:

- Upgrading the adopted technology for the specific product processing.
- Increasing the information and the measurements by properly adding further sensors.
- Applying AI algorithms facilitating the analysis of the hidden information about possible defectivity or failure starting.
- Applying reverse engineering (RE) approaches able to optimize the component design material and processing procedures.

The maintenance strategy is to cut possible costs due to the machine failures. Specifically, predictive maintenance is an excellent tool to contain these costs mainly associated with the machine breakdown or product defectivity. From the diagram in Figure 2.27 it is evident that the vibration analysis is mainly important for the first prediction of malfunctions. Therefore, the analysis

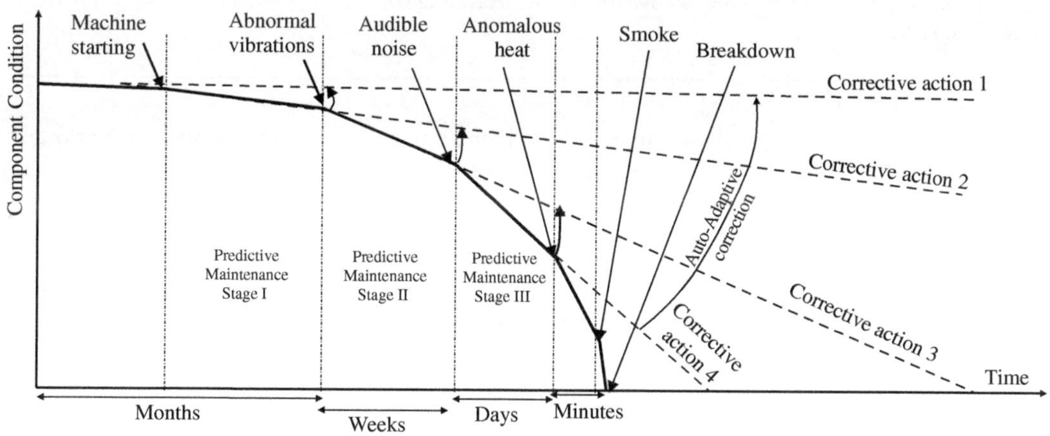

Figure 2.27 Examples of typical alerting signals relating to the failure conditions of machine components, and corrective actions.

Table 2.10 Predictive actions for a polishing process.

Abnormal vibrations (predictive maintenance stage I)	The polishing process of a surface is characterized by vibrations. Vibration analysis is both relevant to the time trend and to the response frequencies calculated by the fast Fourier transform. The anomalous oscillations have different characteristics from the standard of the machinery. The difficulty is precisely in identifying the conditions of normality with respect to those of abnormality. Artificial intelligence classifiers improve the analysis by making it automatic
Accuracy of image processing analysis (predictive maintenance stage I)	Image vision techniques using cameras are useful for analyzing surface wear using appropriate segmentation or classification algorithms. This analysis requires a high resolution and an ideal setting of the operating conditions for image acquisition (positioning for zooming, lighting conditions, protection from dirt, etc.). The use of smart cameras, i.e. cameras equipped with an internal processor, aids the local analysis of the textures and characteristics of the images, but may not be suitable due to the high computational cost (data processing of high resolution images). For this reason, it is preferable to adopt cloud computing solutions. Appropriate classifiers, such as the random forest algorithm [18], simplify the extraction of the features by highlighting the wear regions. In any case, the choice of the suitable algorithm, for the extraction of the characteristics, is "weighted" according to the technology used, and to the training dataset. If anomalous defects or textures are identified, it is necessary to update the machine setting (corrective action 1)
Noise (predictive maintenance stage II)	The noise increase is a degeneration of the polishing process and soon damages the surface to be processed. If the noise increases, it is preferable to change the machine tool to operate directly on the surface (corrective action 2)
Infrared thermography (predictive maintenance stage III)	The thermographic analysis usually helps to understand the effects of anomalous processing, providing information about the correct processing of the surface. In the specific case, infrared thermography provides information about the check of the polishing processing, analyzing the tool processing area by the thermal response, and allowing automatic recognition. Some criticalities of the radiometric analysis may be associated with the orientation of a possible thermal excitation light source enhancing the release of the thermal effect (release of thermal radiation). Appropriate artificial intelligence classification algorithms facilitate this analysis, by highlighting non-homogeneous treatments or processing. Also, in cases of inhomogeneous heated surface, it is preferable to change the machine tool (corrective action 3 and corrective action 4)
Polishing machine lifetime (predictive maintenance stage I, stage II, and stage III)	Other data such as the time of use of the machinery, if properly crossed with the wear analysis data, provide important information about the maintenance plan, adjusting the relative trend in a forecasting key

vibration data, if crossed with other data relating to the application context, allows efficient intervention by avoiding the wear and breakdown events.

Table 2.10 lists some predictive maintenance conditions of a product polishing process, referring to the first three stages of Figure 2.27.

2.7.1 Preventive Maintenance and Predictive Maintenance Operations in the Railway Industry

Predictive maintenance is formulated after preventive maintenance. In the railway industry the planning of the maintenance of wagons, train, and infrastructures are fundamental to avoid serious problems for transport and logistics. The workflow in Figure 2.28 indicates the passage from

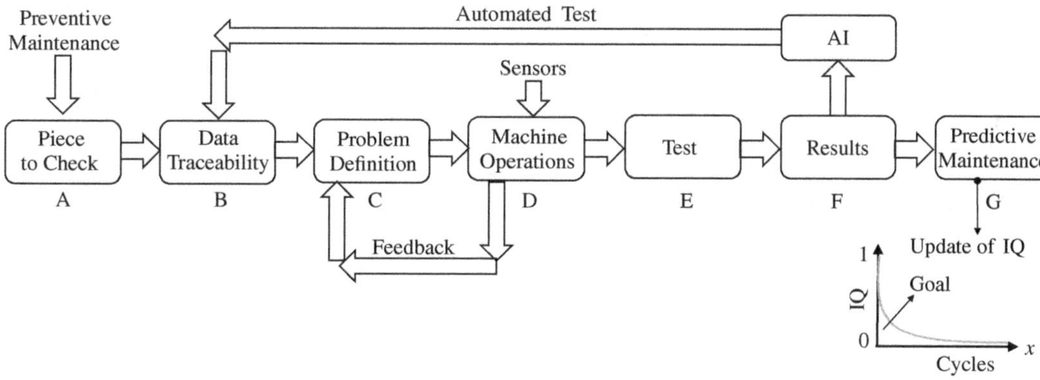

Figure 2.28 Workflow defining predictive maintenance for pieces in the railway industry.

preventive maintenance to predictive maintenance concerning the pieces to process for reuse in wagons or trains. The workflow is structured as follows:

- Step A (piece to check): the preventive maintenance is addressed on the integrity check of a piece (the workpiece is worn and requires maintenance).
- Step B (data traceability): by RFID or QR technologies, the piece to process is registered by enabling the processing traceability (the traceability is also related to registration of the testing results and takes into account the estimation of the residual lifecycle based on the index of quality [IQ]).
- Step C (problem definition): in this step the types and the grade of the out of tolerance, and the activities to perform for the workpiece processing are defined.
- Step D (machine operations): the solution of a problem provides the best parameter setting of machines; the feedback control is provided by means of the sensor monitoring the machine processing and correcting the parameters during the processing.
- Step E (test): different tests are performed according to the previous results.
- Step F (results): the tests are planned dynamically and automatically by the AI engine enabling a continuous variation of the processing parameter setting.
- Step G (predictive maintenance): the predictive maintenance can to predict the next piece processing based on the updating of the IQ characteristic; the ANNs are good candidates for the prediction estimation.

The IQ characteristic is defined by the following equation:

$$\mathrm{IQ}(x) = \exp^{-\lambda x} \tag{2.16}$$

where λ is a coefficient determining the theoretical trend and the piece degradation IQ(x) is a function of the performed processing quality, with x being the number of working cycles. A high value of λ indicates a piece not processable.

The predictive maintenance goal is to decrease the λ value by means of an intelligent and optimized piece processing based on the procedure illustrated in Figure 2.28. The goal is to obtain an updated IQ curve similar to the initial one (Figure 2.29).

By considering the predictive maintenance of train parts and brake quality control made on an ordinary locomotive, the tests are optimized by measuring the parameters in Table 2.11 (best quality check) [19].

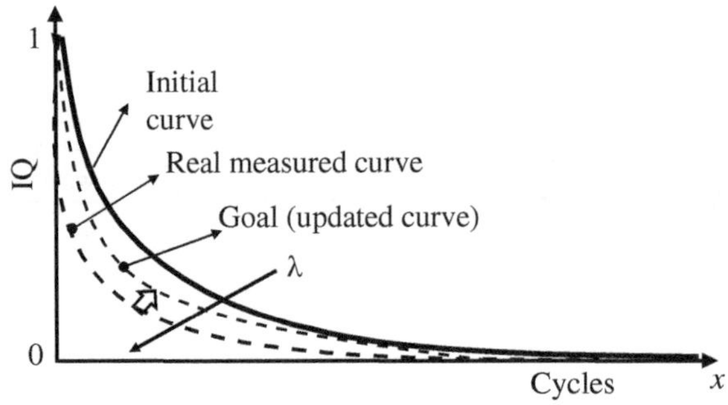

Figure 2.29 IQ trends: initial curve, real measured curve, and updated curve.

Table 2.11 Best measured main parameters for locomotive brake tests.

Parameter	Value
Main tank pressure (bar)	7.5–9
Pipe pressure (bar)	5 ± 0.05
M brake (filling time up to the 95% of the maximum pressure of the brake cylinder) (s)	18–30
V brake (filling time up to the 95% of the maximum pressure of the brake cylinder) (s)	3–5
M brake (brake releasing for a pressure of 0.4 bar at the brake cylinder) (s)	40–60
V brake (brake releasing for a pressure of 0.4 bar at the brake cylinder) (s)	15–20
Pressure decrease for a full breaking (bar)	1.5 ± 0.1
Brake sensitivity	The brake must act before 1.2 s for a pressure decrease of 0.6 bar in 6 s
High pressure stroke	6 bar with a total duration of 2 s. From 6 bar to 5.2 bar in 1 s
Air loss in the generic pipe (bar min^{-1})	≤ 0.1
Loss in the generic pipe	≤ 0.1 bar/5 min
Loss of the brake cylinder	≤ 0.1 bar/5 min
Times of pressure loss in the generic pipe, with pressure of the generic pipe stable at 5 bar, enable a pressure decrease of the generic pipe up to 3.5 bar	Time interval between 3 s and 4 s, with a pressure decrease of the generic pipe from 5 to 3.5 bar
Time of pressure increase in the generic pipe, with stabilized pressure in the generic pipe at 3.5 bar, increasing the pressure inside up to 5 bar	Time interval between 3 s and 4 s for a pressure increase of the generic pipe from 3.5 to 5 bar
Variability of the braking increase and releasing, with a stabilized pressure of the generic pipe at 5 bar, decrease the pressure up to 5 bar, with intermediate levels of 0.2 bar	For a pressure of the generic pipe of 3 bar, verifying for 1 min the stability of the pressure, the variation should be ≤ 0.1 bar
Efficiency of dynamic braking, tested at a constant speed of 20 km h^{-1}, on a planar and linear track	Braking length ≤ 20 m

Source: Modified from [19].

2.8 Examples of Advanced Infrastructures Implementing AI

The information industry infrastructures oriented on AI platforms are an important topic in industrial research. Table 2.12 lists some examples of topics discussed in industry projects.

Table 2.12 Example of advanced information platforms integrating artificial intelligence algorithms.

Industry sector	Description	Reference
Service company	The platform is able to improve the call center service by implementing a multilevel user access (FAQ level, chatbot with AI level, direct operator contact level). The goal of this structuration is mainly to optimize the human resource time. An important issue is to implement a chatbot with an ANN algorithm also able to "understand" colloquial usage forms. The service platform can be also improved by a cybersecurity network managing the communication of the virtual private network channels	[20]
Food trade network	Consulting industries require the evaluation of KPIs. Concerning the food trade network, recurrent neural networks predict KPIs of a company working in global distribution system consulting	[21]
Industry working in realization of electrical, electronic, and energy systems and radio bridge	The data mining engine and augmented reality technology support the human resources KPI evaluation and the decision process, thus consenting to the re-engineering of production processes also predicting the workload	[22]
Store marketing	The AI platform is based on decision tree and XGBoost algorithms providing by an E-commerce platform, a multistore management based on priority rules and sales prediction	[23]
Fashion marketing	A big data platform integrates different AI algorithms improving marketing, business intelligence, and logistics. The platform integrates social text mining and sales prediction tools	[24]
Roadside assistance services	Data mining algorithms facilitate the improvement of business intelligence, strategies, and services	[25]
Document retrieval services	The platform includes the K-means algorithm and support vector machine algorithms for the automatic retrieval of digitized documents. The expert systems are adopted for the recognition of different features of the document such as color, layout, and texture. The content-based image retrieval approach is suitable for the automatic recognition of documents, thus accelerating and optimizing industry services	[26]
Lighting production	Big data and AI are tools added to improve the ESB infrastructure system of the industry. All software exchange data through the ESB which allows data transfer between different database systems. AI tools are linked to the databases by the ESB also managing the data storage in the big data system	[27]
Global distribution store	The platform is based on the execution of multiple ANNs optimizing the planogram and the store shelf layout	[28]
Wheat processing industry	The information system integrates big data and the AI engine improving predictive maintenance, energy consumption optimization, and production machine calibration. Anomalous energy consumption indicates a possible ineffectiveness of production machines	[29]

(*Continued*)

Table 2.12 (Continued)

Industry sector	Description	Reference
Tire industry	The platform integrates the AI image vision technique controlling tire defectivity and quality. The techniques include the combined use of K-means and long short-term memory algorithms applied to the tire image detecting different defect types. The image vision software is embedded into a platform tracing all the production process of tire assembly	[30]
Vegetable production	Image vision and blockchain tools for quality check of the vegetable production phases	[31]
Telemedicine homecare assistance services	Companies working on health assistance and homecare assistance require an information system embedding different software management tools and decision support systems capable of predicting the health status of monitored patients thus automatically updating the assistance plan. Big data systems enforce the AI data processing and the definition of health risk levels contributing to optimizing the health assistance processes. For remote monitoring of patients, wearable devices are used linked in the cloud to platform check in real time the physiological parameters of each patient. The telemedicine platform is also useful to send alerting signals when the physiological parameters pass specified thresholds. Patient data are stored in a big data system by means of crons timing the data storage process	[15, 32, 33]
Telemedicine for laboratory connectivity	Companies performing blood analyses in properly equipped laboratories send results to a platform controlled by the hospital. The ANNs are executed to predict the philological parameters of the patient validating the results of clinical algorithms based on a multilevel risk alerting system	[34]

AI, artificial intelligence; ANN, artificial neural network; ESB, enterprise service bus; KPI, Key Performance Indicator.

2.9 Examples of Telemedicine Platforms Integrating Advanced Facilities

Modern health platforms are upgraded by advanced networks managing patient data in the cloud environment, and by the implementation of AI algorithms able to improve the prognostic and to provide a DSS for doctors and human resources working in assistance. The proposed models are suitable for public hospital and care centers, and in general for companies working in assistance services.

2.9.1 Advanced Telecardiology Platform

Figure 2.30 shows an example of an innovative architecture integrating an advanced information system. The architecture is formulated following a preliminary analysis carried out for a telecardiology control room, by upgrading the network through the introduction of cloud management of a new data flow. The new data flow concerns the audio and video streaming type by the introduction of smart glasses and the electrocardiogram data transmission for 10 experimental nodes. The experimental nodes can improve the first aid center activities supporting the traceability workflow of heart patient care and monitoring. The data flow is managed in the cloud through a DSS

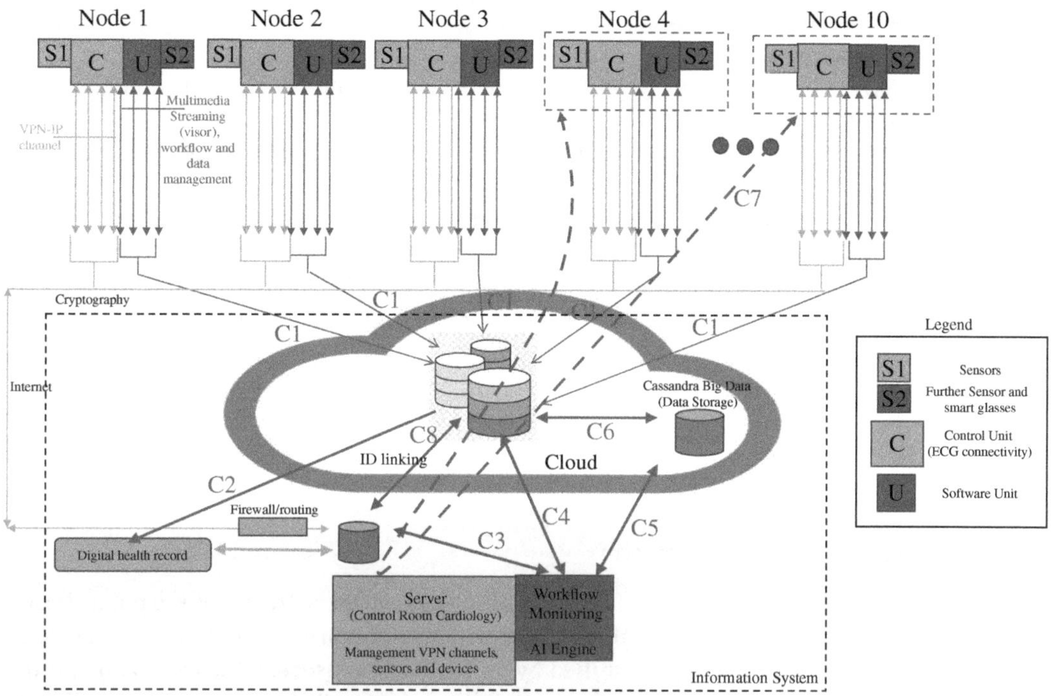

Figure 2.30 System architecture: upgrade of a telecardiology network.

upgraded by a big data system and by an AI engine. The platform can connect the digital electronic file by recognizing the patient's identification number (ID). Table 2.13 lists some of the main network specifications.

Some examples of connectivity (indicated in Figure 2.30 by the letter C), useful for understanding some cases of use of the upgraded network, are:

- C1 connectivity: this connectivity concerns audio and video cloud streaming connection, and sensor data transmission; the hardware device is recognized by the subscriber identification module by activating the registration procedure; the cloud server enables bidirectional transmission.

Table 2.13 Main specification of the telecardiology network.

Network typology	4G/4,5G with dynamic Internet Protocol (IP) and virtual IP, connecting 10 nodes by VPN channels
Network security	Audio and video multimedia streaming with 128 bit Advanced Encryption Standard cryptography, and cyclic redundancy check detecting any errors that may compromise the integrity of the data. Hypertext Transfer Protocol over Secure Socket Layer Protocol
Connected sensors	Electrocardiogram sensors (S1) and other sensors monitoring patient parameters including smart glasses viewing the real patient status (S2)
Estimated additional network load (*Throughput*)	30 megabit for each node channel

- C2 connectivity: the data stored in the cloud is migrated into the digital health record using the Health Level 7 (HL7) Protocol.
- C3 connectivity: the physical server (workflow management) is interconnected to other database systems (bidirectional connection) by a dedicated ESB.
- C4 connectivity: the cloud is interconnected to the physical server managing dynamic workflow and data flow.
- C5 connectivity: the Cassandra big data system is interconnected to the physical server by means of appropriate drivers.
- C6 connectivity: all data stored in the cloud are migrated to the big data Cassandra system.
- C7 connectivity: this link sends a status of request or of starting of a grid interconnection service through the control room.
- C8 connectivity: as an alternative to the C2 link, where data are migrated from the platform database to the digital health record through other database systems interconnected to different software.

2.9.2 Advanced Teleoncology Platform

The telemedicine platform of Figure 2.31 is designed to improve the activities of the public hospital centers, analysis laboratories, general practitioners, and specialists for the care assistance of oncological patients. Each patient is identified by an ID used to associate the care path and the related analyses. Through this platform and a control room, it is possible to have real-time information of the analyzes and to perform AI and clinical algorithms providing important prognostic results. The platform can optimize personalized drug administration enabling the therapeutic stratification, therapy precision. The communication between laboratories, research centers, doctors, and control room, facilitate the prevention of health risk conditions. Industries working in

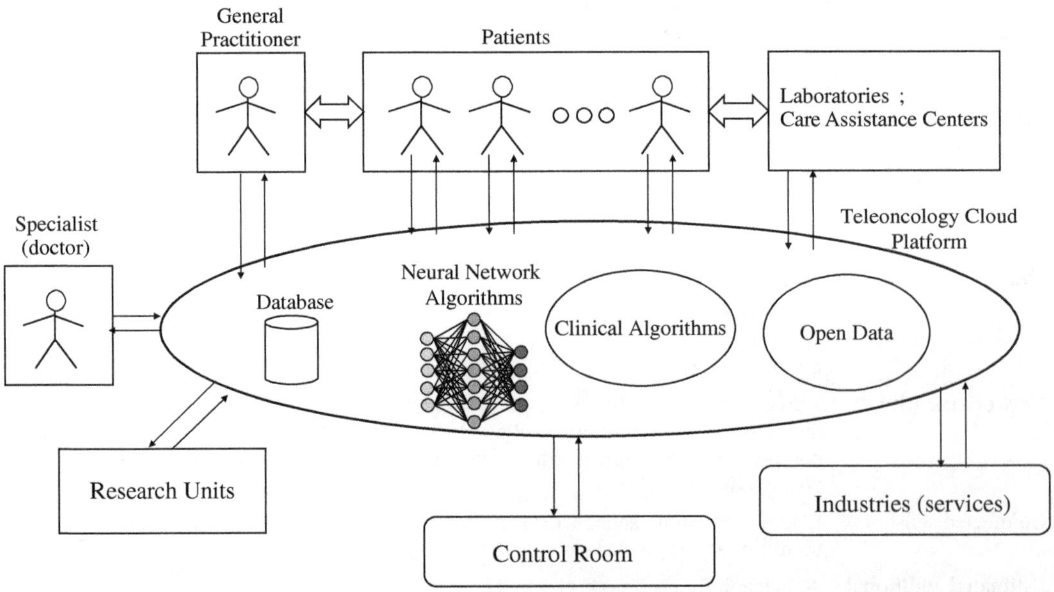

Figure 2.31 System architecture of a teleoncology platform: functional scheme.

software development contribute to the platform maintenance. Important scientific results are in the epidemiological research finding correlations between environment pollution and oncological pathologies. It is necessary to implement a big data system to collect massive data. The actors involved in the architecture of Figure 2.31 are:

- Dehospitalized patients to control.
- General practitioners.
- Laboratories for hematological analyses.
- Control room.
- Research centers analyzing clinical results and information contained in the open data system.
- Industries.

The control room manages all the system actors. The application of neural network algorithms combined with clinical ones, automatize the care assistance procedures [34]: the visit planning and the analyses are automatically planned based on the prediction of the patient's health status. The whole platform provides a DSS for managing the patient, and at the same time suggest medical decisions. The doctor can follow the suggestions or ignore them. In any case, after reading the results, the final decision is made by the doctor. The benefits of the telemedicine platform are the prognostic efficiency, the efficiency in the assistance processes and paths, and the decrease in cost of the hospital due to dehospitalization.

In order to collect many data to process for the formulation of the AI training models, big data is adopted as the database system. The big data is necessary to detect the larger physiological parameters and associated with many patients. The platform of Figure 2.31 can be improved by adding more algorithms to execute in the cloud environment, and by upscaling the access for a large connectivity as for national research networks. The model of Figure 2.31 can be modified to be applied in other industrial telemedicine contexts including further technologies such as E-learning, mobile applications, and other clinical algorithms depending on the pathology. Other AI algorithms to be implemented in the platform are addressed on voice and facial expression recognition during the rehabilitation process.

2.9.3 Multipurpose E-Health Platform

Health platforms are also applied for multipurpose assistance services. Figure 2.32 shows an example of multipurpose assistance service integrating different service functions such as:

- Therapeutic compliance.
- Dietary compliance.
- Psychotherapy.
- Video consultation.
- Diagnostic therapeutic assistance path formulation.
- Rehabilitation path formulation.
- Dietary path formulation.
- Visit scheduling plan formulation.
- AI prediction of the patient health status.

In particular, the prediction of the health status is important to improve assistance services and assistance processes. Different actors are involved in the platform use such as:
System administrator (managing the backend system).

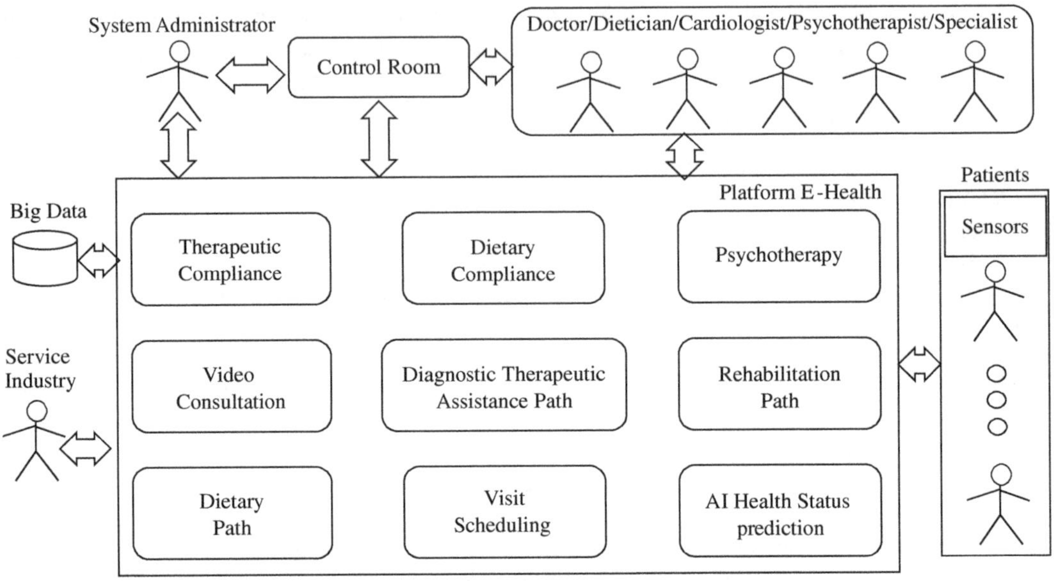

Figure 2.32 E-health system architecture of a multipurpose health platform.

- Doctors.
- Dieticians.
- Cardiologists.
- Psychotherapists.
- Other specialists.
- Patients to be controlled.
- Industries working on software development and software maintenance.
- Control room managing all assistance services and all actors of the system.

All patients are controlled by wearable sensors providing in real time physiological measurements useful for the health status prediction estimated by the AI algorithms. The platform of Figure 2.32 takes into account many care assistance processes. To merge all actors' functions, it is important to map the processes by means of the BPM modeling. A BPM software tool can be programmed to implement these processes by setting a specific checkpoint for the assistance paths to formulate, according to the adopted clinical protocols. For more homecare assistance services, other system actors can be added. The modular design of the workflows automatically implementing processes facilitates the integration of adding services modeled by other workflows. In order to process massive data related to different patient data, the big data is integrated into the advanced information system. The assistance or service workflows to follow can be dynamically and automatically selected by the output of the AI algorithms, predicting patient health status: by mean of a binary logic condition, it is possible to select or to deselect a workflow by automatizing the processes. The process automation is indispensable for the management of many patients or customers requiring a precise assistance correlated to a specific pathology, such as for chronic or older patients. An example of assistance workflow is depicted in Figure 2.33, where different actors such as specialist doctors, psychologists, and dieticians collaborate in a unique care service: the dietary path is parallel to the

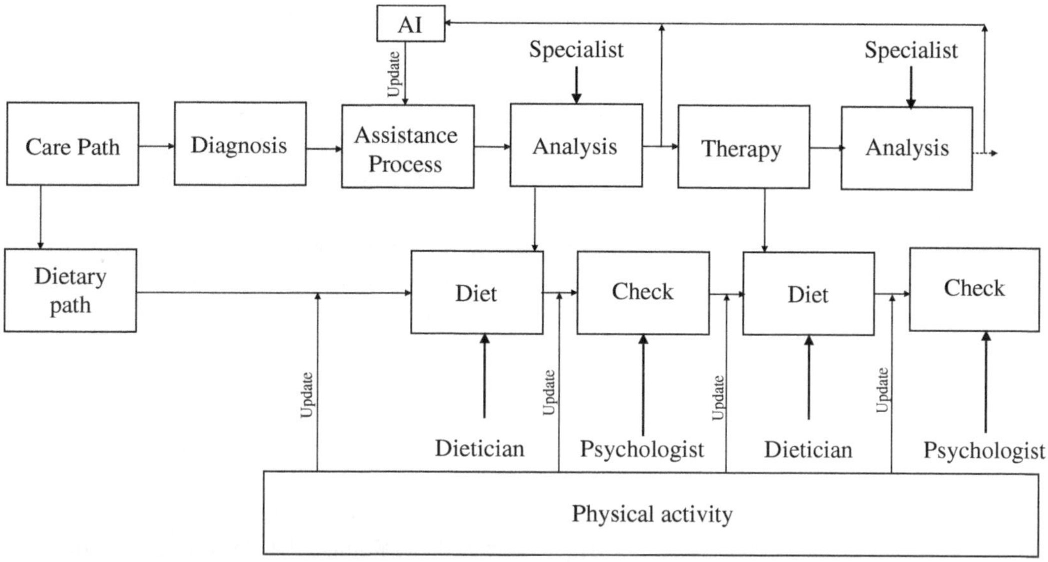

Figure 2.33 Workflow of combined assistance services.

care path, and the actors are synchronized in the workflow optimizing the care process. The assistance process is updated in each step, and the whole process is adjusted by means of AI predicting patient health status.

References

1 Blanco-Novoa, Ó., Fernández-Caramés, T.M., Fraga-Lamas, P., and Castedo, L. (2017). An electricity price-aware open-source smart socket for the internet of energy. *Sensors* 17 (643): 1–34.
2 Martínez, E., Toma, D.M., Jirka, S., and Del Río, J. (2017). Middleware for plug and play integration of heterogeneous sensor resources into the sensor web. *Sensors* 17 (2923): 1–28.
3 Massaro, A., Maritati, V., Giannone, D. et al. (2019). LSTM DSS automatism and dataset optimization for diabetes prediction. *Applied Sciences* 9 (3532): 1–22.
4 Massaro, A. and Galiano, A. (2020). Re-engineering process in a food factory: an overview of technologies and approaches for the design of pasta production processes. *Production & Manufacturing Research* 8 (1): 80–100.
5 Massaro, A., Vitti, V., Mustich, A., and Galiano, A. (2019). Intelligent real-time 3D configuration platform for customizing E-commerce products. *International Journal of Computer Graphics & Animation (IJCGA)* 9 (4): 13–28.
6 Lam, H.K. and Leung, F.H.F. (2007). Design and training for combinational neural-logic systems. *IEEE Transactions on Industrial Electronics* 54 (1): 612–619.
7 Choi, A., Shi, W., Shih, W., and Darwiche, A. (2019). Compiling neural networks into tractable boolean circuits. In: Proceedings of the AAAI Spring Symposium on Verification of Neural Networks (VNN19), Stanford University, USA (25–27 March 2019). AAAI.
8 Vijaychandra, J., Sesha Sai, B., Sateesh Babu, B., and Jaganna, P. (2019). A comprehensive review on McCulloch-Pitts neuron model. *International Journal of Innovative Technology and Exploring Engineering (IJITEE)* 8 (6): 1516–1522.

9 Massaro, A., Maritati, V., Galiano, A. et al. (2018). ESB platform integrating KNIME data mining tool oriented on Industry 4.0 based on artificial neural network predictive maintenance. *International Journal of Artificial Intelligence and Applications (IJAIA)* 9 (3): 1–17.

10 Igel, C., Toussaint, M., and Weishui, W. (2005). Rprop using the natural gradient. In: Trends and Applications in Constructive Approximation. ISNM International Series of Numerical Mathematics, vol. 151 (eds. D.H. Mache, J. Szabados and M.G. de Bruin), 259–272. Basel: Birkhäuser.

11 Massaro, A., Manfredonia, I., Galiano, A., and Xhaysa, B. (2019). Advanced process defect monitoring model and prediction improvement by artificial neural network in kitchen manufacturing industry: a case of study. In: Proceedings of the IEEE International Workshop on Metrology for Industry 4.0 and IoT, Naples, Italy (4–6 June 2019). Piscataway, NJ: IEEE.

12 D'Aloia, M., Russo, R., Cice, G. et al. (2017). Big data performance and comparison with different DB systems. *International Journal of Computer Science and Information Technologies* 8 (1): 59–63.

13 Massaro, A., Galiano, A., Meuli, G., and Massari, S.F. (2018). Overview and application of enabling technologies oriented on energy routing monitoring, on network installation and on predictive maintenance. *International Journal of Artificial Intelligence and Applications (IJAIA)* 9 (2): 1–20.

14 Bastos, P., Lopes, I., and Pires, L. (2014). Application of data mining in a maintenance system for failure prediction. In: Safety, Reliability and Risk Analysis: Beyond the Horizon (eds. R. Steenbergen, P.H.A.J.M. Van Gelder, S. Miraglia and T. Vrouwenvelder), 933–940. Taylor & Francis Group: London.

15 Massaro, A., Maritati, V., Savino, N. et al. (2018). A study of a health resources management platform integrating neural networks and DSS telemedicine for homecare assistance. *Information* 9 (176): 1–20.

16 Massaro, A., Selicato, S., and Galiano, A. (2020). Predictive maintenance of bus fleet by intelligent smart electronic board implementing artificial intelligence. *MDPI IoT* 1 (2): 180–197.

17 Tchakoua, P., Wamkeue, R., Ouhrouche, M. et al. (2014). Wind turbine condition monitoring: state-of-the-art review, new trends, and future challenges. *Energies* 7 (4): 2595–2630.

18 Al Samarraie, M.M., Nordin, J., and Al-Anizy, G.J. (2005). Texture classification using random forest and support vector machines. *Journal of Theoretical and Applied Information Technology* 73 (2): 232–238.

19 Massaro, A., Cannella, E., Dipierro, G. et al. (2020). Maintenance and testing protocols in railway industry. *ACTA IMEKO* 9 (4): 4–12.

20 Massaro, A., Maritati, V., and Galiano, A. (2018). Automated self-learning chatbot initially built as a FAQs database information retrieval system: multi-level and intelligent universal virtual front-office implementing neural network. *Informatica Journal* 42 (4): 515–526.

21 Massaro, A., Dipierro, G., Saponaro, A., and Galiano, A. (2020). Data mining applied in food trade network. *International Journal of Artificial Intelligence and Applications (IJAIA)* 11 (2): 15–34.

22 Massaro, A., Galiano, A., Mustich, A. et al. (2019). A case study of process engineering of operations in working sites through data mining and augmented reality. *International Journal of Data Mining & Knowledge Management Process (IJDKP)* 9 (5): 1–20.

23 Massaro, A., Mustich, A., and Galiano, A. (2020). Decision support system for multistore online sales based on priority rules and data mining. *Computer Science and Information Technology* 8 (1): 1–12.

24 Massaro, A., Vitti, V., Galiano, A., and Morelli, A. (2019). Business intelligence improved by data mining algorithms and big data systems: an overview of different tools applied in industrial research. *Computer Science and Information Technology* 7 (1): 1–21.

25 Massaro, A., Leogrande, A., Lisco, P. et al. (2019). Innovative BI approaches and methodologies implementing a multilevel analytics platform based on data mining and analytical models: a case of study in roadside assistance services. *International Journal on Soft Computing, Artificial Intelligence and Applications (IJSCAI)* 8 (1): 17–36.

26 Barbuzzi, D., Massaro, A., Vitti, V. et al. (2017). Intelligent system for image retrieval (ISIR) based on CBIR technology: industrialization process. *Journal of Next Generation Information Technology (JNIT)* 8 (2): 1–13.

27 Massaro, A., Calicchio, A., Maritati, V. et al. (2018). A case study of innovation of an information communication system and upgrade of the knowledge base in industry by ESB, artificial intelligence, and big data system integration. *International Journal of Artificial Intelligence and Applications (IJAIA)* 9 (5): 27–43.

28 Massaro, A., Vitti, V., and Galiano, A. (2018). Model of multiple artificial neural networks oriented on sales prediction and product shelf design. *International Journal on Soft Computing, Artificial Intelligence and Applications (IJSCAI)* 7 (3): 1–19.

29 Massaro, A., Selicato, S., Miraglia, R. et al. (2020). Production optimization monitoring system implementing artificial intelligence and big data. In: Proceedings of IEEE MetroInd4.0&IoT 2020, Rome, Italy (3–5 June 2020). Piscataway, NJ: IEEE.

30 Massaro, A., Dipierro, G., Cannella, E., and Galiano, A. (2020). Comparative analysis among discrete Fourier transform, K-means and artificial neural networks image processing techniques oriented on quality control of assembled tires. *Information* 11 (5): 1–11.

31 Massaro, A., Maritati, V., Savino, N. et al. (2020). Blockchain and technologies matching with the case of study of vegetables production. *Information Technology in Industry (ITII)* 9 (2): 1–8.

32 Massaro, A., Maritati, V., Savino, N., and Galiano, A. (2018). Neural networks for automated smart health platforms oriented on heart predictive diagnostic big data systems. In: Proceedings of IEEE AEIT 2018, Bari, Italy (3–5 October 2018). Piscataway, NJ: IEEE.

33 Massaro, A., Selicato, S., Ricci, G. et al. (2020). Decisional support system with artificial intelligence oriented on health prediction using a wearable device and big data. In: Proceedings of IEEE MetroInd4.0&IoT 2020, Rome, Italy (3–5 June 2020). Piscataway, NJ: IEEE.

34 Massaro, A., Galiano, A., Scarafile, D. et al. (2020). Telemedicine DSS-AI multi level platform for Monoclonal Gammopathy assistance. In: Proceedings of IEEE Medical Measurements and Applications (MeMeA) 2020, Bari, Italy (1–3 June 2020). Piscataway, NJ: IEEE.

3

Human–Machine Interfaces

The design and the development of human–machine interfaces (HMIs) is fundamental for technological advances in industry. This chapter proposes different practical and common cases where sensor system integrations are required, providing different solutions by means of artificial intelligence (AI) representing a topic of the advances in the passage from Industry 4.0 to 5.0. The goal of this chapter is to provide different schemes and layouts enabling control and actuation, describing local network architectures implementing digital protocols, AI data processing, and data security. The study is focused on the feedback systems and on the implementation of neural networks by means of logic ports. Specifically, some solutions for parameter setting and for production variable classification are discussed. The neural network feedback systems are discussed, explaining how the information networks operate for the classification and for the definition of decision making. Mechatronic interfaces aid the research on the automatisms allowing self-adaptive production processes by means of HMIs implementing intelligent feedback systems, and of AI-based machine-to-machine (M2M) facilities. This chapter discusses advanced feedback controls by neural networks managing programmable logic controller (PLC) units, and supporting product defect classification, machine failure prediction, and system control and actuation. The information infrastructure, enabling the upgraded control, includes different levels, networks elements and protocols suitable for the implementation of advanced HMIs. An important aspect is the formulation of logic conditions controlling production, which can be implemented by logic ports constituting neurons and multilayered perceptron (MLP) neural networks. Digital sensors provide data and information to the interface boards, thus allowing the training of AI models used to improve production quality. The development of the HMI is facilitated by graphical user interfaces (GUIs) structured in operative workflows. Another important aspect of innovative HMIs is cybersecurity. The data security is managed by interfaces controlling priorities, network attacks and threats, queries and machine logs containing the traceability of anomalous access conditions or classifying possible attacks and dangerous network conditions. In this scenario, the AI estimates the classification of dangerous conditions by identifying insecure network accesses and anomalous queries of data access requests, by means of inference engines. The networks managing different communication channels, are improved by AI algorithms and big data systems addressing the network to a communication security system suitable for disaster recovery. This is possible by creating intelligent middleware interfaces usable for communication channel switching and network attack detection. This chapter proposes different examples and architectures with the goal to provide a complete overview of the upgrading of interfaces in industries, by analyzing different application fields involving production quality, intelligent production infrastructures, and cybersecurity. Different examples of architectures interfaced with GUIs complete the overview of possible solutions to adopt for HMIs.

Electronics in Advanced Research Industries: Industry 4.0 to Industry 5.0 Advances, First Edition. Alessandro Massaro.

3.1 Mechatronic Machine Interface Architectures Integrating Sensor Systems

Mechatronic interfaces are implemented to improve automatisms and to integrate sensor systems. This section provides different interface architectures and mechatronic layouts for auto-adaptive control and actuation systems, addressed on Industry 5.0 advanced production systems. The mechatronic tools to integrate in an industry working with old systems induces some technological risks which are listed in Table 3.1.

3.1.1 Multiple Mechatronic Boards Managing Different Production Stages

HMIs represent a group of multiple functions allowing the interaction between operators and production machines. The HMIs are integrated into the whole control and actuation system as illustrated in the example of Figure 3.1, involving complex physical and logical interconnections of the modeling, and the control of a production line constituted by three main processes (three production stages).

Starting with the first stage, it is observed that at the end of each process, the Internet of Things (IoT) sensors for the acquisition of measurements of the raw material check, and of the quality of the semi-finished product are integrated. In detail, these measurements are strictly linked to the quality of the processing carried out by the production stages that precede the sensor itself. These devices transmit the various measurements collected in the central system and processed by the AI engine. The processing parameters, such as the speed or the temperature, are always sampled by the central system through communication protocols inherent to control devices such as a PLC and programmable automation controller (PAC). In particular, the sampling phase happens in asynchronous mode, and is managed by a specific software service, called the Manufacturing Service Bus (MSB) [1], which sequentially retrieves the data from the machines (through requests), and saves the same values into a table of the database (DB) system together with the timestamp attribute. The MSB is a software providing management services for interfacing of production machines in order to perform the acquisition, coordination, and saving of production data. This service is also extended to the IoT system depending on the adopted technologies. In addition to the MSB, the data entry service digitizes further various data, whose acquisition cannot be automated. All data are stored in a DB capable of responding to the specifications required for sampling from multiple connected controlled points: it contains all the numerical values of the production line, and those entered by the technicians by means of properly structured workstations equipped with data entry interfaces. The AI engine extracts and processes data from the DB in order to optimize the efficiency, production, and quality of the product. In practical terms, the algorithms that concern the intelligence engine search for a continuous correlation between the parameters inherent to the best quality of the product and the raw materials processing parameters. Furthermore, the intelligent engine is expected to learn new correlations between data during the processing. The data generated by the AI algorithm, together with the sensor data, are able to examine and to predict the production phases. A bidirectional communication system is able to provide a feedback check when the production cycle is completed (feedback control which involves the use of mechatronic interfaces and PLCs).

3.1.2 Mechatronic Boards Managing Component Processing

Component processing is typically in milling, cutting, additive manufacturing, laser modeling and other advanced processes according to reverse engineering (RE) tools able to optimize the working

Table 3.1 Risks arising from technological innovation.

Type of technical and technological risk	Impact	Probability	Mitigation measures	
			Preventive action	**Corrective action**
Integration of devices to already installed systems	High	High	It is advisable to examine each machine that makes up the production line to determine if the hardware communication interface is accessible	If the machine interface is not present, analyze if the production machines are "open" to perform an integration (feasibility action)
Compatibility and availability of communication protocols with the adopted systems	High	High	For each machine the communication protocol to interact with the communication infrastructure is examined	Analysis of the possibility to actuate integration directly modifying the protocol contents
Installation of new sensors and hardware devices in the production line.	Low	Average	Check the spaces available in the production line according to the characteristics of the sensors (spatial requirements)	Analysis of possible further spaces to achieve according to possible machine displacements
Data integration	Average	High	The current data and other databases are analyzed to understand the data integration procedure to be accomplished	Synchronization of data with the transmission system, to facilitate the new data integration
Internet of Things wireless connection of the production network	Average	Average	Due to the high-power electric drives there may be wireless data transmission problems (interferences). Therefore, a preliminary compatibility check is carried out	In the case of impossibility of transmission by Internet of Things technologies, a minimally invasive wired and shielded network is implemented
Acquisition of data related to quality control tests of the processed product	Average	High	It is verified whether the devices and instruments used in the various control and quality tests can be integrated	If the instruments are not equipped with communication interfaces, specific data entry applications are created
Ineffective optimization of the intelligence algorithms to be implemented for the production upgrade	High	High	Due to some undetected physical or chemical phenomena, artificial intelligence algorithms may not be able to correctly model the new production processes	It is therefore expected to import a lot of information related to each process, through the implementation of additional sensors
Non-interpretability of dashboards for production optimizations to be achieved	Average	Low	The outputs of the artificial intelligence algorithms are related to the optimization to be performed and can be difficult to evaluate. Therefore, a discussion phase is planned with the technical and managerial staff of the company to establish the reading methods of the outputs	In case of misunderstandings, the implemented dashboards are promptly modified (even remotely)

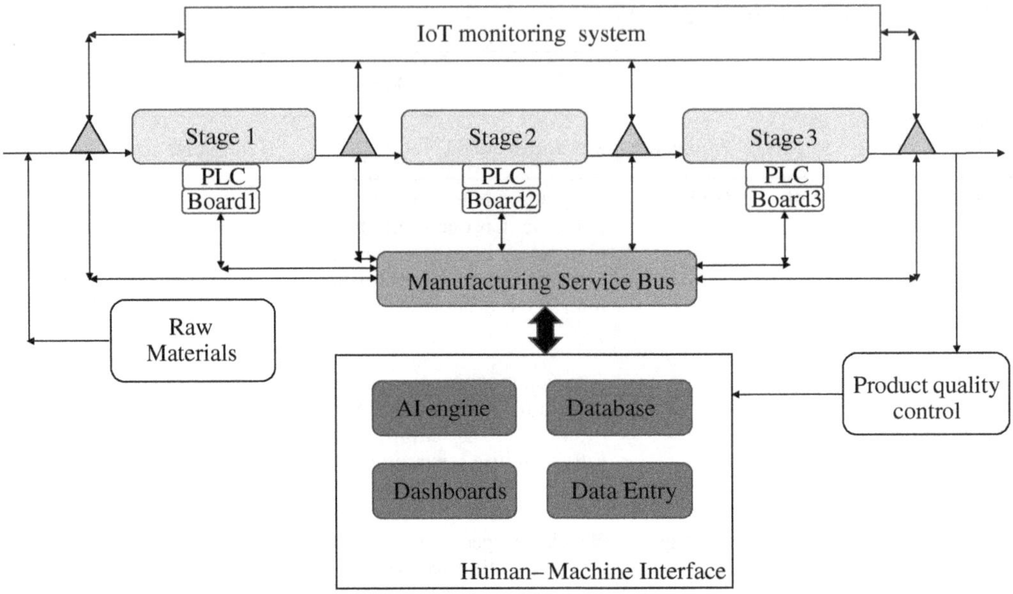

Figure 3.1 Human–machine interface managing multiple mechatronic boards; modeling of a production line constituted by three production phases named stages (stage 1, stage 2, and stage 3).

parameters. Figure 3.2 shows a system integrating mechatronic boards able to control the sensor inputs and the actuators. The automated system of Figure 3.2 considers:

- Sensors (S1, S2, ..., S*n*) for the monitoring of the working processing (temperature, distances, pressures, vibrations, etc.).
- Some sensors are managed locally by a properly designed printed circuit board (PCB) allowing a correct signal detection (PCB1, PCB2, ..., PCB*n*).
- Tools processing a target.
- Three-phase synchronous motors enabling movimentation on the spatial domain, and varying processing parameters.
- A controller collecting and synchronizing the acquisition of the signals of all sensors connected to the system.
- A processor implementing AI algorithms able to predict failures, and, consequently, to optimize the parameter setting such as definition of thresholds.
- Input and output analog units connected to the Ethernet for Control Automation Technology (EtherCAT) linked to the industrial controller.
- An industrial controller able to manage the three-phase motors.

The EtherCAT is deterministic industrial Ethernet technology implementing the international standards IEC61158 and IEC61784.

3.2 Machine-to-Machine Interfaces: New Concepts of Industry 5.0

M2M interfaces are mainly constituted by electronic boards and processors transforming analog signals into digital ones. Particular attention is focused on M2M interfaces adapting a standalone system to digital networks and to messaging mechanisms for the collaborative automation in

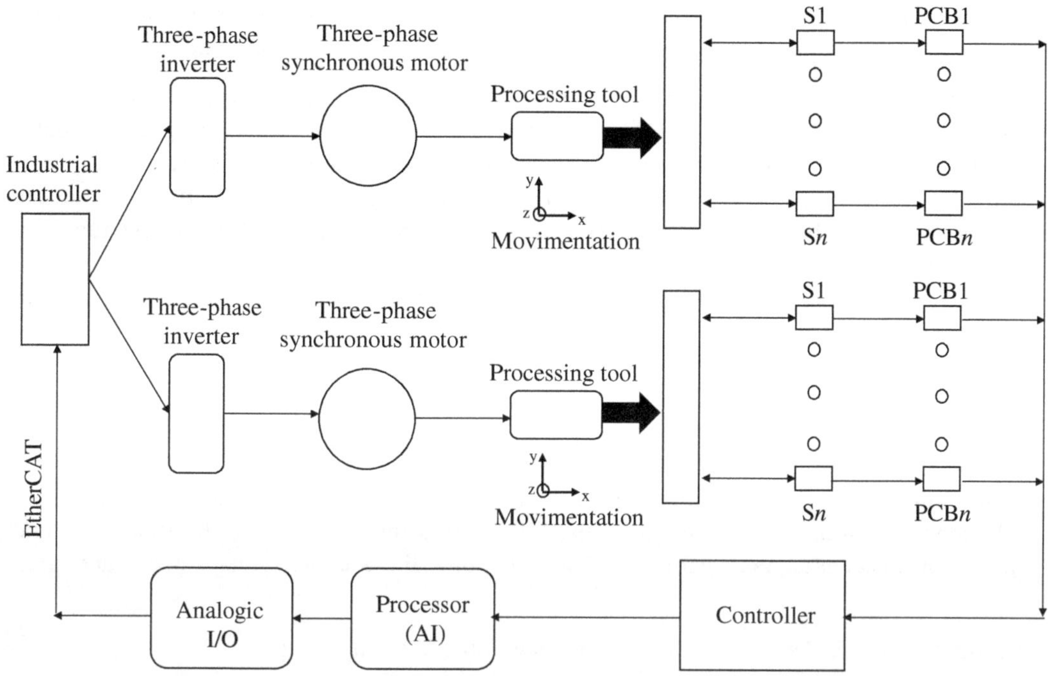

Figure 3.2 HMIs managing multiple mechatronic boards.

industrial networks. The basic concept of M2M is based on the feedback control system processing sensor signals (S1, ..., S*n*) by AI processing (Figure 3.3): the feedback system reduces the error by a neural network processing simultaneously all sensor data as data input. The block diagram of a retro-operated control follows the traditional way by comparing directly the input signals with the feedback signal of the main sensors, or is improved by the AI engine optimizing the calculation of the comparison thresholds, and controlling machine operations.

Figure 3.3 corresponds to a block diagram, where the same blocks represent the components of the control system, while the connections correspond to the signals of different nature and different protocols (in this last case a MSB or an enterprise service bus [ESB] is considered for

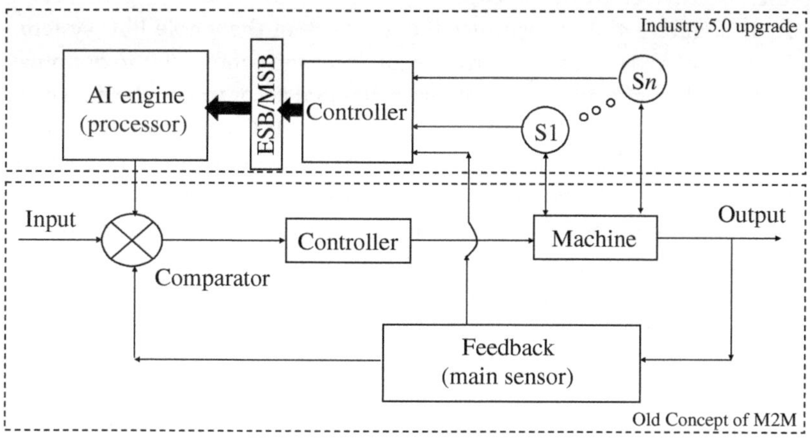

Figure 3.3 M2M concept in Industry 5.0.

simultaneous data processing). Each block takes as input a signal (electrical or mechanical), which is processed (or filtered) to obtain the output signal. In Figure 3.3, the basic blocks are:

- **Machine block of a specific production line**: this represents the part to be controlled such as an electric motor controlling a specific tool; at the input it takes the signal, and at the output it provides the variable to be controlled; the input signal corresponds to the electrical voltage providing the power of the motor, while the output signal corresponds to the rotation speed of the tool, the variable to be controlled or to be kept constant mitigating the disturbances that may occur such as friction.
- **Feedback block**: this is given by the main sensor that measures the control variable in order to have information on how far it deviates from the desired signal; the output signal of this element (feedback signal) is compared with the desired output signal, through a comparator node estimating the error signal; in the case of a mixer example, this feedback element can be given by an encoder that measures the rotation speed of the motor shaft.
- **Controller block**: this corresponds to the electronic control device on which a control algorithm is implemented; it reads the sampled error signal, and processes the signal to be provided to the machine; in the case of the mixer, the controller is given by a PLC or by an inverter and acquires the encoder signal; a control algorithm supplies the electrical voltage power actuation of the motor.

The Industry 5.0 blocks determining the upgrade of the M2M are:

- **Controller managing sensors**: the controller manages different sensors (S1, ..., Sn) controlling production machine processing; it synchronizes all signals to be transmitted in the local network.
- **Bus block**: ESB or MSB bus type is adopted to transport different data protocols in the same local network for the processing of the AI engine.
- **AI engine**: the AI processor provides a further element for the comparator, thus defining dynamically the thresholds for the machine parameter setting.

Given the operation of the industrial control system, the user, i.e. the user of the machinery on which the feedback control is implemented, has the task of setting the reference signal (input signal), which in some cases corresponds to a constant value, and in other cases it may have a defined trajectory to follow (trajectory tracking). The trajectory can be refined by the AI algorithms optimizing the error within the tolerances: a reference signal corresponds to the output of the optimization to be performed, which is provided by the AI engine.

As sketched in the model in Figure 3.4, AI constitutes the feedback of the whole PLC system, where the artificial neural network (ANN) provides further signals for the comparator to optimize and reduce the error. The thresholds of the output signals are calibrated according to the data processing: the digital outputs (D. Output signals) and the analog outputs (A. Output signals) constitute the actual output, which is compared with the ANN output, thus providing the best signal for the controller managing the processing tool. The ANN enables the backward propagation of errors as for feedback systems, improving simultaneously the feedback effect and the comparison of the output signals. The input module can connect the PLC board as in Figure 3.4, where digital input signals (D. Input), analog input signals (A. Input), and timing signals (Timer), also contribute to the training and for the testing of the ANN model. The training process is also optimized by other input signals further improving the training dataset and consequently the training model.

The AI engine is constituted by the AI algorithms, the software interface, and the algorithms adopted for the comparison: the comparator can be implemented by the McCulloch–Pitts neuron

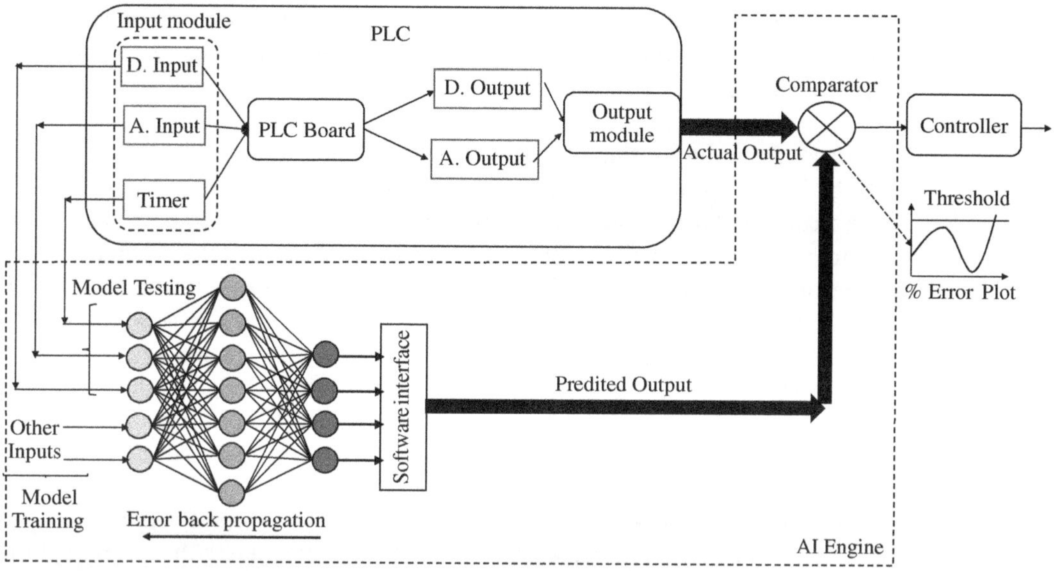

Figure 3.4 Advanced PLC integrated system: feedback system by ANN.

model [2] enabling specific activation functions. The historical data of the sensors are collected into a big data system increasing during the time of the training dataset. The dataset is filtered and pre-processed in order to optimize the output calculation. Different levels are distinguished for the feedback control and actuation system (Figure 3.5): the field level of IoT sensors, the supervisory level integrating the neural networks for defect classification and machine failure prediction supporting decision making, and the control level constituted by the PLC hardware system

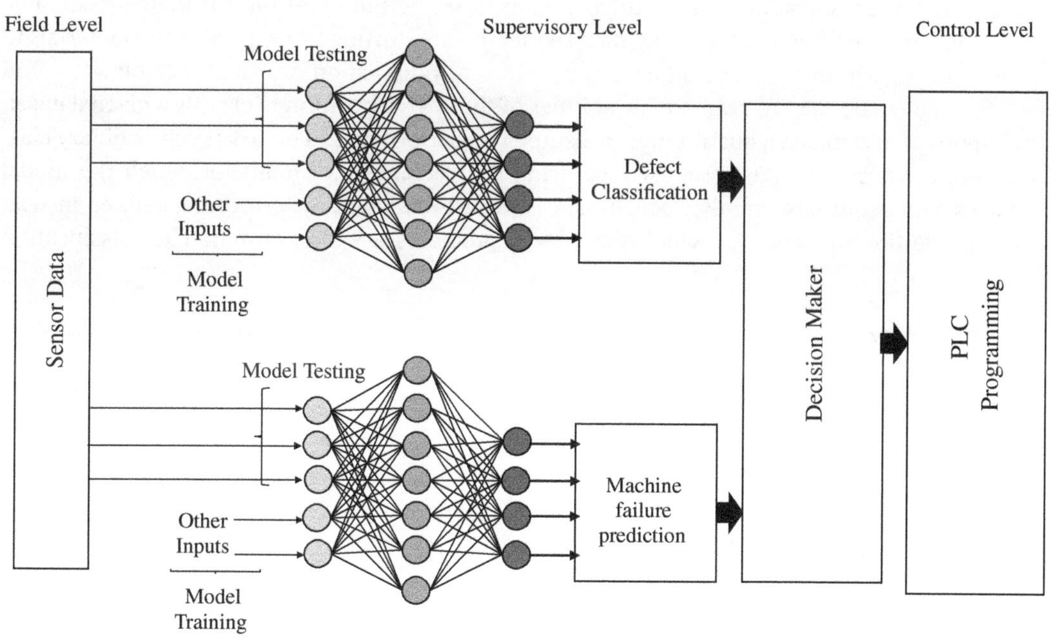

Figure 3.5 Standalone decision maker based on an ANN programming PLC system.

programmed by the setting provided by the decision maker interface. The decision maker generates a code structuring a specific protocol suitable for the PLC programming. The intelligent upgraded PLC can be managed online by a client–server system enabling online control and remote actuation.

Concerning Industry 4.0 communications systems, the Open Platform Communications' Unified Architecture (OPC-UA) [3, 4] is suitable for automatic communication. OPC-UA is appropriate for multiple open and secure platforms, and allows both new and previous generations of PLCs to communicate. Communication based on the OPC-UA protocol has the advantage of being an open standard and connecting multiple devices, machines, or systems in communication with each other. Furthermore, the intrinsic possibility in the protocol to manage a secure communication, guarantees data protection when a machine is connected online. Figure 3.6 shows an example of architecture client–server implementing OPC-UA protocol, integrating AI algorithms which can be implemented in a cloud environment, or in the local machine. The cloud processing is suitable for application where a low computational cost is required, mainly for a quasi real-time feedback response.

Communication via OPC-UA is used to retrieve the samples of the control variables, and to transmit data following a production order to be carried out. By the integrated Ethernet port of the controller, it is possible to use the functions of the OPC-UA protocol with protection, signature and encryption properties through the use of proper certificates. An important advantage of using this protocol integrated in PLCs is that communication between two devices is not based on a simple exchange of bytes, but data are transferred as objects including their properties (data type, date and time of reading, etc.). This aspect is important for the PLC application using an optimized DB for data exchange. The AI engines focus on the analysis of data acquired by the machines, data entry modules, and IoT sensors in order to classify the processed product and to predict machine malfunctions. The ANN algorithms correlate the data related to the raw materials, the machine parameters set, and the data detected by the IoT sensors, with the data of the product quality. The correlation, the classification and the prediction analyses permit to estimate if there are suitable sets of machine parameters, improving the product quality starting from specific raw materials to process: the algorithm processes all data and defines a combination of machine parameters and choices of raw materials. To evaluate the abilities of the ANN algorithm to classify a given dataset, it is important to estimate a quantitative measure of its performance. For tasks such as binary classification, performance is assessed by measuring the percentage of items for which the model produces correct outputs. Another benchmark is the estimation of the error rate, defined instead as the proportion of items for which the system processes a wrong output. The classification

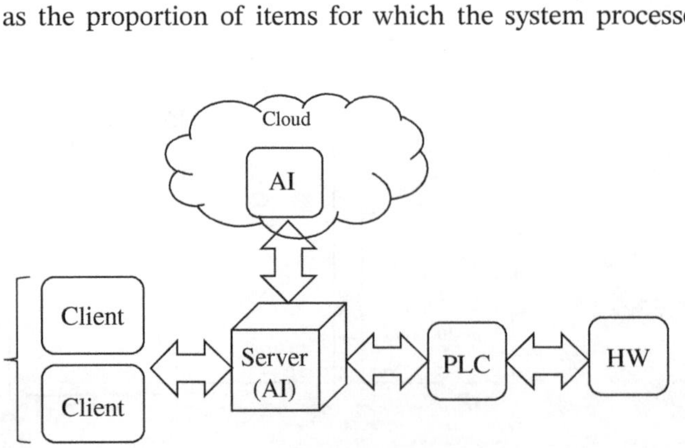

Figure 3.6 Client–server connection scheme via OPC-UA protocol (HW, hardware).

efficiency of the algorithm is also evaluated starting with the receiver operating characteristic (ROC) curve.

The new concepts of Industry 5.0 involve different aspects of the analyzed topics such as:

- A new concept of feedback system based on the implementation of neural network interfaces.
- The possibility to train the AI model with the analog and digital data of the same production line.
- The possibility to adopt mechatronic systems automatizing the production in self-adaptive modality by processing in real time the AI outputs.
- The possibility to predict production failures thus automatically adjusting production machine parameters.
- The redefinition of the threshold values based on the analysis of the AI classifications, automatically adjusting the working conditions of the different tools adopted in each production line.
- The possibility to use cloud computing to optimize AI calculation for real-time data processing enabling real-time production change (change of tool velocity, change of tool type, product refinements, etc.).
- The possibility to integrate other sensor data to optimize the production quality and to decrease the possibility of malfunctions.
- The possibility to integrate RE and rapid prototyping tools for a flexible production following the product processing and the working conditions.
- The possibility to upgrade the production by changing the product to fabricate, and orienting production on innovative devices and components different to the competitor products (nanotechnology implementation, advanced manufacturing, etc.).
- To set automatically machine synchronization by adapting at each production stage the production velocity to achieve a high-quality product.
- To make production flexible for possible changes of production and of products according to the marketing trend.
- To eliminate human errors.
- To drastically decrease the waste materials due to the machine, breakdown, human errors, and machine malfunctions.
- To optimize the raw materials used for the production process.
- To decrease the whole costs associated with the production.

3.3 Production Line Command and Actuation Interfaces in Upgraded Systems

This section analyzes command interfaces for human working stations connecting PLC lines and local area network (LAN), by providing guidelines for the design of synchronized command interfaces working on the production lines. In particular, different schemes are provided of electronic logic circuits implementing adaptive commands of actuation, including the possibility to set thresholds estimated by AI algorithms.

3.3.1 PLC, PAC, Industrial PC, and Improvements

In order to carry out the expected functions correctly, processing machinery must manage the electromechanical actuators according to the detected signals of the installed sensors. This management is entrusted to industrial control systems which, depending on the machinery and on the functions to be executed, is of different types. Therefore, given a production line, each machine provides a control system which, in addition to activating and deactivating the electric drives, generates and processes production data to be analyzed to attain production optimizations. The wired or relay logic, and the management of the various electromechanical actuators, are assigned to a series of electrical circuits that include the use of relays or devices corresponding to switches controlled by voltage signals. These devices are implemented within actuator logics and Boolean functions activating or deactivating electrical drives. Through this approach, it is not possible to reach particularly complex functions, since at the same time the complexity of the wiring is increased. With the advent of today's technology, and especially with the greater availability of semiconductor devices, relays have been replaced with electronic devices as they are much smaller in size and have enormous programming capabilities. In the actual modern industrial scenario, there are three types of control systems:

- PLC
- PAC
- Industrial PC (IPC)

There are some key features that distinguish PLCs from IPCs, microcontrollers, and other industrial control solutions:

- **Input/output (I/O).** The PLC central processing unit (CPU) stores and processes the program data, but the input and output modules only connect the PLC to the rest of the machine; these I/O modules provide information to the CPU and activate actuators for specific tasks. The I/O modules are analog or digital; input devices include sensors, switches, and smart meters, while outputs include relays, lights, valves, and drives. Users match the I/O of a PLC in order to obtain the right configuration for the desired application.
- **Communications.** In addition to input and output devices, a PLC may also need to connect with other types of systems; for example, users may want to export application data recorded by the PLC to a supervisory control and data acquisition (SCADA) system, which controls multiple connected devices. PLCs offer a range of communication ports and protocols to ensure the enabling of the PLC communication with the other systems. In the context of Industry 4.0, this feature assumes a fundamental role, since the interconnection of several production machines is envisaged, not only for monitoring production efficiency but also for managing and activating orders.
- **HMI.** To interact with the PLC in real time, users need HMIs. These HMIs typically are simple displays, with text and keyboard reading, or large touchscreen panels similar to consumer electronics, but in both cases they allow users to review and enter information into the PLC in real time. Also in this case, in the context of Industry 4.0, the HMIs change since the machinery is managed by the network and are integrated to intelligent units constituted by processors.

Today programmable controllers are adopted to communicate data via web browsers, and to connect to DBs via Structured Query Language (SQL), and adopting cloud data via the Message Queuing Telemetry Transport (MQTT) protocol. These innovative features represent new functionalities processing quickly a lot of data for a total monitoring and control of production. In this way, the recording of different data makes it possible to apply predictive maintenance policies (prediction failures) in order to further optimize production. For example, it is possible to store the temperature trend of an engine or of an electric actuator, automatically checking possible anomalies. To date, an All-in-One PLC integrates the controller with the operator panel, creating a compact and easy-to-use automation solution. Technicians no longer need to establish PLC-to-panel communications and program both ladder logic (LL) and HMI design in a single software environment. An All-in-One approach saves time, reduces cabling, and decrease the costs of purchasing multiple devices. A PLC program is usually written by a computer and then downloaded to the controller in order to execute a specified function. Most PLC programming software offers programming in LL or "C" or Python. LL is the programming language most used by most technicians because it is not based on the drafting of a low-level structured code, as for the C language. Furthermore, the programming of a PLC using LL can be performed even by non-highly specialized personnel through learning courses. Its success is because it mimics circuit diagrams with logic read from left to right. Each branch represents a specific action controlled by the PLC, starting with an input or a series of inputs (contacts) that generate an output. It therefore consists of Boolean logic functions, where the output and inputs assume a high or low logic level. The PLC manufacturers provide both the hardware and the software for programming and testing the desired application in a single package. In particular, PLCs are regulated by the open standard IEC 61131, where its third part deals with the basic software architecture and the programming languages of the control algorithm inside the PLC. Other programming languages are:

- Function Block Diagram (FBD), which is also a graphic language showing the connection of functional blocks where each has a specific task (the sequence of connection of the blocks defines a specific function).
- Structured Text (ST), which is a high-level text language, structured in blocks.
- Instruction List (IL), which is a low-level language recalling the assembler machine language.
- Sequential Function Chart (SFC), which is a graphical programming language for processes that are divided into steps or tasks.

The PAC is a type of controller that combines the characteristics of a personal computer-based control system with those of a typical PLC. In particular, it includes a class of industrial controllers proposed by various manufacturers with more advanced communication and control functions and is based on a modular architecture that combines PLC structural foundations, such as the real-time operating system and software control loops, with open programming languages, where the architecture is interfaced with various peripherals. The development framework is based on the IEC61131-3 standard, and a software platform also managing complex control functions, permits the communication to other systems characterized by different protocols. Software tools are also available for the design and management of program flows, and for the execution of multiple tasks in parallel on the same control unit. The PAC connectivity is performed by serial, fieldbus, and Ethernet interfaces. Furthermore, since they communicate with different network technologies and protocols, PACs are able to optimally manage M2M in Industry 4.0 communication systems. Its architecture integrates modules specifically oriented on signal conditioning, with various I/O points able to manage sensors and transducers. For example, special signals conditioning modules are available for sensors such as accelerometers and thermocouples. The difference between PACs and PLCs is found in network connectivity, costs, software needs and applications. Their success is growing, especially considering the growing diffusion of integrated control systems. PAC is therefore seen as an integrated industrial platform capable of meeting functional needs in a single solution that in the past were carried out by different devices. PAC systems are very often used in industrial environments for the control of complex processes, for data acquisition, for remote monitoring equipment, for artificial vision, and for motion control. Unlike devices, the hardware architecture of a PAC is broader and more composite, as it must ensure greater flexibility for appropriate applications. PACs typically include microprocessors with floating point units, safety systems, memory protections, and other different options for operational and mass memories, and for data storage. The software optimizes the performance of the system, also adopting non-traditional techniques (for example, by combining proportional integral derivative [PID] feedback with fuzzy logic control). The software applications are typically developed through advanced languages, or through the configuration of the control system parameters and communication interfaces that use web protocols and Ethernet technologies. The PAC–PLC combination is suitable mainly to manage automated tasks. There are different PAC models and types depending on the use. A software for programming and acquiring the signals inherent to these modules is the Laboratory Virtual Instrument Engineering Workbench (LabVIEW) [5]: it supports the design, development and implementation of industrial systems using a unified graphic programming approach, and being able to directly recall codes from other programming languages, including Dynamic Link Library (DLL). Further, LabVIEW is based on a graphical programming language called "G" in which blocks (in jargon called nodes) are connected following a workflow that goes from left to right depending on the data availability, also including the classic constructs provided in the IEC 61131 standard such as the for loop, the while loop, and the if condition. There are also blocks that enable the integration with other programming languages (such as Python) and with different computing framework environments. This feature allows an easy implementation possible also for unskilled technicians.

Finally, for quite complex applications, such as robotic ones, there are IPCs. These are computers very similar to office ones but with hardware and software characteristics suitable for industrial environments. In these contexts, the system must be reliable enough from the point of view of the tasks to be performed, therefore highly robust processors with Advanced RISC Machine (ARM) and x86 architectures are provided. Being very hostile environments, the IPC cover is reinforced with heavy metal material and in some cases a cooling system using fans to avoid the risk of failures. IPCs typically have small windows for heat dispersion on the cover and on the sides. From a software point of view, IPCs are equipped with real-time operating systems since the active processes for control cannot have delays or interruptions. In detail, a real-time operating system (RTOS) is an operating system intended to serve a real-time application that processes data as it arrives, mainly without buffer delay. Some important reasons for using a RTOS are:

- It offers priority-based scheduling, separating analytical processing from non-critical processing.
- It provides functions enabling an application code.
- The possible implementations of task-based designs.
- It offers modular activity-based development, which enables modular activity-based testing.
- It is characterized by an activity-based feature facilitating the modular development as an activity (it allows teams to work independently on their parts of the project).
- It proposes the event driven programming modality.

The important components of a RTOS are:

- The scheduler (this component tells in what order tasks must be analyzed, generally taking into account a priority).
- Symmetric multiprocessing (SMP) with a series of multiple different activities managed by RTOS to execute parallel processing.
- Function libraries (important elements of RTOS acting as an interface for the connection between the kernel and the application code).
- A memory management system (this element is needed in the system to allocate memory for each program).
- Possibility to use common programming languages such as C or C ++.

There are three types of RTOS:

- Hard real time (deadline is very strictly managed, which means that the assigned task must start running at the scheduled time).
- Firm real time (this system tolerates infrequent deadline misses).
- Soft real time (this system accepts some short period delays from the operating system).

Some important features of RTOS are:

- It requires a low memory.
- It consumes few resources.
- The response times are highly predictable.
- The kernel saves the status of the aborted task announcement, determining which task should be considered next.
- The kernel resets the task state and passes CPU control for that task.

Essential factors to consider when selecting a RTOS are:

- Good performance.
- Middleware interface.
- Error free (it is not possible for an error to occur during the execution of the task).
- Programs are small in size.
- Maximum consumption of computational capabilities.
- Shifting of tasks is minimal.
- RTOS is constituted by an efficient memory protection system.

An example of a RTOS is RTLinux which is a real-time operating system based on a microkernel. The hard real-time system running RTLinux applications can control robots, data acquisition systems, manufacturing plants, and other time-sensitive tools and machines.

3.3.2 SCADA Systems for Centralization of Data Production

Generally, an entire production line is monitored and controlled through a complex system that includes different industrial controllers. This complex system is SCADA suitable for Industry 4.0 applications [6]: in a SCADA system, data acquisition is important for carrying out the functions of observation of the evolution of the controlled process, and functions of control and implementation of actions managing the states and the transitions between states. Generally, a SCADA system is divided into a central component and a peripheral one. The data acquisition systems are placed in the peripheral component, while the monitoring, control and processing of data units are included in the central component. Between the two components lies the data transmission layer which can vary depending on the implementation. Figure 3.7 shows a SCADA system upgraded by an AI engine: the architecture is based on the upgraded control center constituted by the supervisor, the DB, the HMIs, and an AI engine supporting production supervision. The Ethernet bus is connected to the Profibus by an X-Gateway permitting data transmission of the physical processes concerning the PLC and remote terminal unit (RTU) with the supervisor.

On the basis of this reference architecture, considerable integration is often required between the various subsystems. To create a distributed control system, the subsystems must exchange information, at various levels and in different ways, as happens in an ordinary production line. These levels are defined in the computer integrated manufacturing (CIM) [7, 8] pyramid which deals with the coordination and management of the various production stages. This pyramid has a

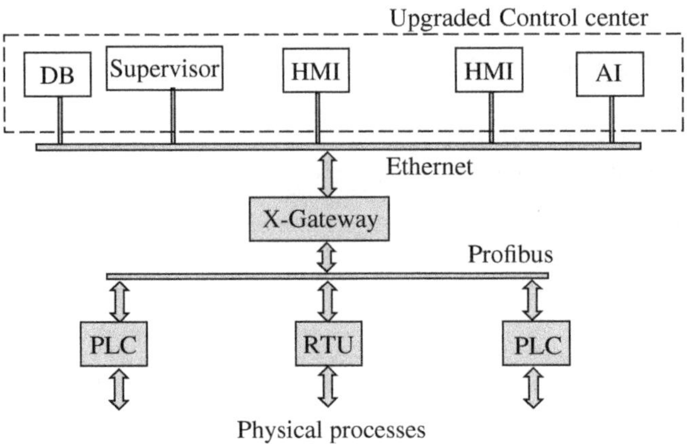

Figure 3.7 Architecture of a SCADA upgraded system.

logic that starts from the production line network and reaches the administration offices. In detail, there are the following levels:

- Level of field sensors for production control.
- Device level suitable for the acquisition and entry of data such as I/O and HMI modules.
- Control level including industrial controllers such as PLC, PAC, and IPC.
- Central company level with PC executing advanced AI algorithms for the supervision of the entire industrial plant.

Moreover, concerning the CIM pyramid:

- It is used to represent information at various hierarchical levels.
- At higher levels, its information is inferior, but the quality is higher and with lower updates.
- For lower levels, its technicality level is high.

For the device level, the control level, and the enterprise level the CIM pyramid is upgraded by the AI pyramid which is embedded as shown in Figure 3.8. The first three most important CIM levels are:

Level of sensors. The data acquisition mechanisms represent the lowest SCADA architecture layer. The actuators of the system, of different technologies, are also directly connected. The sensor level provides digital signal to be processed by the higher layers. At this level different alarm signals are enabled, and different indicators (threshold exceeding alarms, energy efficiency, production efficiency, etc.) are estimated. This layer is a function of the control device and of the communication protocols. Different sensors are connected to monitor the whole production line, and data sensors are stored into a big data supporting the dataset creation for the AI data processing. Sensors are directly connected to the PLC system. Figure 3.9 shows an example of a temperature sensor connected to the PLC.

In order to decrease the degree of complexity of the sensor level, wireless sensors are typically adopted. Table 3.2 lists some sensor wireless protocols typically adopted in monitoring production processes [9].

Level of device acquisition. Particularly important for the implementation of SCADA systems and centralization of data are the methods in which information passes between the various components of the system. In some cases, the choice may be constrained by the technology already

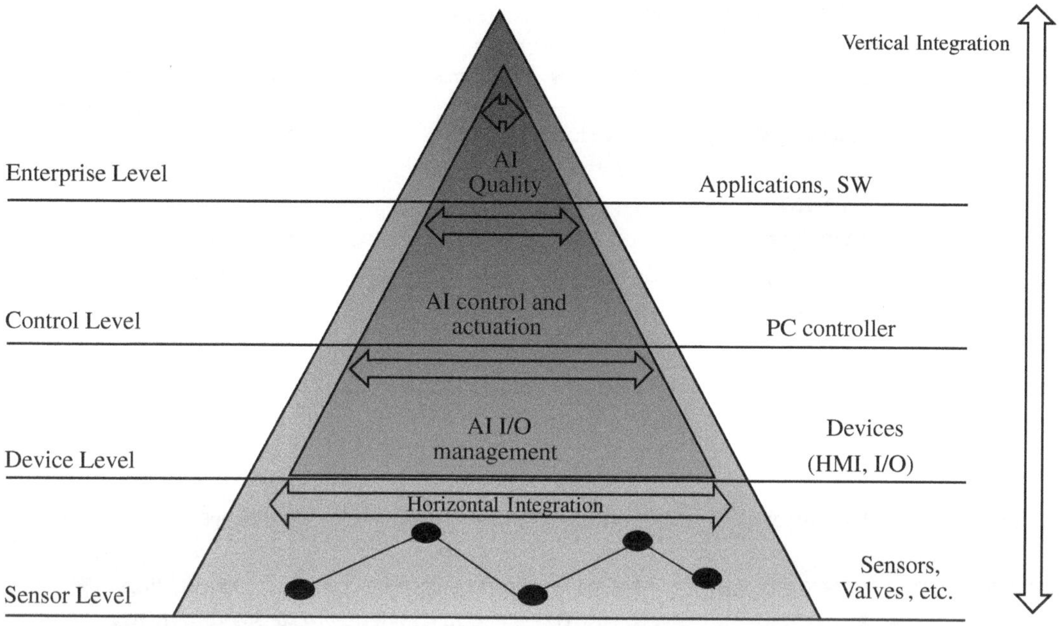

Figure 3.8 CIM pyramid upgraded by AI.

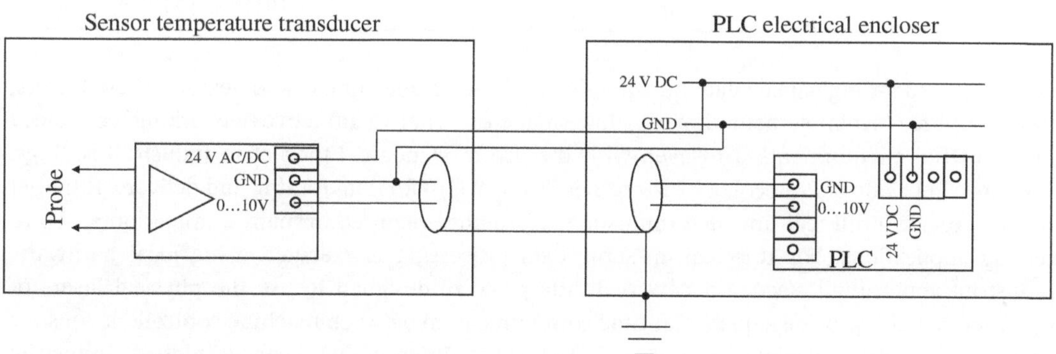

Figure 3.9 Scheme of the connection between the PLC and temperature transducer.

installed, in others it is necessary to methodically analyze the various options, not only according to the system specifications, but also based on future needs. In fact, it is possible to have a system monitoring production machines, which has a type of electronic device that generally have only one access point. Regarding the connection between the second and third level of the pyramid, and between the data storage devices and the central component of the system, many choices can be made based on the transmission speed, spaces, and allocations. The direction in industrial automation is however that of a uniformity in the application of a single technology, which therefore provides the creation of an infrastructure with the possibility of integrating other components. This orientation consists in choosing a network that follows the Industrial Ethernet specifications, which is standardized in the IEEE 802 standard. This allows the use of a series of communications media for simultaneous use. It is also possible to directly interconnect different networks with each other using the internet network, thus obviating the need to interconnect very distant nodes. Ethernet

Table 3.2 Examples of wireless sensor protocols.

Protocol	Standard	Frequency	Transmission range	Data transmission velocity
Bluetooth	Bluetooth 4.2	2.4 GHz	50–150 m	1 Mbps
ZigBee	IEEE802.15.4 ZAD12837	2.4 GHz	10–100 m	250 kbps
Z-Wave	ZAD12837	900 MHz	30 m	9.6/40/100 kbps
WiFi	IEEE 802.11	2.4 GHz	50 m	150–600 Mbps
NFC	ISO/IEC 18000-3	5 GHz 13.56 MHz	10 cm	100–420 kbps
Sigfox	Sigfox	900 MHz	30–50 km (outdoor) 3–10 km (city)	10–1000 bps
Neul	Neul	900 MHz	10 km	10–100 kbps
LoRaWAN	LoRaWAN	Varies	15 km (outdoor) 2–5 km (city)	0.3–50 kbps
Mobile (cellular)	GSM/GPRS/EDGE 2G), UMTS/HSPA (3G), LTE (4G)	900 MHz 1800 MHz 1900 MHz 2100 MHz	35 km (GSM) 200 km (HSPA)	35–170 kbps(GPRS) 120–384 kbps (EDGE) 384 kbps–2 Mbps (UMTS) 600 kbps–10 Mbps (HSPA) 3–10 Mbps (LTE)

was developed for high-speed data transmission and is efficient, simple, and flexible. These factors, together with a low implementation cost, have made Ethernet an attractive networking technology for industrial environments. However, with the use of standard Ethernet in industrial settings, Transmission Control Protocol (TCP)/Internet Protocol (IP) for data routing and delivery, Ethernet cannot provide a full real-time deterministic performance required in many automation and processing applications. For this reason, some data processing is executed using local hardware. Industrial applications needed a new real-time protocol designed to use the physical layers of Ethernet that also provide a quasi real-time communication between machine controllers, sensors, actuators, and other devices. To this end, Industrial Ethernet has been developed. Industrial Ethernet refers to a series of Ethernet protocols based on standard Ethernet hardware (the physical and data link layers) and internet protocols (network and transport layers), along with a proprietary application layer. The supported protocols are: EtherCAT, Ethernet Industrial Protocol, process field net (Profinet), POWERLINK, Sercos III, CC-Link IE, and Modbus TCP/IP (Table 3.3).

Figure 3.10a shows the EtherCAT protocol structure, Figure 3.10b shows the Sercos III protocol layout, and Figure 3.10c shows a typical configuration of the Modbus TCP/IP protocol.

The application layer protocol ensures the correct data transmission and reception in the same way that standard Ethernet allows file sharing and access to network devices. Industrial Ethernet protocols permit controllers to access data within units, PLCs, workstations, and I/O devices. Industrial Ethernet protocols work mainly in the following architectures:

- Standard software and standard Ethernet architecture uses standard Ethernet embedding TCP and IP protocol.
- Open software and standard Ethernet architecture adopting standard Ethernet (POWERLINK architecture).

Table 3.3 Main specifications of communication protocols.

Protocols	Main properties
EtherCAT [10]	• Ethernet-based fieldbus system • Good protocol for protocol is optimized for short cyclic process data • Each EtherCAT slave receives the Ethernet packet sent by the master and it automatically retransmits to the next device • "Pass-through telegram" data transmission • Short cycle times (≤100 μs) • Low jitter for accurate synchronization (≤1 μs) • Low hardware costs • Protocol integrating datagrams • Gigabit Ethernet for data-intensive applications
Ethernet Industrial Protocol [11]	• Communication Profile IEC 61784-1 and EN 61784-1 • Standard capable of handling large amounts of data at speeds of 10 or 100 Mbps, and at up to 1500 byte per packet • Used with IPC, mainframes, robots, I/O devices, PLCs, and other interfaceable devices
Profinet [12]	• Profinet devices: I/O controller, I/O field device, I/O supervisor • Safety system (implementing Profisafe protocol) • Integration of existing field buses like PROFIBUS, AS-Interface, and INTERBUS • Real-time communication • Isochronous real-time communication in motion control applications and in high performance factory automation applications
POWERLINK [13]	• Ethernet real time • High data transmission velocity (of the order of 100 Mpbs) • High synchronization accuracy (±100 ns) • Expansion of ethernet with a mixed polling and time slicing mechanism (technique to implement multitasking in operating systems) • Motion control applications
Sercos III [14]	• Ethernet real time • High data transmission velocity (of the order of 100 Mbps) • Time slot method (bus access) • Vertical integration • Synchronous control of multi-axis systems • Data transfer between decentralized controls • Master/slave communication • Full integration with Ethernet network • Multi-protocol capable (simultaneous transmission)
CC-Link IE [15]	• 1 Gigabit Ethernet-based integrated network • Real-time automation • EEE802.3ab (1000BASE-T) compliant • Compatible with other devices
Modbus [16, 17]	• Implementation of the TCP/IP protocol • Master/slave protocol • RS-232, RS-485 and TCP/IP communication systems

EtherCAT, Ethernet for Control Automation Technology; I/O, input/output; Profinet, process field net.

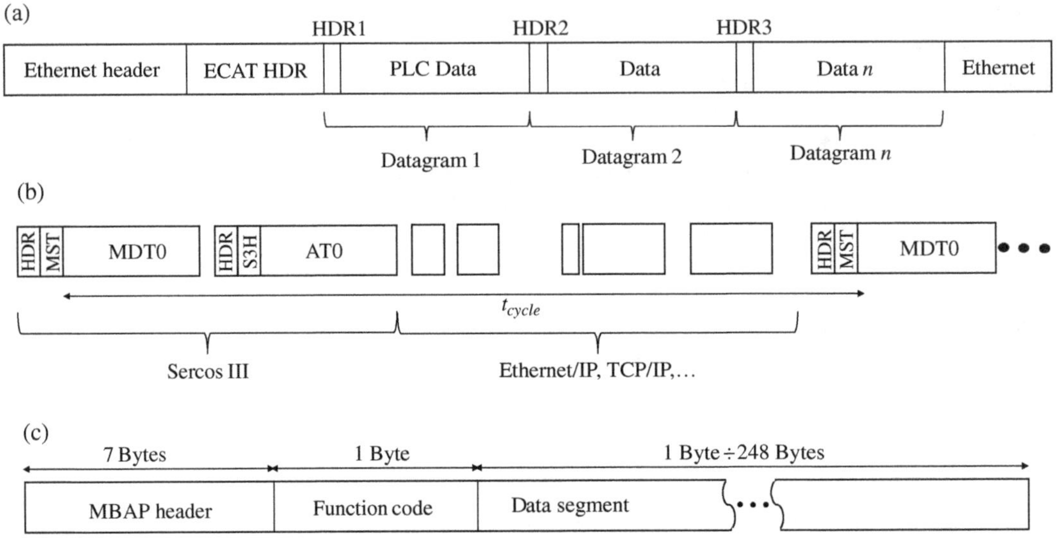

Figure 3.10 (a) EtherCAT protocol specifications (HDR, header). (b) Sercos III protocol (AT, reply telegram; HDR, header; MDT, master data telegram; MST, master sync telegram; S3H, Sercos III header). (c) Application data unit of the Modbus TCP/IP protocol (MBAP header, Modbus application protocol header).

- Open software and modified Ethernet architecture based on standard Ethernet hardware, and adopting complementary hardware, compatible with other protocols (such as EtherCAT, Sercos III, and Profinet).

Concerning an industrial environment, robust hardware is required for the harsh environmental conditions suitable also for electromagnetic compatibility problems.

- **Controller level.** After the detection of data from the various devices, different specifications of the whole system are analyzed. This happens in the controller level with the following main characteristics:
  - Verification of the data availability.
  - Control of the storage process of acquired data.
  - Control and processing of basic and advanced algorithms (such as Boolean logic and AI algorithms managing production).

The controller manages different functions oriented on data management, in particular for data exchange with peripheral units, and for recording data into DB systems accessible by the central system. Concerning data availability, through the controller, the single PLC at the same level ensures the availability of the data to other systems that are suitable for archiving in a DB or for the calculation of representative indicators including production and energy efficiency. The data processing is enabled in the controller level, thus consenting the monitoring of the production process, and checking the production evolution trend. For example, in the food industry in an industrial oven system of a production line, the PLC receives by a thermal sensor the temperature data of the oven chamber. The data are processed to control the cooking temperature of the processed food product. The temperature is also detected on the product at the oven exit. From the measured temperatures, the PLC increases or decreases the thermal power so that the temperature remains at the reference and constant value. Furthermore, the temperature samples are processed to monitor the quality of the processed product, and to check the state of the production machine.

All measured data are stored in a DB for data processing that is carried out by the central system. The storage and the data processing are performed by software protocols based on Industrial Ethernet interfacing the AI engine. Figure 3.11 shows the architecture of the production system based on the application of an industrial oven able to process food. The control and the actuation are accomplished at the control level by means of a logic control of both the measured temperatures T_1 and T_2: an intelligent algorithm, based on a neural network, estimates the threshold condition based on the concept that both temperatures must be higher than fixed thresholds (if the measured values are higher than the thresholds then the process is correct). Table 3.4 gives some logic conditions related to the reading of the temperatures and consequent actions.

Once the control algorithm is developed, and the applications for data storage are implemented, a huge amount of data are processed by AI algorithms and by big data analytics tools by analyzing the historical signals useful for predictive maintenance. The data processing by the AI engine supports:

- Definition of the data type to analyze.
- Data preparation (data filtering and data pre-processing).
- Execution of the supervised and unsupervised algorithms.
- Analysis and interpretation of the achieved results.

From the architectural point of view, the central system generally corresponds to a SCADA server, where both the DB system and then AI algorithms are implanted. Figure 3.12 shows the SCADA central system is interconnected with object linking and embedding for process control (OPC), RTU, PLC, and SCADA clients.

From Figure 3.12 it is possible to observe two different type of network technologies used within the same SCADA system: the LAN Ethernet line, and the field bus line. Modern SCADA systems, in the context of Industry 4.0, activate the data access in real time. This access to information in

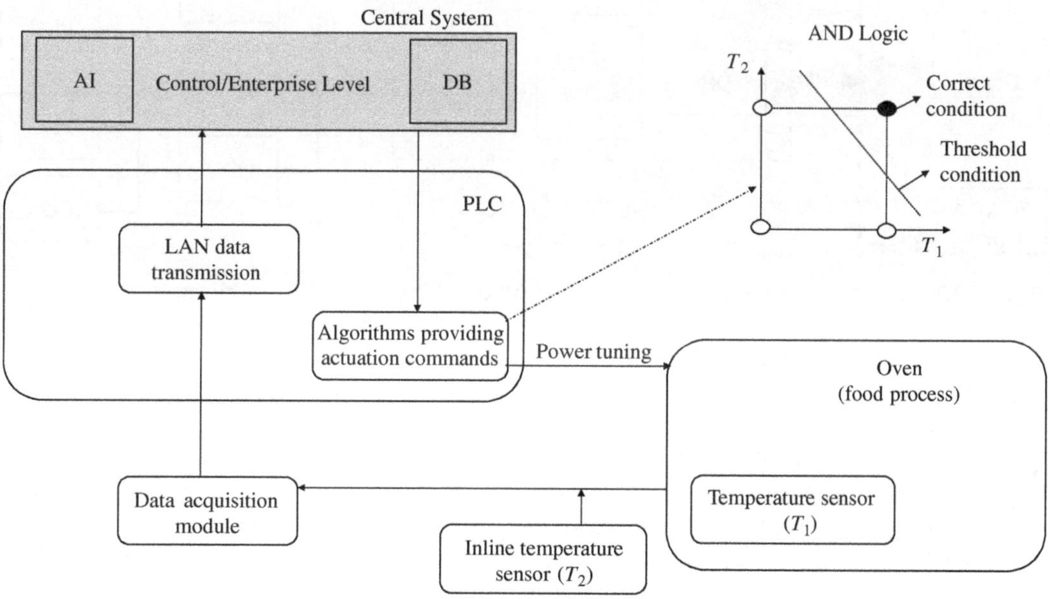

Figure 3.11 Industrial oven temperature control system and automated thermal power tuning: temperature control scheme and display of the acquired data.

Table 3.4 Logic conditions and actuation of the production system illustrated in Figure 3.11.

Variable condition	Description	Process to enable
$T_1 < T_{1\text{Threshold}}$, $T_2 < T_{2\text{Threshold}}$	The oven is out of order or it is turned off for very low values of T_1 and T_2	The oven is adjusted in the case of very low temperatures. For values closest to the thresholds, the oven temperature can be increased gradually (this action is actuated automatically by the PLC system)
$T_1 < T_{1\text{Threshold}}$, $T_2 > T_{2\text{Threshold}}$	The temperature distribution is not homogeneously distributed in the oven chamber, it is focused on the product to process	T_2 is evaluated to make sure that the product does not burn. If the T_2 value overcomes a burning value $T_{2\text{Burn}}$, the process is stopped. It is possible to tune the oven thermal power until $T_2 < T_{2\text{Burn}}$ (new threshold conditions are set according to observation results)
$T_1 > T_{1\text{Threshold}}$, $T_2 < T_{2\text{Threshold}}$	The oven does not work properly because the product is not placed in a good position for cooking	The food movimentation layout inside the oven is revised. If the value is closest to $T_{1\text{Threshold}}$, it can be gradually and automatically increased to obtain good cooking conditions (new threshold conditions are set according to observation results)
$T_1 > T_{1\text{Threshold}}$, $T_2 > T_{2\text{Threshold}}$	The AND logic condition represents good food processing conditions	No action is taken. It is only checked if the T_1 value does not increase drastically by burning the food

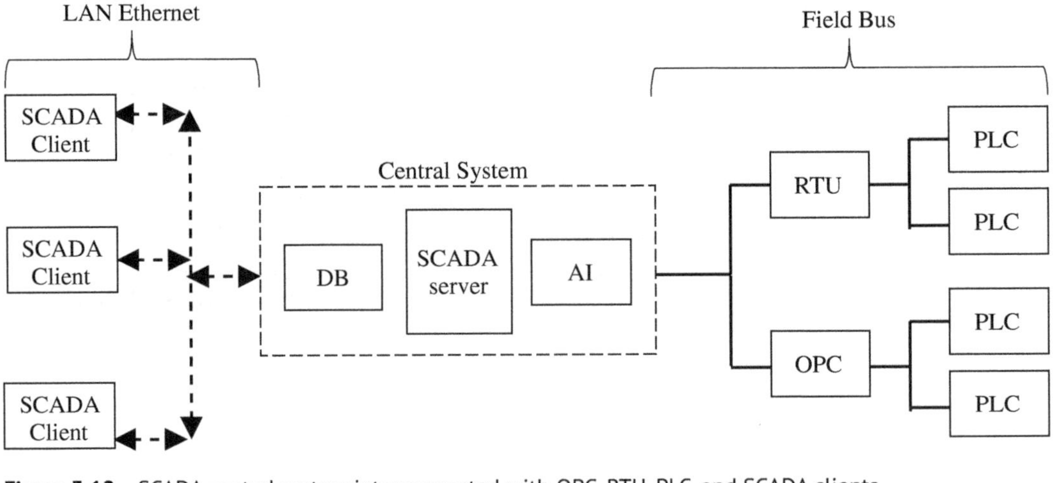

Figure 3.12 SCADA central system interconnected with OPC, RTU, PLC, and SCADA clients.

real time facilitates the business managers to make data-driven decisions to improve their processes. Without SCADA software, it is extremely difficult to collect enough data for consistently well-informed decisions. In addition, most modern SCADA design applications have rapid application development capabilities that allow users to design applications relatively easily, even if they do not have extensive knowledge of software development. The introduction of modern standards and practices of IoT, SQL and web-based applications into SCADA software, have greatly improved the efficiency, security, productivity, and reliability of SCADA systems. A great benefit of using SQL DBs with a SCADA system is that it simplifies integration into existing systems, permitting data to flow in the information system. Historical data of a SCADA server can be also migrated into a not only SQL (NoSQL) big data system.

3.4 McCulloch–Pitts Neurons and Logic Port for Automatic Decision-Making Setting Thresholds

McCulloch–Pitts neurons and ANNs are applied to implement the basic logic gates [18–20] to be used in automatic decision-making systems enabling a specific actuation. The generic structure of a neuron is provided by the model in Figure 3.13, indicating the McCulloch–Pitts neuron. Each neuron is typically linked to other neurons. The output signal of a neuron is provided by the following relationship:

$$y_i = \text{sgn}\left(\sum_{k=1}^{n} X_{k,i} w_k - \tau_i\right) \tag{3.1}$$

with the basic activation function of the signum function defined as:

$$y = \text{sgn}(u) = \begin{cases} 0 & \text{for } u < \tau_i \\ 1 & \text{for } u \geq \tau_i \end{cases} \tag{3.2}$$

The input signals X are weighted by parameters w contributing to the output. By using a signum function the output is binary type. Figure 3.14 illustrates the McCulloch–Pitts neuron model by considering the bias signal.

ANNs are modeled by graphs connecting nodes with specific weights. The following three main architectures are usually considered for ANNs:

- Single-layer feedforward neural networks (as for the single perceptron, the network is structured with one input layer and one output layer with no-feedback linking).
- Multilayer feedforward neural networks (as for the multilayered perceptron, the network is constituted by one input layer, an output layer and one or more hidden layers with no-feedback linking).
- Recurrent neural networks (as for the simple recurrent network, all the networks with at least one feedback linking are included).

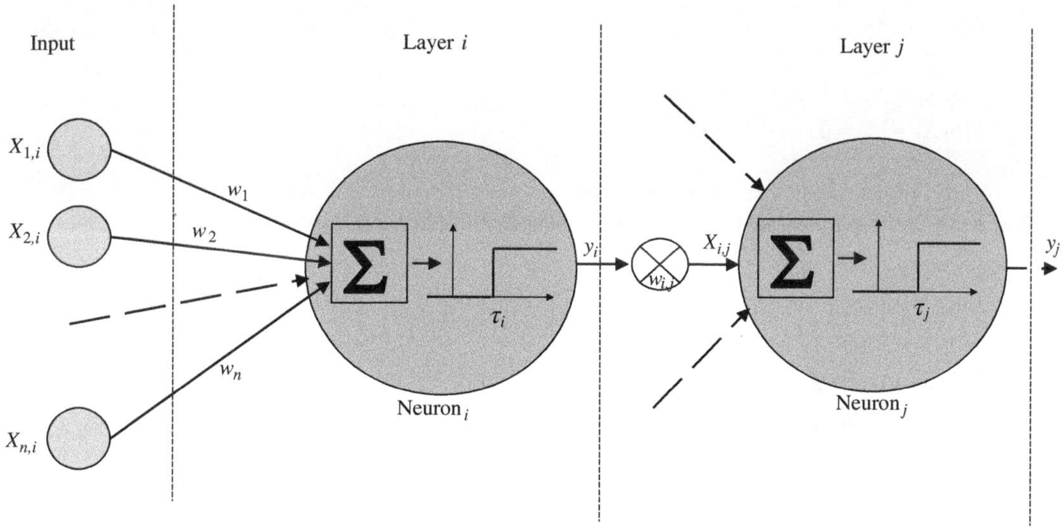

Figure 3.13 Modeling of McCulloch–Pitts neurons.

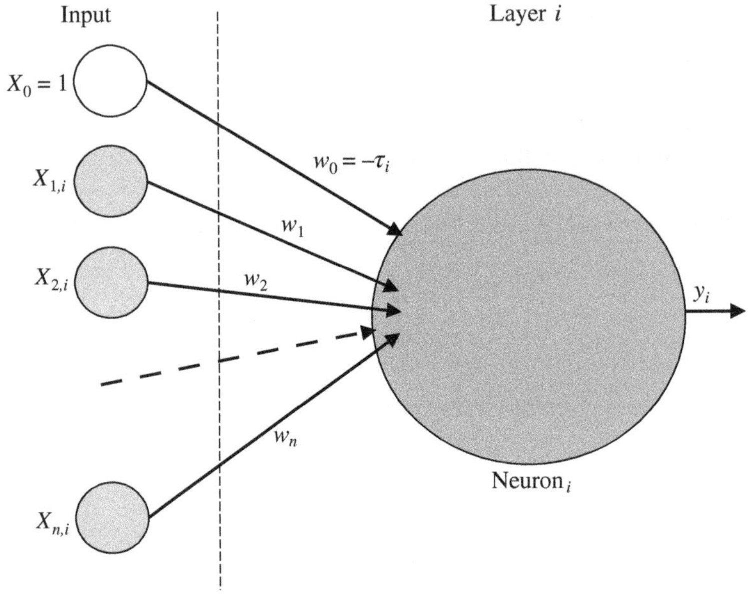

Figure 3.14 Modeling of McCulloch–Pitts neuron by bias signal.

Figure 3.15a–d shows the single-layer feedforward (single-layer perceptron) network, the multilayer feedforward (multilayered perceptron) network, the recurrent (simple recurrent) network, and a generic McCulloch–Pitts neuron network, respectively. For each network the input signals, the weights, and the thresholds τ of the signum functions are defined.

Logic conditions for control and actuation by PLC can be implemented by the networks previously classified. The network can be realized by electronic components, modeling logic port circuits. Figure 3.16a shows an example of NOT logic implementation. According to Eq. (3.1), the threshold and logic conditions for the parameter setting of the network is provided by:

$$X_1 w_1 - \tau_1 = 0 \tag{3.3}$$

$$\begin{cases} \text{for } X_1 = 0, Y = 1: w_1 \cdot 0 - \tau_1 \geq 0 \\ \text{for } X_1 = 1, Y = 0: w_1 \cdot 1 - \tau_1 < 0 \end{cases} \rightarrow \begin{cases} \tau_1 \leq 0 \\ w_1 < \tau_1 \end{cases} \tag{3.4}$$

A solution of the NOT port can be provided by $\tau_1 = -0.5$ and $w_1 = -1$.

The AND logic network in Figure 3.16b is provided by the following conditions:

$$X_1 w_1 + X_2 w_2 - \tau_1 = 0 \tag{3.5}$$

$$\begin{cases} \text{for } X_1 = 0 \text{ and } X_2 = 0, Y = 0: 0 \cdot w_1 + 0 \cdot w_2 - \tau_1 < 0 \\ \text{for } X_1 = 0 \text{ and } X_2 = 1, Y = 0: 0 \cdot w_1 + 1 \cdot w_2 - \tau_1 < 0 \\ \text{for } X_1 = 1 \text{ and } X_2 = 0, Y = 0: 1 \cdot w_1 + 0 \cdot w_2 - \tau_1 < 0 \\ \text{for } X_1 = 1 \text{ and } X_2 = 1, Y = 1: 1 \cdot w_1 + 1 \cdot w_2 - \tau_1 \geq 0 \end{cases} \rightarrow \begin{cases} \tau_1 > 0 \\ w_2 < \tau_1 \\ w_1 < \tau_1 \\ w_1 + w_2 \geq \tau_1 \end{cases} \tag{3.6}$$

A solution of the AND port can be provided by $\tau_1 = 1.5$, $w_1 = 1$, and $w_2 = 1$.

The OR logic port in Figure 3.16c is defined by the following logic conditions:

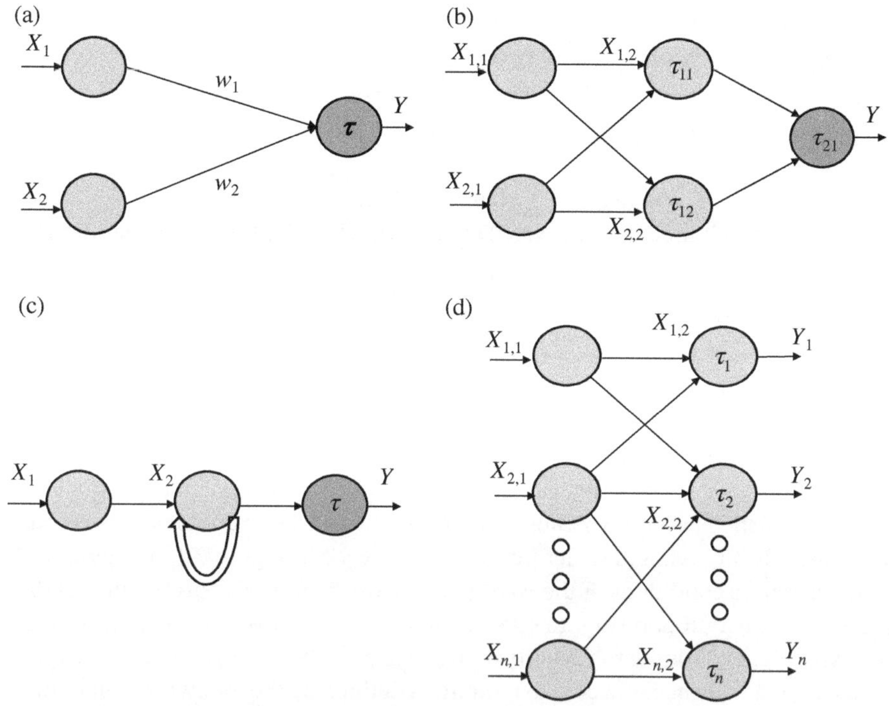

Figure 3.15 (a) Single-layer perceptron network. (b) Multilayered perceptron network. (c) Simple recurrent network. (d) Generic McCulloch–Pitts neuron network.

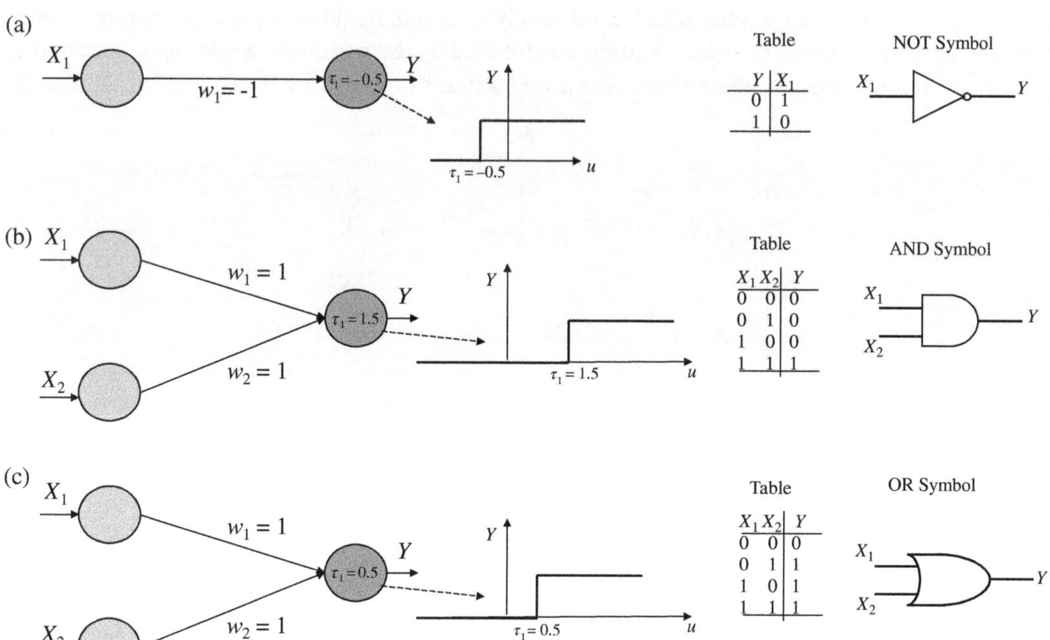

Figure 3.16 (a) NOT logic and its implementation (b) AND port and its implementation. (c) OR port and its implementation.

$$\begin{cases} \text{for } X_1 = 0 \text{ and } X_2 = 0, Y = 0: 0 \cdot w_1 + 0 \cdot w_2 - \tau_1 < 0 \\ \text{for } X_1 = 0 \text{ and } X_2 = 1, Y = 1: 0 \cdot w_1 + 1 \cdot w_2 - \tau_1 \geq 0 \\ \text{for } X_1 = 1 \text{ and } X_2 = 0, Y = 1: 1 \cdot w_1 + 0 \cdot w_2 - \tau_1 \geq 0 \\ \text{for } X_1 = 1 \text{ and } X_2 = 1, Y = 1: 1 \cdot w_1 + 1 \cdot w_2 - \tau_1 \geq 0 \end{cases} \rightarrow \begin{cases} \tau_1 > 0 \\ w_2 \geq \tau_1 \\ w_1 \geq \tau_1 \\ w_1 + w_2 \geq \tau_1 \end{cases} \tag{3.7}$$

A solution of the OR port can be provided by $\tau_1 = 0.5$, $w_1 = 1$, and $w_2 = 1$.

Figure 3.17a shows the network model of the NAND port defined by the following conditions:

$$\begin{cases} \text{for } X_1 = 0 \text{ and } X_2 = 0, Y = 1: 0 \cdot w_1 + 0 \cdot w_2 - \tau_1 \geq 0 \\ \text{for } X_1 = 0 \text{ and } X_2 = 1, Y = 1: 0 \cdot w_1 + 1 \cdot w_2 - \tau_1 \geq 0 \\ \text{for } X_1 = 1 \text{ and } X_2 = 0, Y = 1: 1 \cdot w_1 + 0 \cdot w_2 - \tau_1 \geq 0 \\ \text{for } X_1 = 1 \text{ and } X_2 = 1, Y = 0: 1 \cdot w_1 + 1 \cdot w_2 - \tau_1 < 0 \end{cases} \rightarrow \begin{cases} \tau_1 \leq 0 \\ w_2 \geq \tau_1 \\ w_1 \geq \tau_1 \\ w_1 + w_2 < \tau_1 \end{cases} \tag{3.8}$$

A solution of the OR port can be provided by $\tau_1 = -1.5$, $w_1 = -1$, and $w_2 = -1$.

The parameters of the XOR (or EXOR) port cannot be calculated by the system of inequalities because the system is not satisfied by a proper choice of the parameters. The XOR logic gate is composed of OR, NAND and AND ports; it is structured by the network in Figure 3.17b. Figure 3.18 graphically shows the threshold conditions of the AND port (Figure 3.18a), OR port (Figure 3.18b), NAND port (Figure 3.18c), and XOR port (Figure 3.18d). For a complex network having more input variables, threshold hyperplanes defining the validity of the logic conditions are defined. The hyperplane of $n - 1$ dimension of a neuron connected to n inputs is defined by the following equation:

$$X_1 w_1 + X_2 w_2 + \ldots + X_n w_n - \tau_i = 0 \tag{3.9}$$

For complex networks requiring nonlinearly separable curves and special activation functions, it is important to classify the inputs which are initially encoded, by developing a procedure to estimate the network weights. The classification is estimated by the neural networks able to learn the appropriate weights from a representative training dataset provided by measurement values. In

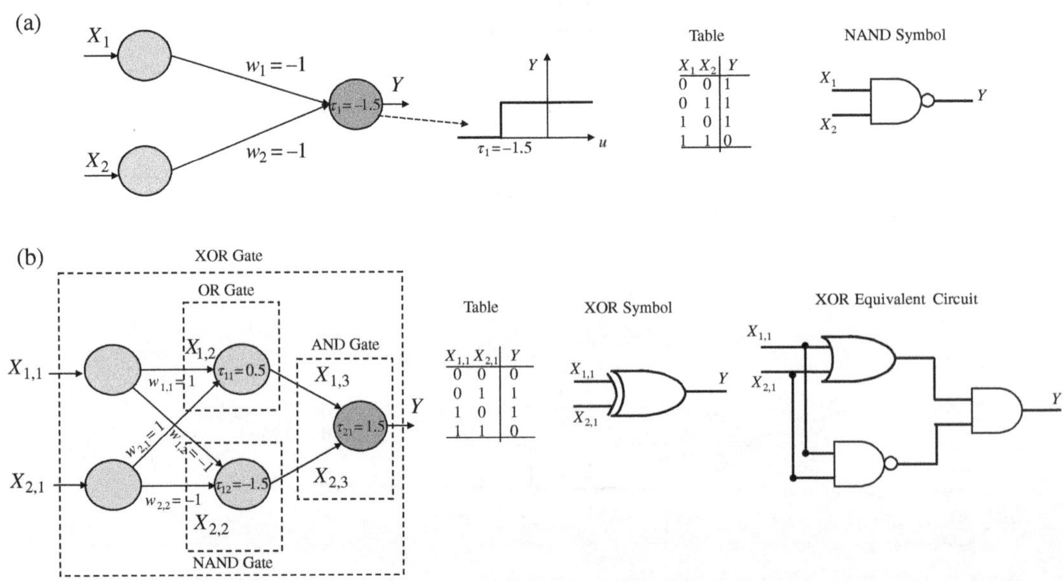

Figure 3.17 (a) NAND Logic and its implementation. (b) XOR logic and its implementation.

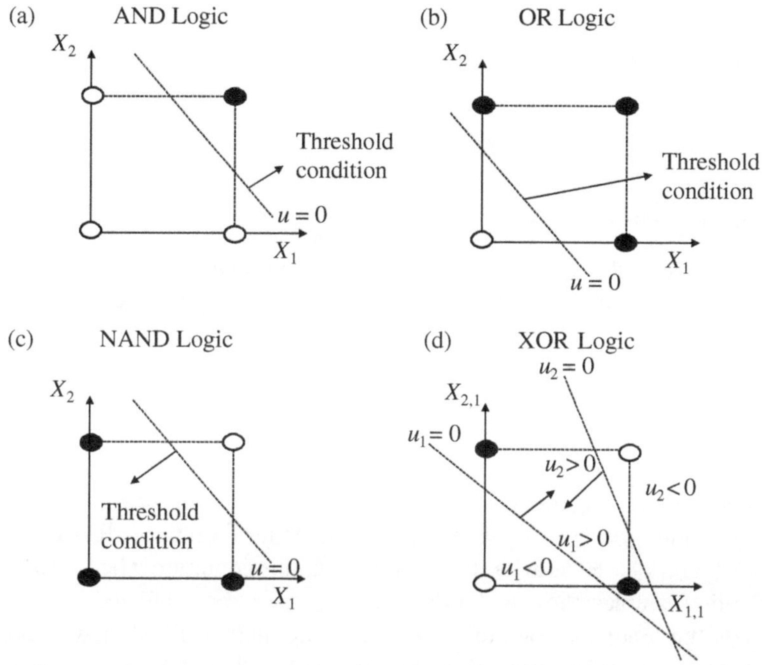

Figure 3.18 Threshold condition of the AND port (a), OR port (b), NAND port (c), and XOR port (d).

any case, the classification is always performed by the classification algorithms, able to automatically define the decision boundaries due a data input classification. Figure 3.19 shows an example of data input classification defining three classes of data input. For the definition of the threshold conditions defining the decision boundaries, a complex classification network is required. Some data mining classification algorithms are:

- k-nearest neighbour (KNN)
- ANN
- Support vector machine (SVM)

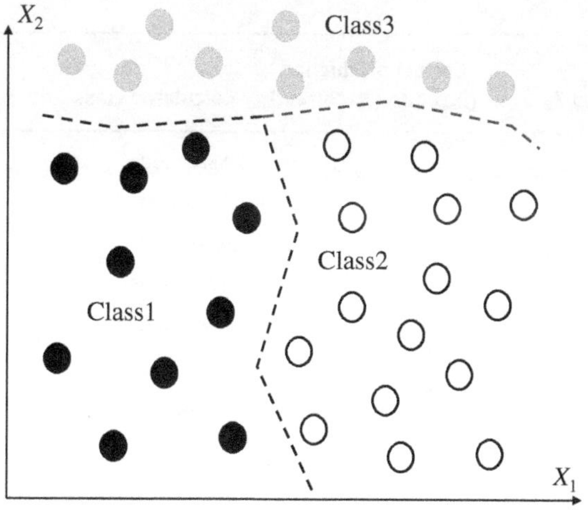

Figure 3.19 General decision boundaries due to a data input classification.

- Decision tree
- Extreme gradient boosting (XGBoost)
- Random forest (RFo)
- Extremely randomized trees
- Gradient boost
- Naive Bayes
- K-means clustering (unsupervised classification)

In order to understand the automated classification mechanism and how a GUI implementing an ANN model works, the following theoretical example summarized in Table 3.5 is proposed: a production line is monitored by two sensors detecting temperatures (temperatures T_1 and T_2) at two different points of two different production stages; the detected temperature are encoded to be processed by the classification algorithm reading the binary input (the active value 1 is assigned for $T_1 > 52\,°C$, otherwise $T_1 = 0$; the active value 1is assigned for $T_2 > 49\,°C$, otherwise $T_2 = 0$); the correct production is observed for the conditions of $T_1 < 57\,°C$ and $T_2 > 50\,°C$ thus structuring the training of the model based on the scheme in Figure 3.20. Observing the first seven dataset records used for the ANN model training, the condition $T_1 = 58\,°C$ and $T_2 = 58\,°C$ is estimated to be the failure condition: this allows the testing condition of $T_1 = 60\,°C$ and $T_2 = 60\,°C$ to be classified as the failure condition. The last three rows of Table 3.5 refer to the testing samples: the class for these samples has to be calculated.

The network able to estimate the predicted class of the last measurements is illustrated by the workflow in Figure 3.21 related to the GUI implementing a MLP ANN-based model. The Konstanz Information Miner (KNIME) graphical workflow is structured as follows:

- An input node reads data from an external data source imported into a DB or into a file (node 1).
- A block interface able to partition data input as a training dataset partition and a testing dataset partition (node 2).
- A learner block able to learn the model by means of a resilient back propagation (RProp) algorithm of the multilayer feedforward networks, accomplishing a local adaptation of the weight updates according to the error function trend (node 3) [21].
- The MLP predictor block predicting class (node 4).

Table 3.5 Example of a dataset used for the automated classification of the input data (the gray shading indicates the testing samples).

T_1 (°C)	T_2 (°C)	Encoded T_1	Encoded T_2	Correct production (0, failure; 1, correct)	Calculated class
51	51	0	1	1	Assigned
52	52	0	1	1	Assigned
50	54	0	1	1	Assigned
53	49	1	0	0	Assigned
48	55	0	1	1	Assigned
56	49	1	0	0	Assigned
58	58	1	1	0	Assigned
49	58	0	1	?	1
60	60	1	1	?	0
51	54	0	1	?	1

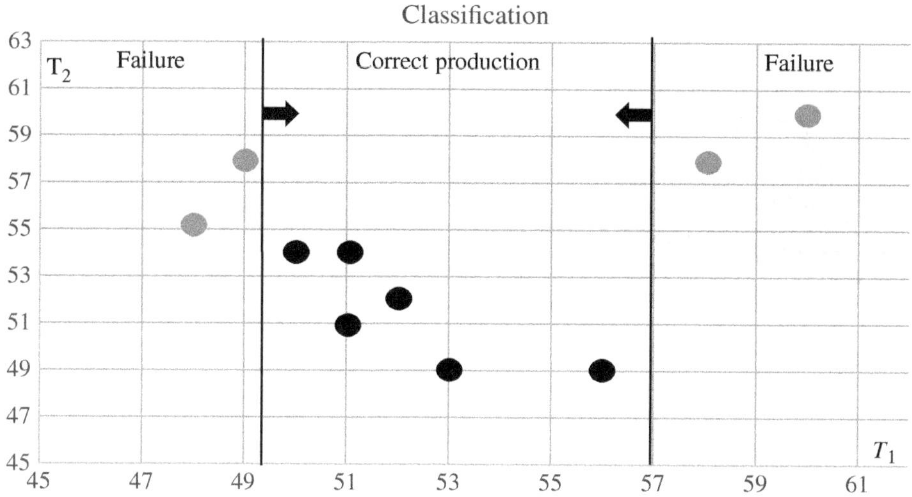

Figure 3.20 Data input mapping and correct production or production failure classification.

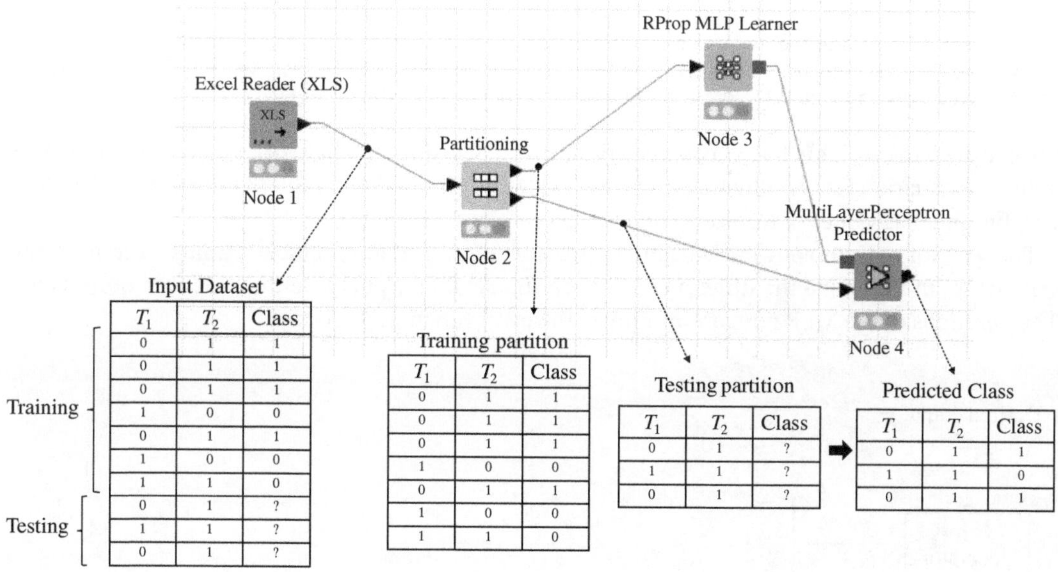

Figure 3.21 KNIME workflow implementing MLP model predicting correct production or production failures, and related data flow at each stage of the workflow.

The partitioning object is able to fix the percentage of records of the dataset to choose for the learning and for the testing. By the configuration of the learned block, it is possible to fix some algorithm parameters such as the number i of neurons of the hidden layer, and the maximum number of iterations (epochs). The MLP network in Figure 3.22 is constituted by bias nodes: the bias is a constant value that is added or subtracted to each unit of a specific layer, in a consistent way; it is possible to find the hidden bias node as another level of input unit, similarly, the output bias unit behaves as an additional unit of the hidden layer; these units do not receive input from the input level units; since there is no back link for the bias units, the error cannot propagate

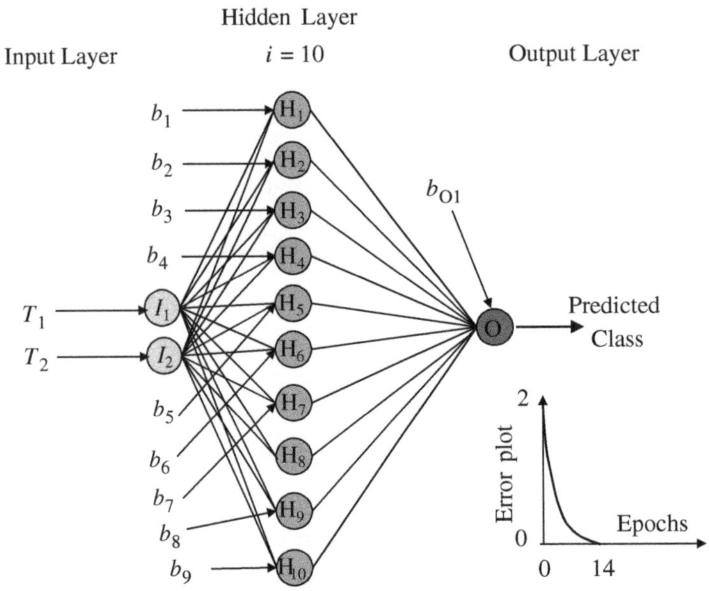

Figure 3.22 MLP network adopted for the temperature classification (the number of neurons of the hidden layer is $i = 10$, and the maximum number of iterations is equal to 100). A few epochs are enough for the decrease of the calculation error and for the correct estimation of the network weights. The input nodes are indicated by I_1 and I_2, and the only output neuron is indicated by the letter O.

further once the weights have been automatically adjusted by the network. Each hidden neuron defines a generic class as illustrated by the model in Figure 3.23. Table 3.6 lists the calculated weights and bias values of a convergent calculation.

The general requirement for learning is that the algorithm automatically adjusts the network weights w_{ij} to minimize the difference between the actual outputs y_j and the desired outputs y^d_j. The iterations of the MLP network minimize the error function defined as:

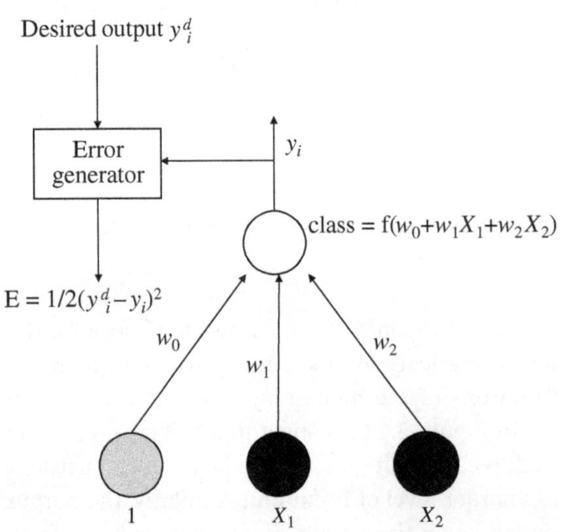

Figure 3.23 Generic class defined by two input neurons and one bias input.

Table 3.6 Calculated bias and weight values for each neuron of the hidden layer.

Hidden neuron	Bias b_i	w_{i1} (from I_1 to H_i)	w_{i2} (from I_2 to H_i)	w_{io} (from H_i to output node O) b_{o1} = 0.42
H_1	−11.82	57.51	−12.50	−5.05
H_2	−4.17	79.29	−35.08	−1.63
H_3	4.38	79.35	34.80	12.74
H_4	−0.99	54.88	−14.65	−1.6
H_5	−11.34	65.69	−14.08	−12.55
H_6	9.13	−82.29	29.43	12.66
H_7	−1.30	12.58	−2.33	−4.03
H_8	−10.18	81.82	−28.32	−1.67
H_9	2.17	−55.82	14–80	12.90
H_{10}	−12.36	66.99	−12.42	−12.53

$$E(w_{ij}) = \frac{1}{2}\sum_{p}\sum_{j}\left(y_j^d - y_j\right)^2 \tag{3.10}$$

where the index p indicates the training patterns; Eq. (3.10) is the total squared error summed over all the j outputs and all the p training patterns. The error function is estimated at each step (Figure 3.23). The weight function is defined by the following equation:

$$\hat{w}_i = w_i + \eta \nabla E(w) \tag{3.11}$$

where η is the learning rate. The learning rate is a tuning parameter adopted for the training model: it is a small positive value (defined in a range between 0 and 1), and its function is to control how quickly the model is adapted to the classification or to the prediction to solve (low learning rates require more training epochs, whereas larger learning rates require fewer training epochs). The error function can be traced on a surface as sketched in Figure 3.24, where the new weights during the iteration processes are obtained by moving in the direction of the negative gradient along the multidimensional error surface. A pseudocode describing the error calculation for a single layer feedforward network is as follows:

1. *it is defined the set of training patterns to learn* $\{X_i^p, y_j^p, i=1...n, j=1...m, p=1...o\}$;
2. *set up of the network model defining the n input units fully connected to the m output units by the weights* w_{ij};
3. *generate random initial weights;*
4. *choice of an appropriate error function* $E(w_{ij})$ *and a learning rate* η;
5. *apply the weight update:*

$$\Delta w_{ij} = -\eta \partial E(w_{ij}) / \partial w_{ij} \tag{3.12}$$

to each weight w_{ij} *and for each training pattern p. A step of updates of all the weights for all the training patterns is named one epoch of the training process.*

6. *The last step is repeated until the error function is minimized.*

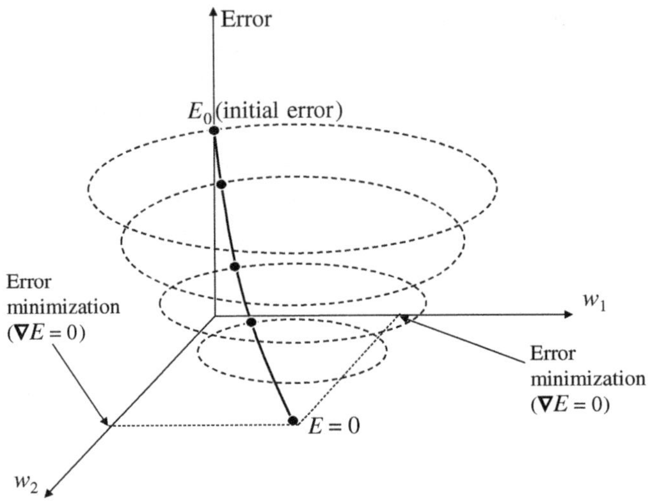

Figure 3.24 Multidimension error surface.

3.5 Programmable Logic Controller I/O Ports Interfacing with AI Engine

Digital values of sensors coming from the PLC's I/O ports can be processed as parallel values. The shift registers are adopted to convert parallel values to serial values allowing data storage. Consider a production line composed of two stages (Figure 3.25a), where digital sensors from 1 to n are adopted for the monitoring of the machine in stage 1, with sensors from $n+1$ to m used for the monitoring of the machine in stage 2. As illustrated in the block diagram of Figure 3.25b, digital

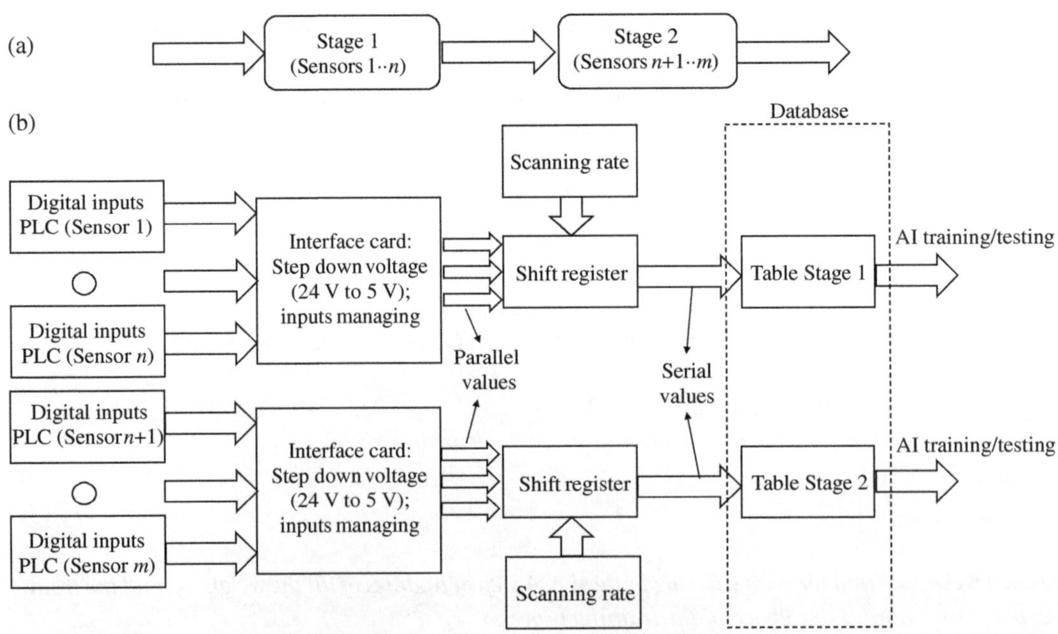

Figure 3.25 (a) Production line composed of two main stages monitored by sensors. (b) Digital I/O PLC values using shift registers storing data into a DB addressed for AI data processing.

values on PLC ports go through two stages, the step-down voltage converter and the shift registers, to store them into the specific table of the DB dedicated for AI data processing (AI engine). Usually, PLCs operate on 24 V. Many shift registers work with circuits at low power with low input voltage (5 and 3.3 V). The first step to construct a digital I/O from the PLC is to design and implement an electronic interface (electronic card) between the PLC and shift registers, decreasing the voltage.

Figure 3.26 shows a possible configuration of a parallel to serial register constructed by flip-flop D elements.

The system in Figure 3.25a can be full synchronized by storing all data sensors into a unique table. This is the case when a product is first processed in stage 1 and then processed in stage 2 (stage 2 waits for the end of the first processing). Figure 3.27 shows a configuration where the PLC data are synchronized by shift registers storing data into a unique table to be processed by AI.

By considering analog inputs, it is necessary to apply at the output of the interfaces named interface cards in Figure 3.27, the analog-to-digital converter (ADC). An example of an ADc is the flash converter shown in Figure 3.28a. Figure 3.28b shows an example of a 3-bit quantized output, obtained by implementing seven comparators in the flash converter configuration.

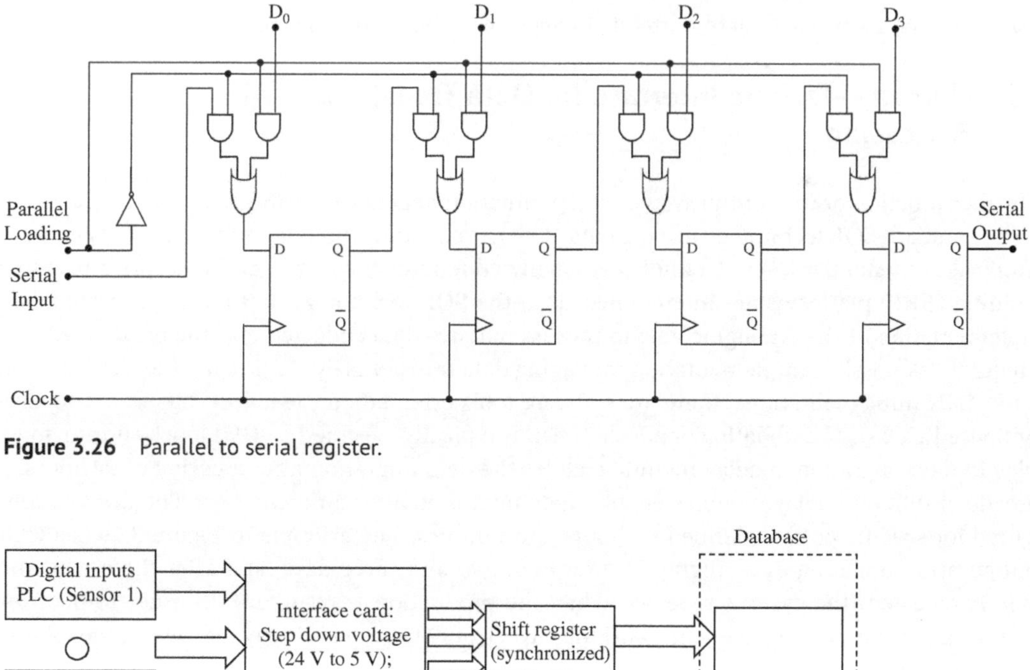

Figure 3.26 Parallel to serial register.

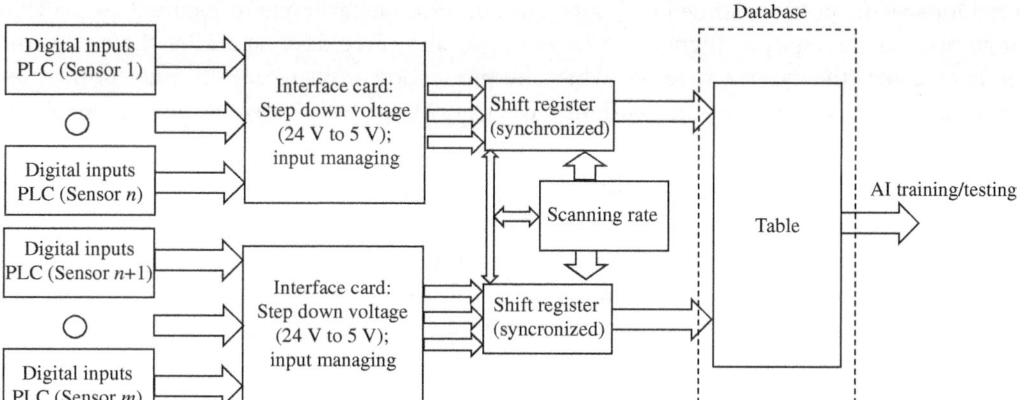

Figure 3.27 Digital I/O PLC values using synchronized shift registers storing data into a unique table addressed for AI data processing.

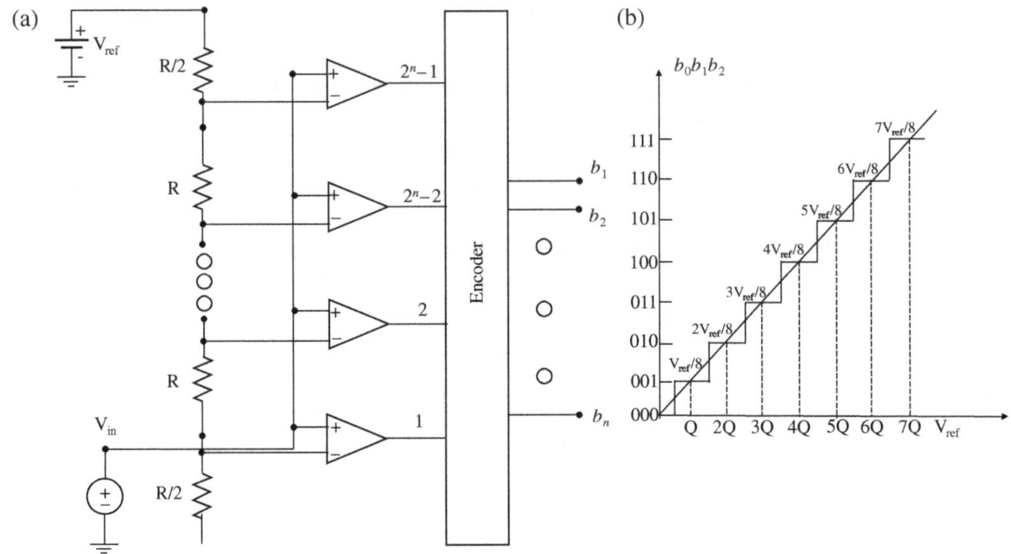

Figure 3.28 (a) Scheme of flash converter. (b) Example of a 3-bit quantized output.

3.6 Human–Machine Interface for Data Transfer and AI Data Processing

A further function of the central system of a production network infrastructure is data migration from a standard SQL technology to a big data NoSQL one. Many industry information systems are equipped with standard SQL technologies: many commercial software and enterprise resource planning (ERP) platforms are interconnected to the SQL technology, but for a system upgrade implementation of the AI engine able to process massive data collected into the big data system is required. A possible implementation managing data transfer is by GUIs using the KNIME tool: as for ANN implementations, there are software tools oriented on the object (block) to properly configure [22, 23]. The updating of a NoSQL DB is typically executed by GUI blocks improving a delay in data migration: a delay module enables the data migration after a period of 24 hours; a cascade of different delay modules enables data migration after different days. The delay is configured for specific periods defined by hours and minutes. The example in Figure 3.29 is a GUI configuration updating data migration after one, two and three days, and after three days the cycle is repeated (the cycle is repeated when the production is resumed). In order to plan for example a data migration during the working days (from Monday to Friday) of a week, the following configuration is designed:

- A single delay module allowing data migration after Monday.
- Two delay modules allowing data migration after Tuesday.
- Three delay modules allowing data migration after Wednesday.
- Four delay modules allowing data migration after Thursday.
- Five delay modules allowing data migration after Friday.
- All the delay modules are executed again after Saturday and Sunday by continuing the automated data migration process.

A possible scheme of NoSQL and AI integration in an industry network, able to control and actuate the production machine, is illustrated in Figure 3.30 where some GUIs running on the central

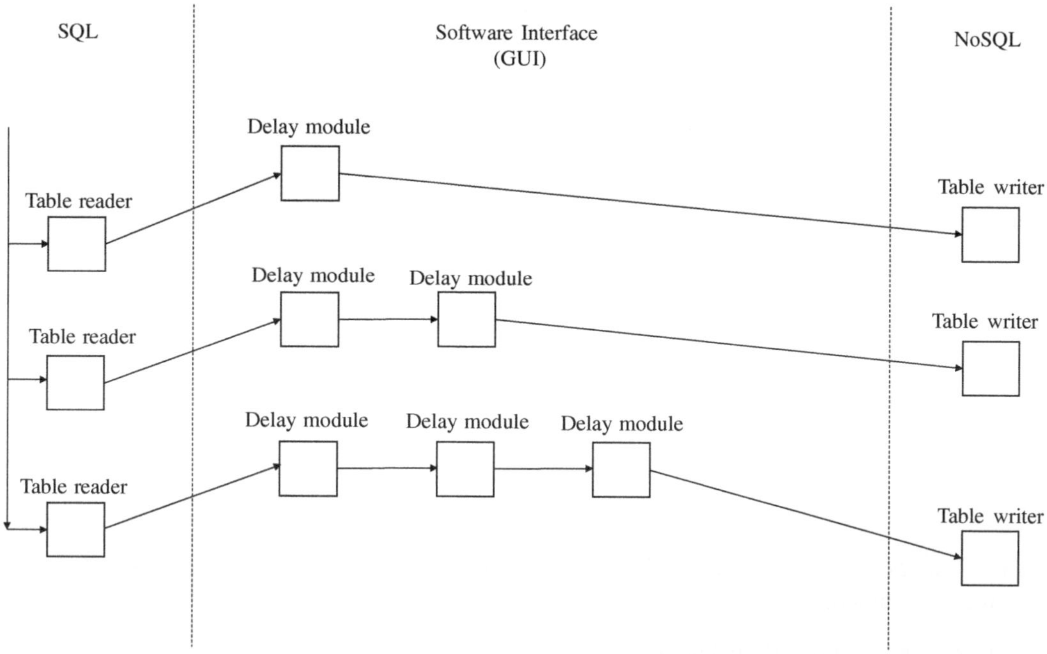

Figure 3.29 GUI implementing delay modules for data migration from SQL to NoSQL technology.

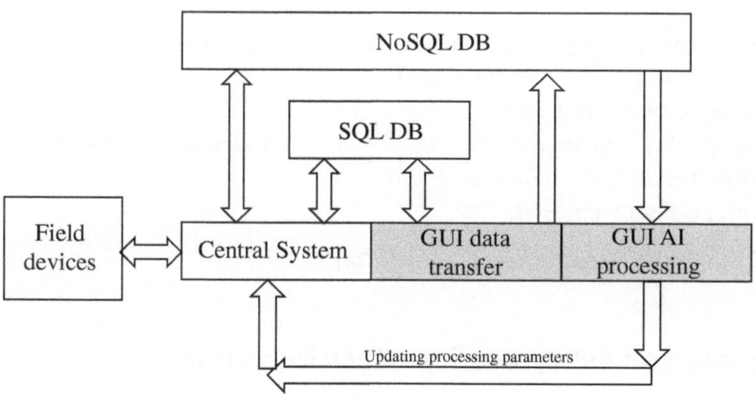

Figure 3.30 Network integrating NoSQL and AI technologies.

systems are dedicated for SQL to NoSQL data migration, and other GUIs are used for the implementation and execution of AI algorithms. The output of the AI algorithms is adopted to configure the new parameter setting according to failure and defect prediction.

3.7 Example of Interface Configuration of Temperature Control

Some tools are used for data recovery and data traceability concerning sensor data detection. Node flow programming tools [22] are useful for wiring sensors applied in the local information infrastructure by application programming interfaces (APIs) and online services: it consists of a series of interconnected nodes (wired nodes) tracing sensor data flow. Figure 3.31 shows an example [22]

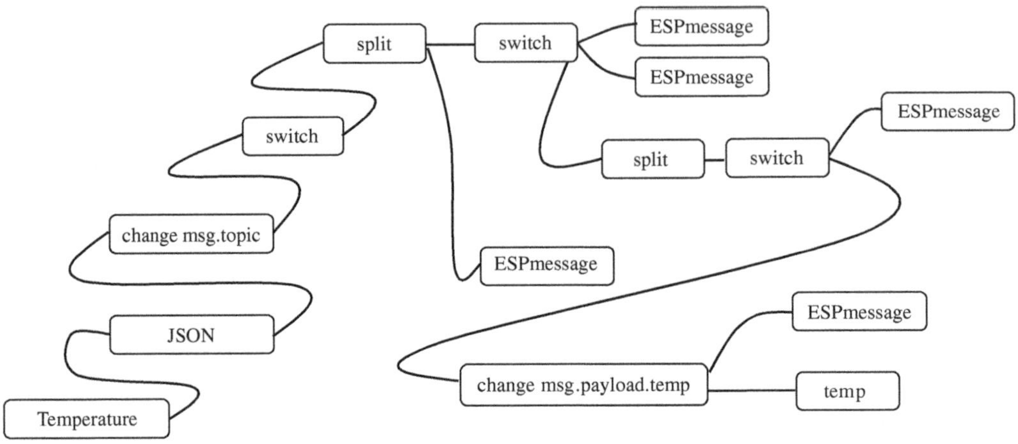

Figure 3.31 Data flow inherent data recovery from an ambient temperature sensor.

of data flow inherent data recovery from an ambient temperature sensor, where the following functions are defined:

- Temperature field indicating data flow source corresponding to the digital output of the sensors.
- JavaScript Object Notation (JSON) field: this field represents a JavaScript object as a string, used by web APIs to return data (if a message property contains a JSON string it must first be parsed to its equivalent JavaScript object).
- Message topic field: defines the message topic that can be accessed by a variable assignment.
- Switch: message switching (switching to the user interface).
- Split: split a message into a sequence of messages.
- ESPMessage: encapsulating security payload data protocol for data confidentiality, data source authentication, data integrity checking, and data protection.
- Change message: changes properties of a message.

An example of a node flow tool is the Node-Red program [22, 24].

3.8 AI Interfaces Oriented on Cybersecurity Attack Detection

AI is increasingly seen as the new frontier of solutions for the defense of companies: AI in cybersecurity enhances the predictive capabilities of control systems, and therefore protects data of public and private organizations from increasingly sophisticated cyberattacks. It is essential to respond in real time to any threats. Machine learning (ML) tools are increasingly used to detect and mitigate threats, but the real, great potential of AI in cybersecurity lies in its ability to make realistic predictions before the attack. Traditional security systems recognize the threat as pre-classified threat infection but are not oriented on the analysis of multiple attributes, and to a dynamic self-learning processing of threat classification. The infection happens:

- In memory: residing in the random access memory (RAM) of computers by executing malicious code directly in the memory; infections caused by attacks of this type are generally not persistent.
- In persistent modality: the attacker obtains persistence on compromised systems by loading a payload into the memory so that the infections are persistent.

- By the use of dual use tools (tools natively installed in the operating system with a high risk and difficult to trace).

However, new technologies alone are insufficient to counter any type of cyberattack. In fact, there is a gradual change of the environments and cloud frameworks in which companies operate, requiring a change in the logic of protection during the time. However, the application of ML algorithms is limited by the type of attack to which the networks may be subject. For example, it might be useful to implement ML techniques to be able to detect viruses by analyzing code, but it is quite difficult to prevent sniffing and spoofing attacks since it is difficult to understand this type of attack as it is necessary to distinguish legitimate purposes (such as identifying connection problems) from illegitimate purposes (in sniffing, the attacker "listens" into a data transmission and acquires data packets adopting packet sniffers; in spoofing, the attacker pick up the credentials of a user to legitimate the user connection). In contrast, AI is useful in being able to counteract insidious exploits. To date, not many antivirus platforms have adopted AI systems.

All machines (personal computers, servers, tablets, etc.) that must be protected in some way also have installed the so-called "agent,0" i.e. the antivirus software that carries out protection activities and constantly communicates with the control center: it is essential, however, that this agent is not too "heavy" or does not need too many resources of the host machine. Furthermore, the control center where the "intelligence" of the antivirus resides must be fast and reactive.

Neural networks are applied in the prevention of certain types of attacks such as Denial of Service (DoS) and Distributed Denial of Service (DDoS): DoS attack is a service attack where a personal computer is used transmitting TCP, and User Datagram Protocol (UDP) packets. A DDoS attack consists of an attempt to overload a website or an online service with traffic from different sources to make it unavailable to users. Concerning threat detection, it is important to set a data traffic threshold, because traffic peaks are not always suspicious to be classified as DoS attacks. A neural network potentially prevents worms, viruses, Trojan horses or phishing scams, using large amounts of data (collected into a big data system) to classify any viruses or to identify anomalous patterns traced by the user connection and user data processing. With regard to intrusion detection system (IDS) type attacks, neural networks are suitable to analyze incoming communications, in order to possibly identify intrusions within the server. A good design of the calculus AI engine must guarantee a system that does not affect machine performance. However, the idea remains that AI should be a support to any system administrators. In order to partially overcome the problem associated with the computational cost of threat detection and data processing, the AI calculation can be executed in the cloud. As illustrated in the architecture in Figure 3.32, cloud computing is related to:

- Big data storage.
- Big data analytics.
- Data clustering.
- Data classification.
- Prediction (predictive maintenance and defect classification).
- Data security (threat detection).
- Graphical data reporting.

AI models to predict threats of data access are calculated by means of a learned network such as the Tree Ensemble algorithm model [25]. The ensemble approach combines a set of classifiers learning a target function by combining prediction results.

The workflow implementing the Tree Ensemble algorithm is structured by the following nodes:

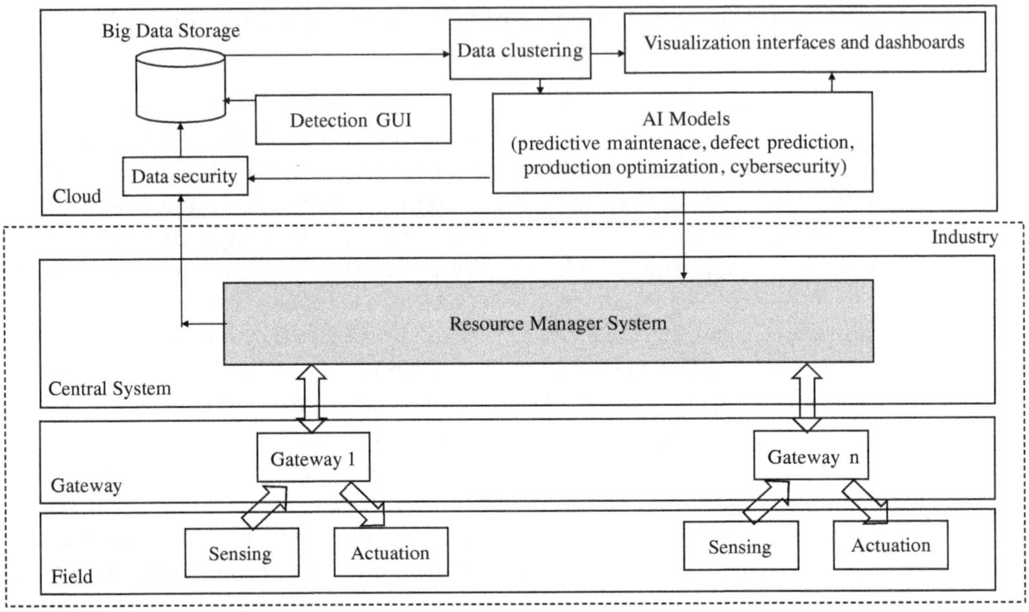

Figure 3.32 Advanced architecture linking the cloud environment with local company infrastructure, where AI algorithms are executed in the cloud.

- "Cache" (node 1): node reading input data such as log data extracted from server machine (the input data are stored as time stamp, request type coding, codified message, IP request, IP arrival, IP request port, IP arrival port, and other information concerning server status).
- "Partitioning" (node 2): the input scheduled data are split into the two dataset partitions (the train dataset at the "Tree Ensemble Learner" node input and testing dataset addressed to the input of the "Tree Ensemble Predictor".
- "Tree Ensemble Learner" (node 3): this node learns the model to apply for the log classification.
- "Tree Ensemble Predictor" (node 4): this node predicts patterns of attribute information by means of the grouping of predictions of the RFo model of each tree.

Figure 3.33a shows the KNIME workflow implementing the Tree Ensemble algorithm suitable for anomalous server accesses. Figure 3.33b shows the log matrix related to the Tree Ensemble output determining the attack risks by a range of tree values: "No Risk," "Low Attack Risk," and "High Risk." The log protocol to analyze at the input of the classifier is constructed by the following fields:

- Timestamp (log time registered in the server; each record is structured after a period).
- Request type (as access port; it is the labeled attribute for the prediction results).
- Message (returned codified message subsequent to the request).
- IP request (origin of the request IP).
- IP arrival (information on the IP destination).
- IP request port (information about the origin port).
- IP arrival port (destination port concerning IP arrival address).

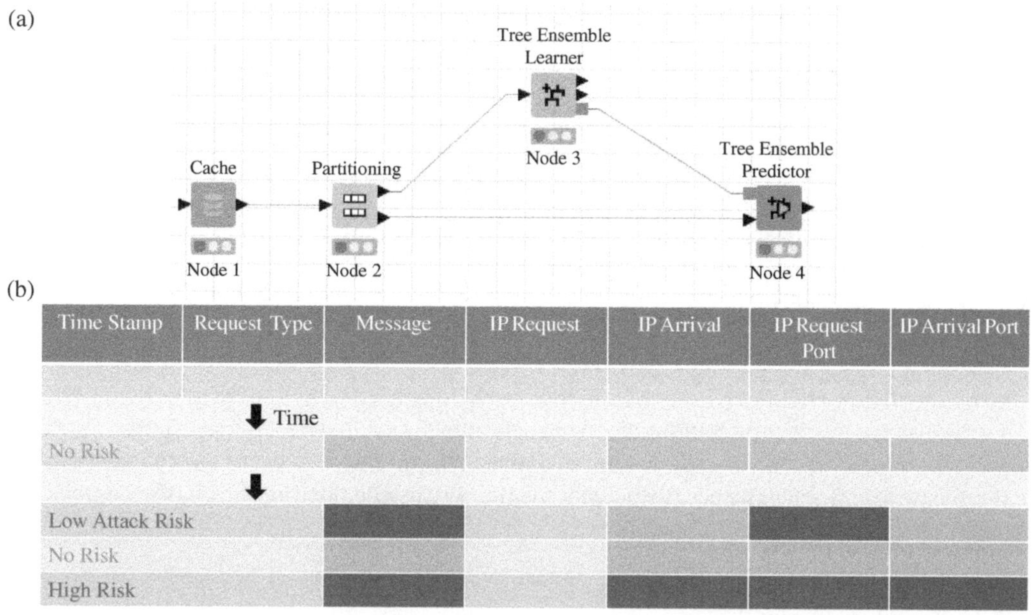

Figure 3.33 (a) KNIME workflow implementing the Tree Ensemble algorithm classifying risks. (b) Classification of the attack risk levels.

The pattern to analyze is constituted by the message, IP arrival, IP request port and the IP arrival port fields. In this case, the analyzed pattern is classified by the Tree Ensemble algorithm as an attack or as a normal system log.

3.9 AI Interfaces Oriented on Database Security

Most companies and organizations nowadays are oriented to safeguard sensitive data. IDSs are important components for a secure framework, and AI algorithms provide an important contribution to data protection. The IDS using the ML approach is gaining more and more attention in the field of DB anomaly detection due to high detection accuracy, efficiency, and automation features. In particular, ANNs are useful to model intrusion behaviors based on misuse. Furthermore, data mining is adopted to detect SQL-injection attacks [26, 27]. An IDS based on a ML technique using pattern classifiers to detect injection attacks and to protect web applications and the backend system can identify queries that do not match with typical and usual query patterns. The usual query patterns are adopted as a training dataset [28] to construct an AI model classifying and detecting cyberattacks. Concerning backend systems, a relational database management system (RDBMS) is typically used for high performance data storage. Attacks on a RDBMS are classified as insider and outsider attacks. Attacks from the outside, such as SQL intrusion, are usually mitigated with defensive programming techniques. Some classifiers for query classification are:

- Learning classifier system (LCS)
- Convolutional neural network (CNN)
- Convolutional neural-based learning classifier system (CN-LCS)

The CN-LCS technique combines the conventional LCS with the CNN enabling the DB intrusion detection system. The access control of DB can be managed by the following access models [29–32]:

- Discretionary access control (D-AC)
- Mandatory access control (MAC)
- Identity-based access control (I-BAC)
- Role-based access control (R-BAC)
- Task-based access control (T-BAC)
- View-based access control (V-BAC)
- Task-mandatory access control (T-MAC)
- Organization-based access control (Or-BAC)

There are also a number of combinations of these models such as T-MAC and T-BAC, to refine and to adapt the access control logic according to the environment in which it is implemented, and the company's safety policies. Security models are mainly classified in three levels of access control:

- the access control policy
- the access control model
- the access control mechanism

An access control policy defines the rights, prohibitions, verified information and persons authorized to use the applications and data, or, even, the modification of the access control policy itself. The security policy aims to ensure confidentiality, integrity, and availability.

The access control model is the intermediary for joining the gap between policy and implementation.

The access control mechanism permits implementation of the default security policy. To determine whether a subject can read the information contained in an object, the mechanism checks if the authorization is included in the access control list.

The D-AC model can restrict access to objects based on the identity of subjects and groups to which they belong. The controls are optional: a subject with a specific access permission can pass the permission to other subjects.

The MAC model is a system-enforced access control model based on a subject's permission and on the object's labels. Subjects and objects have authorizations and labels classified as confidential, secret, and top secret.

The I-BAC model introduces the fundamental concepts of subject and object together with the action. The subject is the active entity and most often refers to a user or an application that runs for the benefit of a user. The object is the passive entity that refers to information or resources that a subject can access to perform an action. The action means a desired effect when a subject accesses an object (read, write, modify, etc.). The purpose of the I-BAC model is to control any direct access of subjects to objects through the use of actions. This check is based on the identity of the subject and object. However, this model has limitations because when creating a new subject or object, the authorization policy must be updated to define new permissions associated with that subject or object.

The R-BAC model can execute an action on an object, while the subject must first have a role with adequate permissions. Roles are assigned to subjects according to the function assigned to them in the organization. The basic principle of the R-BAC model is to consider that permissions are directly linked to roles and not to the individual.

Compared with the I-BAC model, the management authorization policy is simplified as it is not necessary to update the policy when a new entity is created.

Actions generally correspond to elementary commands, such as reading the content of an object or writing an object, but in recent applications there is a need to check the results of composite actions, called tasks or activities.

The T-BAC model introduces the task concept. In other words, access is granted depending on the task, so the access check compares this task with the roles that contain it. In this manner, tasks can be considered as subroles.

The V-BAC model is a security model proposed for the SQL relational DBs, able to structure the objects of a security policy on the concept of view: a view is a set of access rights, which are permitted or prohibited for operations on objects.

The T-MAC model is developed to control access to collaborative activities. In this model, permissions are associated with roles associated with teams, where a team is an abstraction that encompasses a set of users with specific roles in order to accomplish a specific task or goal in the collaboration activity. The permissions of a subject are obtained by combining the permissions associated with the roles on the team and the permissions on the team to which the subject belongs.

The Or-BAC model is indicated for the modeling of a variety of security policies, and is based on the following four fundamental principles:

1. The organization is the central entity of security policies and models.
2. There are two levels of abstraction:
   - a concrete level (made by the subject, the action, the object);
   - an abstract level (made by role activity).
3. The possibility of expressing permissions, prohibitions, obligations, and recommendations.
4. The ability to express contexts.

Subjects (users) and objects (e.g. files) are classified into the following levels (access classes):

- levels for subjects (clearance levels)
- levels for objects (sensitivity levels)

The Bell–La Padula (BLP) model [33] is designed to achieve security mainly in a military environment, ensuring the confidentiality of information. There are four levels of document sensitivity:

- not classified
- confidential
- secret
- top secret.

For continuous forthcoming information, the AI classification automatizes the priority levels indicated in Figure 3.34.

People are assigned to levels according to the documents to review. The classification supports this assignment for different labeled documents.

The security roles establish how information can circulate. There are two main classifications:

- Simple security property (a process running at a specific security level, only read objects at its level or lower).
- Ownership (a process running at a specific security level, only write objects at its level or above).

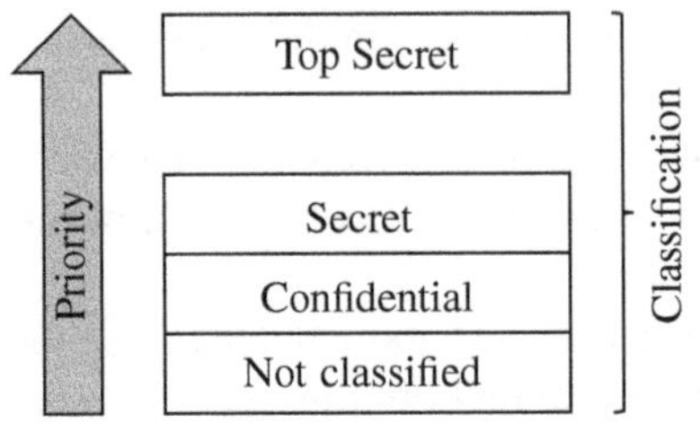

Figure 3.34 Priority levels of information.

To these roles are added the security roles specified by the access matrix. Processes read down and write up in the level scale: the flow of information is from the bottom up and not vice versa.

The BLP model was designed to keep secrets and therefore confidentiality, not to ensure data integrity. In fact, it is possible to overwrite the information belonging to a higher level.

The goal of the Biba model [33] is data integrity. The simple security property in this case believes that a process running at a specific security level can only write objects at its level or lower (no writing upwards is possible). While the integrity property believes that a process running at a specific level can only read objects at its level or above (no reading down it is allowed).

The BLP and Biba model conflict with each other, so it is not possible to use them at the same time.

The term "administration" means the management and control of all components of the security policy such as users, actions, objects, roles, permissions, etc.

The specifics of the security policy and its updating are the two most important administrative tasks. Security models are not all accompanied by a model for administration. The best-known model of administration is the flexible model of Harrison, Ruzzo, and Ullman [34]. A particularly important aspect for data security is access control.

The design of a general architecture for access control is composed of three basic blocks:

- DB
- Access Control Module (ACM)
- Inference Control Module (ICM)

Concerning inference control architectures, an ACM is always present as the first line of defense in the system. This is a DAC or a MAC access control module able to adjust the user's access privileges and permissions before starting any inference checking. The AI interface can improve the inference model able to identify the anomalous user accesses. An architecture integrating AI is shown in Figure 3.35.

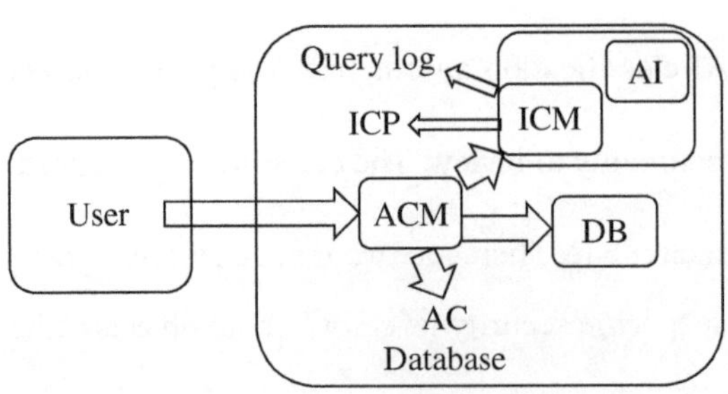

Figure 3.35 Access control interface model.

Any user request to the server is always first routed to the ACM, where the user is checked verifying his/her legitimacy and authorization for a specific request. The access control DB stores for each user his/her credentials and permissions which are regulated by a server administrator. After the user's request (i.e. query) is verified by the ACM, it is forwarded to the ICM for further inspection of any inference processing about indirect malicious requests. In the model in Figure 3.36, the ICM consists of two blocks, namely an inference control policy (ICP) and a query log (log of each SQL query received from a client). Each query that is routed to the ICM for inspection is first saved in the query log and then checked for threats using the ICP. Depending on the type of ICP, different inference checking strategies are used, analyzing all previously saved queries in the query log. If the ICP finds no inference treatment, the user's query is marked as safe.

Many companies store and keep continuously multiple information (customer data, information on company production, etc.), often using a RDBMS. One of the first levels of security involves the IDS. Attacks on an RDBMS are classified as internal and external. External attacks, such as SQL injections, are prevented through appropriate security systems; internal threats are much more difficult to detect and are the most damaging. Some IDSs are unable to detect and manage some classes of intrusion, in particular, attacks that the system has not previously seen can cause a serious security breach in the IDS. The ML-based approach allows algorithms to be able to learn and adapt and remedy this limitation of the IDS.

In [35] a CN-LCS system is proposed as a new model of the IDS system. The architecture of the CN-LCS is composed of two components: a rule discovery component to select and filter characteristics from queries based on a learning classification system, and the query intrusion detection component based on a CNN to classify the role of each query.

During the training phase, n queries are collected from the DB log with their role in order to feed the algorithm, and features are extracted from the queries to form a vector of characteristics $\mathbf{F} = (x_1, y_1), \ldots, (x_n, y_n)$. Figure 3.36 shows an example of architecture integrating a CNN network in the IDS structure.

The selected features are the CNN input, where convolution operations are used to extract threat characteristics. The trained CNN classifies if the query is abnormal, so the response engine alerts the administrator and does not execute the query, otherwise it is performed in the DB.

The increasing presence of DBs on the internet means that the need for security of the information exposed on the network increases. Some DBs are very important and their security is of crucial importance; these can contain financial information, banking, personal data, medical data, etc. Often the managers of online services report the theft of their users' data through their websites:

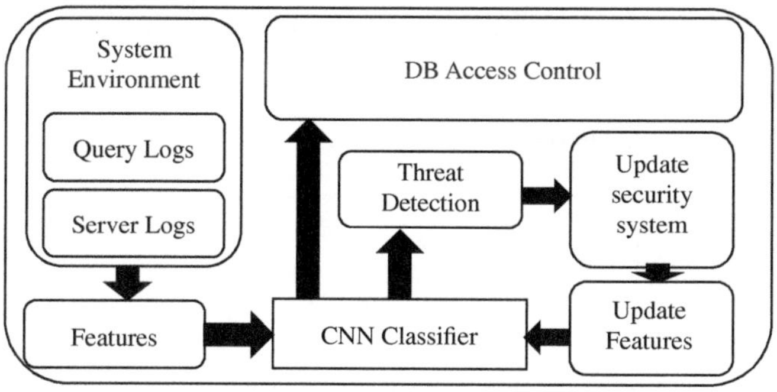

Figure 3.36 IDS architecture based on a CNN classifier.

hackers use the flaws of poorly written web applications and exploit, when there is a DB as a data source, the techniques of SQL injection.

Today, because of the internet, E-commerce works due to the presence of a DB where information is stored about articles, customers, purchases, orders, transactions, payments, etc. The safeguarding of the DB information by IDS is intended to detect intrusions as soon as possible so any damage caused is reduced to zero or otherwise minimized. Intrusion detection is a research topic focused on the network and server's security. The detection of network traffic intrusions is done by monitoring and analyzing network traffic; host intrusions are detected by analyzing the server logs. Both systems look at the "signature" of the attack because each attack has specific patterns facilitating the discrimination of malicious actions from non-malicious ones. These two types of IDS do not work on the application layer offering much more potential and accuracy in detecting suspicious actions. Figure 3.37 shows an architecture for detecting DB access intrusion based on a signature-based IDS system capable of learning and recognizing new access signatures.

The data flow of the architecture in Figure 3.37 is described as follows:

1) A service request is enabled by the application user by means of a transaction in the application server (the transaction may or may not be legitimate).
2) The application server formulates the necessary SQL statements and sends them to the DB server via the DB user.
3) The user accesses the DB, the DB session is traced, and the SQL message is processed by the AI interface module.
4) In the intrusion detection module, the received SQL statements are matched with the DB legitimated transaction set.
5) The anomalies or intrusions are then channeled to the module which decides which are the most appropriate actions to take (an alarm is transmitted to the administrators in the case of the presence of a possible attack).
6) The output is returned to the application user.

The system in Figure 3.37 uses the following three components:

- A tool to log all SQL statements submitted to the DB server (this is used to capture SQL queries and transactions comparing them with the allowed signatures).
- A process to derive the SQL query signature to legitimate transactions.
- A signature database (SDB) used to detect intrusions (each submitted SQL statement is compared with the SDB, and the intrusion is detected if there is no matching).

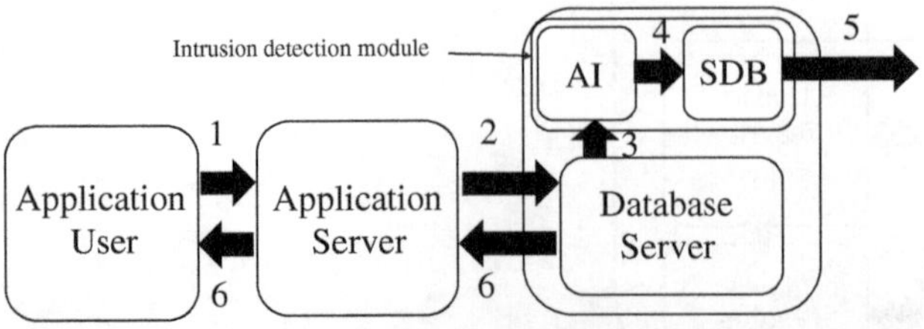

Figure 3.37 DB access intrusion system based on attack signature (SDB, signature database).

Hidden Markov model (HMM) [36] is applied to build DB security models trying to capture changes and to detect patterns that an operator is unable to recognize. The HMM uses the relationships between similar past events and current ones, by recognizing anomalous patterns.

To manage all the transactions of an entire DB it is necessary to use data reduction techniques, for this reason a clustering algorithm can be used, where each cluster is built on a set of attributes representing a DB region. The HMM is used to model the behavior of the DB clusters. Value changes in records with different time granularity, and this is one aspect that is processed by the HMM. The HMM provides the acceptance probability of an unknown sequence. If the variation of the acceptance probability is outside the established threshold, a system of acceptance of the results deriving from different training strategies generates a possible alerting condition. The HMM consists of three main modules:

- The clustering module pre-processing data to obtain a reduction operation of the clustering process.
- The modeling module constructing the HMM model (for each cluster the parameters and recognized pattern classes are initialized; there are several methods for training a HMM with a sequence of data; the model is trained incrementally with data sequences that grow over time).
- The validation module identifying anomalous or valid sequences (this control is based on the probability of accepting a sequence at time $t+1$ in relation to the probability of accepting the sequence at time t).

The encryption represents a further approach to ensure DB access. Using cryptography a text is transformed providing an incomprehensible version (encrypted text). To achieve this transformation, it is necessary to apply an encryption algorithm and a key able to encrypt the data. The encryption operation only makes sense if the following conditions are satisfied:

- **Reversibility:** given an algorithm and a key to decrypt the message, it is always possible to obtain the original starting text.
- **Security:** the ciphertext (result of encryption performed on plaintext using an algorithm, named the cipher) is not understandable to anyone who does not have the key.

The symmetric encryption approach utilizes a single key to be shared by users who need to receive the message, while asymmetrical encryption operates with a pair of public keys and a private key to encrypt and decrypt messages during the communication.

The symmetric key encryption algorithm uses the same key to encrypt and decrypt data. The most used approaches are the so-called "block ciphers," in which the text to be encrypted is divided into blocks of a predetermined size, which are then subsequently encrypted. If the last data block is smaller than the predetermined size, further bytes are added until a full block is obtained. This technique is called the padding technique.

For its importance, the key must be chosen and protected with care. A key of a few bytes or containing only some types of values is not robust and can be easily cracked by a cryptanalyst. Many symmetric encryption algorithms use keys of at least 128 bits, a length considered safe for commercial use. Unfortunately, it is often not possible to guarantee that a key is chosen accurately, so there are methods to overcome this problem by creating a robust key from the original one. These methods, called "key derivation," use hashing functions repeated several times to generate the new key starting from the original one.

An initialization vector containing a set of random values is used to make sure that a cryptanalyst cannot decrypt multiple texts encrypted with the same key algorithm. The initialization vector must be unique.

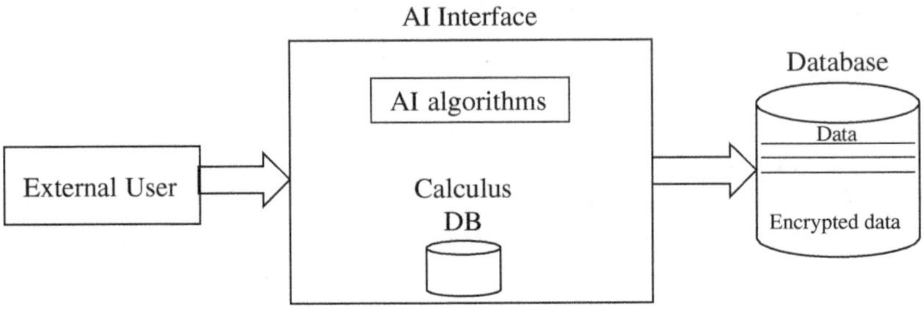

Figure 3.38 DB security model based on an AI interface.

In the encryption process, the individual blocks are "linked" to each other in different ways. The most common methods for linking are:

- Electronic Code Book (ECB): each block is encrypted independently and is not linked to the others (this mode is considered safe only if it is used to encrypt a single block of data; using it on several blocks generates a risk compromising the security of the encrypted text).
- Cipher block chaining (CBC): each block is first combined with the previous encrypted block and then is encrypted.
- Cipher feedback (CFB): this processes one portion of the block at a time; like CBC, CFB also includes a block chaining mechanism.

Common encryption algorithms are the Dynamic Encryption Algorithm for Real-Time Applications (DEA-RTA) algorithm, and the AES-Rijndael algorithm [37].

The DB security model illustrated in Figure 3.38 is an architecture considering the AI interface as the DB access control. Specifically, an external user is interfaced with a DB by means of an AI interface implementing algorithms for:

- Classifications (classifications of queries, threats, attacks, anomalous patterns, etc.).
- Risk prediction of possible external user behavior (comparison of protocols containing external user information).
- Communication switching.
- Access roles (definition of user type accesses, administration roles, etc.).
- Other functions based on the correlation analysis of different attributes.

The data collected into the DB are stored in a multilevel partition, where the levels represent the user authorization levels. A partition of the same DB is dedicated for secret data, which can be encrypted to furthermore increase the security of the whole system.

A more complex DB security system model provides a higher security level. Figure 3.39 shows an example of an architecture based on the simultaneous adoption of an inference engine with the AI classifier. Specifically, the model is structured by:

- A user interface managing the DB accesses of internal users and external ones.
- An AI interface classifying queries and access information patterns of all users.
- An access control unit managing the outputs, and making the decision to authorize the DB access.
- A central DB to control.

In order to improve security, all data of the central DB are copied periodically into another DB as for disaster recovery systems.

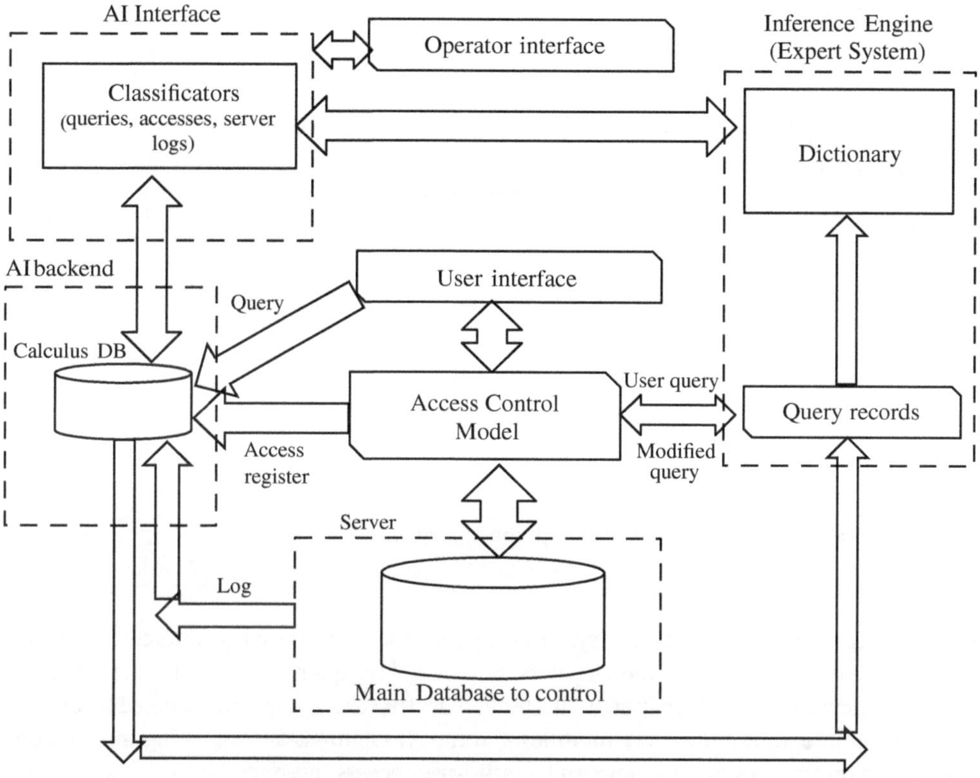

Figure 3.39 Architecture of a DB security system constituted by an inference engine and an AI classifier.

The inference control [38, 39] consists of the following three modules [39]:

- The knowledge acquisition module extracting knowledge of data dependency, knowledge of data structure, and semantic knowledge of the domain (based on the DB scheme and data sources, the dependency of data is extracted between attributes within the same entity and between entities; semantic knowledge of the domain can be derived from semantic links with specific constraints and condition).
- A semantic inference module (SIM) which can be built based on the gained knowledge.
- A violation detection (VD) module.

The SIM is a data model that combines data structure, dependency, and semantic knowledge. This module links the attributes and related entities, as well as the semantic knowledge required for data inference. The SIM represents all possible relationships between data source attributes. A semantic inference graph (SIG) is constructed by instantiating entities and attributes in the SIM. For a given query, SIG provides inference channels to infer sensitive information.

Based on the inference channels derived from the SIG, the VD module combines the new query request with the request log, and checks whether the current request exceeds the predetermined information threshold.

To reduce the computational complexity for inference, the SIG is mapped into a Bayesian network, so that available Bayesian network tools are used to evaluate the probability of inference along the inference channels.

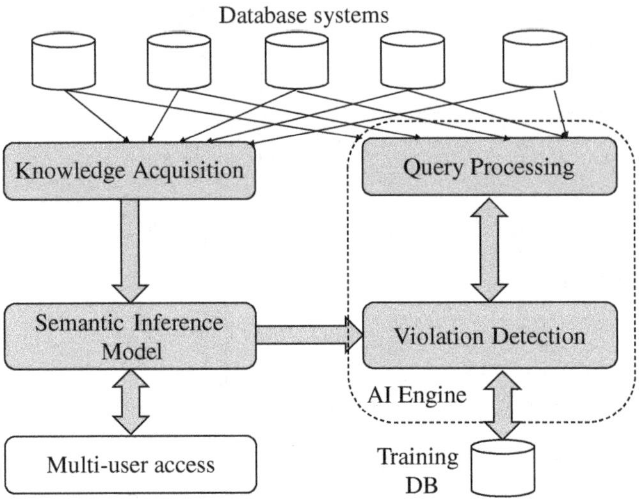

Figure 3.40 Inference detection system coupled with the AI engine.

When a user poses a query, the tracking system examines the past query logs and calculates the likelihood of inferring sensitive information from the response of the query request. The query request is denied if it can infer sensitive information with a probability that exceeds the prespecified threshold.

A possible model integrating the three modules and the AI engine is shown in Figure 3.40: the scheme includes a multidata source system and a multi-user access interface.

3.10 Cybersecurity Platform and AI Control Interface

Cybersecurity tools for the control and implementation of intelligent switching of virtual private network (VPN) channels is an important issue in industrial research [40]. Through real-time application of AI algorithms able to identify and classify cyber threats in self-learning mode, and to check the intrusions in VPN channels dedicated to bidirectional communications, and through the analysis of traffic logs, it is possible to improve the security of the whole network system. The same cybersecurity network is structured to support disaster recovery, to synchronize data migration processes, and to enable virtualized switching functions, also using supernodes and big data systems [40], capable of saving switching configurations, and to dynamically redirect communication into more secure VPN channels. The AI algorithms are of greater efficiency for the automatic classification of the attacks. The following can be adopted as classification algorithms: XGBoost, RFo, CNN, logistic regression, nearest neighbors, SVM, and others. The architecture in Figure 3.41a is structured to predict cyberattacks through the ability to identify characteristic patterns of normal data and network accesses from the anomalous and risky accesses. An AI software interface, behaving as a middleware control room, detects and classifies in real time the attacks in the cloud environment. After the identification of a possible risk, the interface enables the channel switching from a node to another one, assuring the communication security. The supernodes can reconfigure the communication to

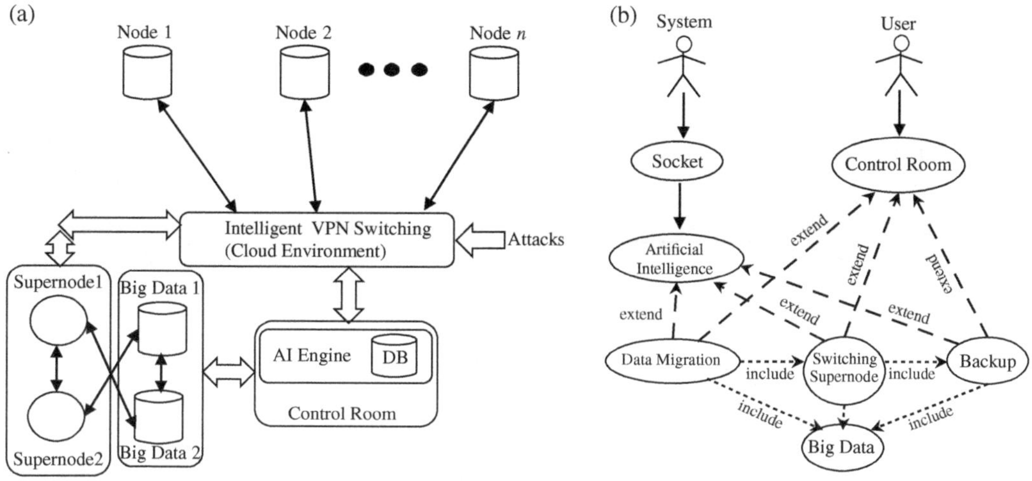

Figure 3.41 (a) Cross platform architecture oriented on cybersecurity, virtual connectivity, big data and artificial intelligence control. (b) Diagram explaining the main system use cases and actors.

another node thus virtualizing the communication system. All data concerning reconfiguration are stored into big data suitable also for disaster data recovery. The main actors of the network are the platform system and the user. Figure 3.41b shows a diagram including the main functions of the platform system:

- node switching by AI detection
- data migration
- data backup (disaster recovery function)

The applications that can be implemented by the network in Figure 3.41a are:

- Gaming network.
- Videocall platform managing teams.
- E-learning platforms for bidirectional communication.
- Augmented reality real-time communication systems.
- Other systems involving cloud communication and internet network.

Figure 3.42 shows all the functions involved in the simplified scheme in Figure 3.41a. The described flowchart traces the data flow occurring in the platform during a communication. In particular, the following main phases are distinguished:

- Traffic logs are pre-processed and clustered to be processed by the CNN algorithm recognizing attacks and managing the traffic.
- Traffic is managed by a "Traffic Manager" interface enabling traffic virtualization switching VPN channels.
- The virtual control room visualizes the switching operations, manages data migration and data storage, and permits the node configuration setting to be memorized.

The simultaneous use of more big data systems is useful for the management of complex networks based on multimedia platforms.

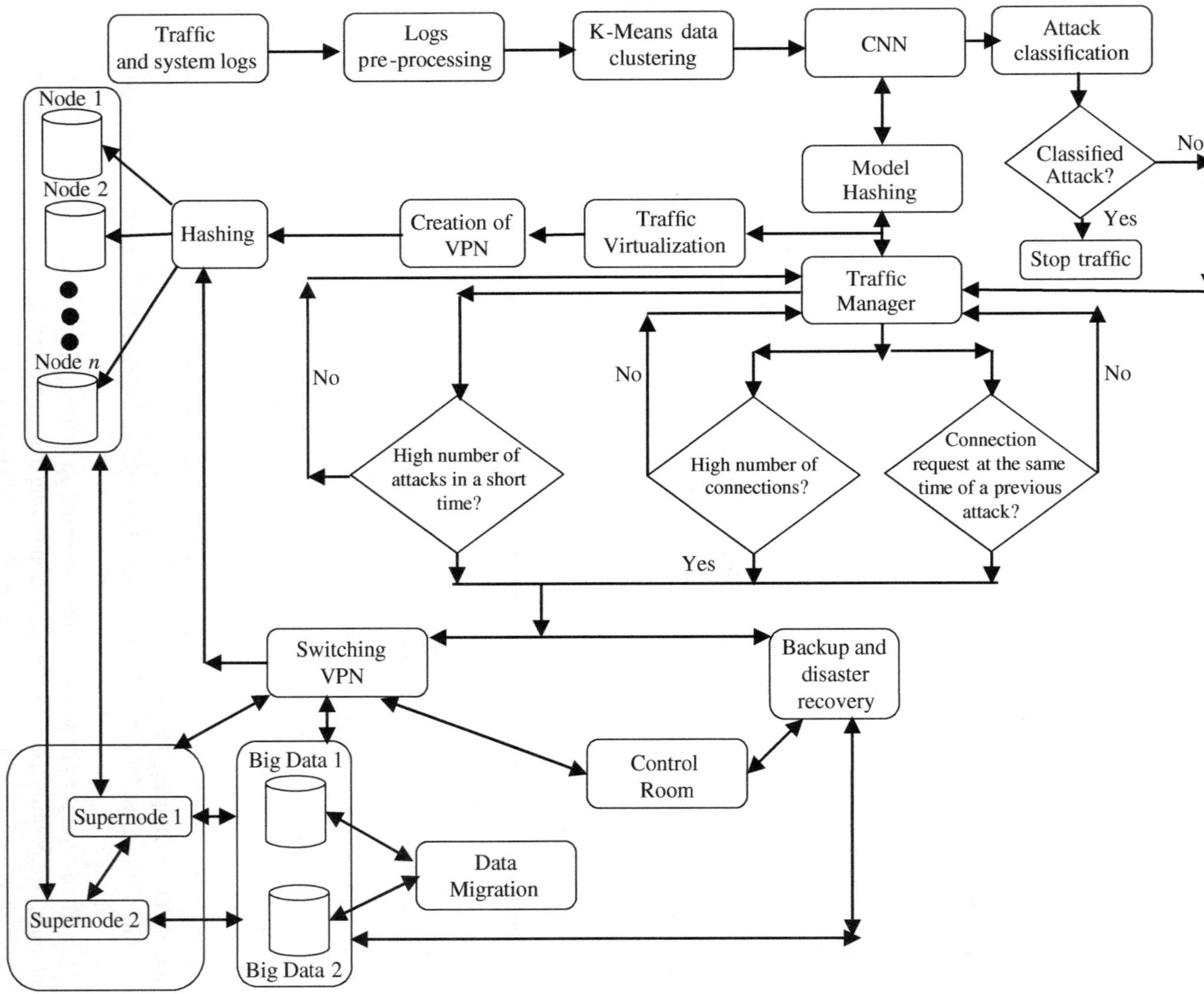

Figure 3.42 Platform functions involved in Figure 3.41a.

References

1 Schela, D., Henkela, C., Stocka, D. et al. (2018). Manufacturing service bus: an implementation. *Procedia CIRP* 67 (1): 179–184.

2 Vijaychandra, J., Sesha Sai, B., Sateesh Babu, B., and Jaganna, P. (2019). A comprehensive review on McCulloch-Pitts neuron model. *International Journal of Innovative Technology and Exploring Engineering (IJITEE)* 8 (6): 1516–1522.

3 Nikiel, P.P., Farnham, B., Filimonov, V., and Schlenker, S. (2015). Generic OPC UA server framework. *Journal of Physics: Conference Series* 664 (082039): 1–8.

4 Fraile, F., Sanchis, R., Poler, R., and Ortiz, A. (2019). Reference models for digital manufacturing platforms. *Applied Sciences* 9 (20): 1–25.

5 Barsoum, N.N. and Chin, P.R. (2011). Ethernet control AC motor via PLC using LabVIEW. *Intelligent Control and Automation* 2 (4): 330–339.

6 Godoy, A.J.C. and Pérez, I.J. (2018). Integration of sensor and actuator networks and the SCADA system to promote the migration of the legacy flexible manufacturing system towards the Industry 4.0 concept. *Journal of Sensor and Actuator Networks* 7 (23): 1–21.

7 Nagalingam, S.V. and Lin, G.C.I. (2008). CIM – still the solution for manufacturing industry. *Robotics and Computer-Integrated Manufacturing* 24 (3): 332–344.

8 Saadaoui, S., Tabaa, M., Monteiro, F. et al. (2019). Discrete wavelet packet transform-based industrial digital wireless communication systems. *Information* 10 (3): 1–17.

9 Oppitz, M. and Tomsu, P. (2018). *Inventing the Cloud Century: How Cloudiness Keeps Changing Our Life, Economy and Technology*, 1e. Transworld Research Network.

10 Jansen, D. and Buttner, H. (2004). Real-time Ethernet: the EtherCAT solution. *Computing and Control Engineering* 5 (1): 16–21.

11 Davies, S. (2007). Industrial ethernet – the fundamentals of ethernet/IP – EtherNet/IP has reached the million-node landmark, but what is making this protocol so attractive to industrial control engineers? *Computing & Control Engineering Journal* 18 (1): 42–45.

12 Ferrari, P., Flammini, A., and Vitturi, S. (2006). Performance analysis of PROFINET networks. *Computer Standards & Interfaces* 28 (4): 369–385.

13 Barrios-Avilés, J., Rosado-Muñoz, A., Iakymchuk, T., and García-Chulbi, G. (2017). POWERLINK and Ethernet/IP comparison as robust industrial Ethernet protocols. *IFAC-PapersOnLine* 50 (1): 363–368.

14 Hou, C.Y., Jiang, H.H., Yang, Y. et al. (2010). An improved synchronization mechanism of SERCOS III in strong EMI environment. *Applied Mechanics and Materials* 29–32: 246–251.

15 Zurawski, R. (2015). *Industrial Communication Technology Handbook*, 2e. CRC Press.

16 Bai, Q., Jin, B., Wang, D. et al. (2018). Compact Modbus TCP/IP protocol for data acquisition systems based on limited hardware resources. *Journal of Instrumentation* 13 (4): T04004.

17 Jaloudi, S. (2019). Communication protocols of an industrial Internet of Things environment: a comparative study. *Future Internet* 11 (66): 1–18.

18 Solairaju, A. and Periyasamy, R. (2011). Effect of linear threshold gate on the countenance of artificial neural networks. *International Journal of Open Problems in Mathematics and Applications* 1 (1): 51–55.

19 Singh, V.K. (2016). Proposing solution to XOR problem using minimum configuration MLP. *Procedia Computer Science* 85 (2016): 263–270.

20 Sylvester, E.I. and Adesola, W.A. (2015). Artificial neuron network implementation of Boolean logic gates by perceptron and threshold element as neuron output function. *International Journal of Science and Research (IJSR)* 4 (9): 637–641.

21 Riedmiller, M. and Braun, H. (1993). A direct adaptive method for faster backpropagation learning: the RPROP algorithm. In: *Proceedings of the IEEE International Conference on Neural Networks (ICNN)*, San Francisco, USA (28 March–1 April 1993). Piscataway, NJ: IEEE.

22 Massaro, A., Galiano, A., Savino, N. et al. (2020). Systems for an intelligent application of automated processes in industry: a case study from "PMI IoT Industry 4.0 Project". *Proceedings of IEEE MetroInd4.0&IoT 2020*, Rome, Italy (3–5 June 2020). Piscataway, NJ: IEEE.

23 Massaro, A., Maritati, V., Giannone, D. et al. (2019). LSTM DSS automatism and dataset optimization for diabetes prediction. *Applied Sciences* 9 (17): 1–22.

24 Saravanakumar, G., Baalachandar, C.V., and Mohaidden, A.K.K. (2019). Environmental condition monitoring system of textile industry for sustainable development goal. *International Journal of Innovative Technology and Exploring Engineering (IJITEE)* 9 (2S2): 986–991.

25 Massaro, A., Panarosa, G., Savino, N. et al. (2020). Advanced multimedia platform based on big data and artificial intelligence improving cybersecurity. *International Journal of Network Security & Its Applications (IJNSA)* 12 (3): 23–37.

26 Jeet, R.I. and Deshdeepak, S. (2012). Data mining based database intrusion detection system: a survey. *International Journal of Engineering Research and Applications (IJERA)* 2 (4): 1752–1755.

27 Pinzón, C.I., De Paz, J.F., Herrero, Á. et al. (2013). idMAS-SQL: intrusion detection based on MAS to detect and block SQL injection through data mining. *Information Sciences* 231 (1): 15–31.

28 Cheon, E.H., Huang, Z., and Lee, Y.-S. (2013). Preventing SQL injection attack based on machine learning. *International Journal of Advancements in Computing Technology (IJACT)* 5 (9): 967–974.

29 Mudarri, T. and Al-Rabeei, S.A. (2015). Security fundamentals: access control models. *International Journal of Interdisciplinarity in Theory and Practice* 1 (7): 259–262.

30 Muntean, M. and Dijmărescu, L. (2018). Sustainable implementation of access control. *Sustainability* 10 (6): 1–9.

31 Kashyap, K.I. (2013). Database security & access control models: a brief overview. *International Journal of Engineering Research & Technology (IJERT)* 2 (5): 743–751.

32 Debbech, S., Bon, P., and Collart-Dutilleul, S. (2019). Toward semantic interpretation of goal-oriented safety decisions based on foundational ontology. *Journal of Computers* 14 (4): 257–267.

33 Dholakia, Y. (2017). Mandatory access control – problems in it and propose a model which overcomes them. *International Research Journal of Engineering and Technology (IRJET)* 4 (4): 2031–2035.

34 Harrizon, M.A., Ruzzo, W.L., and Ullman, O. (1976). Protection in operating systems. *Communications of the ACM* 19 (8): 461–471.

35 Bu, S.-J. and Cho, S.B. (2017). A hybrid system of deep learning and learning classifier system for database intrusion detection. In: HAIS 2017: Hybrid Artificial Intelligent Systems, Lecture Notes in Computer Science, vol. 10334 (eds. F. Martínez de Pisón, R. Urraca, H. Quintián and E. Corchado), 615–625. Cham: Springer.

36 Mor, B., Garhwal, S., and Kumar, A. (2020). A systematic review of hidden Markov models and their applications. *Archives of Computational Methods in Engineering* https://doi.org/10.1007/s11831-020-09422-4.

37 Omari, O.H., Al-Kasasbeh, B.M., Al-Qutaish, R.E., and Muhairat, M.I. (2009). DEA-RTA: a dynamic encryption algorithm for the real-time applications. *International Journal of Computers* 3 (1): 191–199.

38 Marks, D.G. (1996). Inference in MLS database systems. *IEEE Transactions on Knowledge and Data Engineering* 8 (1): 46–55.

39 Chen, Y. and Chu, W.W. (2008). Protection of database security via collaborative inference detection. In: *Intelligence and Security Informatics. Studies in Computational Intelligence*, vol. 135 (eds. H. Chen and C.C. Yang). Berlin: Springer https://doi.org/10.1007/978-3-540-69209-6_15.

40 Massaro, A., Gargaro, M., Dipierro, G. et al. (2020). Prototype cross platform oriented on cybersecurity, virtual connectivity, big data and artificial intelligence control. *IEEE Access* 8 (1) https://doi.org/10.1109/ACCESS.2020.3034399.

4

Internet of Things Solutions in Industry

The design of Internet of Things (IoT) systems is a function of the industry activities. Sensors interact with the cloud for data computing accomplished by artificial intelligence (AI) engines supporting the full production automatisms which characterizes the evolution of Industry 4.0 to Industry 5.0. The automatism is managed by intelligent units, controllers, agents, or in general processors executing AI algorithms. All the technologies are designed to provide the data flow suitable for auto-adaptive machine tuning, thus improving production quality. After a description of basic sensor configurations usable in industrial applications for real-time control and actuation, different architectures integrating advanced technologies such as big data, augmented reality (AR), and blockchain are discussed. Starting with the basic principles of the main sensors used in industry, the mechanism allowing the feedback control improved by AI is introduced. Concerning examples of automatisms, intelligent motion can be implemented by means of sensors scanning a surface or an object, adapting the detection according to the object to be detected. The sensor control is characterized by different time delays due to the adopted technology and to the data processing in a "quasi" real-time detection and actuation system. The time delay is a limit in industrial robotic applications requiring a quick response according to the production velocity and represents an important factor in auto-adaptive applications. The role of AI engines is mainly to support the auto-adaptive processes managing the control and actuation by properly designed mechatronic interfaces implemented in production lines. The AI algorithms are executed within decision support systems (DSSs), preferably working in cloud environments. Particular attention is focused on energy monitoring by the DSS, where the neural networks are suitable for the energy consumption prediction. The real issue in Industry 5.0 is to increase the computational performances of the AI engines using graphics processing units (GPUs) or cloud computing approaches. Different computational basic examples are provided for GPUs, comparing performances with central processing units (CPUs). Another useful aspect for innovative production is the implementation of traceability technologies improving warehouse product allocation, quality check points and intra-logistic operations. The whole production, the company information system, and the automatized processes can be managed by the online configuration of the DSS. Different applications in engineering services are enabled by IoT and DSS, with the possibility to configure online machine parameters. Specifically, services improved by DSS for bridges, railway and road non-invasive diagnostics, machine predictive maintenance, platforms controlling machine processing, platform controlling quarry production, and telemedicine are discussed. This chapter also discusses some architectures of quasi real-time data processing in the cloud environment for industrial applications, including multi-visor AR connection, blockchain integrated in information systems, and dynamic production infrastructures

Electronics in Advanced Research Industries: Industry 4.0 to Industry 5.0 Advances, First Edition. Alessandro Massaro.

integrating AI and robotics. The goal of this chapter is to provide an overview of industrial applications using sensors transmitting in the cloud environment and linked in a local area network (LAN). The proposed examples help the reader to have a general overview of how IoT systems are integrated in advanced infrastructures improved by AI facilities.

4.1 Cloud Computing IoT

Cloud computing is an important issue in modern research. AI algorithms can be executed in the cloud environment by improving the production management. IoT agents support the resource management and enable data processing oriented towards production optimization.

4.1.1 IoT Agent

IoT interfaces applied in automatic processes are classified as IoT agents. An IoT agent is a tool component that manages groups of IoT devices and aggregates the data to transmit to the application layer. IoT agents are usually components of a broader IoT platform and are typically managed by IoT gateway devices placed on the network edge. The network edge is defined as virtual network services integrated into a modular infrastructure platform and optimized for real-time development and interconnection of industry network services, including control and automation. In this scenario, the IoT agent is part of the industry information system [1]. The system architecture in Figure 4.1 is representative of a typical data workflow involving an IoT agent, AI, and human–machine interfaces (HMIs). The IoT sensor network represents the interface with the production line through field IoT sensors. The use of IoT devices allows data real-time access from a monitored site, checking for example the measured energy, ambient temperature and humidity, temperature of the machinery and of products, any temperature of liquids or gases, number of products per unit of time, and other production parameters. This level forwards the real time information to the higher level. The gateway level represents the "embedded" hardware module capable of supporting data collection, compatible with the selected IoT and traditional industrial protocols working in cloud platforms. IoT data are typically processed by AI algorithms for predictive maintenance and for the extraction of knowledge on the state of production. The AI algorithms are possible to integrate with other company software such as enterprise resource planning (ERP), product lifecycle management (PLM), customer relationship management (CRM), etc. The protocols adopted in the information systems are wired protocols (Modbus, OPC) or wireless protocols (LoraWan, ZigBee, Bluetooth, WiFi). More applications are addressed on the use of ZigBee protocol. Zigbee defines the highest levels of the protocol stack, based on the 802.15.4 standard. As a main function of the gateway is the aggregation and orchestration of industrial data by making them available in an integrated system able to:

- Receive data from the set of IoT field sensors characterized by different communication protocols.
- Forward such data to a cloud layer for data storage and processing.

The sensors are suitable for the implementation of predictive maintenance and production optimization policies. The resource manager (RM) behaves as a collector of packets coming from the sensors deployed in the field. Communication takes typically place through IEEE 802.11 (WiFi/5 GHz) or ModBus standards, following the most common mechanisms for asynchronous communication, in particular, the packets typically are encapsulated for example according to the Message Queuing Telemetry Transport (MQTT) protocol. The IoT agent layer behaves as an intermediate level: the information retrieved from the field is translated by the messaging services, and is exchanged with the middleware representing the cloud interface. The data acquired from the field through the sensors are sent to the RM through representational state transfer (RESTful) endpoints. The RM server is able to communicate with the gateway through calls to web services, or through data exchange according to the mechanisms of the MQTT protocol. Moreover, the RM is able to pass the data to the storage systems of the cloud level for the application of AI algorithms. An example of architecture integrating AI cloud computing managed by the IoT agent is shown in Figure 4.1. In this system architecture the following different levels concerning data flow are distinguished:

- The field level constituted by the IoT field network including sensors and machine-to-machine (M2M) devices interfacing production machines.
- The gateway level enabling data transmission to the information network according to the adopted protocols.
- The "factory" level made by a RM named IoT agent managing cloud data transmission and data switching.
- The cloud level where data are collected into a Database (DB) providing data to the AI engine interfaced with HMI monitoring production activity.

In the architecture of Figure 4.2 different actors and roles are distinguished. Two main actors are the domain expert and the production manager. The domain expert is addressed on the data organization process and data storage, and the production manager manages all data flowing in different fields. Both actors are linked by the modeler as indicated in Figure 4.2, which is a software engine. The modeler creates a predictive model used to generate the roles to be applied in the streaming analytics phase, directly on the data received from the sensors located along the

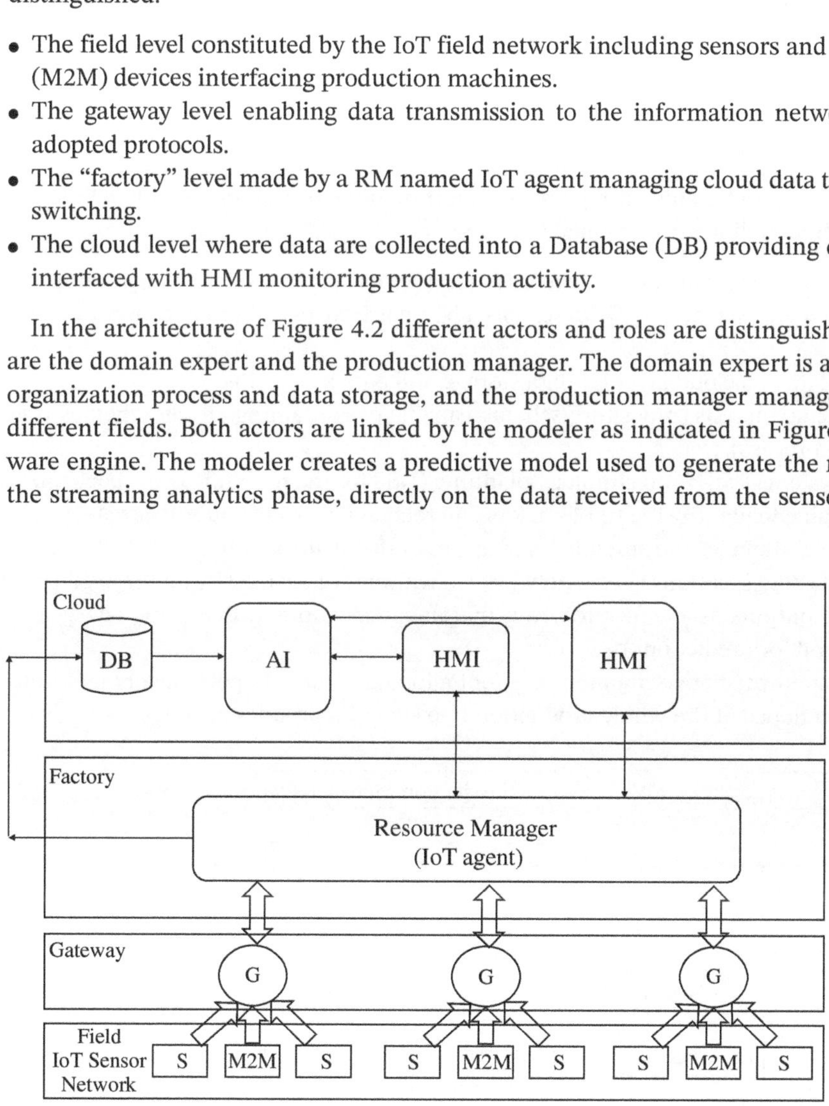

Figure 4.1 Architecture model of a company information system integrating IoT agent (S, sensor; M2M, machine-to-machine; G, gateway; DB, database; AI, artificial intelligence; HMI, human–machine interface).

production line. Typically, the modeler is based on Python code performing data processing for training, validation, and testing of the production line model. The modeler is used essentially for two different use cases:

- **Clustering:** the modeler is adopted to extract hyperparameters suitable for grouping a dataset detected by sensors; the data for processing are labeled or unlabeled.
- **Classification:** the modeler is used to extract roles (in the form of IF–THEN–ELSE roles) for the specific dataset constructed by sensors.

The modeler calls to itself the preparatory procedures for data processing, such as ones concerning data reading and pre-processing.

Figure 4.3 shows a diagram of an example of a production manager process related to the enabling of an actuation command: the process involves message broker, IoT agent and IoT device actions.

4.1.2 IoT Gateway in Smart Environments

Figure 4.4 shows a conceptual architecture of a IoT smart manufacturing framework, where it is possible to distinguish the following elements:

- IoT gateway: a hardware/software system capable of implementing the following operations:
  - Receive data from the set of IoT field sensors characterized by different communication protocols.
  - Forward sensor data to a cloud layer for data storage and data processing.
- DB connector for data analysis (repository scheme capable of containing all the schemes used for the data storing process).
- Decision support system (DSS) for technology planning (DSS for the monitoring of specific production processes, able to identify the right sensors and related hardware and software technologies typical of the IoT systems, and suitable for smart manufacturing features).
- Neural network data classification (neural network algorithms able to classify all the aggregated data and to provide outputs usable for predictive maintenance, anomaly detection, energy efficiency, and in general for predictions).
- Specific and structured predictive maintenance tool adopted to update periodically and automatically the maintenance of the whole production line avoiding production stops.

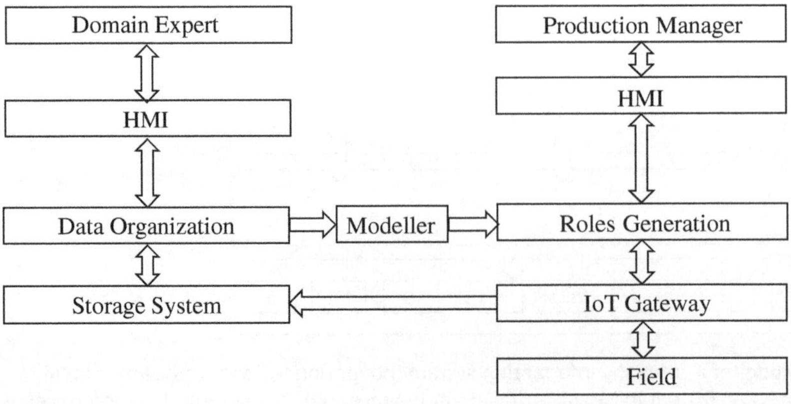

Figure 4.2 Domain expert and production manager roles and relationships in an IoT agent system.

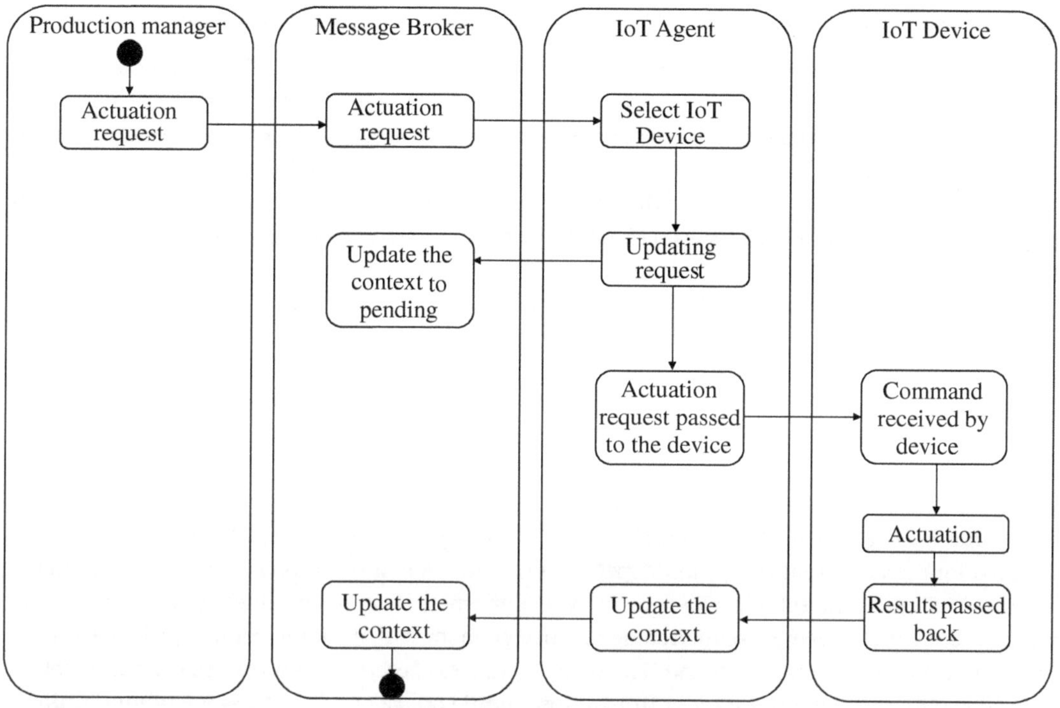

Figure 4.3 Relationship between production manager, message broker, IoT agent and IoT devices for an actuation command.

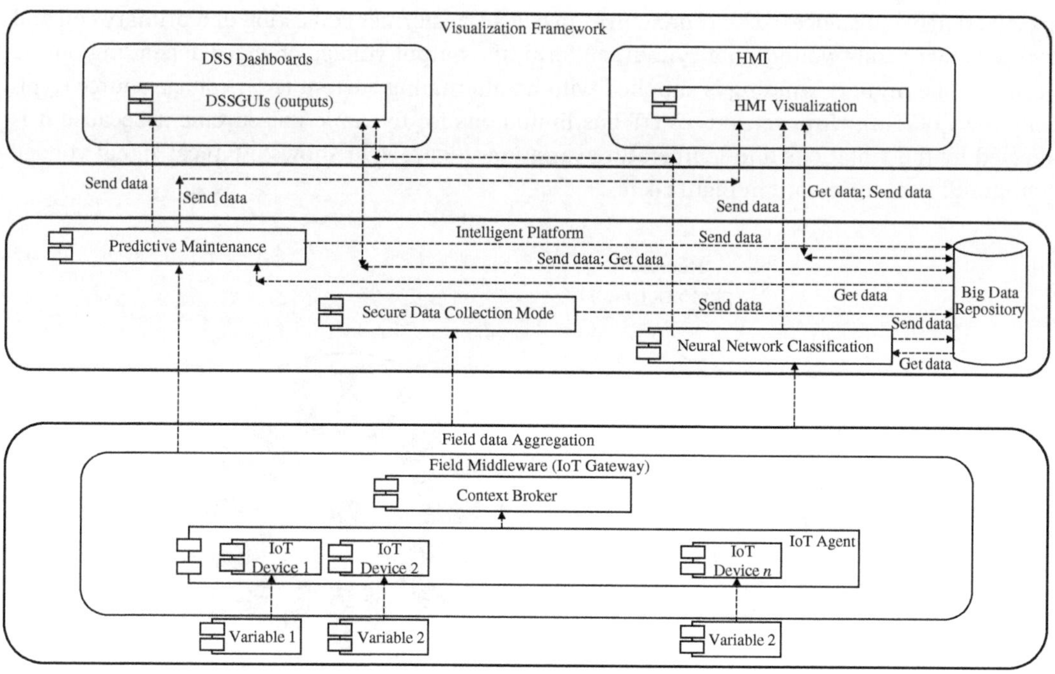

Figure 4.4 Architecture of a IoT smart manufacturing framework integrating an IoT agent.

4.1.3 Basic Elements of a Smart Industry Environment Controlling Production

Encoders are the IoT devices detecting the angle or position, generating pulse signals as the position output. Figure 4.5 shows a sketch of a rotary encoder detection system which is categorized into two forms: (i) the incremental encoder, which produces digital pulses as the measurement of relative position; and (ii) the absolute encoder where a unique digital word corresponds to the absolute position. Figure 4.5a shows the principle of the rotary encoders: a stationary light-emitting diode (LED) emits a light focused through the disk with windows permitting the light to pass through (Figure 4.5b); a phototransistor is placed on the other side of the disk detecting the LED light. The incremental encoders have two sets of windows in many cases, which are offset from each other. The optical disk of the absolute encoder is designed to produce a digital word that distinguishes distinct positions. The encoder signal is sent to a controller, which monitors motion parameters such as speed, rate, direction, revolutions per minute (RPM), acceleration, distance, inclination, or position. Encoders are suitable for precise position determination in industrial automation. While rotary encoders measure rotational movements, linear encoders check the position or the synchronization in linear movements.

Another basic element of the smart industry are potentiometers. Potentiometers are typically adopted for direct current (DC) motor speed control, direct control of power, control of inputs of electronic circuits, and robotic arm control. A potentiometer is an adjustable resistor (Figure 4.6), which is constructed using a semicircular resistive element with a sliding contact (called a wiper), and can be tuned by an AI command. The potentiometer is described as a voltage divider to obtain a variable output voltage at the slider from a fixed input voltage applied across the input terminals of the device.

The linear variable differential transformer (LVDT) is a device for measuring linear distance (a three-dimensional [3D] reconstruction is shown in Figure 4.7a). As illustrated in Figure 4.7b,c, the LVDT is a position sensor whose output is proportional to the position of a moving core over a specific range of distance. As the core moves inside a transformer consisting of a primary coil and two secondary coils wound on a cylindrical form, the output voltage changes depending on the position. The primary winding is supplied with an alternating current (AC) voltage source (typically several kHz). However, the LVDT has limitations for dynamic measurement because it is affected by the vibrations and temperature variation. Figure 4.7d shows a typical signal voltage configuration of the circuit in Figure 4.7c.

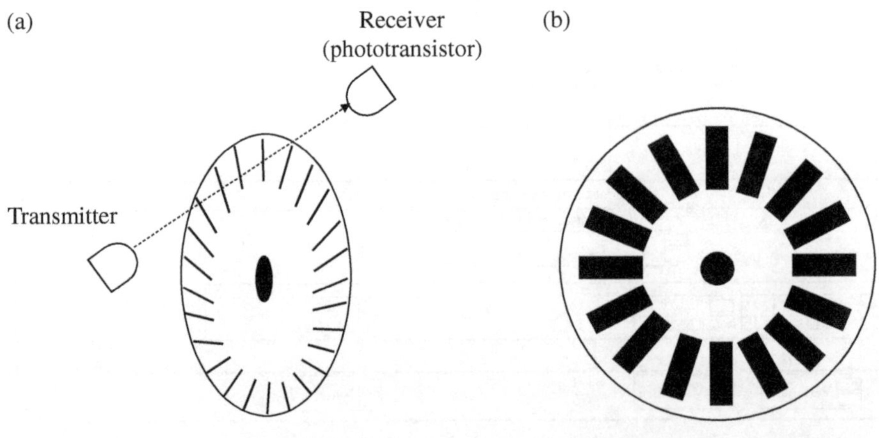

Figure 4.5 (a) Rotary encoder detection system. (b) Front view of a basic encoder layout.

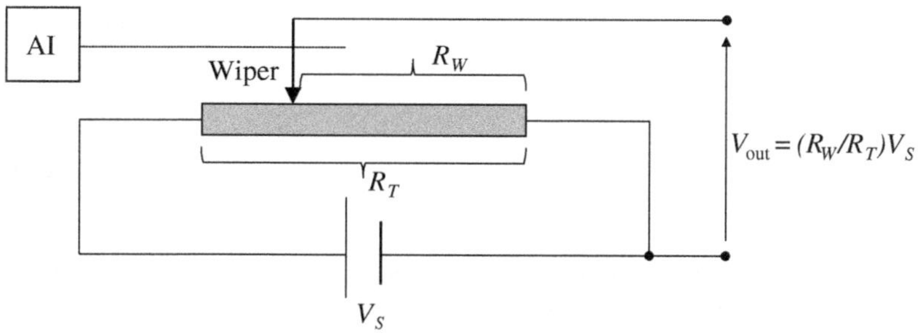

Figure 4.6 Basic principle of potentiometer and AI wiper control.

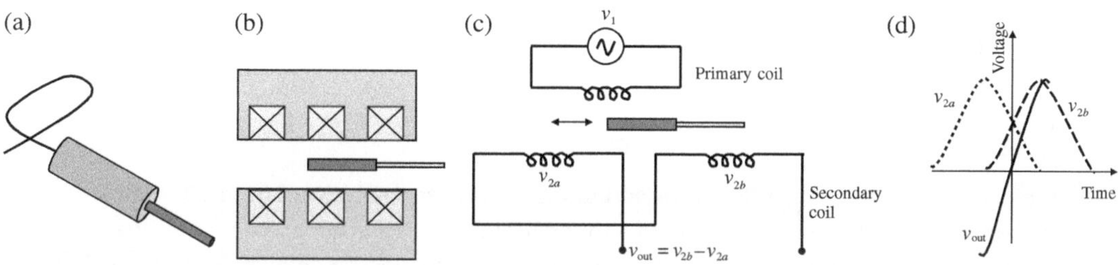

Figure 4.7 (a) LVDT 3D configuration. (b) Structure and basic principle of the LVDT. (c) Equivalent LVDT circuit and (d) signal characteristics.

Strain gauge is used to measure a small-scale strain on an object. In an elastic body subject to an applied force F, the resulting strain ε is proportional to the force, i.e. $\varepsilon = \Delta L/L \propto F$, where L is the original length, and ΔL is the increase in length. The applied force is estimated when the strain is known. There are several methods to measure the strains, and the LVDT represents a typical solution. A widely used sensor is the strain gauge in Figure 4.8.

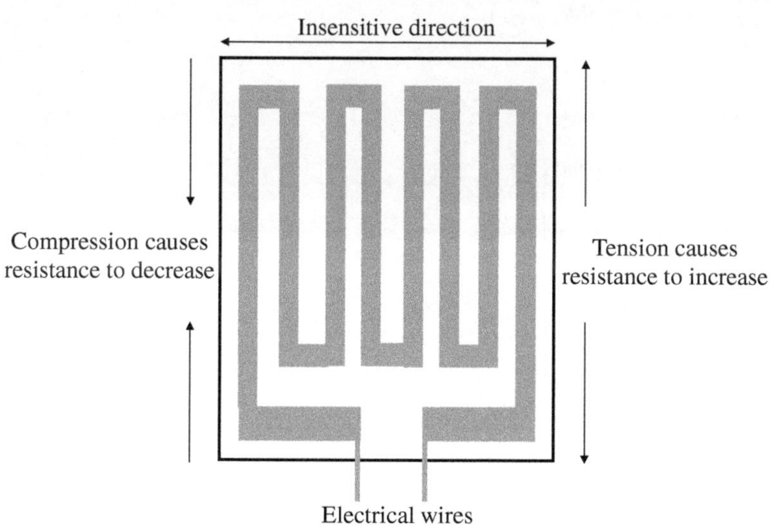

Figure 4.8 Metallic strain gauge.

The strain gauge is attached directly to a test sample subjected to an applied force: the strain changes the electrical resistance of the strain gauge R. The ratio of the change of the resistance ΔR to R is given by:

$$\frac{\Delta R}{R} = \frac{\Delta \rho}{\rho} + \frac{\Delta l}{l} - \frac{\Delta A}{A} \tag{4.1}$$

where A is the resistivity, l is the total length of metallic wire, and A is the wire cross section. In a metallic material, $\Delta\rho/\rho$ is negligible, and $\Delta A/A$ is expressed using the Poisson ratio σ, then the ratio is estimated as:

$$\frac{\Delta R}{R} \cong (1+2\sigma)\frac{\Delta l}{l} \tag{4.2}$$

The gauge factor or strain factor F_G is defined as:

$$F_G = \frac{\Delta R / R}{\Delta l / l} \cong 1 + 2\sigma \tag{4.3}$$

The typical value of the gauge factor for actual metallic strain gauges is around 2.

Other elements usually adopted in sensing applications in production processes are load cells. A load cell refers to a device that converts an applied force into electric signals. Figure 4.9 shows a load cell suitable for measuring forces with higher accuracy, in which the spring element has a structure called a Roberval mechanism. Strain gauges are located on the load cell to detect tensions and compressions.

Assuming the applied force is on the right side of the spring element in Figure 4.9 with the left side fixed, the strain gauge resistances R_1 and R_3 increase while R_2 and R_4 decrease. The resistance increases with the generated tension force, while the resistance decreases with the compression force. The changes in the electrical resistances are detected by the bridge circuit in Figure 4.10. The output voltage is given by:

$$V_o = \left(\frac{R_2}{R_1+R_2} - \frac{R_3}{R_3+R_4}\right)V_S = \frac{R_2R_4 - R_1R_3}{(R_1+R_2)(R_3+R_4)}V_S \tag{4.4}$$

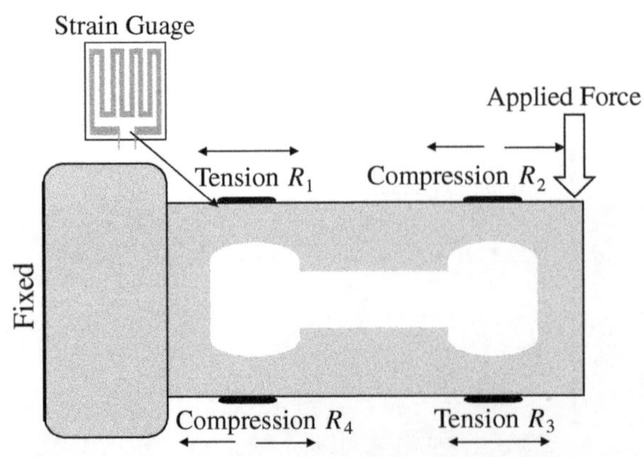

Figure 4.9 Load cell with a Roberval mechanism.

Assuming that $R_1 = R_3 = R + \Delta R$ and $R_2 = R_4 = R - \Delta R$ where ΔR is the change in resistance, the output voltage is defined as:

$$V_o = \frac{(R-\Delta R)^2 - (R+\Delta R)^2}{(R+\Delta R + R - \Delta R)} V_S = \frac{-\Delta R}{R} V_S = -F_G \varepsilon V_S \tag{4.5}$$

where ε is the strain.

Triangulation laser scanners are adopted to control defectivities in production lines. In triangulation laser scanners the emitting and receiving devices are separated by a fixed and calibrated distance. The position in space of each measured point is determined by a trigonometric calculation performing submillimeter accuracies and providing information on the three-dimensionality. Normally, detection stability is inversely proportional to the scanning speed. The laser sensor must be interfaced to a processor via the Ethernet line and Internet Protocol (IP), through touch screen panels or similar devices, on which the measurement software is executed. The laser detection system is composed of:

- A controller, which communicates with processors, programmable logic controllers (PLCs), encoders, sensors, triggers, and other components, through Ethernet, RS-232C, input/output (I/O) cables and other protocols.
- A display output unit, which communicates with the console, with RGB (red, green, and blue) monitor and HMI touch panel.
- A communication unit, which is linked typically via an Ethernet/IP or Profinet protocol with PLC and processor.

Figure 4.11 shows the basic scheme of a laser system detecting a target line region: the system is composed of a laser emitter and a lens focusing the reflected wave onto a photomatix reconstructing the scanned region.

Another sensor typically used for the monitoring of machine vibration is the accelerometer. The analyses in the time domain, and in the frequency domain (Fourier transform), provide important information about anomalous vibrations. A vibration analysis of a production machine in the (x,y) plane is described in Figure 4.12, where a few measurements over a threshold indicate possible

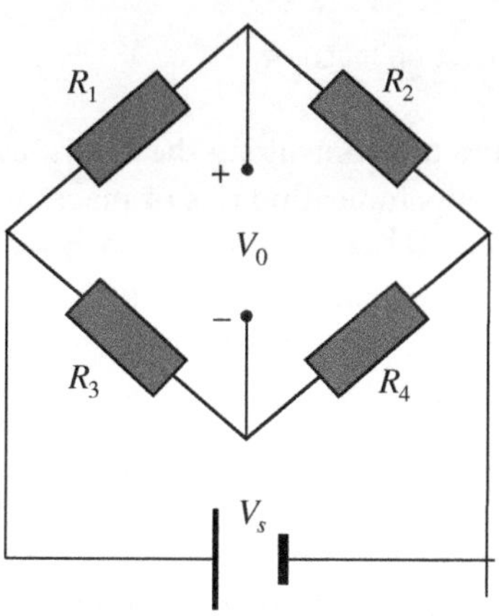

Figure 4.10 Wheatstone bridge for load cell.

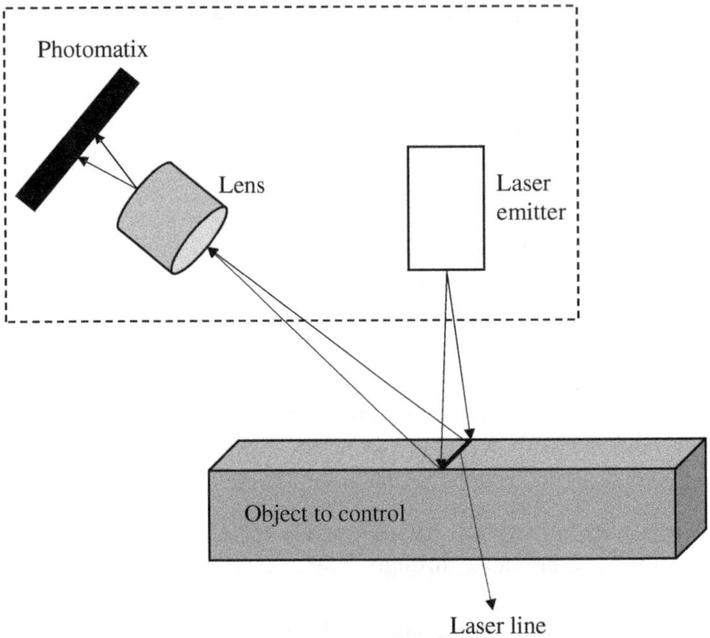

Figure 4.11 Basic scheme of a laser detector.

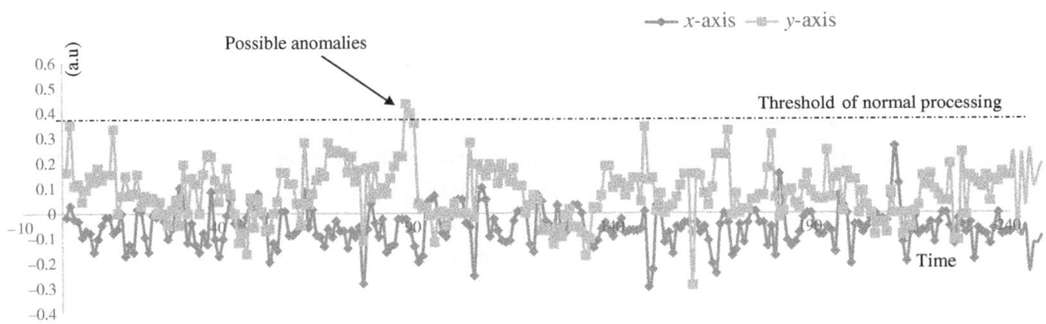

Figure 4.12 Measured accelerations in the (x,y) plane of a vibrating production machine.

anomalies. The stronger vibration is mainly registered for the same machine along the z-axis: as illustrated in Figure 4.13, different values are over the threshold limits indicating risk of machine rupture.

Furthermore, the fast Fourier transform (FFT) potentially provides more information about anomalies determined by specific frequencies. The prediction of possible ruptures can be estimated by the workflow in Figure 4.14 described by the following blocks:

- Input file reader loading accelerometer data (*Excel Reader*).
- *Table view* block visualizing data on a table loaded in a local repository.
- *Normalizer* block normalizing input data.
- *Partitioning* block splitting data input into the model learner (*RProp MLP Learner*) and into the *Multilayered Perceptron Predictor* block (neural network algorithm).
- *Numeric scorer* block estimating algorithm performance by estimating different indicators such as the mean absolute error (MAE), mean square error (MSE), and root mean square error (RMSE).

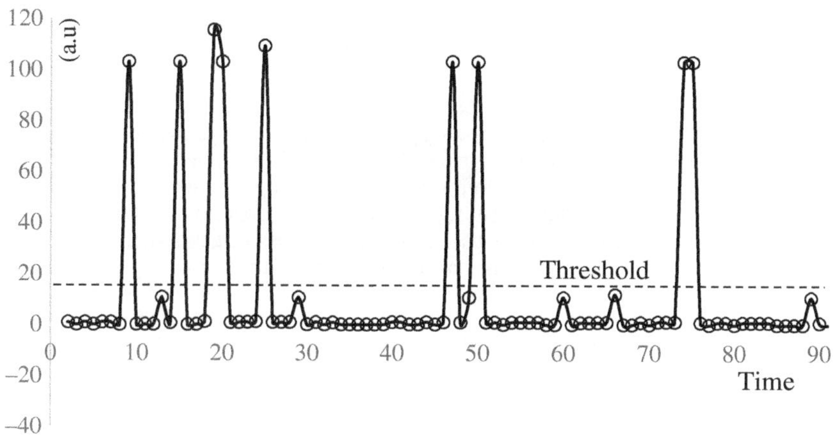

Figure 4.13 Measured accelerations along the *z*-axis of a vibrating production machine.

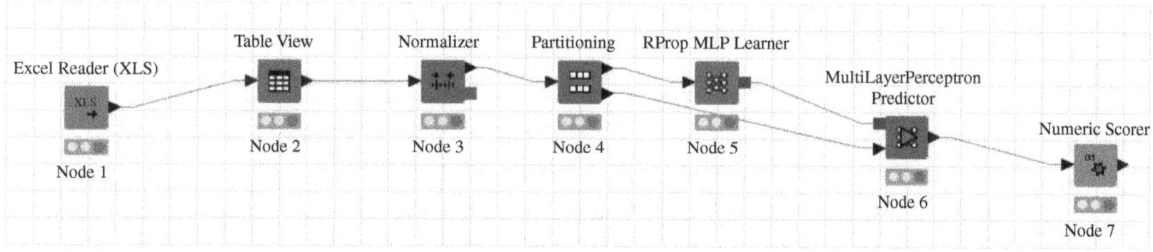

Figure 4.14 Konstanz Information Miner (KNIME) workflow predicting accelerations.

An example of application of accelerometers is illustrated in Figure 4.15, where n accelerometers are placed in critical parts of a production line, transmitting each signal to a receiver electronic board behaving as an interface of a neural network algorithm for cloud damage risk prediction. A graphical dashboard is suitable to represent the risk map as a layout of production line status. The risk map is useful for predictive maintenance.

Table 4.1 lists some applications of standard sensors in *in-line* monitoring processes.

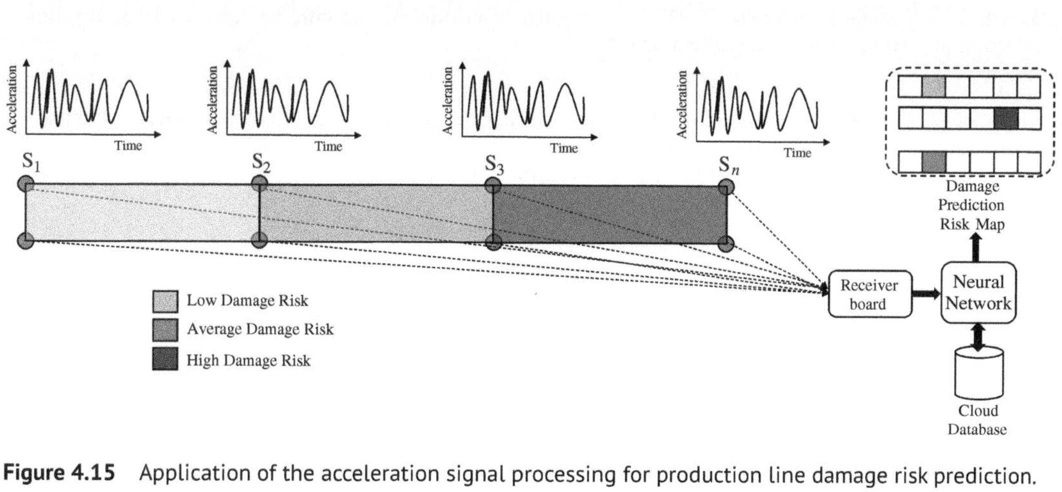

Figure 4.15 Application of the acceleration signal processing for production line damage risk prediction.

Table 4.1 Application of basic sensors in industry.

Basic sensing	Importance
Temperature	A basic parameter indicating the main processing parameters. In order to control the temperature of a product having large dimensions, it is preferable to use a thermal infrared camera. The temperature measurement is important for processing monitoring when tools are used contacting the product (blades, drilling processes, etc.), or for non-contact processing approaches (by lasers, plasma). The temperature monitoring is combined in advanced mechatronic systems for real-time corrective actions of the processing tools
Humidity	As for temperature, the humidity is an important parameter to control for specific production processes where it potentially damages the product (as for pasta production processes). Typically, food production rooms are controlled for temperature and humidity. For this purpose, an automatic controlling system provides a real-time alert of the production.
Weight	Load cells allocated in specific regions of the production line provide information on product weight supporting quality check. The *in-line* product weight monitoring is important to check possible failure of the machines, dosing products into packages. Industrial load cells are typically adopted for weight check
Surface defectivity	Laser scanner and image processing provide information about product defect on its surface. Image vision techniques are mainly suitable for surface defectivity detection. For micro-roughness defectivity, use of an optical microscope is required. The control and monitoring of the surface defectivity is important for the improvement of quality processes
Count	Laser sensors are usually adopted for the counting of products moving on a conveyor belt. The counting provides information about production volumes and production velocity
Traceability	Tag detection systems are used to trace all the production processes. In particular, tagging by RFID and QR codes are adopted to collect information about all the production phases by tracing quality. The traceability can be improved by a blockchain network
Vibration	Accelerometers detect vibrations on different parts of the production machines predicting possible breakages or failures. The vibration analysis can be supported by a thermal one, because a strong vibration generates an accented local heating. Accelerometers are also adopted to predict breakages of production machines or of movement parts. The artificial intelligence engines are useful to predict machine malfunctions by the classification of anomalous machine behaviors. The prediction is performed by analysis in the time domain and in the frequency domain
Pollution detection in products	Metal detectors are applied to recognize impurities in a specific product (useful application mainly for food production)
Product alignment	Optical sensors are useful in order to check the correct alignment of a product during its processing on the production machine. This detection is important in manufacturing industries for example for the cutting of a wood piece, checking the alignment of the piece to cut.
Defect recognition by image processing	Smart cameras are applied to locally check the product defects by means of image processing. The image processing includes image pre-filtering, image segmentation, and defect classification. Artificial intelligence supports automatic defect identification. Defects are detected also for packages
Image processing for warehouse patterns recognition	Image tracking of movements inside the warehouse layout is appropriate for intra-logistic design and optimization: the detected patterns and the estimated times for operation processes provide important information about optimized allocations of products inside the warehouse

4.1.3.1 Feedback Control: Basic Concepts

Sensors are implemented in industrial processes mainly for the control of production machines or robots. Specifically, feedback control is a fundamental concept for actuation processes. A basic scheme of feedback control is provided by a single-input and single-output system described by:

$$\begin{aligned}\dot{x} &= Ax + bu \\ y &= cx\end{aligned} \tag{4.6}$$

where the factors (A,b) are controllable, and the pair (A,c) observable. Figure 4.16a illustrates the concept of the feedback control, which generates the control input u by the system state vector x multiplied by the feedback gain vector k. This system establishes a regulator system so that the state variable is always moved back to the origin wherever the current state is located. The AI controller processes different data inputs tuning the gain value, thus optimizing the trajectory as shown in Figure 4.16b. The trajectory diverges for systems without the control.

The trajectory adjusted by AI allows to achieve the same results of the feedback control with a better precision and converging in minor time, if compared with the feedback control time without AI correction. A typical feedback control system is shown in Figure 4.17 representing a loop shaping in the frequency domain. Performance of feedback systems is characterized by the frequency characteristic of the open loop transfer function named the loop gain $G(s) = P(s)C(s)$. The closed transfer loop function $T(s)$, the transfer function $G_d(s)$ describing the ratio between $W(s)$ and $Y(s)$, and the transfer function $G_n(s)$ indicating the ratio between $N(s)$ and $Y(s)$ are expressed by:

$$\begin{aligned}T(s) &= \frac{Y(s)}{R(s)} = \frac{P(s)C(s)}{1+P(s)C(s)} = \frac{G(s)}{1+G(s)} \\ G_d(s) &= \frac{Y(s)}{W(s)} = \frac{P(s)}{1+G(s)} \\ G_n(s) &= \frac{Y(s)}{N(s)} = \frac{G(s)}{1+G(s)}\end{aligned} \tag{4.7}$$

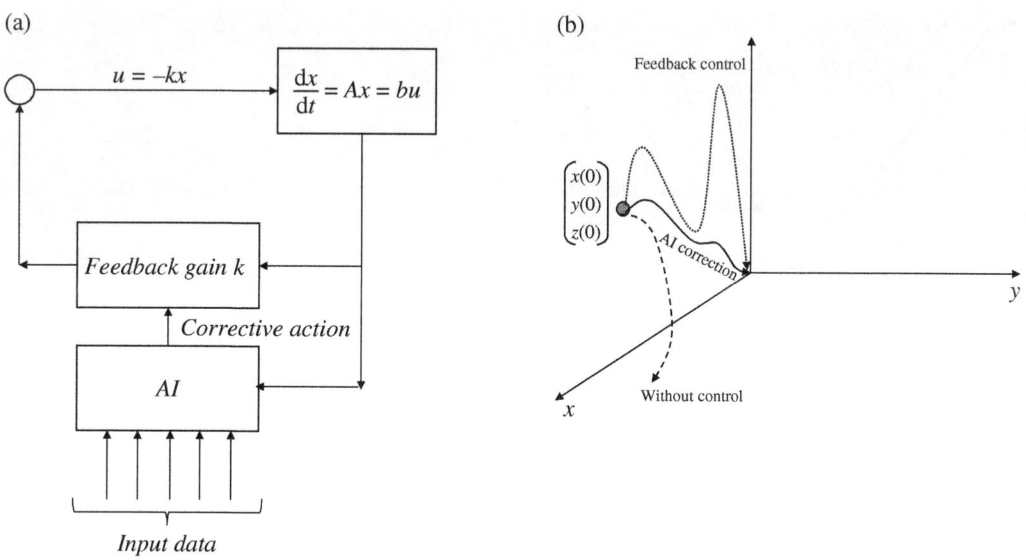

Figure 4.16 (a) Feedback control and AI corrective action. (b) Trajectory of point with and without control action. The vector $(x(0),y(0),z(0))$ represents the initial state.

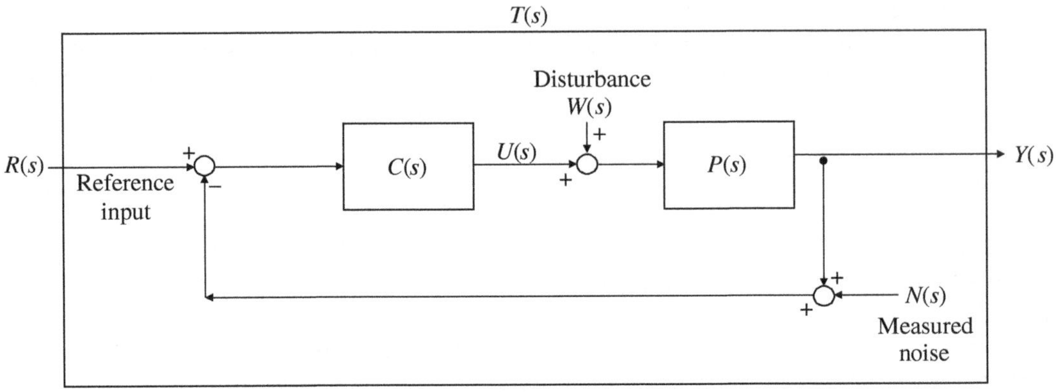

Figure 4.17 Basic feedback control system.

The reference signal and the disturbance one usually appear in the lower frequency range, while the measurement noise is at higher frequencies. The Bode sensitivity is the influence of the change of $P(s)$ on the closed loop transfer function, and is defined by:

$$S(s) = \frac{\Delta T / T_v}{\Delta P / P_v} \tag{4.8}$$

where $P_v = P + \Delta P$ and $T_v = T + \Delta T$. By using Eq. (4.7) the Bode sensitivity becomes:

$$T_v = T + \Delta T = \frac{(P + \Delta P)C}{1 + (P + \Delta P)C} \tag{4.9}$$

Figure 4.18 shows an example of desirable gain plot versus the frequency range.

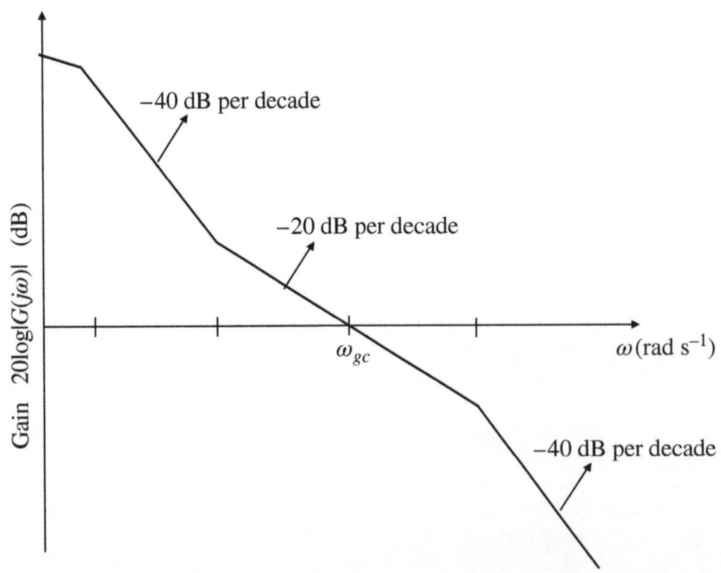

Figure 4.18 Desirable gain characteristic (ω_{gc}, desirable gain crossover frequency).

4.1.4 Augmented Reality Hardware and Cloud Computing Processing

AR multiple systems, working in the cloud environment, are emerging research topics. The aim of the research is to overcome AR limits upgrading at the same time the technology. In this context, a fundamental requirement of an innovative solution is the possibility to use in real time all the data collected in different DBs by AR systems. To meet this need, the solution of Figure 4.16 intends to use a virtual and AR device, wearable without cables, which allows to enrich the real field of view with the display of digital contents, acquired from heterogeneous DB sources containing production monitoring data. Access, viewing and manipulation of digital data is possible through voice interfacing and head movement, thus leaving the operator's hands free. Figure 4.19 summarizes the logical architecture of the entire multivisor AR system, where all the visors are connected simultaneously to an integration layer. The system is characterized by the following three logic levels:

- **Integration layer:** it is the middleware that allows the interoperability of data processing with heterogeneous data sources belonging to the company information system; the data processing activities of the various contents is achieved in this level; the integration layer is installed once on a dedicated server.
- **Application programming interface** (API) **layer:** it contains the API and web socket modules that, respectively, allow the management of requests for access to external data, and the management of dynamic content such as machine parameters continuously monitored during the production processes; the API layer is installed on a dedicated server.
- **Client:** it is the mixed reality device (visor) used by the worker; the client application is installed on this device once, and allows access and use of all available contents.

The real-time use of content on the viewer is possible due to careful management of computational resources on both the server and client sides. Simple calculation operations are accomplished directly on the visor hardware unit to avoid network-related latency. The complex

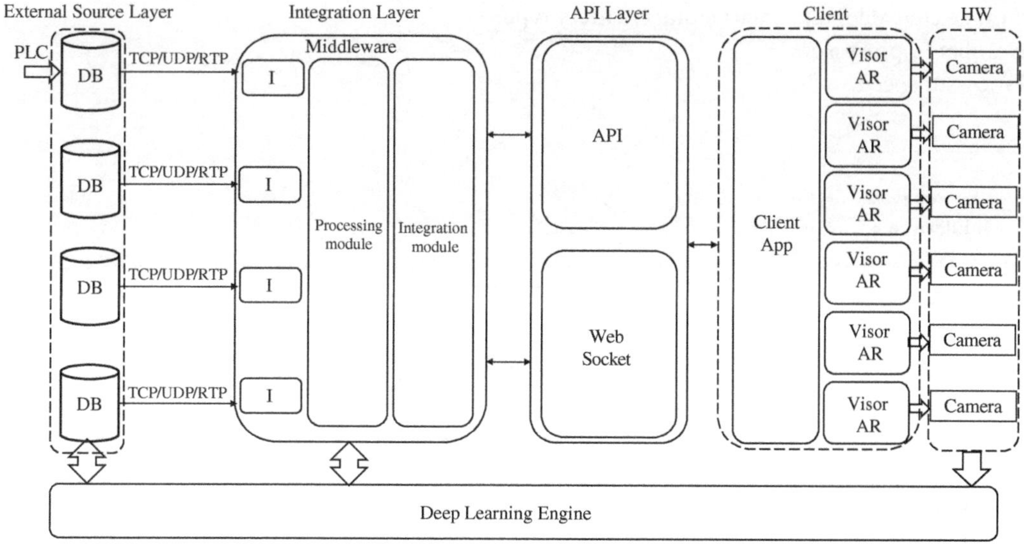

Figure 4.19 Architecture of a multivisor AR architecture.

calculation operations, including deep learning (DL) data processing, are performed on the server to avoid saturation of the client's resources. On the server side, the computational performances are further optimized due to the design and the implementation of a scalable architecture distributing the workload in parallel across multiple server machines. GPUs can be used for local data processing, by improving the computational cost.

The multivisor AR system in Figure 4.16 is suitable for various frameworks involving multi-users. Some working scenarios are:

- Multiple workers operating in the same site (the coordination between workers is improved by the simultaneous use of more visors).
- Multiple workers operating in different sites (the work in different sites must be synchronized, or the work performed at a site demonstrates the work to be executed in the other sites).
- Multi-user training (the interaction between the members of a team potentially is improved by means of the AR interactivity).

Furthermore, the data that are visualized and imported on the visors can be of external sources or DB collecting production machine data, transferred by PLC systems. The physical architectural scheme in Figure 4.20 describes the workload distribution considering a parallel data flow performed on multiple servers, where the Load Balancer element manages the balancing of the workload on the various nodes of the server cluster, and each server has high performance characteristics suitable for the specific application.

The specifications for a server machine are:

- Core number of the processor core.
- Processor cache capacity.
- Memory.
- Network controller type.
- System fan features.
- Power type.
- Expansion slot number.
- Storage controller and smart storage battery type.
- Number of processors.
- Processor speed.

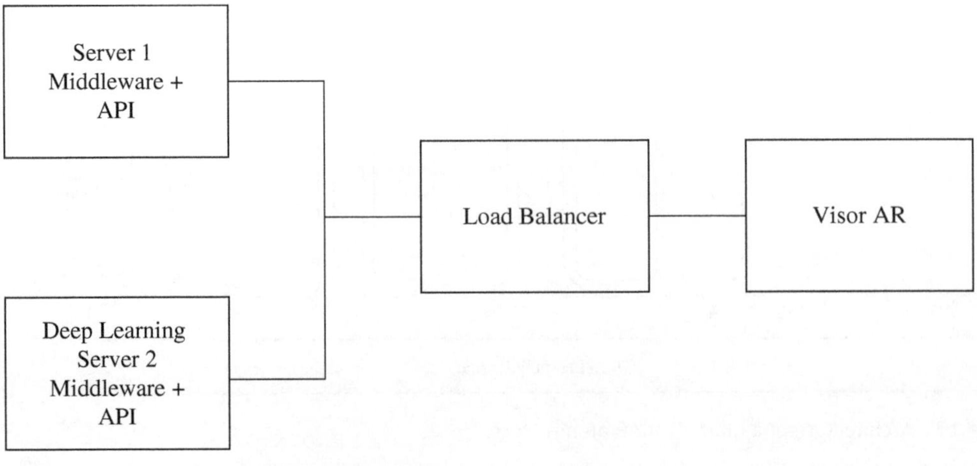

Figure 4.20 Load balancing of the workload by parallel data flow involving two servers.

The protocols used for the DB importing data for processing are:

- **Transmission Control Protocol** (TCP) is a packet network protocol of the transport level, and within the implementation of the ISO-OSI protocol stack, it is based on the IP. Consequently, it is known as TCP/IP. TCP provides a reliable, orderly and error-controlled framework for delivery of data between applications running on host communicating systems across an IP network. Major internet applications such as the World Wide Web (WWW), emails, remote administration, and file transfer make use of TCP. TCP enables a service for communication at an intermediate level, between the application and the IP, permitting host-to-host connections at the transport layer. At the transport layer, the protocol handles all the details, presenting an abstraction of the connection to the application. At the lower levels of the stack, due to network congestion and other unpredictable behavior, some IP packets can be lost, duplicated, or delivered out of order. TCP detects these problems, requires retransmission and helps to minimize problems related to network congestion. TCP is optimized to ensure accurate delivery rather than delivery sometimes involving long delays while fixing errors due to transmission. This makes it unsuitable for real-time applications, and for this type of application other protocols are better suited, such as the Real-time Transport Protocol (RTP) which uses the User Datagram Protocol (UDP).
- **RTP** is a network protocol used for broadcasting audio and video over IP networks. RTP is used extensively in communication and entertainment systems that involve streaming, such as telephony, video conferencing applications, television services, etc. RTP is used in conjunction with the RTP Control Protocol (RTCP). RTP takes care of the streamed data, while RTCP is used to monitor the statistics of the transmission and the quality of service (QoS). RTP is designed to allow real-time end-to-end streaming. The protocol provides compensation for out of sequence data reception, a common phenomenon in transmission IP networks. RTP is a standard for transporting audio and video data across IP networks. Real-time streaming applications require fast delivery information, and, for this purpose, packet losses are accepted. Although TCP is standardized to be used for RTP, it is not usually used in RTP applications, because it promotes reliability at a certain speed. Most RTP implementations are built on UDP.
- **UDP** makes use of a simple transmission without connection. It does not provide particular mechanisms of transmission between hosts, and therefore exposes the final application to many problems due to unreliability of the network. UDP provides only a mechanism to ensure data integrity. Using UDP, applications send messages (datagrams) to other hosts through IP networks. UDP is suitable for purposes that do not require error checking. Time-sensitive applications often use UDP, because, in these cases, losing datagrams is preferable to waiting for retransmitted packets.

DL algorithms can be applied in order to check the skills of the workers: by analyzing the AR experience of each user, the DL engine defines the worker behavior and class appertaining to a specific skill level. Some parameters to process for the formulation of human resources skills are difficulty of the work, time to conclude the work, operation steps, and movements. The same architecture is implemented for users with smart glasses, properly equipped by camera, working in image recognition: in this case the DL engine is adopted for object classification and recognition. By considering this last case, the DL training model is constituted by features of objects to detect, or of obstacles, or of other objects indicating a risk or a dangerous condition for the worker.

4.1.5 Real-Time Control and Actuation

Real-time control and actuation is an interesting topic in the research industry. Concerning sensing and actuation, it is correct to speak of quasi real-time data processing for the different time delays occurring in a complete process of sensing and actuation. The model in Figure 4.21

indicates the time delays expressed in Eq. (4.10), where ΔT_{se} is the sensor detection time, ΔT_d is the computational time of sensor data pre-processing, ΔT_n is the latency due to the data transmission, ΔT_a is the AI data processing time, and ΔT_{sy} is the synchronization time necessary to synchronize sensors and actuators.

$$\text{Time delay} = \Delta T_{se} + \Delta T_d + \Delta T_n + \Delta T_a + \Delta T_{sy} \tag{4.10}$$

A term which contributes to a significant delay is ΔT_a related to the AI data processing. The intelligent actuation is activated when the AI controller provides corrective actions. For a fast response, it is necessary to use high-performance hardware technologies oriented towards a decrease in the time ΔT_a. Many auto-adaptive applications require fast data processing to refine the sensing pattern during the measurement and consequently a low value of ΔT_a. In the era of quantum computing, the time delays associated with data processing will be negligible.

Concerning optical detection of shapes and colors by means of the optical probe in Figure 4.22a [2], AI data processing can optimize the detection pattern as for the layout in Figure 4.22b, described by the following steps:

- The detection patter layout follows an initial setting.
- When the initial part of an object is detected, the AI feedback adjusts the pattern gradually.
- When the detection is terminated, the recognition pattern follows the initial setting to increase the detection velocity, especially when more targets must be detected on a large surface.

The AI delay can increase when the pattern recognition is refined or changed, and the sampling is fitted. The sampling is a function of the sensor resolution, of the dimension of the object to detect, and of the admissible tolerance.

The basic network of an intelligent real-time sensing and actuation system is represented in Figure 4.23, where field sensors transmit data in the cloud where a DSS integrating AI algorithms processes data and decides to enable electro valves. The architecture explaining this application is illustrated in Figure 4.24 where:

- Field sensors transmit data ZigBee protocol.
- A coordinator electronic board receives all the signal transmitted by the field sensors.
- Data are transmitted by the coordinator to the external servers adopting the General Packet Radio Service (GPRS) protocol for communication with a cloud server.
- Data are digitally processed by the DSS/AI engine and stored in a DB system.
- The output of the DSS/AI engine is transmitted as an actuation command to the coordinator behaving as an I/O interface by adopting the GPRS protocol.
- The electro valve receives actuation commands by again using the ZigBee protocol.

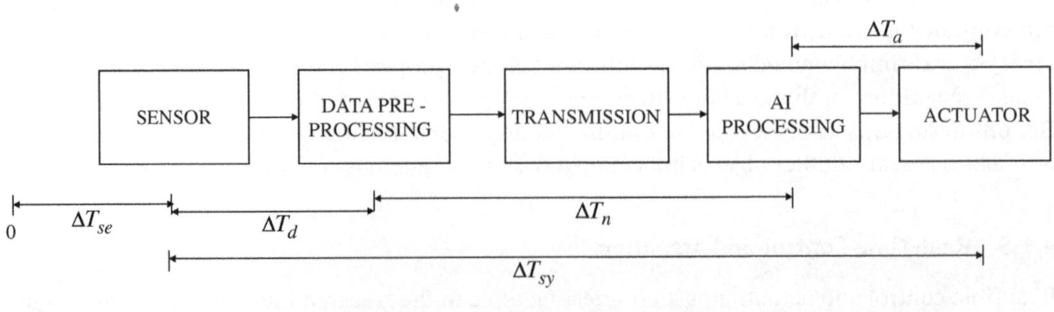

Figure 4.21 Model representing quasi real-time data processing involving sensing and actuation.

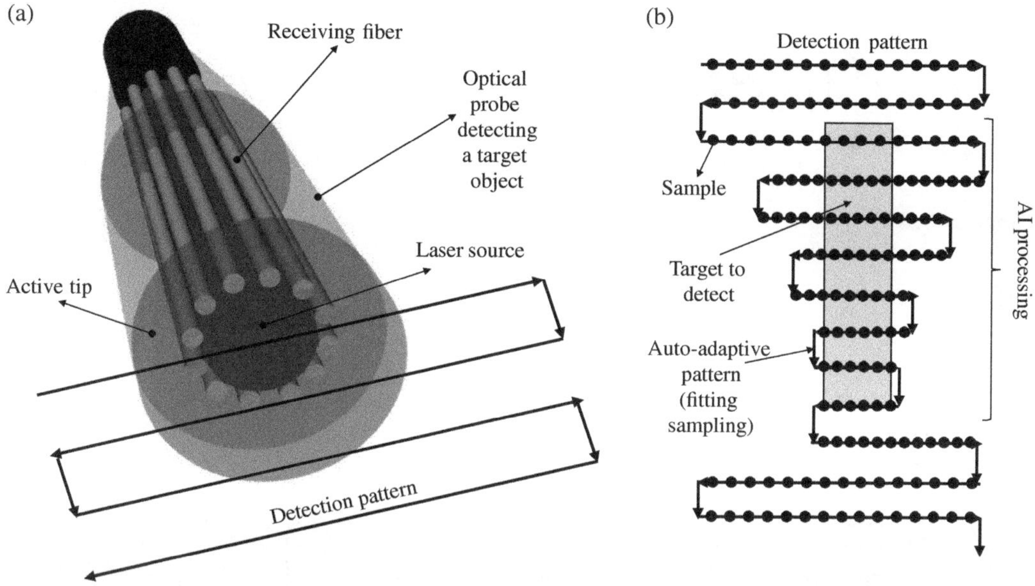

Figure 4.22 Automatism in sensor detection: (a) nanocomposite optical probe detecting object profile by moving the tip following a detection pattern. (b) AI auto-adaptive detection pattern improving acquisition time and sampling.

The DSS system in Figure 4.23 is suitable in agriculture for the fertigation process: the matching between meteorological parameters and predicted ones facilitates the decision to activate the irrigation network according to the analysis of other parameters such as terrain humidity, evapotranspiration (ET), multispectral plant signals, and presence of chemical elements such as nitrogen (N_2). The DSS is able to detect hydric stresses by activating the irrigation by electro valves, and determines conditions to operate with other fertilizations. In advanced systems, the DSS operates for "agriculture of precision" by mapping through a geolocalized matrix all the health status regions, and acting on the hydric contribution for the single plant. Sensors and unmanned aerial vehicles (UAVs) contribute to improve the DSS system by increasing data collecting. The systems in Figure 4.23 and Figure 4.24 are standalone if a GPRS transmission and a photovoltaic energy source are considered. The DSS, working in the cloud, is able to activate and deactivate a complex hydric network by automatisms provided by logic functions and prediction results through a supervised approach. The DB system can be big data collecting massive data to process, thus optimizing the fertigation process: data of multiple data sources are collected in the big data and preprocessed by means a data fusion (DF) approach, indexing all data into a unique DB system, where raw data and non-structured data are transformed into useful correlated information, useful for the analysis of growth factors, and for the formulation of relationships between plants and environments.

The scheme in Figure 4.24 is suitable for precision agriculture, where the Coordinator node manages all the geolocalized nodes displaced in the field to monitor. Each node constitutes an element of the spatial matrix representing a region or a plant. Each node transmits to the Coordinator by ZigBee protocol, and typically is composed of a microcontroller managing field sensors and electro valves.

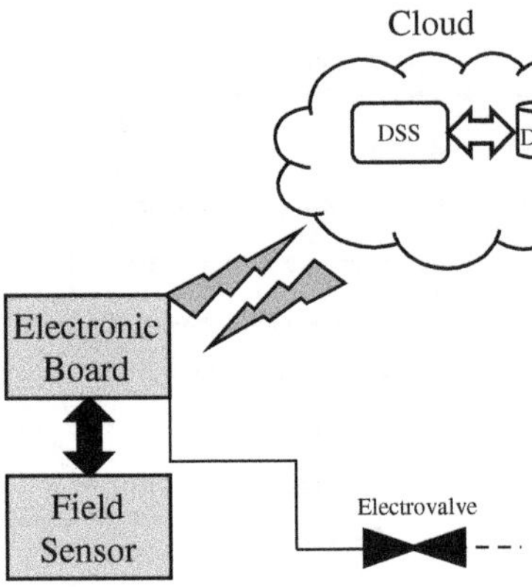

Figure 4.23 Quasi real-time data processing of a sensing/actuation process managed by a DSS.

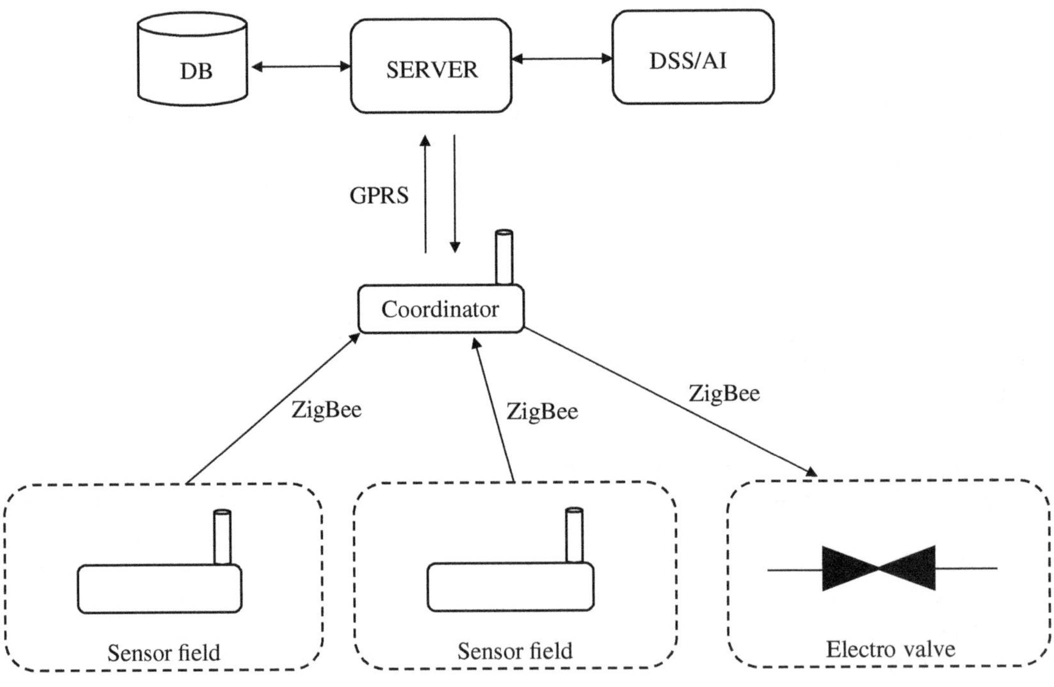

Figure 4.24 Architecture of quasi real-time data processing involving cloud DSS/AI data processing.

The UML (Unified Modeling Language) sequence diagram in Figure 4.25 describes the automation process involving sensing and actuation, where the actuation command is enabled by the DSS. The data flow starts sending data from the sensor field constituted by two sensors (sensor 1 and sensor 2) to the Coordinator board. The Coordinator sends the measurements to the server providing all data to the DSS algorithm for data processing by means of a DB system. The results of the DSS algorithm enables the electro valve actuation by specific commands.

The sequence diagram in Figure 4.25 follows the scheme in Figure 4.21, representing a case of quasi real-time control and actuation.

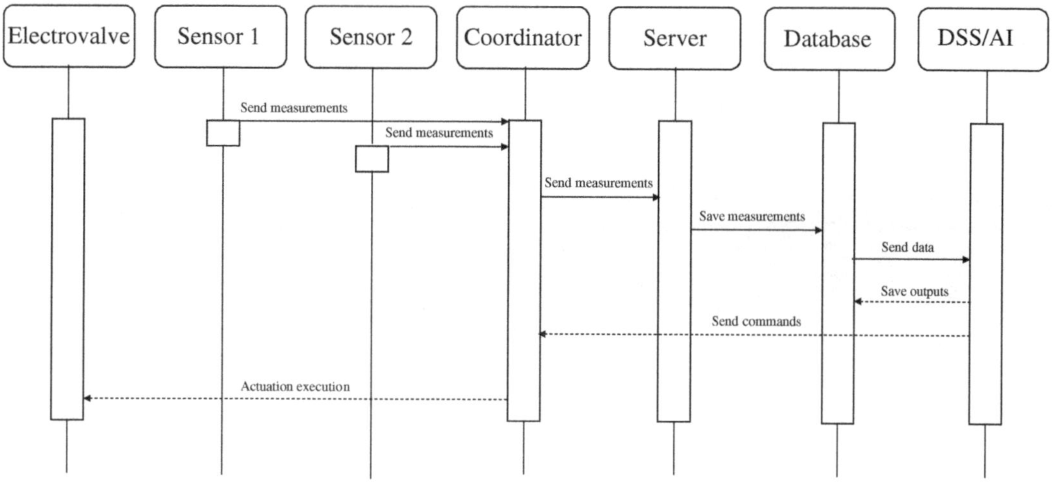

Figure 4.25 UML sequence diagram describing the sensing and actuation processes in Figure 4.24.

4.1.6 Localization Technologies in an Industrial Environment

Production checks can be facilitated with the use of drones equipped with optical or radiofrequency tracking and sensing. This function is important mainly for the monitoring of production which requires an accurate control in a particular dangerous industrial environment. The traceability and the monitoring of products can be accomplished by an UAV or remotely operated video enhanced receiver (ROVER) type drones, by automating the inspection activity to program. In addition, the drone unit can be equipped not only with an RFID/barcode/QR tag reader for product traceability, but also with a data detector interconnecting with sensors allocated in particular stages of the production line, useful for the risk evaluation of production quality. Other technological equipment involves cameras and infrared cameras. The image processing possibly provides further information about the industrial environment temperature and humidity, estimation of production volumes, pollution, dangerous gases, and possible product defects. Indoor problems related to drone movement mainly derive from obstacles [3, 4] such as protruding elements. As for indoor positioning, some studies have evaluated the possibility of using ultrasound and radio frequency (RF) technologies in fixed stations [4, 5] to improve UAV position accuracy. The drones are suitable to check production over short times, by moving in locations where workers are not present, or in fully automated production systems as depicted in Figure 4.26a, by collecting more data and information, and updating the ERP. The drone is able to automatically read, at a specific stage, different information by transmitting all data to a DSS checking possible production anomalies, or executing image processing algorithms suitable to classify defects or dangerous working conditions. The signal transmission layout of Figure 4.26b indicates that RF is used for UAV control and simultaneously for UAV positioning. The traceability technology, including camera detection systems, can be also applied to analyze worker movements in the warehouse. In this case the DSS, by analyzing all data, is able to optimize the production patterns and the intra-logistic layout according to the security policies and procedures. As a basic application, a microcontroller board, installed on the UAV, can acquire the temperature and humidity of the local environment. Checking the industrial environment conditions ensures the integrity of products during the entire processing. The combination of the indoor geolocalized data and the detected values of sensors are mapped by the DSS providing possible predicted alerting conditions for possible product defects or failures.

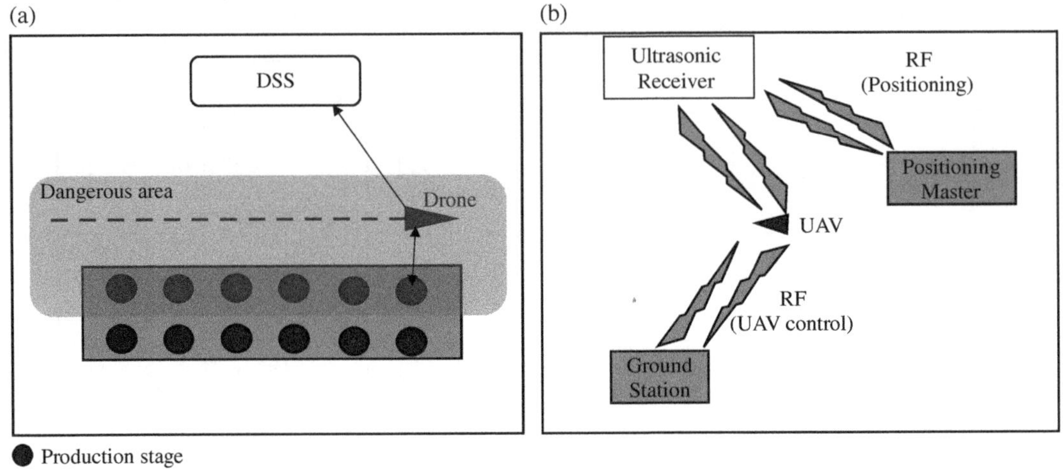

Figure 4.26 (a) Production line layout and drone monitoring in dangerous areas. (b) Indoor transmission system and UAV positioning.

The UAV and the ROVER units can be adopted also for outdoor industrial environments. In this case, the drone positioning characterization may not be crucial.

4.1.7 GPU Processing Units

GPUs are technologies supporting real-time data production processing. The optimization of the computational cost addresses the self-adaptive production from quasi real-time to a real-time sensing and actuation. In order to analyze the potential of the GPUs, different examples oriented towards the comparison with CPUs are provided.

4.1.7.1 Performance of GPUs by Processing Binary Matrices

In order to check the GPU performance, it is necessary to define a procedure suitable for the comparison of the computational cost between different technologies. The calculation associated with an elementary binary matrix is a good way to estimate performance. A binary matrix is a matrix whose elements assume values equal to 0 and 1, respectively, "False" and "True." Binary matrices, and more generally binary functions, are basic elements in computer science, because all implementations are based on functions operating on sequences of bits, therefore on binary sequences. These functions are processed by hardware computing units, consisting of several cores (processing cores) that work in parallel. There are two types of hardware computing units: CPUs and GPUs. The CPU directs and controls every task of the computer and coordinates the activities of memory and peripheral units as well as performing all the arithmetic and logical operations relating for example to an executing program. All operations are made in bits (0 and 1). The GPU is designed specifically for creating digital graphics and is characterized by a very different structure from the CPU. The GPU is built to process large blocks of data in parallel, which allows it to execute the tasks faster than the CPU [6, 7]. In fact, modern GPUs are high-performance multicore processors capable of achieving high floating point operations per second (FLOPS). The General-Purpose Graphics Processing Unit (GPGPU) refers to the use of a GPU for the following purpose:

it is used to speed up those parts of applications that require intensive numerical computations, since they have a much better floating point operation rate than the CPU. The most striking feature of this type of use of GPUs is the enormous theoretical power offered by these solutions and, consequently, the reduced processing time. A typical GPU architecture is the Compute Unified Device Architecture (CUDA). The CUDA parallel programming architecture has three main key features: a hierarchy of thread groups, shared memories, and synchronization management. This model requires that the problem to be solved is partitioned into tasks to be executed independently and in parallel by blocks of threads, and each task is further divided into more specific parts of code, to be executed in parallel by threads within the block. Each block of threads is scheduled to run on one of the available processor cores, independently of the other blocks, and this allows a CUDA program to be run on different processor cores. According to advanced applications, CUDA is adopted to simulate nanostructures performing with a good computational cost compared with the CPU data processing [8]. Different tools can be adopted for GPU testing. NumPy is an open source library for the Python programming language that adds support for calculation of large arrays and multidimensional arrays, having a large collection of high-level mathematical functions. Numba is a just-in-time compiler for Python suitable for the execution of NumPy arrays and functions. It does not require pre-compilation and when a new code is executed, it compiles each annotated function specifically for the available hardware. Numba has the advantage of doing all the operations automatically. Additionally, Numba supports CUDA GPU programming, directly compiling a limited subset of Python code into the CUDA kernel, and device functions following the CUDA execution model. NumPy arrays supplied as arguments to the kernel are automatically transferred between the CPU and GPU. In the CUDA system, the written code is executed by multiple threads simultaneously, and the solution is modeled by defining a hierarchy of grid, block, and threads. A program is broken down into functions to be executed independently. The single CUDA kernel is executed, in logical parallelism, by a grid (a set of thread blocks). Thread blocks can run on the same multiprocessor in parallel with other threads on a CUDA device, and can therefore share memory. The Shared Memory serves as the basis for inter-thread communication where the threads in a block cooperate by exchanging data stored in it. The other memory managed and used in the script is the Global Memory, which is the largest and most used memory of the GPU. This can also be accessed by threads belonging to distinct blocks. Figure 4.27a is representative of the CPU layout blocks. The structure of the blocks in the GPU structure of Figure 4.27b,c highlights the differences between the two technologies.

The algorithm used to compare performances between CPU and GPU considers matrices loaded in the Global Memory. The block structure is based on the number of threads per block (TPB). The first example takes into account the exponentiation of a binary data matrix. Exponentiation is achieved through a Phyton script implementating a FOR loop that multiplies the matrix by itself n times, where n is the exponent. Furthermore, in this example the value of TPB (between 4, 8, 16, and 32) that ensures the shortest execution time is investigated, fixing the matrix dimensions to 256×256: by observing Figure 4.28, it is noted that the best TPB in terms of execution is 8. In particular, it is noted that, by providing less data to be processed by the single CUDA, the execution time decreases reducing the entire calculation time. Furthermore, with a very low value of TPB equal to 4, there is a time consumption worsening due to the loading of the submatrices (threads) to be processed by the single CUDA. Therefore, a value of 8 is an excellent compromise between memory traffic and computational complexity.

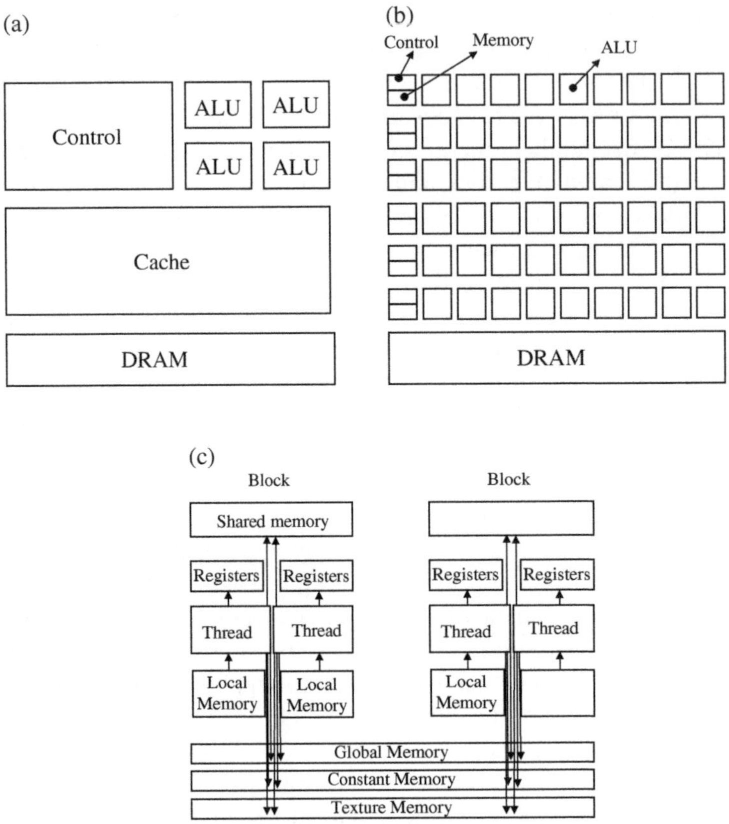

Figure 4.27 (a) CPU structure. (b) GPU structure and (c) related functions (DRAM, dynamic random access memory; ALU, arithmetic logic unit).

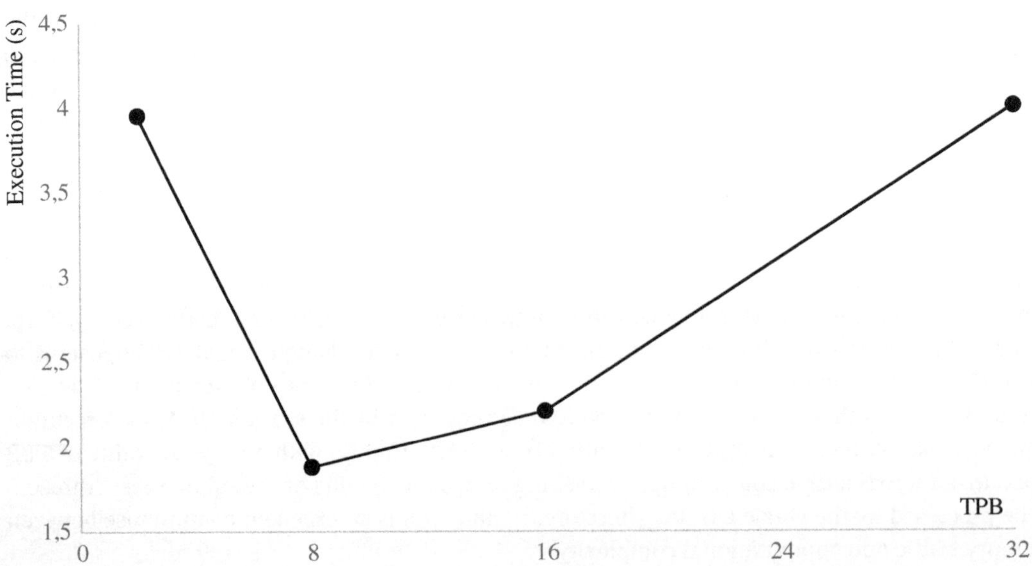

Figure 4.28 Execution time versus TPB for the exponentiation of a binary data matrix.

The comparison between CPU and GPU execution time for the exponentiation case is shown in Figure 4.29 highlighting the best performance of the GPU.

The comparison case concerning the multiplication between two binary matrices is illustrated in Figure 4.30, representing the execution time versus the matric dimensions: it is observed how much more efficient the GPU calculation is compared with the CPU as the size of the matrices increases.

By considering the elevation of a binary matrix to a power, two other cases are shown in Figure 4.31 and Figure 4.32 proving best performance of the GPU technology in terms of execution times.

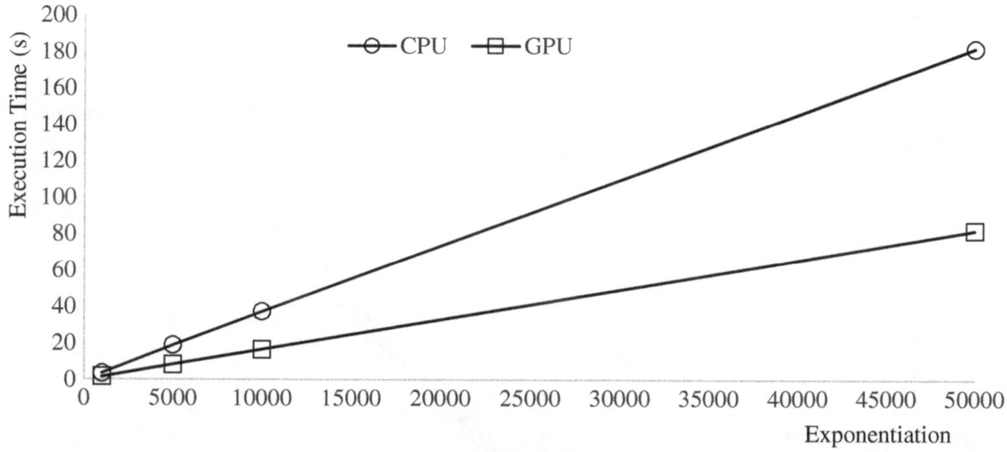

Figure 4.29 Execution time versus the exponentiation n of a Float64 data matrix.

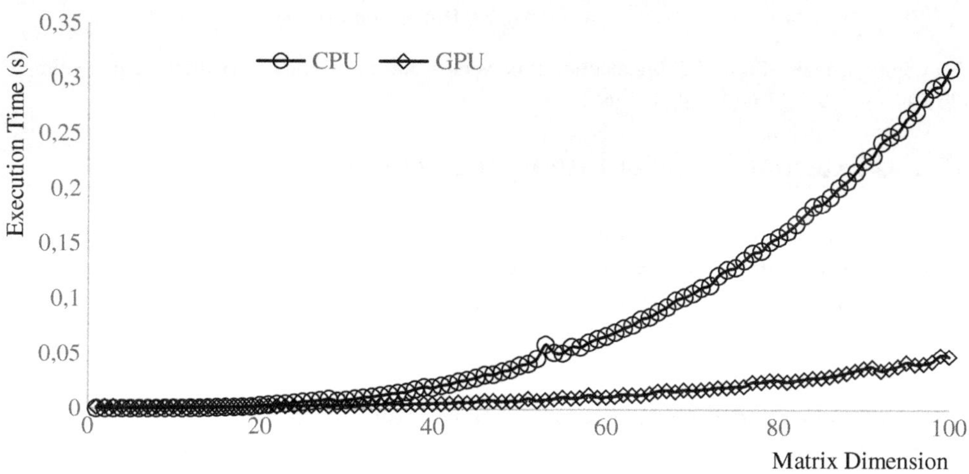

Figure 4.30 Execution time versus the binary matrix dimension.

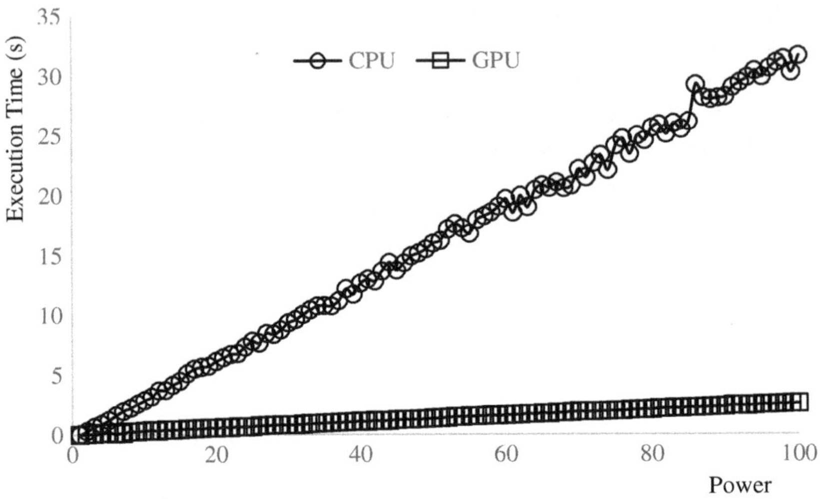

Figure 4.31 Comparison of GPU and CPU execution time versus the power of the 100×100 binary matrix.

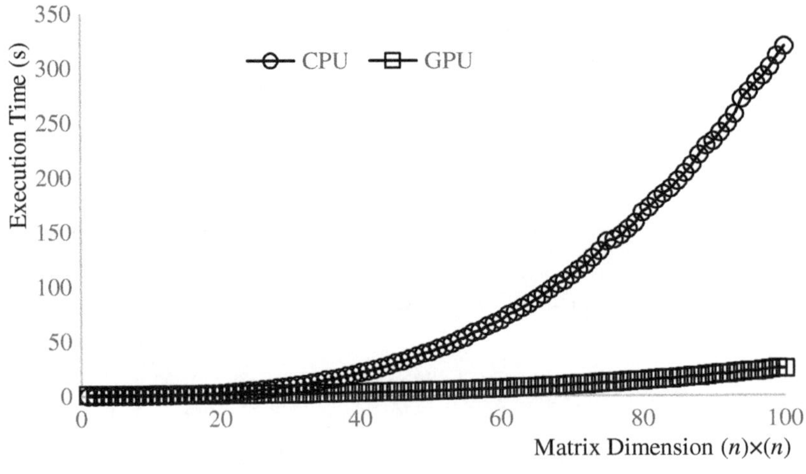

Figure 4.32 Comparison of GPU and CPU execution time versus the size of the binary matrix dimension increasing by considering a power fixed at 1000.

4.2 IoT and External Artificial Intelligence Engines

IoT devices are important for control and adjustment of production, possibly improved by AI engines. AI algorithms are typically executed locally as in smart devices locally embedding sensors and processor units, but a powerful calculation is executed on external server machines integrating big data systems.

4.2.1 Artificial Engines and Server Location: Artificial Intelligence and Adaptive Production

Servers located on the cloud environment simplify the production management, and are appropriate to improve the calculation of different engines suitable for production and quality monitoring, and auto-adaptive production systems, automatically setting production parameters. The server system processes all data coming from the company field including IoT, production machine enabling Industry 4.0 functions, and other sensors integrated in the production lines, and improving the control and the actuation. The architecture of Figure 4.33 is representative of a complete scheme of a cloud server implementation. The architecture is composed of the following elements:

- **System integration module:** other sensors are integrated in the production lines enforcing the production control; a mobile app is interfaced with an API web service (Jason request and Ack response), thus providing a further system to collect data into a big data system containing all historical production data; a buffer memory of a Structured Query Language (SQL) DB supports the data transmission and data migration to the big data.
- **Industry 4.0 module:** the IoT devices, I/O, and the production machines enabled to transmit digital information work in quasi real-time configuration; all these units are connected to a cloud controller supported by the AI engine.
- **Backend module:** all data are managed in the backend system where all the algorithms managing sensing and actuation functions are executed; the AI engine predicts possible machine failures or product defects, supporting the activity of the controller to adjust production parameters; the predictive analysis is achieved by constructing training models using historical production data collected into the big data repository; alerting signals are sent to dashboards of the frontend system.
- **Frontend module:** this includes HMI and dashboards running on applications software which are connected to ERP and to other tools controlling the production.
- **Industry 5.0 module:** the machine controller is activated by the predicted results, and the machine parameter setting is self-adapted simultaneously by the actual sensor monitoring, and by the failure predictions.

The improvement of the Industry 4.0 production is in the AI engine which is the intelligence of the new production. In the proposed architecture, the AI engine is located on a cloud server, allowing cloud computing and self-adaptive production, thus optimizing the production quality.

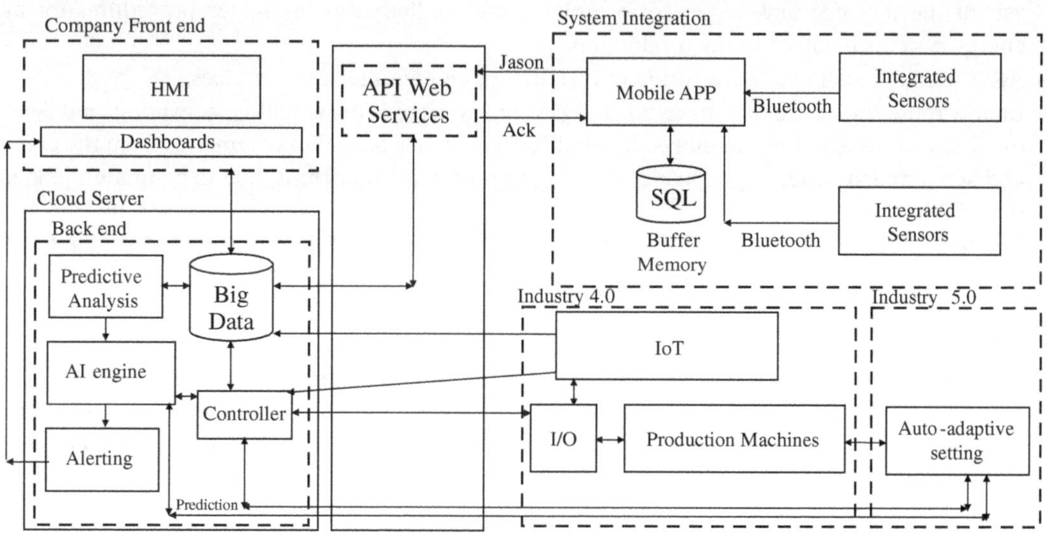

Figure 4.33 Architecture integrating AI cloud server and IoT device: Industry 4.0 and system upgrade in Industry 5.0.

4.2.2 IoT Security Systems in the Working Environment and Implementation Aspects

IoT sensors are also adopted to improve security. Table 4.2 lists some examples of IoT sensors describing implementation aspects.

4.2.3 Example of Energy Power Control and Actuation: Energy Routing and Priority Load Management for Energy Efficiency

Energy routing is an important issue for industries having renewable energy systems used for production. Figure 4.34 shows the architecture of an intelligent energy router system based on AI prediction and on intelligent load switching. In particular the architecture is composed of:

- Sensors (S1, S2, . . ., S*n*) capable of transmitting data to a data processing unit interconnected with the energy routing prototype hardware; the sensors provide data such as lighting intensity, temperature, energy consumed, humidity and other data useful for the estimation of the load curve, for its prediction, and for the intelligent management of load activations and deactivations based on priority roles; the sensor data are transmitted in the cloud.
- Renewable energy devices such as micro-wind, or photovoltaic or geothermal units interfaced with smart energy meters capable of measuring the current produced by renewable energy and suitable both for the energy storage, and for its input in grid networks; the measurement data are sent to the cloud and are processed by AI algorithms predicting future energy generation, and correlating the analyses with the weather forecasts; the predictive calculation of energy generation, combined with the prediction of the load curve, dynamically defines the threshold curves related to the activation and deactivation switching of the loads connected to the prototype routing network.
- Thermography is applied to check the efficiency of the micro-wind and photovoltaic modules, and to measure energies of the production lines, thus providing important indications about any energy leaks or anomalous electrical consumptions; thermography is also adopted for indoor temperature and leakage control, and as a monitoring tool for machinery maintenance; the analysis of the thermograms serves to formulate new predictive maintenance procedures for all energy systems involved in the production.
- AR is used to facilitate the execution of installation and maintenance services.
- Energy router module behaving as an energy gateway capable of switching, activating, and deactivating various load lines managed by intelligent switches receiving commands from the cloud and activating or deactivating loads following appropriate algorithms able to compare the load thresholds.
- The AI engine able to process, through innovative algorithms, the data collected in a suitable DB and to provide the following outputs:
  - Prediction of load curves.
  - Optimal and dynamic thresholds setting for load activation and deactivation.
  - Predictive maintenance of renewable energy devices based on predictive calculations.
  - Intelligent classification of priority and non-priority loads (implementation of association roles according to the application context and the data collected by the various sensors).
  - Switching, activation and deactivation commands of the electrical loads of the production lines.

Table 4.2 IoT sensors in a secure working environment.

IoT sensor type	Implementation aspects
Gas sensor	Gas sensing is important for all environments characterized by gas emissions. A gas threshold value is detected by a microprocessor emitting an acoustic alerting signal. Smart gas sensors can be embedded into worker clothing with the smart microprocessor connected to an acoustic transducer. All the detected signals are stored into a database system useful to trace during the time the risk map of the environment: the indoor geolocalization by RFID, iBeacon or GPS signals can be combined with the localized indoor and outdoor gas measurements to map the dangerous areas during the time. Gas sensors are also suitable to detect chemical risks in research and development laboratories. AI engines are able to predict dangerous areas by analyzing historical geolocalized data
Thermal sensors	Temperature risk is typical for industries working with very high temperatures, such as in foundry industries. Fixed infrared thermography devices define and control high temperature zone risks. Thermal sensors and microprocessor units can be embedded into worker clothing as standalone devices advising the worker when a high temperature is checked. In order to optimize data processing, all data are transmitted to a cloud server and processed in cloud computing modality by an AI engine able to provide temperature risk maps (maps predicting dangerous working areas)
Face recognition	A camera recognizes the face of a worker thus ensuring authorized access. This function is mainly useful for the control of indoor access. The biometric information of the face is processed by an AI classification algorithm such as k-nearest neighbors [9, 10]
Dangerous worker position	The proximity of a worker near a dangerous work site is classically detected by proximity sensors using RFID, or iBeacon technologies: when a tagged place detects the proximity of the worker, an acoustic or luminous alarm indicates the danger. A dangerous position is also an area dedicated to the movement of a robotic arm or of a vehicle. Dangerous areas are also all zones with vibrating infrastructures. In this case, different accelerometers placed on the infrastructures detect the danger conditions
Electromagnet emission	Electromagnetic sensors capable of detecting an electromagnetic field are important for the worker operating in high voltage environments. Antenna sensors detect the electric and magnetic field intensities. The measured intensity values must be under a threshold value defined generally by laws and regulations
Radioactive emission	Natural radioactivity is exhibited by several elements, such as radium, uranium, and other members of the actinide series, and by some isotopes of lighter elements, such as carbon-14, used in radioactive dating. Furthermore, radioactivity may also be induced, or created artificially. The radiation produced during radioactivity is mainly of three types, designated as alpha, beta and gamma rays. Radioactive sensors detect the radioactivity thus enabling the early warning conditions especially in laboratories with radioactive elements
Laser rays	Lasers can be dangerous for the eyes of a worker in manufacturing additive production processing. For this reason, lasers are mounted and sheltered by specific elements to avoid the risk. Sensors support the verification of the laser beam focus supporting risk control. High power lasers are also used in industry for the cutting of metallic product. In this instance, the laser is dangerous for all body parts, and sensors could advise the worker if he/she is in a dangerous region

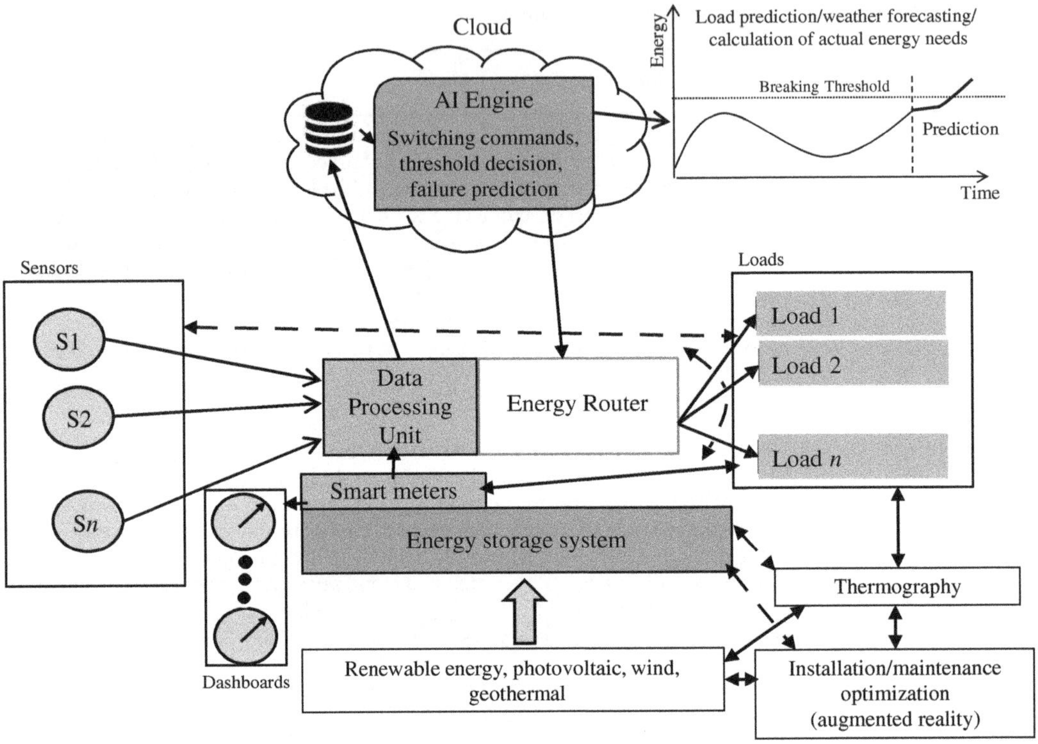

Figure 4.34 Architecture of an energy router system.

The architecture therefore is oriented to create a device, the energy router, for the management and the optimization of the complex of energy flows entering or leaving the distributed machines, implementing software tools operating on the cloud for analysis of production data, and for the optimization of energy resources. By means of the energy router, it is possible to achieve a small intelligent distribution system, and therefore a smart grid which generates economic benefits in terms of increasing self-consumption, increasing energy efficiency, increasing in renewable production, reduction of losses, and sale of adding services. The system architecture is hybrid and consists of two main modules:

- The field section, installed in the industry and able to control energy generation, accumulation (storage), and energy flows.
- The cloud section implementing advanced software solutions for supervision and system control with the use of forecasting and optimization models, detecting input parameters (production from renewables, loads, etc.), and optimizing the use and the management of distributed energy resources.

This hybrid architecture in Figure 4.34 requires continuous exchange of information between the cloud and the production areas, by solving security and data quality control problems. However, the cloud intelligence system allows to decrease the computational cost for algorithms requiring high calculation resources. Cloud computing and GPU technologies are addressed to improve the computational performance of the software used for the calculations. The architecture in Figure 4.34 is representative of an innovative system for intelligent distribution of electricity within the industry production area through the intelligent energy router, by accomplishing:

- Optimization of charge and discharge cycles, in the electrical storage systems.
- Exchange of data in real time between the production systems, storage devices, sensors (S1, S2, . . ., S*n*), and local control center (control room).
- The control of cloud tools by a middleware communication system.
- The exchange of data in secure modes between the field and the cloud system.

Furthermore, the AI engine provides:

- A decision-making system for the best routing of energy in accordance also with predictive maintenance procedures.
- The forecasting of the electrical consumption (energy consumption optimization thus decreasing production costs, and prediction of anomalous consumption indicating a possible breakage).
- Innovative strategies and techniques facilitating design, installation, and machine maintenance.
- The intelligent distribution of energy within the industry framework, by optimizing the use of storage systems and ensuring their optimal use.

The energy routing system involves the creation of a series of elements including sensors and control hardware for production, consumption and storage systems. The system routes energy based on control provided by the AI tool, providing directives to the production field system, identifying in advance the better routing of electricity between the elements of production, and maximizing the use of stored energy. The whole production can be monitored in the cloud by MQTT, Advanced Message Queuing Protocol (AMQP), or other protocols. By remote control, a user activates or deactivates electric loads predicting the total power consumption by AI [11]. In this scenario, the loads are controlled by analyzing prediction results using different methodologies [12] (Table 4.3).

The LSTM can be developed by adopting Python language and different libraries such as:

- Numpy (library useful for the management of large multidimensional arrays processed by complex mathematical functions).
- Matplotlib (graphical library).
- Seaborn (graphical library mainly used for drawing statistical plots).
- Pandas (library for data analysis processing DataFrame objects as SQL tables).
- Scipy (open-source library of mathematical functions including linear algebra, integrals, FFT, and image processing).

Table 4.3 Requirement for intelligent load switching.

Requirements	Method description
Electrical load trend prediction	Linear regression predicting the slope behavior of the energy consumption (Figure 4.35a)
Accurate electrical load prediction	Application of LSTM. Training and testing of the LSTM model (Figure 4.35b)
Definition of priority roles and enabling/disabling actions based on priority	The load priority is described by a flowchart describing different priority levels of electrical loads and criteria to deactivate and activate loads, by means of the decision support system and of the prediction results
Combination of the slope consumption trend and LSTM prediction	The comparison between the two approaches (Figure 4.35c) optimizes the decision making about load switching

LSTM, long short-term memory.

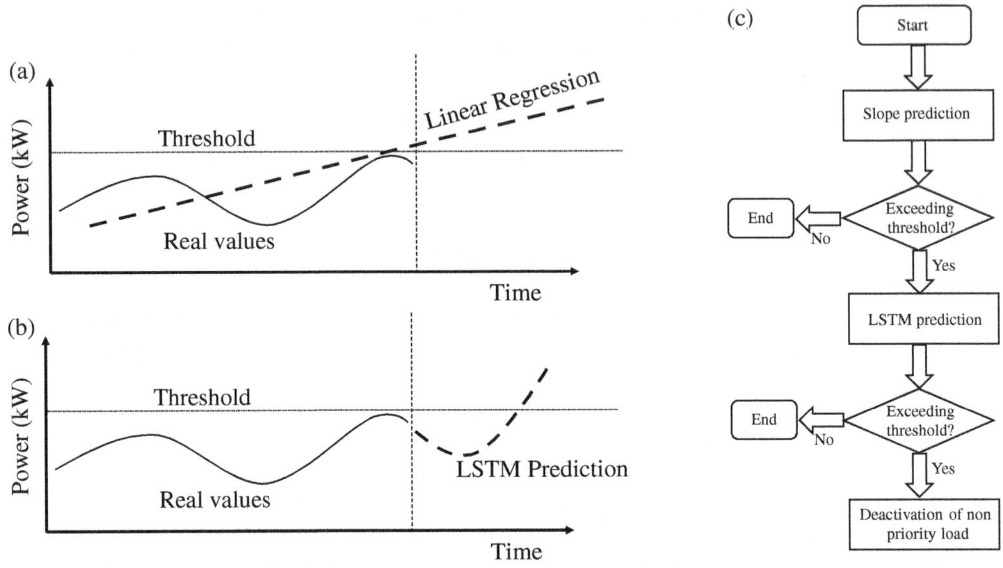

Figure 4.35 (a) Linear prediction of energy consumption by linear regression analysis. (b) LSTM approach to predict energy consumption. (c) Flow chart combining slope prediction and LSTM approach predicting the energy consumption over a specific threshold.

- Scikit-learn (machine learning open-source library typically used for classification, regression, clustering, and other data AI algorithms).
- Keras (neural networks and DL libraries supporting convolutional neural networks [CNNs] and recurring neural networks [RNNs], and matching with parallel computing).

The energy consumption can be abnormal: the smart meters provide information of load consumption and about electric load trend of the whole production line, thus enhancing real-time alerting conditions in cases of incorrect machines working. In this way, infrared thermography is appropriate to detect further electrical and mechanical abnormalities, considering the reference values in Table 4.4:

Consumption prediction is estimated for active and reactive power by applying DL algorithms suitable for energy management systems [13, 14].

4.2.4 Online Configurators: Cloud DSS

The DSS is constructed to manage all the production. For an efficient production system, the DSS is implemented online in the cloud environment: this choice is justified because cloud computing is typically efficient in terms of computational costs, and because it is easy to control a production system remotely. The online configuration concerns different aspects of the production such as:

- Setting the machine parameters (velocity, temperatures, etc.) according to the checked quality.
- Telemetry parameters for production line monitoring.
- Thresholds enabling alerting conditions or specific electrical loads.
- Risk level conditions (depending on the case study).
- Geolocalization of the monitored parameter (outdoor applications).
- Setting and calibrations of sensors by remote control.

Table 4.4 Electrical abnormalities.

Element	Abnormality
Speed variators	Temperature > 50 °C
PLC	Temperature > 45 °C
Contactors	Temperature > 50 °C
Switches	Temperature > 50 °C
Power supplies	Temperature > 40 °C
Fuses	Temperature > 50 °C
PCB	Temperature > 30 °C
Power supplies	Temperature > 40 °C
Batteries	Temperature > 30 °C
Telecommunication devices	Temperature > 55 °C
Power factor correction capacitors	Temperature > 55 °C
Mechanical heating	Temperature > Reference temperature
Hot defect points (critical points)	Region with a major mechanical stress. Points with abnormal vibrations generating abnormal local heating
Bolts subject to strong mechanical stress	They can generate mechanical misalignments producing local heating
Failed interconnections	Detected cold parts
High voltage power transformer	Abnormal voltage peaks

- Setting of the actuators.
- Possibility to control and manage simultaneously different production lines localized in different places.
- Informatic infrastructure network setting.

Table 4.5 (machine processing), Table 4.6 (railway infrastructure monitoring), Table 4.7 (road and bridge infrastructures), Table 4.8 (quarry production), and Table 4.9 (telemedicine) report different application fields of online parameter configurations through the use of a DSS cloud platform.

As an example of bridge control, Figure 4.36 shows a bridge thermogram detected by UAV. The real-time acquisition is pre-processed by the DSS checking all the temperatures over a threshold as highlighted as in Figure 4.37. Further data processing actions of the DSS are the reading of a range of temperatures, and the comparison between hot and cold regions, indicating possible infiltrations in infrastructures (Figure 4.37). Further data output of the DSS is the temperature prediction by means of the AI models trained by historical thermograms. As an example of a railway temperature check, Figure 4.38 shows a thermogram indicating the temperature distribution of the infrastructure.

Furthermore, a non-invasive method suitable for a bridge diagnostic is the GPR approach. The GPR detects the internal part of the bridge infrastructure as sketched in Figure 4.39. The DSS in this case indicates the materials, and, for high resolution radargrams, recognizes infrastructure cracks. DF techniques are adopted to simultaneously collect information about infiltrations by thermograms and the internal diagnosis of the bridge structures.

Table 4.5 Example of online configuration for a platform controlling machine processing.

Parameter to configure	Description
Working parameter of a controlled chamber	In closed controlled production systems, it is necessary to set production parameters such as temperature or humidity (as in pasta production or in food processes in general)
Motor velocities	The production velocity must follow the permissible processing times according to the machine or robot characteristics and velocities. In the whole production line, all the motor velocities must be synchronized. The online control of the production allows the best velocities of all the machines to be calibrated. Sensors installed on the production line, and connected in cloud, provide a synoptic picture of production
IoT sensor sampling	The sensor sampling must follow the production velocities and, consequently, the machine synchronization
Thresholds activating or deactivating specific electrical loads	The threshold estimation is useful to activate or deactivate specific loads thus ensuring a decrease in energy consumption. The loads control is also related to the stop production conditions following an alerting condition (for example by measuring an anomalous vibration or a different spectrum of resonance frequencies). Thresholds are updated automatically by the AI engine predicting failures or dangerous conditions. Alerting conditions are also important for the automatic update of the predictive maintenance. The DSS is suitable to activate and to select remotely the accelerometer sensors by setting its parameters (sampling, noise level, amplification, etc.), and according to the phenomenon to check, where a strong acceleration means that a load is to deactivate thus avoiding possible breakdown risks
Data migration	The data provided by the IoT sensors are collected in a unique database system. In order to allow data migration from a specific database to the big data system, it is necessary to program by crons data transfer. Typically, this operation is executed during the night when the production is stopped. Data migration is performed also in security systems and for disaster recovery actions
AR setting	AR devices are applicable to monitor the production: a worker by his/her smart glass sends the image online to another specialist, activating an interaction by properly using markers on the visualized image. The marker highlights the part to focus attention on for a possible intervention. The parameters to configure are related to software managing interactive contents and the audio/video bidirectional communications. The parameters can be set by means of a cloud platform controlling all the communication channels and managing all the audio and video calls
DSS parameters	AI algorithms are executed by setting different calculation parameters, according to the best DSS model to implement. The parameters are fixed after checking the performance of the algorithm (by analyzing algorithm accuracy or other performance indicators)

AR, augmented reality; DSS, decision support system, IoT, Internet of Things.

Table 4.6 Example of online configuration for a platform controlling a railway infrastructure.

Parameter to configure	Description
Geometrical parameters of railway infrastructure	Laser scanner technology is applied to detect geometrical characteristics of the railway infrastructure including the possibility to check gallery status. The program to execute the measurements can be performed online by a DSS suggesting the measurement resolution and the velocity for the data acquisition of a laser scanner allocated on a railway test carriage or on a UAV (light laser scanner). The DSS, by remote control, permits all the parameters of the laser scanner to be set, also allowing the inspection procedure to be planned
Hydrogeological behavior of the ballast	The DSS platform can update the hydrogeological risk by crossing hydrogeological maps with meteorological data and sensor measurements detecting a bad ballast drainage. A high resolution georadar measures by FFT the effective interspace between ballast particles (average estimated condition by knowing the typology of the ballast used). The setting of the measurements to acquire can be done online by a continuous measurements system: the measurement is adjusted step by step in order to obtain the best raw radargram to process by FFT. A standard photo camera helps the expert user to remotely adjust the georadar parameters by observing the ballast quality. The DSS can overlap the geolocalized maps of the hydrogeological risks with ones of the inspection, thus defining updated risk levels of the whole railway infrastructure to control
Geomorphological environment (georadar setting)	The geomorphological analysis of the infrastructure soil is very important to define possible potential structural failures or breakdown. A useful tool to detect the geomorphological characteristics is the georadar. The possibility to set online the georadar parameters could automatize the detection procedure. The georadar can be installed on a diagnostic railway carriage, or on an UAV unit also used for photogrammetry. The DSS suggests the setting of the parameters of the georadar antenna by planning the inspection procedures. The geormorphological reconstruction is optimized by the data fusion of information extracted from laser scanner, georadar and other digitized maps. The DSS is appropriate to facilitate the three-dimensional model reconstruction
Alerting conditions (railway temperature check)	The temperature check is important to reduce risks in railway transportation. A continuous high temperature over a threshold potentially deforms the railway line (buckling effect occurring when the temperature exceeds a threshold value for different consecutive days). In addition, a very low temperature possibly damages the railway line by internal cracks, increasing the risk of breakage. Temperature standalone units controlled by a microprocessor are indicated to transmit data with a timing depending on the DSS setting. The DSS crosses the temperature data together with the meteorological data in order to predict dangerous conditions for a specific geolocalized point of the infrastructure. The temperatures are also detected by an infrared thermal camera installed on a UAV unit. All the parameters of the UAV and of the fixed stations can be configured online. The main parameter to control is the temperature threshold limit

DSS, decision support system; FFT, fast Fourier transform; UAV, unmanned aerial vehicle.

Table 4.7 Example of online configuration for a platform controlling a road and bridge infrastructures.

Parameter to configure	Description
Car velocity	A car equipped with sensors (accelerometers, infrared thermal camera, georadar, and camera) can send online data with high noise level. In this scenario, the DSS suggests the true car velocity for the best conditions to monitor roads thus decreasing the noise. Car velocity is also a function of the frame rate and of the technology used. The DSS automatizes the sensors and the image acquisition by detecting for each measurement the localization coordinates, thus accomplishing an automatic reconstruction of an inspection map indicating data detection coordinates
Threshold levels of the infrared images	The infrared thermal camera is adopted to check road drainage or water infiltration on bridges. The DSS, based on preliminary image processing, automatically suggests the threshold conditions to discriminate better the humidity in different road zones when the car is in motion. A UAV unit can be equipped with an infrared camera to accelerate the inspection of large areas to check. The DSS is appropriate to set the temperature threshold value for real-time processing and is suitable to process data in post-processing modality to extract hidden information from the radiometric images. Possible parameters to set are the minimum temperature, the maximum temperature, and the temperature range for image filtering. Furthermore, the DSS is useful to control the UAV and the thermogram acquisition process, also planning the inspection flight
UAV detection mode (equipment with georadar and thermal camera setting)	The UAV favors the detection of radargrams and thermograms. The parameter setting of these sensors is managed directly and in real time by an operator controlling online the flying together with the operator locally driving the UAV unit. The parameters to set are mainly the focus for the thermal camera and the scanning depth for the Ground Penetrating Radar system. The height of the UAV is a function of the technology resolution. The GPR system can also be adopted for the check of the first asphalt layer, and for the iron reinforcement analysis of the bridge structure (check of the iron inside the concrete)
Accelerometer	Anomalous vibration detection can represent irregularities of the asphalt layer: when a car travels on the road, the asphalt irregularities provide anomalous vibrations which can be detected remotely by the DSS. The detection of geolocalized anomalous accelerations allow a road census using geographic information system technology to be mapped. The DSS configurator in this case is able to automatically trace the road status indicating the status by risk colors. The reconstructed map allows road maintenance to be planned
Pipeline leakage monitoring	UAV or fixed station controls possible water, oil or gas leakages. LSTM and CNN algorithms are used to support the data post-processing, by highlighting leakage region and estimating the related plume extension and losses. It is preferable to perform the execution of LSTM/CNN algorithms and the parameter setting in cloud computing modality by the DSS. The DSS calibrates the LSTM/CNN parameters according to the estimated probability of leakage classification. By analyzing the plume extension during the time, the DSS provides information about leakage as the estimation of volume losses. AI algorithms are also applied for radargram analysis to distinguish pipeline under roads, by means of image segmentation clustering and image classification
Bridge monitoring	Continuous bridge monitoring involves a seismic oscillation check; single inspections are mainly related to water infiltrations. The oscillations are measured by accelerometers, and the infiltration by infrared thermography detecting a temperature gradient. For continuous monitoring, the sensors are fixed on efficient and strategic positions. Moreover, georadar is functional to check the quantity of iron rods in the reinforced concrete and bridge pillar status. All sensors, including cameras, can be integrated into the UAV. The sensors can be monitored and controlled remotely by the DSS

CNN, convolutional neural network; DSS, decision support system; GPR, Ground Penetrating Radar; LSTM, long short-term memory; UAV, unmanned aerial vehicle.

Figure 4.36 Thermogram of a bridge acquired by UAV. Inset: UAV adopted for thermogram detection.

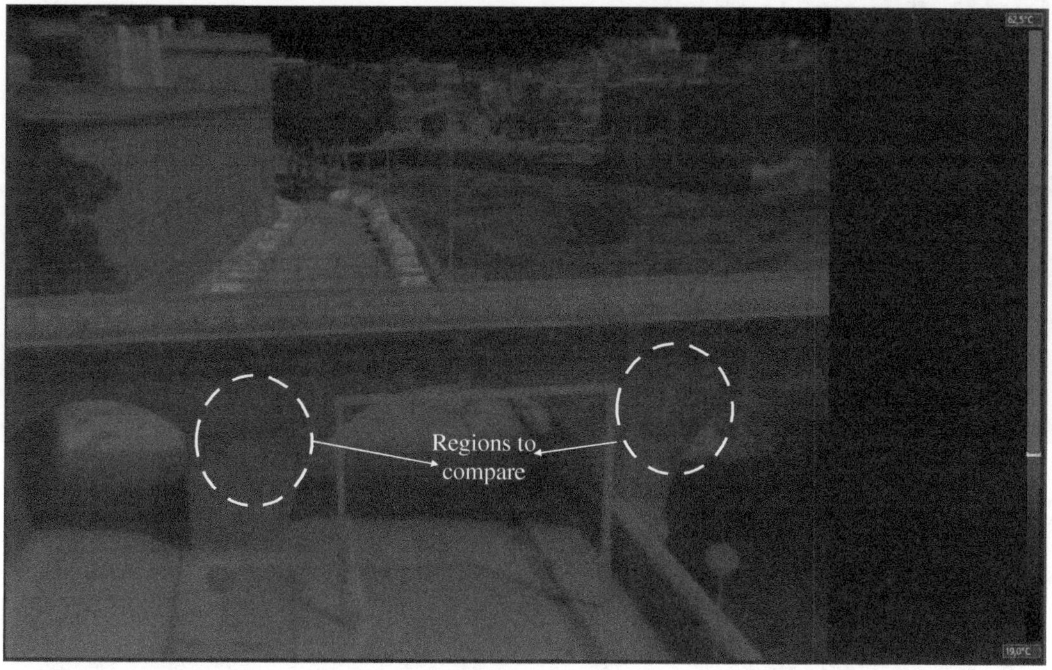

Figure 4.37 Thermogram of Figure 4.36 setting over a threshold of 32 °C (DSS setting).

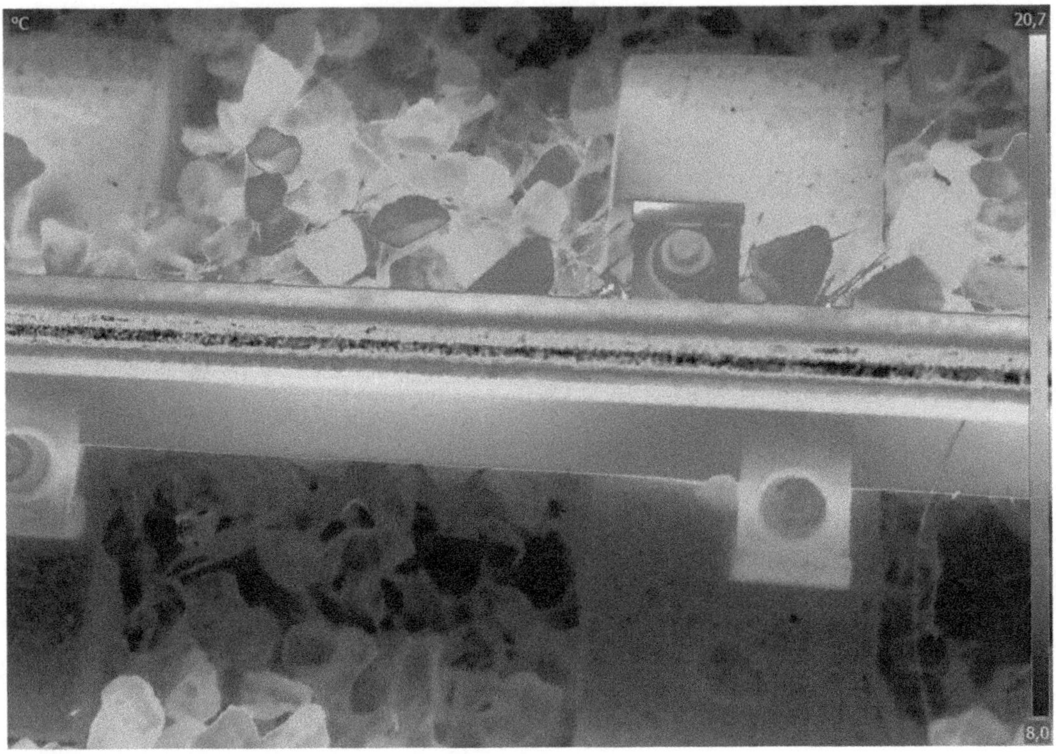

Figure 4.38 Thermogram of a railway infrastructure.

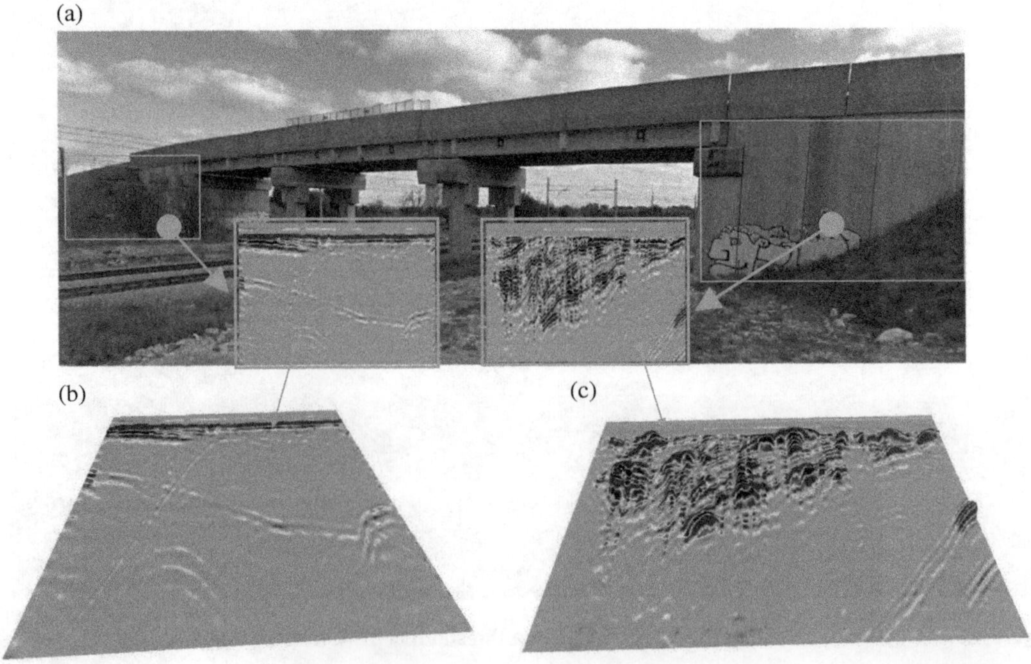

Figure 4.39 (a) Radargram of part of a bridge detected by a UAV equipped with a GPR. (b) and (c) 3D image processing of the analyzed parts with possible different materials used for the bridge construction (iron content).

Table 4.8 Example of online configuration for a platform controlling a quarry production.

Parameter to configure	Description
Accelerometers for crushing plant	Vibration monitoring provides important information about internal breakage prediction of quarry crusher machines. The vibration measurements are performed using accelerometers working at low frequencies. The online connection is useful in this case for sensor calibration and data transmission modality. The DSS cloud platform is structured to estimate spectra in order to find anomalous frequency, by indicating a possible malfunction (breakage prediction). The accelerometers can also be installed on other mechanical parts of the production line such as conveyor belts
Conveyor belt cracks	An image vision technique, by smart camera executing LSTM classification algorithm, detects the cracks on the carpet of the conveyor belt. A camera installed on the conveyor belt detects possible cracks as defects processed by an image segmentation algorithm. The LSTM model is trained by analyzing the homogenous region and anomalous ones of each line of the detected image of the conveyor belt carpet. The image acquisition process is executed when no stones are transported by the conveyor belt, that is when the production is stopped. The data are transmitted in a cloud repository to be processed. The acquisition time and the image segmentation algorithm parameters can be set online by accessing the DSS. AI models are trained to distinguish and classify different cracks by defining a prediction of the crack degeneracy
Map risk update (dangerous area)	An unmanned aerial vehicle is adopted for photogrammetry acquisition, and for georadar inspection allowing the analysis of the nature of the rock to process. The 3D quarry reconstruction combined with the geomorphological analysis is useful to formulate a map indicating the best area to work. The DSS supports the reproduction of the 3D quarry environment, suggesting the areas to dig. The parameters to configure are the efficiency indicators processing data
Pollution detection	The workers are subject to risks from fine dust pollution. Air quality sensors are useful to report risk conditions. The DSS is used to periodically check the pollution sending alerting messages to the project manager of quarry activities

3D, three-dimensional; DSS, decision support system; LSTM, long short-term memory.

Table 4.9 Example of online configuration for a telemedicine platform.

Parameter to configure	Description
Wearable sensors for homecare assistance	Wearable sensors measure in real time physiological data of patients. This technology is useful for homecare assistance processes and for dehospitalization processes. All sensors transmit data in the cloud where it is implemented on a server of the DSS. The DSS telemedicine platform is set to verify threshold values and to send alerting alarms when a health risk is detected. Thresholds can be set online. The artificial intelligence algorithm is also adopted for the prognostic. The DSS platform can be "open" to different wearable technologies, and to the data collection into a big data system
Big data migration	All patient data stored into different database systems can be migrated into the big data system by scheduling crons enabling data transfer. It is preferable to execute this operation directly in the backend system automatizing data migration. The DSS manages data transfer and data migration also considering other external database sources
Hub unit configuration	All sensors or point of care measuring medical parameters are connected to a hub located in a laboratory, in a hospital room, or at home for dehospitalized patients. The hub is configured online to make the data transmission system standalone and user friendly
Health status prediction	Artificial neural networks, long short-term memory, and big data are applied for health care status prediction, thus supporting the care assistance process. All the algorithm parameters are fixed by the DSS. The DSS allows the health risk level to be defined remotely through the definition of alerting thresholds (threshold of physiological parameters)

DSS, decision support system.

4.3 Blockchain and IoT Data Storage Systems

Blockchain is an important tool for security production check [15], and in general for cybersecurity [16, 17]. The data are collected in secure blocks ensuring data integrity. The blockchain is also suitable to enable automatic smart contracting processes [18]. All data are stored into blocks managed by specific rules enabling data transactions. The main function of the AI controller is to check and correctly store IoT sensor data by avoiding incorrect information about the production process. The use of the blockchain enforces the bargaining power thus certifying the quality of the production process.

4.3.1 Blockchain Implementation Rules

Figure 4.40 shows the UML design of the sequence diagram implemented in the prototype blockchain network, where Actor X indicates a generic element of the supply chain inserting data or information. The diagram is structured as follows:

- An IoT sensor transmits the signal to an AI engine behaving as a controller.
- The AI controller checks the production by processing the signal level, and predicts the machine behavior for possible failures.

- If the failures are not predicted and the production machine is under control, the AI controller enables the Actor X to start a new transition request.
- The Actor X enables the blockchain node (Node N), by adding the transaction to the local list of new transactions, thus creating a new block.
- The linked Node N2 is then automatically updated.
- The Node N2 performs the blockchain copy validation and the local update by accepting the transaction.

The blockchain architecture of Figure 4.41 enhances the relationships between the blockchain nodes, IoT sensors, data entry interfaces, and the AI controller managing information and reporting.

An example of a JSON script of the transaction structure is as follows:

```
transaction = {
'Id': integer
(identification number of the actor enabling the event to be stored
into the transaction)
       'id_device': integer
(dentification number of the IoT device generating the transaction)
       'information'  : json information
(data collected in the transaction following a precise protocol)
       'timestamp': integer
(time traceability of the data stored into the transaction)
}
The single block is structured as follows:
block = {
(block index)
'index'     : integer,
'timestamp' : integer,
'transactions'  : list,
(Proof of Work -PoW-)
'proof_of_work' : integer,
(block hash preceding the unique id)
'previous_hash' : string,
(node identification number)
'node_id' : integer
}
```

The blockchain architecture of Figure 4.41 represents the block structure of nodes constituting a part of a blockchain network which is structured by a sequence of blocks (Node 1, Node 2, ..., Node *n*). Each node has a local own copy of the blockchain, and contains the transaction data (Transaction 1, Transaction 2, ..., Transaction *n*). The blockchain can be interfaced with ERP software, mobile applications and IoT devices. The AI controller can enable the transactions according to the sequence diagram. Table 4.10 shows the blockchain class structure.

IoT sensors are primary actors of blockchain data processing: all the storing processes are synchronized by the AI controller. The continuous detection of sensors allows the AI controller to define the timing to transfer data to the blockchain blocks.

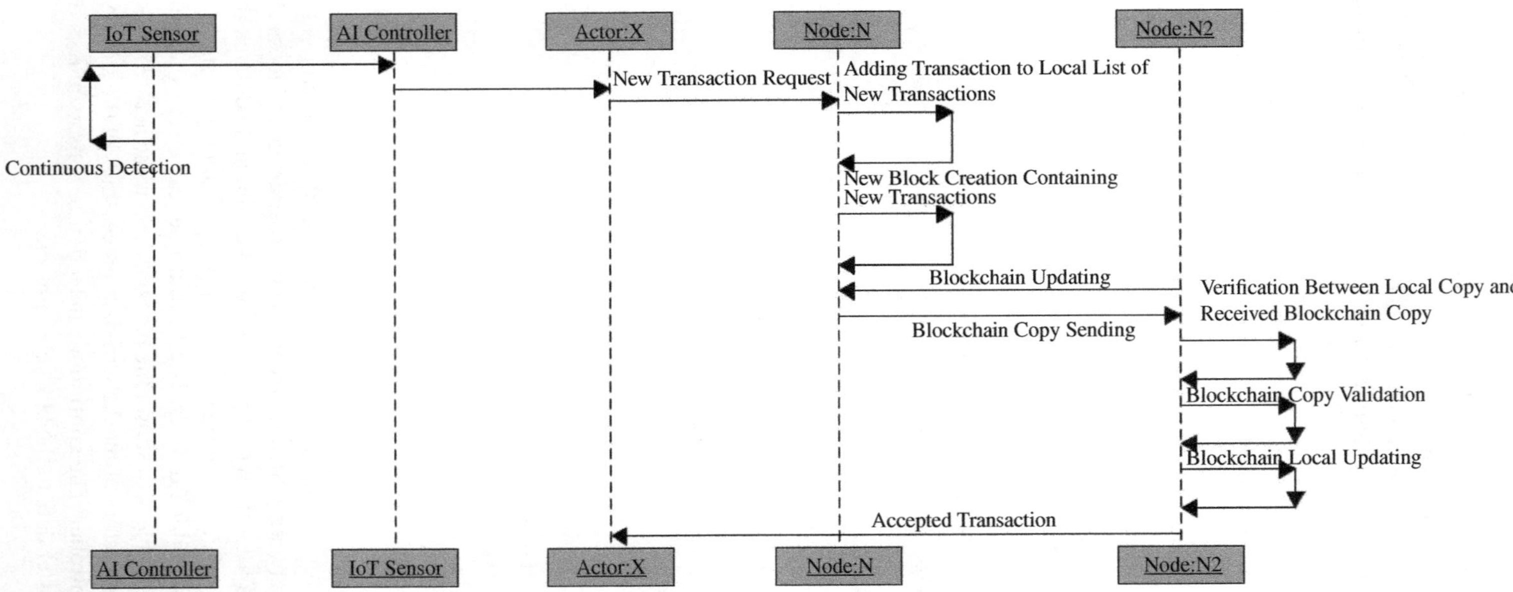

Figure 4.40 Blockchain model and transactions integrating an AI controller.

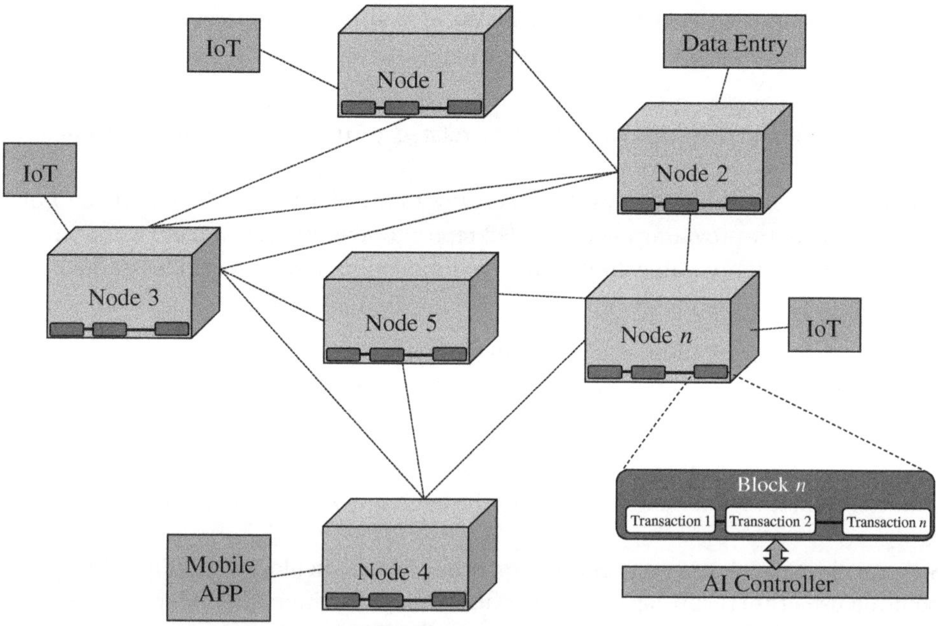

Figure 4.41 Blockchain architecture model.

Table 4.10 Blockchain class definition.

Blockchain class
-transaction: list
-nodes: set
-chain: list
“property”+last_block: string
+ id: integer
+add_node(address: string)
+chain_validation(chain:list):Boolean
+ AI approval(): boolean
+add_block (proof:string, previous_hash: string): string
+add_transaction (idtrans: integer, id_device:integer, information: string, timestamp: integer): string
+last_block(): string
+last_blocks (from_id: integer = 0): list
+hash(string): string
+proof_of_work(string): integer
+proof_of_work_approval (actual_proof, last_proof, last_hash: Boolean

4.3.2 Blockchain and IoT Production Traceability

The blockchain network can be integrated into the company information system improving production traceability. An example of blockchain integration is provided in Figure 4.42, where different workers provide production data by means of a data entry application suitable to digitize data of n production lines. Data entry is possible by implementing a mobile app suitable to add further data not provided by production machines of each production line. All data are collected into a big data repository containing all the information of production, including IoT data, thus ensuring the correct product processing. The data processing platform is loaded by different digitized data such as:

- Project status (project managers trace digitally in the platform the phases of a new production).
- Research and development data (production data related to the research and development sector).
- IoT sensor data (an AI controller defines the procedure to transfer IoT production data to the block of the blockchain, by providing an automated report to send to the company managers).
- Other data and information coming from different production sectors enforcing production traceability.

The access to the whole information system is managed by the system administrator.

Blockchain is applied in different application fields such as vegetable production [15], involving the traceability of the following phases:

- **Sanitization processing:** the information to include in blockchain blocks is the vegetable treatment procedures, combining methodologies about the adoption of different liquids and gaseous substances.
- **Pretreatment process and post-harvest process:** the blockchain blocks contain all the information about the use of inert substances and surfactants, by discriminating the procedures for organic and integrated cultivation; moreover, the blocks also contain information about the absence of contamination (fungi and phytophagous insects).

For wine production blockchain can collect information about all the production phases such as harvesting, crushing and pressing, fermentation, clarification, aging, and bottling. In telemedicine

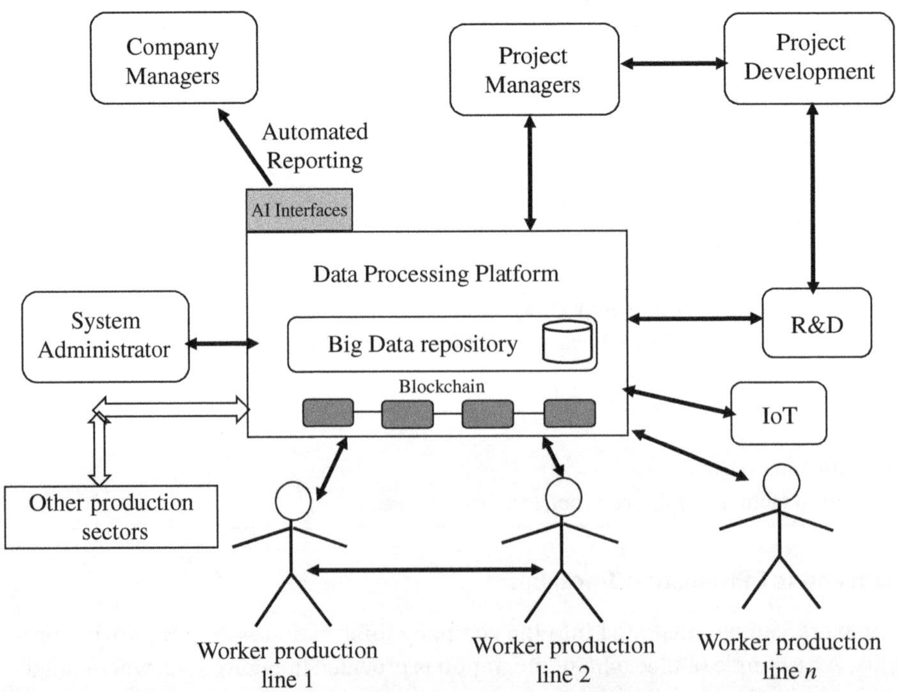

Figure 4.42 Blockchain implementation concerning a full production process data processing involving different actors.

applications particular laboratory analyses can be certified through the blockchain, tracing the adopted procedures and verifications of the obtained results. This application is important for all private laboratory centers having specific facilities. All data stored in the blockchain increase the contract bargaining power.

4.4 Mechatronic Machine Interface Architectures Integrating Sensor Systems

Mechatronic machine interface integrating sensors are fundamental for a fully automatized production. The integration of the AI engine, the AI controllers and other advanced technologies improves the new dynamic production by addressing self-adaptive systems. The self-adaptive production is typical for the evolution of the Industry 4.0 scenario to Industry 5.0. Figure 4.43 shows an example of an advanced architecture integrating different technologies suitable for Industry 5.0 implementation. The adopted technologies include:

- Information infrastructure, typically bus/enterprise service bus, connecting different hardware and software components.
- Data sources such as machine logs, data sensors, laboratory data, ERP data, process/production data, and energy consumption data.
- DBs collecting different data related to a particular data source, and a big data system storing all data and performing analytics.
- Blockchain storing production data and saving information about product quality and phase processes.
- Robotic elements controlled by a PLC standard (exoskeleton and CoBot systems).
- Rapid prototyping and reverse engineering facilities.
- AI engines supporting energy consumption optimization, rework minimization, process optimization, predictive maintenance and autocalibration of production line machines.
- AI controller connected to the AI engine supporting feedback control of motors.

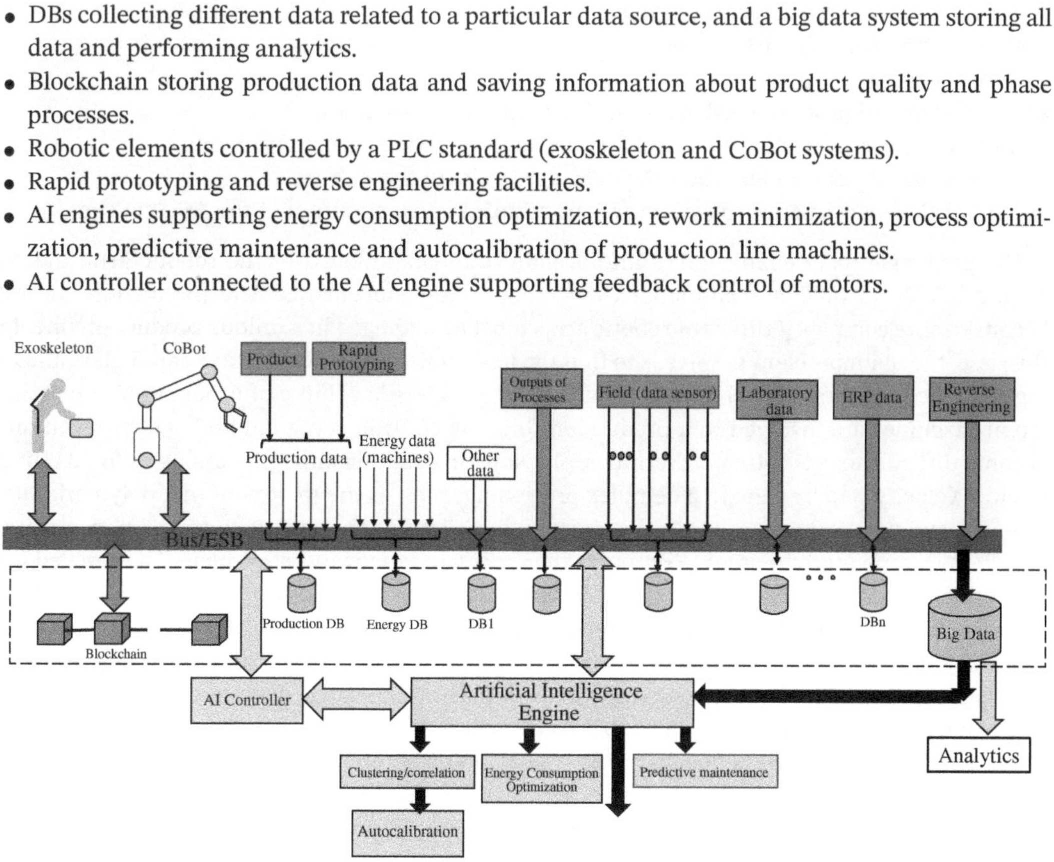

Figure 4.43 Example of architecture implementing facilities improving dynamic production.

In particular, energy consumption data are important to define the correct use of machines: the AI engine is appropriate to analyze the anomalous electric machine consumptions, by predicting possible machine failures or incorrect working conditions. The AI autocalibration of the machine parameters adjusts the production process thus optimizing energy consumption: the analysis of the data clustering, machine parameters, and correlations of variables facilitate the choice of the processing conditions.

4.5 Multiple Mechatronic Boards Managing Different Production Stages

Multiple mechatronic boards manage each stage of the whole production. Each motor machine or robot can be controlled by a properly designed interface. The board interfaces control the movements in 3D space: the typical robotic arm control in Figure 4.44 of a motor in the x, y, and z directions is composed of the following elements:

- Input sensors including a thermal camera monitoring the production process (the infrared thermal camera checks if a high temperature influences the production quality such as in cutting processing with cutting tools or lasers excessively heated).
- Output interface transferring commands to the motor.
- Motor moving an arm in 3D space.
- A robotic arm.
- A PLC standard protocol used for control and actuation processes.
- An AI engine supporting the control and the adaptive robotic arm movements.
- A DB saving all sensor data (the DB can be used to learn the AI model).
- A predictive maintenance scheduling of the robotic system as output of the AI algorithm.

The predictive maintenance takes into account the data analysis of the robotic arm: the AI engine is able to predict malfunctions detecting wrong movements. The production control becomes more complex if different robotic arms must be managed in a unique production line. In this case, the main problem to solve is to find the best choice of the production parameters allowing the synchronization of all robotic arms by fixing a specific volume of production. When different machines are involved in a production line, the control, actuation and synchronizations become difficult to set. Infrared thermography supports the dynamic processing auto-adapting the tool velocity step by step: in a complex processing, such as the treatment of a polymeric surface, the radiometric image is indicative of spatially following the surface to treat, by measuring locally the temperature in order to optimize the successive applications of the tool designated for the surface treatment. For a production machine in cascade configuration, such as the layout in Figure 4.45, the mechatronic boards are installed in each production stage: raw material is processed by the mechanical components, and the control and the actuation are managed by the mechatronic interfaces. For the cascade layouts, two main managing stages are defined. The first managing level of the control is related to the raw materials. The second managing level considers the synchronization of all the mechatronic boards, and is more difficult to implement: typically, the production times of the different stages are considered by defining the true timing and delays according to the production velocity of each stage. This estimation guarantees avoidance of time losses and, consequently, a decrease in the whole production volume. The AI engine optimizes the production activity, also defining the best choice of the production parameters, including motor velocities, thus avoiding product defects. The self-adaptive production in the Industry 5.0 scenario, takes into account the goal of "zero defect" production: the AI could adjust and tune automatically all the production parameters, by setting the maximum production velocity for a high product quality. When some parameters are uncontrollable it is preferable to apply predictive maintenance avoiding a cost increase due to waste material, or to a degradation of the standard quality of the product.

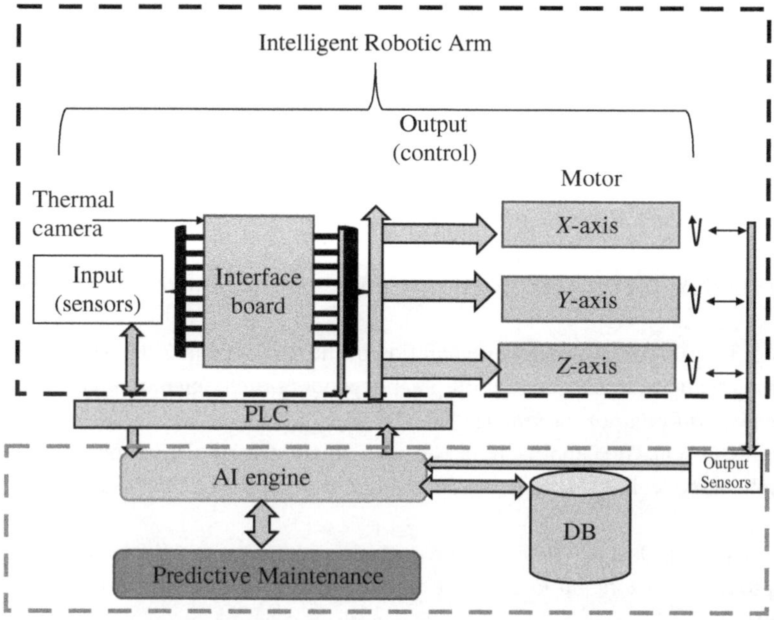

Figure 4.44 Mechatronic interface board controlling three motor axes of a robotic arm.

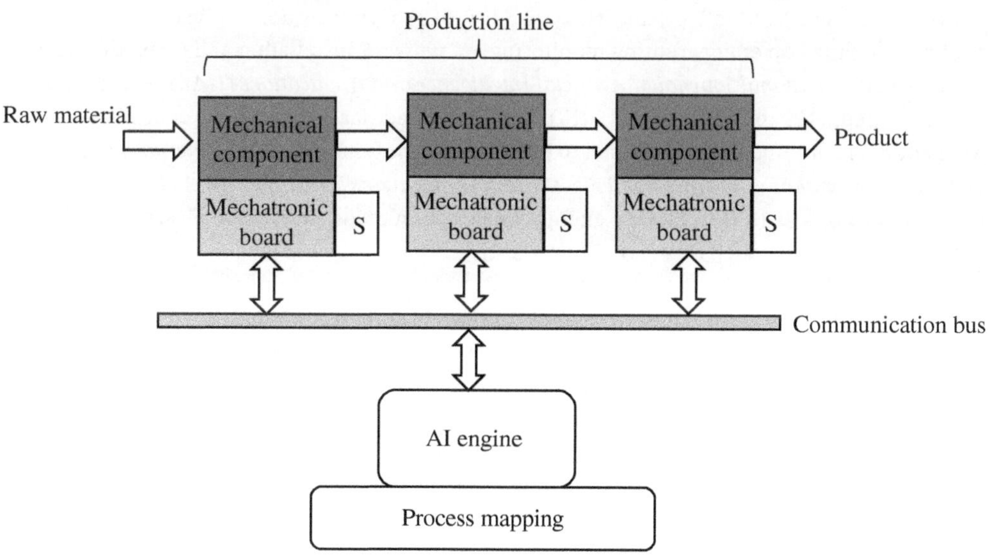

Figure 4.45 Multiple mechatronic boards managing a production line layout characterized by different production stages connected in cascade configuration (S, sensor allocation).

References

1 Massaro, A., Galiano, A., Savino, N. et al. (2020). Systems for an intelligent application of automated processes in industry: a case study from "PMI IoT Industry 4.0 Project". In: *IEEE Proceedings of MetroInd4.0&IoT 2020*, Rome, Italy (3–5 June 2020). Piscataway, NJ: IEEE.

2 Massaro, A., Giannoccaro, N.I., Spedicato, L. et al. (2014). Characterization of an innovative like-eye sensor for feature detection and robot sensing. *Measurement* 55 (1): 153–167.

3 Beul, M., Droeschel, D., Nieuwenhuisen, M. et al. (2018). Fast autonomous flight in warehouses for inventory applications. *IEEE Robotics and Automation Letters* 3 (4): 3121–3128.

4 Khosiawan, Y. and Nielsen, I. (2016). A system of UAV application in indoor environment. *Production & Manufacturing Research* 4 (1): 2–22.

5 Rudol, P., Wzorek, M., and Doherty, P. (2010). Vision-based pose estimation for autonomous indoor navigation of micro-scale Unmanned Aircraft Systems. Proceedings of the IEEE International Conference on Robotics and Automation (ICRA 2010). In: , Anchorage, USA (4–8 May 2010). Piscataway, NJ: IEEE.

6 Owens, J.D., Houston, M., Luebke, D. et al. (2008). GPU computing. *Proceedings of the IEEE* 96 (5): 879–899.

7 Syberfeldt, A. and Ekblom, T. (2017). A comparative evaluation of the GPU vs. the CPU for parallelization of evolutionary algorithms through multiple independent runs. *International Journal of Computer Science & Information Technology (IJCSIT)* 9 (3): 1–14.

8 Tartarini, D. and Massaro, A. (2011). GPU approach for Hertzian potential formulation tool oriented on electromagnetic nanodevices. *Progress In Electromagnetics Research (PIERM) Journal* 17 (1): 135–150.

9 Galiano, A., Massaro, A., Barbuzzi, D. et al. (2016). Face recognition system on mobile device based on web service approach. *International Journal of Computer Science and Information Technologies* 7 (4): 2130–2135.

10 Massaro, A., Barbuzzi, D., Galiano, A. et al. (2017). Simplified and efficient face recognition system on real image set and synthesised data. *International Journal of Biometrics* 9 (2): 143–162.

11 Massaro, A., Galiano, A., Meuli, G., and Massari, S.F. (2018). Overview and application of enabling technologies oriented on energy routing monitoring, on network installation and on predictive maintenance. *International Journal of Artificial Intelligence and Applications (IJAIA)* 9 (2): 1–20.

12 Massaro, A., Meuli, G., and Galiano, A. (2018). Intelligent electrical multi outlets controlled and activated by a data mining engine oriented to building electrical management. *International Journal on Soft Computing, Artificial Intelligence and Applications (IJSCAI)* 7 (4): 1–20.

13 Kim, J.-Y. and Cho, S.-B. (2019). Electric energy consumption prediction by deep learning with state explainable autoencoder. *Energies* 12 (4): 1–14.

14 Ullah, I., Khan, R.U., Yang, F., and Wuttisittikulkij, L. (2020). Deep learning image-based defect detection in high voltage electrical equipment. *Energies* 13 (2): 1–17.

15 Massaro, A., Maritati, V., Savino, N. et al. (2020). Blockchain and technologies matching with the case of study of vegetables production. *Information Technology in Industry (ITII)* 9 (2): 1–8.

16 Taylor, P.J., Dargahi, T., Dehghantanha, A. et al. (2020). A systematic literature review of blockchain cyber security. *Digital Communications and Networks* 6 (2): 147–156.

17 Mathew, A.R. (2019). Cyber security through blockchain technology. *International Journal of Engineering and Advanced Technology (IJEAT)* 9 (1): 3821–3824.

18 Nzuva, S. (2019). Smart contracts implementation, applications, benefits, and limitations. *Journal of Information Engineering and Applications* 9 (5): 63–75.

5

Advanced Robotics

The basic principles of mechatronic are "expanded" by designing artificial intelligence (AI) improvements in sensing and actuation. Different working processes can be achieved by industrial robots and Collaborative Robots (CoBots). The robot actions are executed by means of specific protocols and auto-adaptive controlling systems upgraded by the AI controllers. AI driven robots and exoskeletons represent the new advances of Industry 4.0, addressing industry production in the era of Industry 5.0. The analysis of motor control by signal processing is the basic study to design an advanced controller able to auto-adapt a robotic actuator, and to decrease disturbance effects during the actuation. The AI control can be implemented in many industrial processes including advanced manufacturing and additive manufacturing (AM). Starting with the main classification of industrial robots and protocols, this chapter analytically describes the movements of a robotic arm managed by intelligent control systems constituted by sensors and actuators working in a feedback system. The understanding of the programmable logic controller (PLC) programming functions, together with sensor and protocol integration, facilitate the comprehension of mechanisms addressing the upgrade of self-adaptive robotic motion using AI controllers. Another important aspect is the comprehension of the principles of actuators and motor actuations which is the basis to improve intelligent control of robots, and in general production machines. Different schemes integrating AI for motor control enhance the possibility to act directly on the voltage or electrical signals. The opportunity of an automatic control optimized by AI, managing variable electric signals traveling into a motor circuit, allows to adapt the actuation process into adaptive movements or regulations. AI is suitable for a self-learning data process orienting the production to full and adaptive control, according to the measurements provided by the sensors allocated on the production line. An important aspect is also the possibility to directly control signals in inverter circuits by providing the possibility to optimize the control, acting directly in the time domain of the source signal. The supervised and unsupervised AI algorithms provide a further possibility to optimize the traditional feedback systems for collaborative robotic and controlled AM. The advances in controlling robotic systems represent a new research topic for the upgrade of industry production in the era of Industry 5.0.

Electronics in Advanced Research Industries: Industry 4.0 to Industry 5.0 Advances, First Edition. Alessandro Massaro.

5.1 Collaborative Robotics in Industry and Protocols

Different applications are executed by CoBots oriented towards Industry 4.0 production systems such as:

- Polishing (the robot polishes product raw surfaces tuning the applied force).
- Injection molding.
- Laboratory analysis (automation in testing).
- Tightening (repeating the same movement several times with the same precision and speed to improve processing quality).
- Packaging and palletization.
- Gluing and dosing.
- Machine tending.
- Assembly.
- Picking and positioning.
- Milling.
- Cutting.
- Welding.
- Quality control (image processing, sensor data detection, etc.).

A preliminary robot classification is provided in Table 5.1.

Industrial robot identification is mainly based on a kinematic classification represented by the degrees of freedom of the arm. In particular, the following classes are distinguished:

- Robot with Cartesian geometry (arm along three orthogonal axes, according to the Cartesian coordinates x, y, z, having as features a high precision mainly for assembly and measuring processing, strong load capacity, but usually expensive and characterized by having considerable size).
- Robot with cylindrical geometry (arm moving within a cylindrical working volume, having a high speed, a strong loading capacity, and small size, but can suffer from poor precision and complex programming).
- Robot with polar or spherical geometry (arm moving within a spherical volume, having the advantages of a good working speed, small size, and simple programming, but sometimes can have poor accuracy).
- Robot with articulated geometry (arm moving along three rotation axes having high working speed, high precision, small size, and being inexpensive, but programming can be complex requiring the SCARA system to be included).

All the possible advantages and disadvantages are functions of the proposed technology. Technological advances have recently solved a lot these disadvantages by achieving a high standard precision for the specific applications. Intelligent industrial robots and controlled systems are representative of the fourth industrial revolution of Industry 4.0 [1, 2]. The evolution of the control systems is fundamental for upgrade of production according to quick production variations following market trends. Other classes of CoBots involve Automated Guided Vehicles, and exoskeletons supporting human resource activities in production. Possible common tasks of CoBots are summarized in Table 5.2, indicating a description of the specific robot activity.

Table 5.1 The main classifications for industrial robots.

Robot class	Characteristics
Linear robots	Applications: • handling for plastic molding • sealing • laser welding • pressing Cartesian robot has arms with three prismatic joints (axes coincident with a Cartesian coordinate system)
Articulated robots	Applications: • handling for metal casting • welding • painting • packaging • palletizing • handling for forging This robot is characterized by an arm having at least three rotary joints, high payload capacity, and flexible mounting possibilities by permitting a combination with SCARA systems
Parallel robots	Applications: • picking and placing • assembly • handling Robots have concurrent prismatic or rotary joints
Cylindrical robots	Robot with axes controlled in a cylindrical coordinate system
SCARA robots	Applications: • product assembly • packaging This robot is characterized by two parallel rotary joints to provide compliance in a plane
Other robots	Applications: • Robots in hazardous environments. • Operations under water. • Operations in atmospheres containing combustible gases • Operations in space with Automated Guided Vehicles
Collaborative and auto-adaptive robots	Application in all production processes in manufacturing, including worker security and load assisted systems (collaborative exoskeletons). The robots can be interfaced by artificial intelligence engines

SCARA, selective compliance articulated robot arm.

Kinematic schemes and configurations are related to six and seven axes concerning spherical wrists, parallel axes, offset wrists, and spherical joints layouts [14]. Some commercial light robots are discussed in [15].

Table 5.2 Robot task classification.

Working process	Task	References
Co-manipulation	Object handling by means of haptic interaction, collaborative energy injection, and interpretation of flexible and rigid object swinging. The co-manipulation minimizes the effect of overloading joint torque, maximizes the human manipulation capacity, adapts the human physical fatigue, involves force exchange between the human and the robot (the force exchanged is not usually constant), facilitates the grasping, and generally helps operators in decreasing the weight and the repetitiveness of their activities. Furthermore, co-manipulation robots improve strength enhancement, weight compensation, and movement guidance	[3]
Positioning	Positioning of the product in a preferred position. Computer vision techniques determine the object's position and orientation. The positioning function also concerns object localization and object alignment	[4]
Robotic handover	Handover assistance. Handover involves the transfer of an object from a giver to a receiver. For this function, joint actions enabling physical collaborations are important	[5]
Assembly	Assembly in production processes. Assembly actions include screwing, insertion, fixing and other functions	[6]
Pick and place	Pick and place according to object distance. The pick and place automated process accelerates the process of picking up parts or items and placing them into different places	[7]
Fetches	CoBot fetches parts for assembly. In this class are included CoBots fetching the assembled parts and placing them at a number of specified assembly stations. CoBots for fetching accelerate the assembling process	[8]
Soldering	Wire and point soldering. The soldering operations are executed by high temperature thermal processes in which the metals are joined by melting or fusing their contacting surfaces	[9]
Illumination	CoBot directing illumination to a workspace. CoBot vision inspection systems are important for object recognition	[10]
Inspection	Screwing check and other kinds of inspection	[11]
Drilling/surface finish	Drilling holes. The drilling robotic system can adopt visual inspection	[12]
Screwing	Bolt insertions into holes. Screwing for assembly operations and unscrewing for disassembly	[13]

CoBot, collaborative robot.

5.1.1 Data Protocols

Robots in industry are managed by data protocols traveling in the information system. Table 5.3 lists some protocols adopted for industrial robots.

5.1.2 Basic Concepts of Robotic Arms and Control Improvement

Universal Robots (UR) such as UR3, UR5, UR10, and UR16 (payloads of 3, 5, 10, and 16 kg, respectively) are economically suitable and represent a flexible technology. All the UR have the same Linux-based operating system, a controlled arm with six degrees of freedom, and different input/output (I/O) ports including USB, Ethernet, Modbus, digital, and analog. Open Platform

Table 5.3 Robot protocols.

Protocol	Properties
PROFIBUS	RS-485 cables, fiber optic cable or MBP
EtherCat	Ethernet 10/100/1000 Mbps
CC-Link	RS-485 coaxial cable, 10 Mbps
CC-Link LT	Dedicated flat cable, 2.5 Mbps, 625 Kbps, or 125 Kbps
CC-Link Safety	Based on CC-Link
CC-Link IE Control	Ethernet 1 Gbps (fiber optic cable)
Modbus-TCP/IP	Ethernet 10/100/1000 Mbps
Modbus RTU	Serial cable (RS-232, RS-485)
DeviceNet	CAN bus with twisted pair cables, 1 Mbps
ControlNet	RG-6 coaxial cables, 5 Mbps
EtherNet/IP	Ethernet 10/100/1000 Mbps
FlexRay	Serial bus (data rates up to 10 Mbps)

Communications' Unified Architecture (OPC-UA) information protocol are adopted for robotized manufacturing applications [16]. A robot operating system combines different frameworks to program a robotic system. The mechanical structure generally used in robots consists of an articulated chain of mechanical elements connected to each other by means of joints. The degree of freedom D of a mechanical part is estimated by Grübler's formula:

$$D = 6n - \sum_{i=1}^{j} (6 - f_i) \tag{5.1}$$

where n is the number of moving links, j is the number of joints, and f_i is the degree of freedom of the ith joint. The most common kinematic link is constituted by elements in series, where the elements of the chain are connected to each other with rotoidal couplings (relative motion of rotation) and prismatic ones (relative motion of translation). The rigid body in space is centered on its own center of gravity (o, x', y', z'). With translations along the three axes, the rigid body carries out movements in space, moving with respect to a fixed Cartesian triple (x, y, z). The free rigid body in space is said to have six degrees of freedom, three of which are translation and three of rotation. Figure 5.1a shows a configuration of six degree of freedom of a robotic arm, and Figure 5.1b shows the movement configuration in the (x, y) plane.

Concerning a point moving in a plane, a basic transformation is described by the following rotation matrix R around the z-axis (axis perpendicular to the plane (x,y) of Figure 5.2a):

$$\begin{bmatrix} V_x \\ V_y \end{bmatrix} = R \begin{bmatrix} V^N \\ V^O \end{bmatrix} = \begin{bmatrix} \cos(\theta) & -\sin(\theta) \\ \sin(\theta) & \cos(\theta) \end{bmatrix} \begin{bmatrix} V^N \\ V^O \end{bmatrix} \tag{5.2}$$

Both rotation and translation are described by the following equation referring to Figure 5.2b:

$$\begin{bmatrix} V_x \\ V_y \end{bmatrix} = \begin{bmatrix} P_x \\ P_y \end{bmatrix} + \begin{bmatrix} \cos(\theta) & -\sin(\theta) \\ \sin(\theta) & \cos(\theta) \end{bmatrix} \begin{bmatrix} V^N \\ V^O \end{bmatrix} \tag{5.3}$$

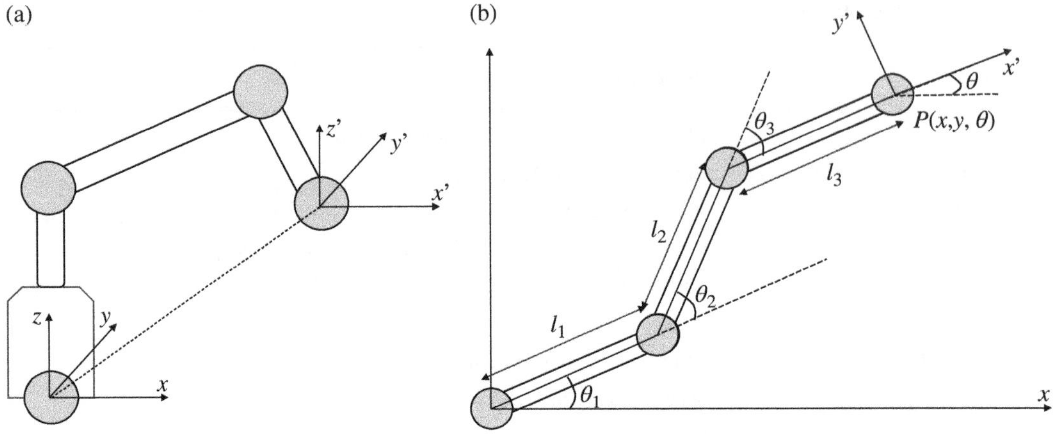

Figure 5.1 (a) Robotic arm with joints connected relatively to each other. (b) Kinematic model in two-dimensional space.

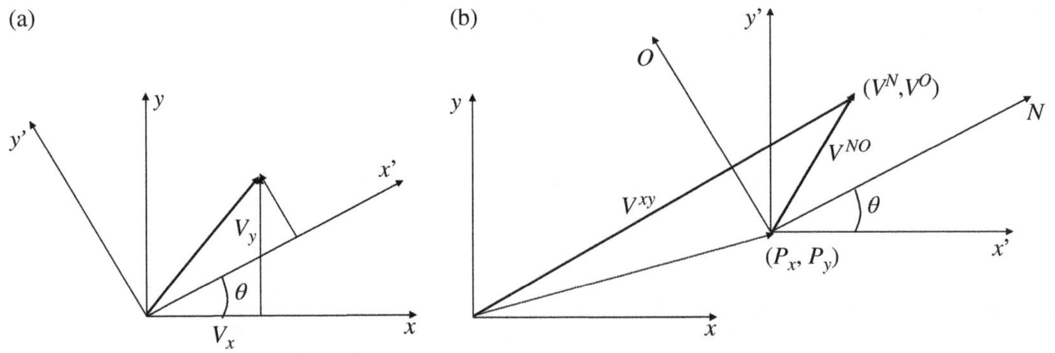

Figure 5.2 (a) Coordinate system determining the rotation matrix in the (x,y) plane. (b) Coordinate system defining translation and rotation.

Concerning the rotation matrix in three-dimensional (3D) space around the *z*-axis (R_z), the *y*-axis (R_y), and the *x*-axis (R_x), respectively:

$$R_z = \begin{pmatrix} \cos(\theta) & -\sin(\theta) & 0 \\ \sin(\theta) & \cos(\theta) & 0 \\ 0 & 0 & 1 \end{pmatrix} \tag{5.4}$$

$$R_y = \begin{pmatrix} \cos(\theta) & 0 & \sin(\theta) \\ 0 & 1 & 0 \\ -\sin(\theta) & 0 & \cos(\theta) \end{pmatrix} \tag{5.5}$$

$$R_x = \begin{pmatrix} 1 & 0 & 0 \\ 0 & \cos(\theta) & -\sin(\theta) \\ 0 & \sin(\theta) & \cos(\theta) \end{pmatrix} \tag{5.6}$$

In the 3D space a 4 × 4 matrix is able to describe both rotation and translation. By considering the coordinate system in Figure 5.3a, the following H matrix indicates translation without rotation:

$$H = \begin{pmatrix} 1 & 0 & 0 & P_x \\ 0 & 1 & 0 & P_y \\ 0 & 0 & 1 & P_z \\ 0 & 0 & 0 & 1 \end{pmatrix} \tag{5.7}$$

Rotation without translation in 3D space is configured by the coordinate systems in Figure 5.3b, and by the following generic 4×4 matrix representing rotation around the z-axis, x-axis, and y-axis, or a combination of the three:

$$H = \begin{pmatrix} n_x & o_x & a_x & 0 \\ n_y & o_y & a_y & 0 \\ n_z & o_z & a_z & 0 \\ 0 & 0 & 0 & 1 \end{pmatrix} \tag{5.8}$$

The combination of rotation and translation is described by the following matrix equation:

$$\begin{bmatrix} V_x \\ V_y \end{bmatrix} = H \begin{bmatrix} V^n \\ V^o \\ V^a \\ 1 \end{bmatrix} \begin{pmatrix} n_x & o_x & a_x & P_x \\ n_y & o_y & a_y & P_y \\ n_z & o_z & a_z & P_z \\ 0 & 0 & 0 & 1 \end{pmatrix} \tag{5.9}$$

The rotation matrix around the x-axis in 3D space is provided by:

$$R_x = \begin{pmatrix} 1 & 0 & 0 & 0 \\ 0 & \cos(\theta) & -\sin(\theta) & 0 \\ 0 & \sin(\theta) & \cos(\theta) & 0 \\ 0 & 0 & 0 & 1 \end{pmatrix} \tag{5.10}$$

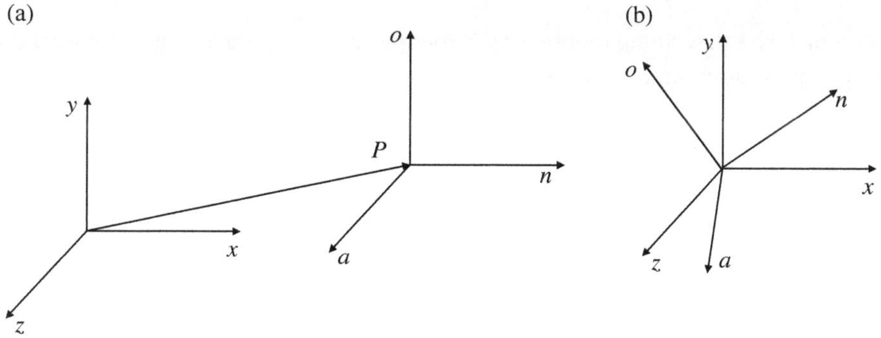

Figure 5.3 (a) 3D coordinate system describing 3D translation. (b) 3D coordinate systems indicating rotation in 3D space.

The translation along the x-axis in 3D space is expressed by:

$$T_x = \begin{pmatrix} 1 & 0 & 0 & P_y \\ 0 & 1 & 0 & 0 \\ 0 & 0 & 1 & 0 \\ 0 & 0 & 0 & 1 \end{pmatrix} \quad (5.11)$$

The following matrix indicates rotation around the z-axis:

$$R_z = \begin{pmatrix} \cos(\theta) & -\sin(\theta) & 0 & 0 \\ \sin(\theta) & \cos(\theta) & 0 & 0 \\ 0 & 0 & 1 & 0 \\ 0 & 0 & 0 & 1 \end{pmatrix} \quad (5.12)$$

The translation along the z-axis is instead defined by:

$$T_x = \begin{pmatrix} 1 & 0 & 0 & 0 \\ 0 & 1 & 0 & 0 \\ 0 & 0 & 1 & P_y \\ 0 & 0 & 0 & 1 \end{pmatrix} \quad (5.13)$$

By considering the two-dimensional (2D) space, the equations describing the vector algebra kinematic solution of a robot end effector are:

$$\begin{aligned} x &= l_1 \cos\theta_1 + l_2 \cos(\theta_1 + \theta_2) + l_3 \cos(\theta_1 + \theta_2 + \theta_3) \\ y &= l_1 \cos\theta_1 + l_2 \cos(\theta_1 + \theta_2) + l_3 \cos(\theta_1 + \theta_2 + \theta_3) \\ \theta &= \theta_1 + \theta_2 + \theta_3 \end{aligned} \quad (5.14)$$

The same set of equations, referred to a system composed of only l_1 and l_2 segments (without segment l_3), is:

$$\begin{aligned} x &= l_1 \cos\theta_1 + l_2 \cos(\theta_1 + \theta_2) \\ y &= l_1 \cos\theta_1 + l_2 \cos(\theta_1 + \theta_2) \\ \theta &= \theta_1 + \theta_2 \end{aligned} \quad (5.15)$$

The transformation matrix T describing motion of the robotic arm in Figure 5.1b is given by the translation matrix $T_{x,y}$ and rotation matrix $R_{x,y}$ as:

$$T = R_{x,y}(\theta_1) T_{x,y}(\theta_1) R_{x,y}(\theta_2) T_{x,y}(\theta_2) R_{x,y}(\theta_3) T_{x,y}(\theta_3) \quad (5.16)$$

The matrix T is then defined by the following blocks T_{ij}:

$$T = \begin{pmatrix} T_{11} & T_{12} & T_{13} \\ T_{21} & T_{22} & T_{23} \\ T_{31} & T_{32} & T_{33} \end{pmatrix} \quad (5.17)$$

$$T = \begin{pmatrix} \cos(\theta_1 + \theta_2 + \theta_3) & -\sin(\theta_1 + \theta_2 + \theta_3) & l_2 \cos(\theta_1 + \theta_2) + l_1 \cos(\theta_1) + l_3 \cos(\theta_1 + \theta_2 + \theta_3) \\ \sin(\theta_1 + \theta_2 + \theta_3) & \cos(\theta_1 + \theta_2 + \theta_3) & l_2 \sin(\theta_1 + \theta_2) + l_1 \sin(\theta_1) + l_3 \sin(\theta_1 + \theta_2 + \theta_3) \\ 0 & 0 & 1 \end{pmatrix} \quad (5.18)$$

where the submatrix is composed of the elements T_{11}, T_{12}, T_{21}, and T_{22} representing the rotation, and T_{13} and T_{23} characterize the translation in the (x,y) plane. Focusing attention on the translation part, the coordinates x and y are assigned to T_{13} and T_{23}, respectively:

$$\begin{aligned} x &= l_2 \cos(\theta_1+\theta_2) + l_1 \cos(\theta_1) + l_3 \cos(\theta_1+\theta_2+\theta_3) \\ y &= l_2 \sin(\theta_1+\theta_2) + l_1 \sin(\theta_1) + l_3 \sin(\theta_1+\theta_2+\theta_3) \end{aligned} \tag{5.19}$$

By estimating the derivatives, the following 2×3 Jacobian matrix groups the equation in the following form:

$$\begin{pmatrix} \dot{x} \\ \dot{y} \end{pmatrix} = \begin{pmatrix} -l_2 \sin(\theta_1+\theta_2) - l_1 \sin(\theta_1) - l_3 \sin(\theta_1+\theta_2+\theta_3) & -l_2 \sin(\theta_1+\theta_2) - l_3 \sin(\theta_1+\theta_2+\theta_3) & -l_3 \sin(\theta_1+\theta_2+\theta_3) \\ l_2 \cos(\theta_1+\theta_2) + l_1 \cos(\theta_1) + l_3 \cos(\theta_1+\theta_2+\theta_3) & l_2 \cos(\theta_1+\theta_2) + l_3 \cos(\theta_1+\theta_2+\theta_3) & l_3 \cos(\theta_1+\theta_2+\theta_3) \end{pmatrix} \cdot \begin{pmatrix} \dot{\theta}_1 \\ \dot{\theta}_2 \\ \dot{\theta}_3 \end{pmatrix} \tag{5.20}$$

The rotation matrix is generalized as:

$$R = \begin{pmatrix} T_{11} & T_{12} \\ T_{21} & T_{22} \end{pmatrix} = \begin{pmatrix} \cos(\theta_1+\theta_2+\theta_3) & -\sin(\theta_1+\theta_2+\theta_3) \\ \sin(\theta_1+\theta_2+\theta_3) & \cos(\theta_1+\theta_2+\theta_3) \end{pmatrix} = \begin{pmatrix} \cos(\theta) & -\sin(\theta) \\ \sin(\theta) & \cos(\theta) \end{pmatrix} \tag{5.21}$$

with

$$\theta = \theta_1 + \theta_2 + \theta_3 \tag{5.22}$$

The derivative of the variable θ is expressed in terms of joint velocities as:

$$\dot{\theta} = \dot{\theta}_1 + \dot{\theta}_2 + \dot{\theta}_3 = (111) \begin{pmatrix} \dot{\theta}_1 \\ \dot{\theta}_2 \\ \dot{\theta}_3 \end{pmatrix} \tag{5.23}$$

All the equations are combined by means of the following 3×3 Jacobian matrix by providing the spatial velocity in the plane:

$$\begin{pmatrix} \dot{x} \\ \dot{y} \\ \dot{\theta} \end{pmatrix} = \begin{pmatrix} -l_2 \sin(\theta_1+\theta_2) - l_1 \sin(\theta_1) - l_3 \sin(\theta_1+\theta_2+\theta_3) & -l_2 \sin(\theta_1+\theta_2) - l_3 \sin(\theta_1+\theta_2+\theta_3) & -l_3 \sin(\theta_1+\theta_2+\theta_3) \\ l_2 \cos(\theta_1+\theta_2) + l_1 \cos(\theta_1) + l_3 \cos(\theta_1+\theta_2+\theta_3) & l_2 \cos(\theta_1+\theta_2) + l_3 \cos(\theta_1+\theta_2+\theta_3) & l_3 \cos(\theta_1+\theta_2+\theta_3) \\ 1 & 1 & 1 \end{pmatrix} \cdot \begin{pmatrix} \dot{\theta}_1 \\ \dot{\theta}_2 \\ \dot{\theta}_3 \end{pmatrix} \tag{5.24}$$

The trajectory in the plan can be programmed and adjusted during the robot processing: AI algorithms can change the coordinates and velocities after data processing, thus improving the work. The self-adaptive solution is appropriate to change rotations and translations without human intervention. In this way, all the variables defined in the kinematic analysis can vary according to the AI commands. A particular AI application driving a robotic arm is based on the image vision technique: the velocity and the position can change after image processing, and data analysis of positioning sensors provides the new trajectory to execute. Figure 5.4 shows an example of a functional scheme controlling trajectory and tasks of a robotic arm by:

- Pressure force detection.
- Torque detection.
- Other sensor data acquisition (positioning, temperature, etc.).
- Image vision technique (object recognition by image processing technique and defect classification by AI algorithms).

All data are useful to optimize the grasp control (force alignment and object reach direction control), and to automatize the automatic rejection of defective piece following *in-line* processing.

The positioning functions of robotic arms are usually performed with direct current (DC) motors and closed loop regulations. For positioning applications, it is necessary, at each step, to compare the position and the speed reached with the reference ones.

5.1.3 Collaborative Exoskeleton Communication System Protocols

Exoskeletons [17–19] are important robotic systems supporting human activity in different production activities. The exoskeletons are classified mainly as (Figure 5.5):

- arm support (Figure 5.5a)
- back support (Figure 5.5b)
- leg support (Figure 5.5c)

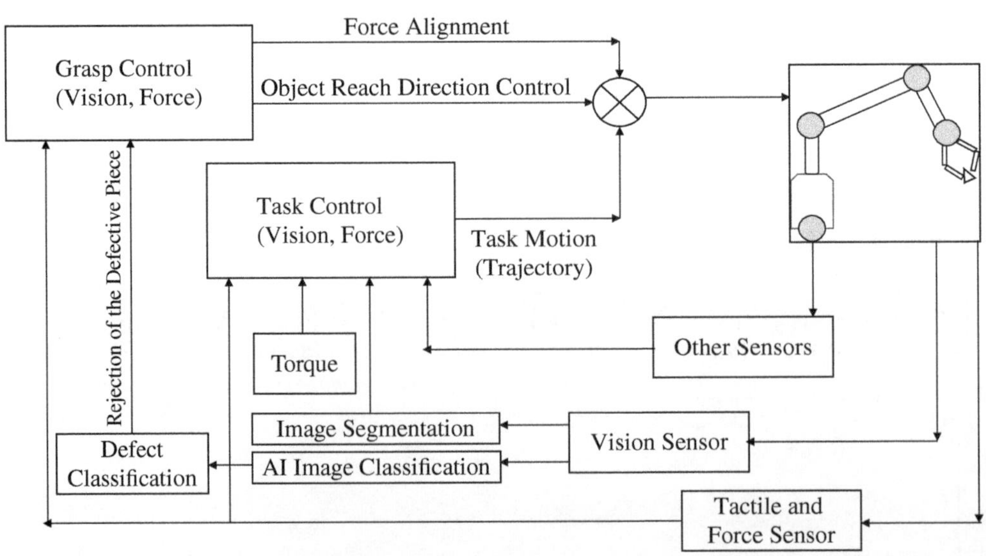

Figure 5.4 Functional scheme of a robotic arm controlled by sensors, image vision and AI supporting defect recognition and automatic defective piece rejection.

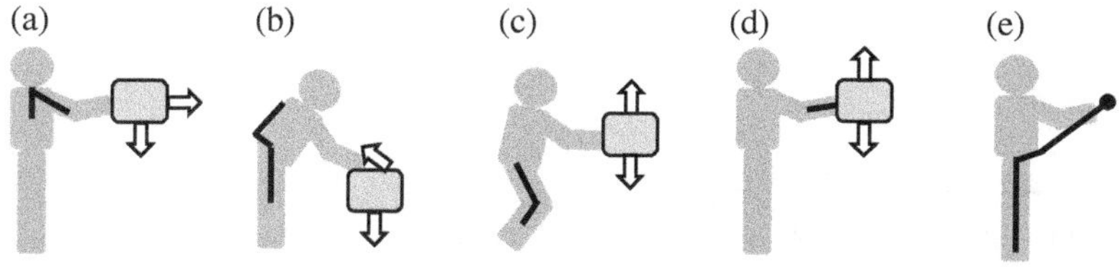

Figure 5.5 Exoskeleton configurations in industry and applied forces: (a) arm support, (b) back support, (c) leg support, (d) power glove, and (e) tool holding.

- power glove (Figure 5.5d)
- tool holding (Figure 5.5e)

The exoskeleton classification is based on interaction with the human body according to the parts of the body involved. Two main classes are defined:

- Active exoskeletons, which use mechanical transmission components to provide additional force to the movement of the worker, thus increasing performance.
- Passive exoskeletons, which are limited to supporting the movement of the worker helping the worker to maintain the same position for a long period.

A further classification is related the parts of the body involved (upper or lower part).

In accordance with intelligent exoskeleton systems, Figure 5.6 shows a scheme integrating AI in the control of the robotic system, by analyzing three different level of data flow: level 1 is related to the sensor and actuator system of the working parts of the exoskeleton; level 2 is of the working framework; level 3 takes into account the data processing enabled by a properly designed user interface. At least one sensor for the motor and one for the actuator can be adopted for output-angle measurements and calibration. More complex actuators integrate force and torque measurement units. The control unit behaves as a sensor and actuator gateway between sensors, and as a safety system with real-time reaction time. The AI processor checks and identifies possible errors such as motor current, actuator angles, joint limit positions, and other mechanical errors including kinematic anomalies. The "quasi" real-time response can automatically correct the errors and the human working process. The training of the AI model allows to adopt the exoskeleton in self learning modality: the exoskeleton analyzes and classifies the human movements, by comparing them with movements and trajectories contained in the training dataset and, consequently, accompanies human movement during the operations. All the central units are linked in order to construct and also measure complex movements by guiding synchronized movements according to the specific work to perform. For the quasi real-time data processing, the AI calculation can be performed by cloud processing, by local powerful server machines adopted for data processing, or by powerful processors as exoskeleton equipment.

5.1.4 Advanced Robotics and Intelligent Automation in Manufacturing: Logic Conditions and PLC Programming

In order to analyze how the intelligence acts in robotic manufacturing systems, PLC programming approaches are discussed by analyzing different examples implementing logic conditions. The logic conditions can be derived by the AI algorithms defining automatically coded instructions as commands for the actuators enabling robotic actions.

A modality to program the PLC system is provided by ladder logic (LL) diagrams. LL is a simple graphical programming language that is used to program a PLC, expressing logic operations with

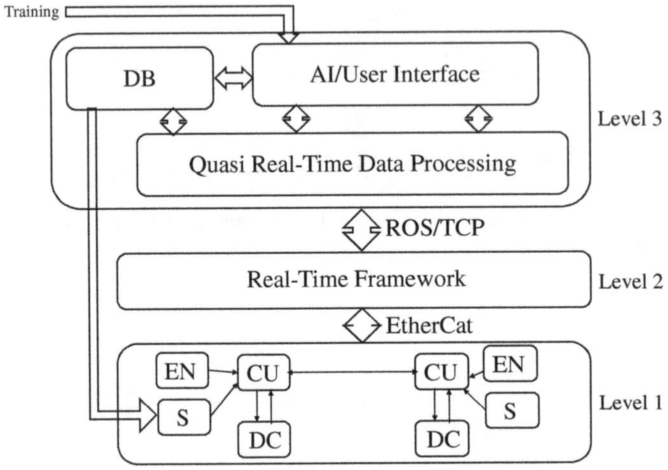

Figure 5.6 Exoskeleton communication model integrating AI (S, pressure and torque sensor; DC, DC motor; CU, control unit; EN, encoder; DB, database).

symbolic notation. The LL diagrams are able to execute logic and sequential tasks modeling industrial automation applications and are suitable for programming.

A control program in a LL diagram generally is characterized by the following main steps:

1. Determine the sequences of the control system.
2. Select the PLC input and output devices (sensors and actuators).
3. Draw a control diagram, using the LL diagram symbols (the data process must be structured in the correct sequence).
4. Code the diagram symbols in a format suitable for insertion into the central processing unit (CPU).
5. Transfer the coded chart (program) to the CPU.
6. Execute the program.
7. Check if the program contains any errors.
8. Correct errors and save the finished program.

The primary evaluation of the operations to be controlled by the PLC is very important. The operations are represented by the I/O variable definition. The input data are mainly provided by sensors, whilst the outputs are actuators, servos, display systems, etc. The number of entry points, added to the number of exit points, gives the total number of I/O points, which must never exceed the maximum capacity of the controller. Depending on the I/O modules selected, the input and output signals can have voltages or different currents, thus achieving an optimal adaptation of the PLC for the best processing conditions. Many programmable controllers use the concept of I/O definition. A design table is appropriate to summarize the relationships between all the PLC system signals, where the table is representative to determine the sequence of events. It is necessary to find the relationship of each controlled device with the others linked to it. Then, it is necessary to code the LL. The logic conditions are related to the I/O ports of the PLC system illustrated in Figure 5.7, where the PLC hardware represents the control system to which sensors and actuators are then connected for the processing of related information. The implementation of the AI algorithms is mainly from the evolution of Industry 4.0 to Industry 5.0 by introducing the new concept of self-adapted production. The main limits to solve are in finding technologies to be integrated in the hardware systems, providing a good performance for quasi real-time production processes: the

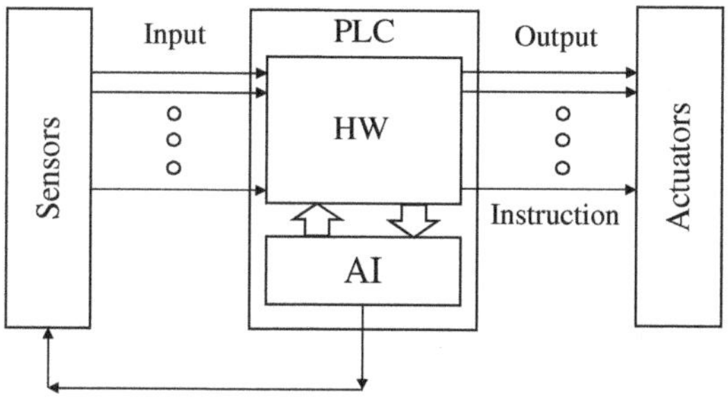

Figure 5.7 PLC scheme enabling AI instructions (HW, hardware).

computational cost must follow the production time, and must provide a quick robot response for rapid production, by generating high production volumes and preserving quality.

The LL diagrams can translate the outputs of the AI algorithms executing specific instructions. The diagram is typically structured by simple symbols indicating a satisfied, or an unsatisfied logic condition. All the logic conditions are summarized into a table indicating the devices detecting the signal (sensors), or actuating an instruction (actuators), the memory address (I 0.00, I 0.01, etc.), and the description of the actions of each element of the diagram. The AI algorithms provide threshold values to control the robotic arm or to set processing parameters according to the sensor signals monitoring the production and possible anomalies (defective products, misalignments, etc.).

In the example of Figure 5.8, three sensors are placed to monitor a robotic process (sensor 1, sensor 2, sensor 3). The self-adaptive classification is performed by an AI algorithm providing the following AND condition of an incorrect robotic operation:

- Sensor 1 detects a high value (address input value I 0.00).
- Sensor 2 does not detect a high value (corresponding to the detection of a low value and to an address input value of I 0.01).
- Sensor 3 detects a high value (address input value I 0.02).

If the three conditions are satisfied then the actuator decreases the motor speed (output address O 100.00), thus ensuring the production quality (instruction transmitted to the motor actuator).

The high and the low values are estimated by comparing threshold values provided by AI processing. The three conditions can refer to the simultaneous measurements of temperatures detected in different points, thus defining the correct robotic processing condition for cutting or drilling.

The logic condition can be related to OR logic conditions. In the example in Figure 5.9 the following case is programmed:

- Sensor 1 detects a low value (address input value I 0.00).
- Sensor 2 detects a low value (address input value I 0.01).
- Sensor 3 detects a high value (address input value I 0.02).

If at least one condition is satisfied, then the actuator moves the robotic arm by rejecting a defective piece. The example can be referred to an application involving image processing of different parts of a produced piece. Three image segmentation algorithms can be applied for defect classification:

- The first algorithm indicates too low or too high pixel intensity of a pixel cluster.
- The second algorithm indicates low or too high pixel intensity of a pixel cluster.
- The third algorithm indicates a high probability of detection of a defect by applying a classification algorithm based on a convolutional neural network (CNN).

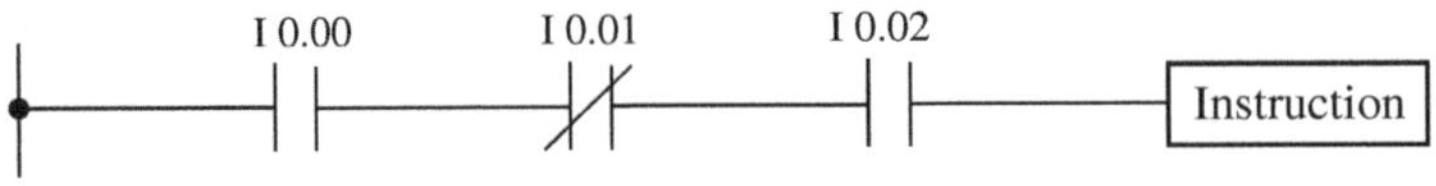

Device	Address	Description
Sensor 1	I 0.00	High value of Sensor 1
Sensor 2	I 0.01	AND low value of Sensor 2
Sensor 3	I 0.02	AND high value of Sensor 3
Actuator	O 100.00	Actuator decreases the motor speed

Figure 5.8 PLC basic program and related table description: AND logic implementation.

The indication of anomalous pixel intensity denotes a possible inhomogeneous welding region as an anomalous colder part when good welding is characterized by a homogeneous hot part. The CNN algorithm function is to complete the analysis, by proving the first segmentation results.

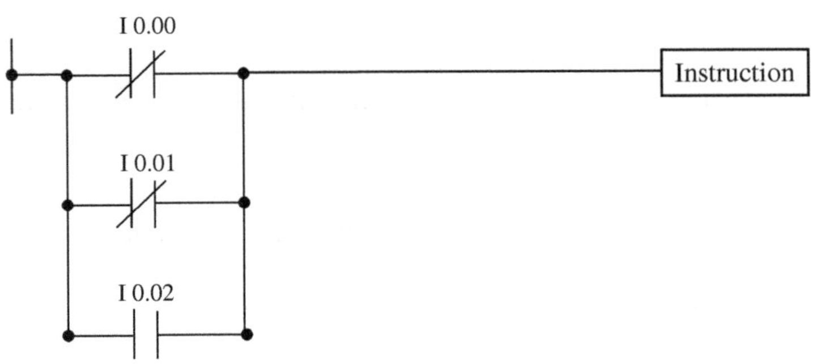

Device	Address	Description
Sensor 1	I 0.00	Low value of Sensor 1
Sensor 2	I 0.01	OR low value of Sensor 2
Sensor 3	I 0.02	OR high value of Sensor 3
Actuator	O 100.00	Actuator moves the robotic arm

Figure 5.9 PLC basic program and related table description: OR logic implementation.

A more complex program is structured by hybrid logic conditions joining AND/OR logic, as in the example in Figure 5.10 where the following conditions are programmed:

- Sensor 1 detects a low value (address input value I 0.00).
- Sensor 2 detects a low value (address input value I 0.01).
- Sensor 3 detects a high value (address input value I 0.02).
- Sensor 4 detects a high value (address input value I 0.03).

The following pseudocode is defined:

```
if the first AND the second conditions are satisfied;
OR
the third condition is satisfied;
AND the fourth condition is satisfied;
then execute the instruction to decrease the speed processing.
```

Finally, in the example in Figure 5.11 a switching condition is represented by implementing the following pseudocode:

```
if the S1 value is low AND the S2 value is high AND the S3 value
is high;
OR
the S1 value is high AND the S2 value is high AND the S3 value is low;
OR
the S1 value is high AND the S2 value is high AND the S3 value is high;
then the circuit is switched enabling a motion.
```

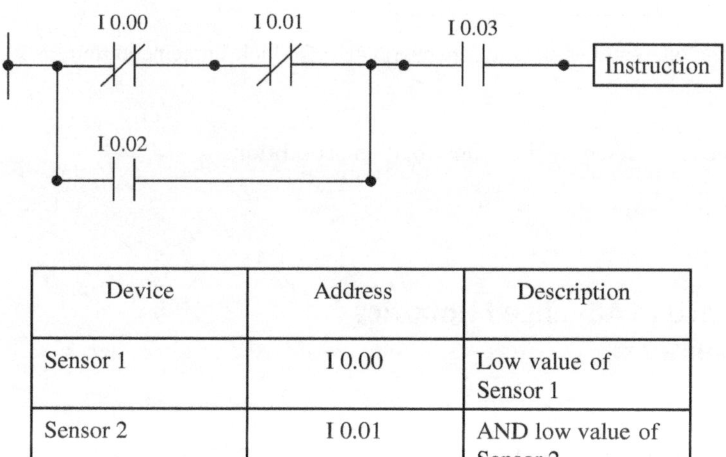

Device	Address	Description
Sensor 1	I 0.00	Low value of Sensor 1
Sensor 2	I 0.01	AND low value of Sensor 2
Sensor 3	I 0.02	OR high value of Sensor 3
Sensor 4	I 0.03	AND high value of Sensor 4
Actuator	O 100.00	Actuator decreases the processing speed

Figure 5.10 PLC basic program and related table description: hybrid AND/OR logic implementation.

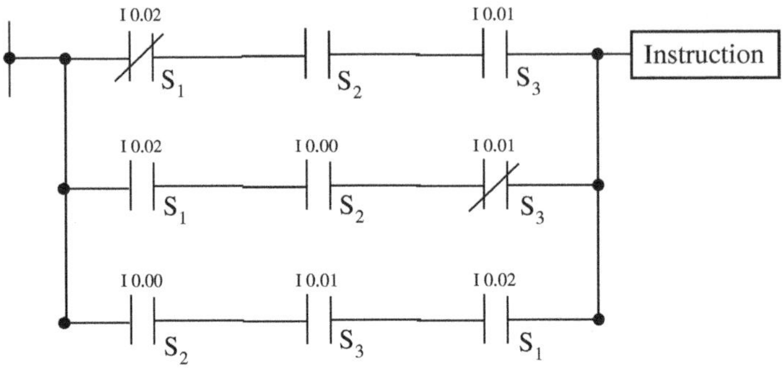

S_1	S_2	S_3	Instruction (1:on; 0:off)
0	0	0	0
0	0	1	0
0	1	0	0
0	1	1	1
1	0	0	0
1	0	1	0
1	1	0	1
1	1	1	1

Figure 5.11 PLC basic program and related table description: hybrid AND/OR logic implementation.

The output of the pseudocode is modeled by the following logic condition:

$$\text{Instruction} = \overline{S_1}S_2S_3 + S_1S_2\overline{S_3} + S_1S_2S_3 \tag{5.25}$$

5.2 Artificial Intelligence in Advanced Robotics and Auto-Adaptive Movement

AI will be used to improve adaptive control of robot movement in the future. The auto-adaptive motion in advanced robotics is achieved by AI actuator control. A full overview of actuation principles and related technologies facilitates the comprehension of the intervention in advanced robotics.

5.2.1 General Technological Aspects about Auto-Adaptive Motion in Advanced Robotics

Technological aspects are modeled by a circuital approach highlighting how electric signals interact with the physical part of the actuators in embedded systems.

5.2.1.1 Main Aspects of Electrostatic Actuators

General technological aspects of advanced robotic systems are derived from the basic principle of actuation. Electrical actuators are adopted for motion control thus generating mechanical motion through electrical energy conversion to mechanical energy (Figure 5.12). There are two main categories of electrical actuators involving the application of an electric field, or of a magnetic field, generating mechanical forces.

Some common cases of actuators take into account:

- Hydraulic pumps (electric motors that increase or decrease the flow rate of a liquid).
- Solenoid valves (opening or closing the passage of a liquid).
- Servo motors (moving part of the machine to a desired position).

Concerning electrical actuators using an electric field, the generated force is restricted by the polarization charge process of a dielectric. Figure 5.13a shows the basic principle of an electrostatic actuator indicating how polarization charges are generated in dielectric materials.

Figure 5.13b shows the application of the polarization charge principle by implementation in a typical electrostatic actuator: when voltage is applied to the capacitor, consisting of electrodes and dielectric materials, polarization charges are induced in the dielectric so that the true charge on the electrode and the polarization charge have opposite charges. The attractive forces between opposite charges are adopted for producing actuation forces: a torque, due to the electric charges, is generated by the rotor when the voltage is applied to the pair of electrodes. To continue the motion in the same direction, the source voltage is then switched step by step to another pair of electrodes.

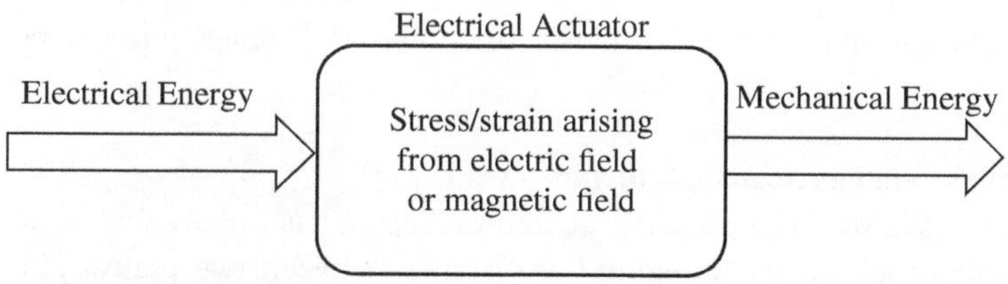

Figure 5.12 Block diagram of an electrical actuator.

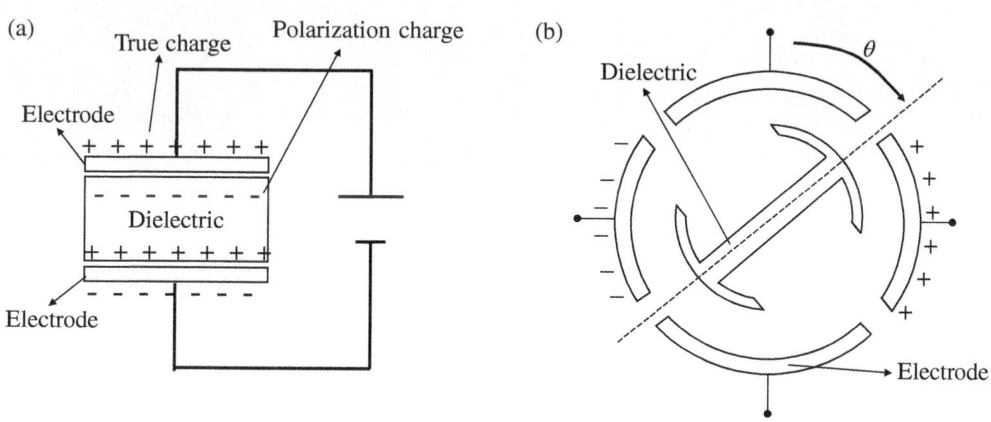

Figure 5.13 (a) Polarization charge of a capacitor. (b) Implementation of the capacitor in an electrostatic actuator by means of a dielectric rotor.

Let W_{el} be the electrostatic co-energy stored in the actuator air-gap between electrodes and dielectric material. This energy is expressed as a function of the applied voltage V and the displacement angle of the mover θ (rad). The torque **T** (Nm) of the actuator is given by the following equation:

$$\mathbf{T} = \frac{\partial W_{el}(V,\theta)}{\partial \theta} \tag{5.26}$$

where $W_{el}(V, \theta) = VQ - W_{el}$, where Q is the electric charge stored in the system. In the example in Figure 5.14 an electrostatic actuator is shown where the two capacitances individuated between the terminals A-A′ and B-B′ are approximated by the following equations (expressed in farads):

$$C_A(\theta) = C_0 + C_1 \cos 2\theta \tag{5.27}$$

$$C_B(\theta) = C_0 - C_1 \cos 2\theta \tag{5.28}$$

By applying a two-phase AC voltage expressed in volts by the voltage generators in Figure 5.14:

$$V_A = V_m \cos(\theta + \delta) \tag{5.29}$$

$$V_B = V_m \cos\left(\theta + \delta - \frac{\pi}{2}\right) \tag{5.30}$$

the electrostatic co-energy is expressed in joules as:

$$W_{el} = \frac{1}{2} C_A V_A^2 + \frac{1}{2} C_B V_B^2 \tag{5.31}$$

Equations (5.26) and (5.31) highlight that torque control is a function of the voltage phase to be controlled.

5.2.1.2 Microelectromechanical System Electrostatic Actuators

Microelectromechanical system (MEMS) electrostatic actuators behave as micro-actuators. A MEMS actuates by a capacitor of area S by changing the distance between two plates. The

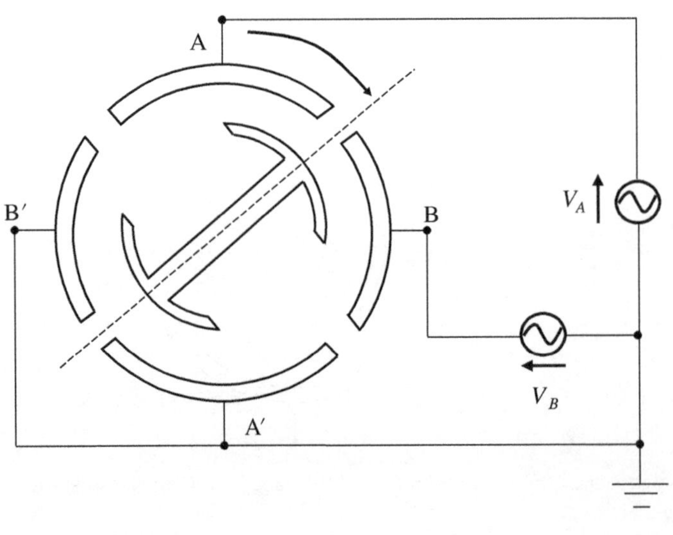

Figure 5.14 Implementation of the electrostatic actuator.

MEMS model is constituted by a movable plate with a mass M (kg) subject to the electrostatic force, which is denoted as F (N), by a viscous force with coefficient b (Ns m^{-1}), and by a restoring force with the spring constant k (N m^{-1}). The capacitance between the plates is provided by the relationship:

$$C(r) = \frac{\varepsilon_0 S}{r} \tag{5.32}$$

where ε_0 is the vacuum permittivity ($8.8541878128 \times 10^{-12}$ F m^{-1}). The electrostatic co-energy is expressed as a function of voltage V and the displacement r:

$$W_{el}(V,r) = \frac{1}{2}C(r)V^2 = \frac{1}{2}\frac{\varepsilon_0 S}{r}V^2 \tag{5.33}$$

The force is then provided by:

$$F = \frac{\partial W_{el}(V,r)}{\partial r} = -\frac{\varepsilon_0 S}{2r^2}V^2 \tag{5.34}$$

The analytical model describing the electrostatic MEMS actuation is provided by the following differential equation related to the circuit in Figure 5.15:

$$M\frac{\mathrm{d}^2 r}{\mathrm{d}t^2} + b\frac{\mathrm{d}r}{\mathrm{d}t} + kr = \frac{\varepsilon_0 S}{2r^2}V^2 \tag{5.35}$$

So, the voltage control of the source V changes the mass M position, actuating a control signal.

5.2.1.3 Piezoelectric Actuators

Piezoelectric actuators are characterized by the piezoelectric effect: when some materials are subjected to a mechanical force, they become electrically polarized; conversely, when an electric field is applied to the materials, physical deformation is observed to cause forces. These two phenomena are called the piezoelectric effect and the inverse piezoelectric effect, respectively. Some commercial piezoelectric materials are:

- silicon dioxide (SiO_2)
- lead zirconate titanate (PZT)
- barium titanate ($BaTiO_3$)
- polyvinyledene fluoride (PVDF)

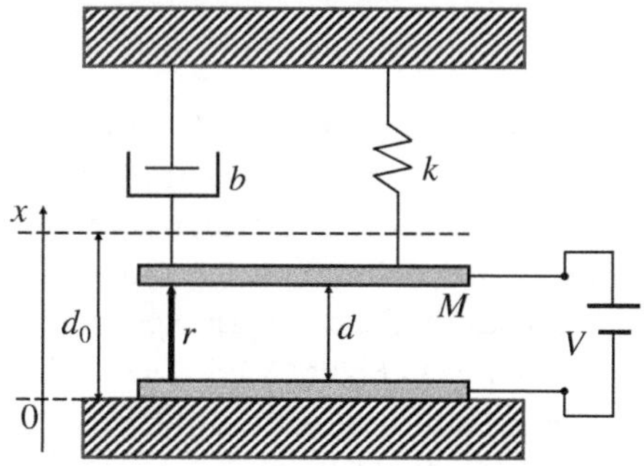

Figure 5.15 Electrostatic actuator in MEMS configuration.

Figure 5.16a shows the basic principle of the piezoelectric actuator system: the material is exposed to an electric field generating an electric polarization **P** coinciding with the applied field. The deformation includes the thickness mode (called the d_{33} effect), which increases its thickness in the direction of the applied electric field, and the planar mode (called the d_{31} effect), which is defined in-plane. The strain tensor is given by:

$$\varepsilon_k = d_{ik}E_i + S_k^E \sigma_k \tag{5.36}$$

where ε_k is the strain tensor and d_{ik} is the matrix of the piezoelectric constants. By considering for example the strain effect generated in the circuit in Figure 5.16b relating to a PZT plate, the following parameters can be calculated:

- PZT plate thickness l.
- Electrical field intensity $\mathbf{E}_3$ inside the plate.
- Source voltage V.

A displacement due to the strain of Δl = 60 nm is obtained for d_{33} = 400×10^{-12} m V^{-1} (PZT value), l = 1.5 mm, V = 100 V, with the electric field defined as $\mathbf{E}_3 = V/l$ (V m^{-1}). In order to obtain a stronger deformation, the scheme in Figure 5.17 is considered, representing a large number of thin piezoelectric layers with electrodes attached so that neighboring polarity appears alternately along the stacking direction. By assembling a multilayer structure with more layers, the total deformation is increased for a fixed source voltage, while decreasing the thickness of each layer, a lower source voltage is required for a strain. The actuators potentially exhibit large force and high speed of motion and return very quickly to their neutral position when the source is disabled. Consider now the plate geometry stack configuration in Figure 5.18, supposing N = 80 thin plates with l = 3 mm, w = 2 mm, and h = 65 μm. By applying a source voltage of 150 V, the displacement is:

$$\Delta h = N \cdot d_{33} V = 80 \times 400 \times 10^{-12} \times 150 = 480 \times 10^{-8} = 4.8 \mu m \tag{5.37}$$

The area of the of each plate is:

$$A = 3 \times 10^{-3} \times 2 \times 10^{-3} = 6 \times 10^{-6} \, m^2 \tag{5.38}$$

The total height of the actuator is:

$$H = 6.5 \times 10^{-5} \times 80 == 520 \times 10^{-5} \, m \tag{5.39}$$

The generated force expressed in newtons (N) is then provided by:

$$F_{33} = \frac{Y \cdot A \cdot \Delta h}{H} \tag{5.40}$$

5.2.1.4 DC Motor Actuation

Electric actuators are also enabled by magnetic fields. Actuators using magnetic fields have a wide variety of uses ranging from home appliances to industrial machines. Two main classes are defined in terms of the voltage source: DC machines and alternating current (AC) ones. Furthermore, AC machines are divided into two main subcategories: asynchronous and synchronous. The induction motor is a typical form of asynchronous machine. Figure 5.19a illustrates the basic principle of the electromagnet constituting an element of the whole DC motor sketched in Figure 5.19b. Due to the function of the brushes, and of the commutators connected to the armature coils, the armature coil currents produce a stationary magnetomotive force mismatched by 90 electrical degrees to the

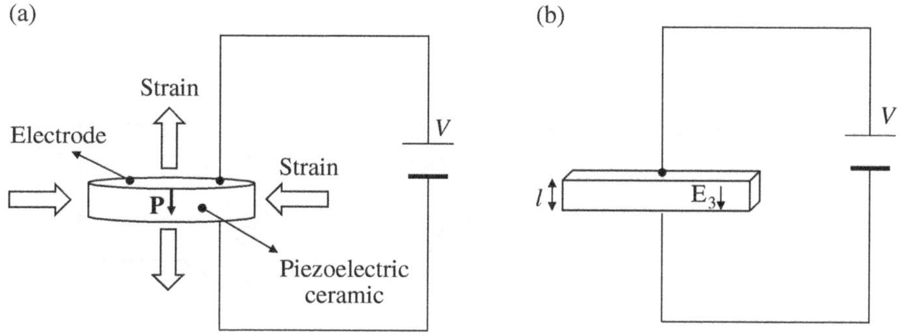

Figure 5.16 (a) Basic principle of piezoelectric actuation. (b) Strain caused in a piezoelectric plate.

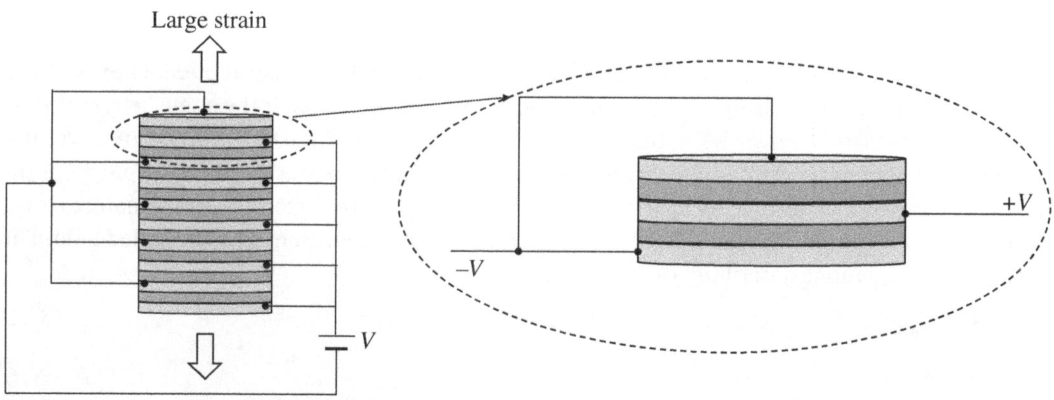

Figure 5.17 Schematic of a piezoelectric actuator for large strain effect.

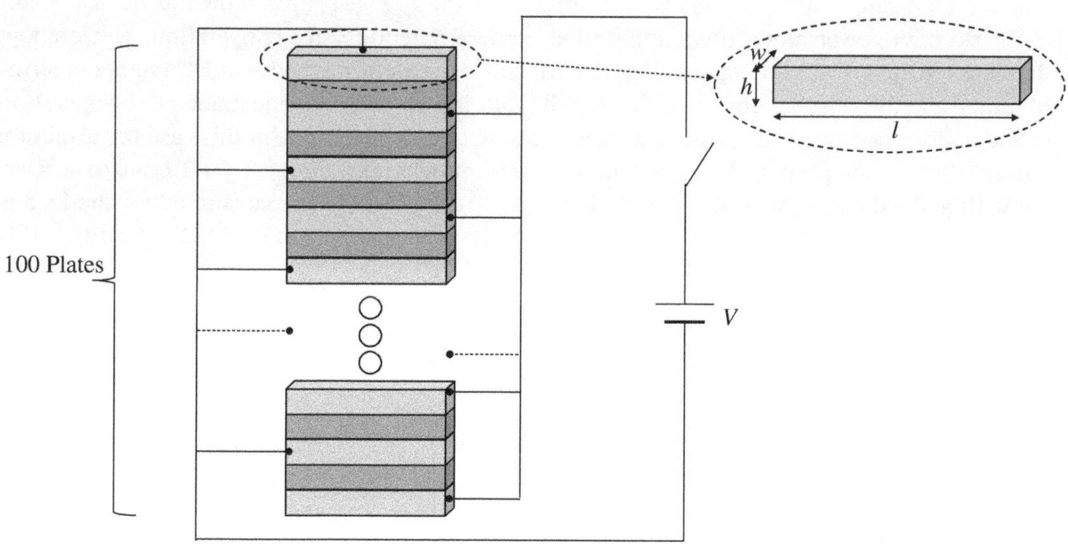

Figure 5.18 Multi-layer plate piezoelectric actuator model.

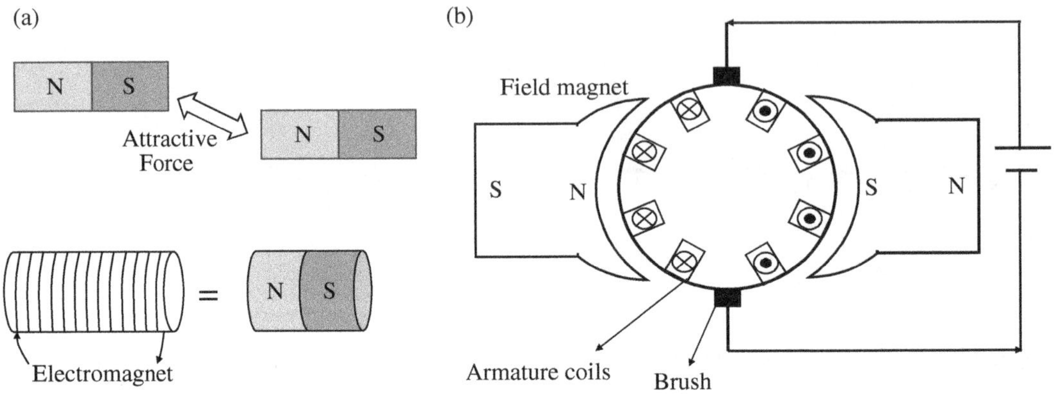

Figure 5.19 DC motor by magnetic field: (a) basic principle of electromagnet equivalence and attractive forces; (b) scheme of an actuator controlled by a magnetic field.

magnetomotive force related to the field magnets. This relationship makes it possible to establish the generated torque proportional to the value of the armature currents. If the angle between those two magnetomotive forces varies in time, the torque is not proportional to the armature current.

Figure 5.20 illustrates the equivalent circuit of a DC motor, where v is the applied voltage, **E** the induced voltage (in terms of electromagnetic field), i_a the armature current, R_a the armature resistance, L_a the armature inductance, **T** the produced torque, and **F** represents the generated field. The fundamental equations describing the model are:

$$\mathbf{E} = K_a \Phi \omega_n \tag{5.41}$$

$$\mathbf{T} = K_a \Phi i_a \tag{5.42}$$

where K_a is a constant and Φ is the magnetic flux. In a normal operation, the applied voltage v is larger than **E** so that the armature current i_a flows in the same direction as the applied voltage. This condition indicates that the energy is transferred from the voltage source to the mechanical load.

Typical high-power motor drive applications generally require steady operation, acceleration, deceleration, and reverse operation. Figure 5.21 shows the working modes of DC motors in terms of torque and the velocity. Condition I (forward motoring operation) is the fundamental operation of the motors: assuming the positive direction of the circuit clockwise, in this case the armature current flows in the positive direction with v positive, producing a torque $\mathbf{T} = KT\, i_a > 0$ to accelerate with a speed $\omega_n > 0$, with $KT = K_a\Phi$, $\mathbf{E} = K_E\omega_n > 0$, $\omega_n T > 0$ (power consumed for the load is positive and the electric power is transferred from the power supply to the load). Condition II is

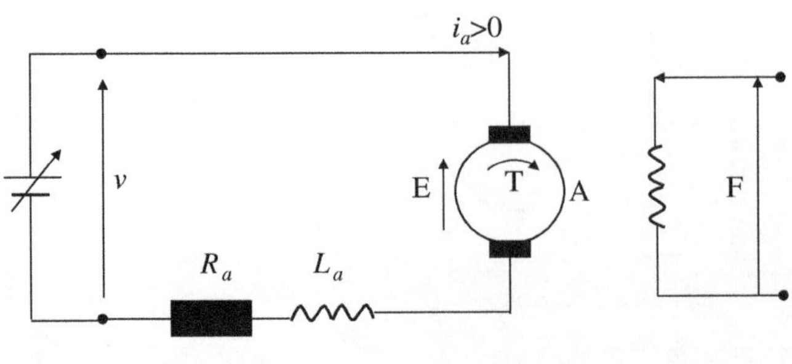

Figure 5.20 Equivalent circuit of a DC motor.

named forward regeneration (second quadrant of Figure 5.21). In this case the applied voltage is $v < \mathbf{E}$, the motor drive is in a forward condition (ω_n>0), and the armature current flows in the reverse direction (i_a<0, the torque $\mathbf{T} = KT\, i_a < 0$, producing braking torque, and $\omega_n T < 0$ indicating the conversion of the mechanical power into electrical power as for electrical and hybrid vehicles). The reverse of the running direction is achieved with $v < 0$. The working condition of quadrant III is the same as the forward motoring, except for the directions of the physical variables. Finally, quadrant IV indicates the reverse regeneration as the quadrant is the counterpart of the forward regeneration of the second quadrant.

The induction motors use the physical behavior of the eddy currents, induced in conductors by the time varying magnetic field, producing electric forces on the currents (the induced currents are also modeled by equivalent magnets). When eddy currents are induced in the motor conductors by the rotating magnetic field, which is caused by the stator AC currents, then the equivalent magnets located at both the stator and the rotor produce forces generating a torque. The rotor conductor is like the shape of a squirrel cage (Figure 5.22a). A three-phase voltage source is typically used in industrial applications: typical structures are shown in (Figure 5.22 b–e). In the configuration in Figure 5.22b, on the stator are the primary coils having structures of distributed windings generating a sinusoidal magnetic field along the rotor surface. Moreover, on the rotor are the secondary conductors where the electromagnetic field is induced so that the secondary currents flow. Indicating the frequency of the source voltage by f and the number of magnetic poles generated by the stator currents by p, the rotating speed of the rotating magnetic field (synchronous speed) is given by:

$$N_s = \frac{2f}{p} \tag{5.43}$$

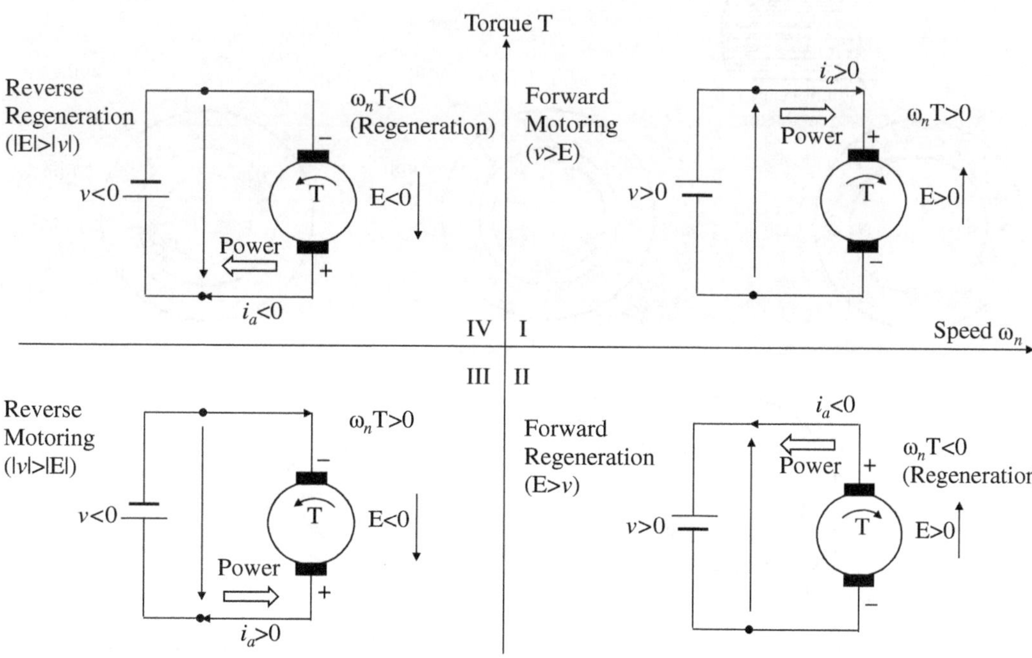

Figure 5.21 Operation mode of a DC motor in four quadrants.

measured in radians per second (RPS). The formula is transformed to revolutions per minute (RPM) as:

$$N_s = \frac{120f}{p} \tag{5.44}$$

The rotor speed must be lower than the synchronous speed to generate torque, i.e. $N_r < N_s$. The slip parameter s is defined as:

$$s = \frac{N_s - N_r}{N_s} = \frac{\omega_s - \omega_r}{\omega_s} \tag{5.45}$$

where ω_s is the synchronous angular speed (rad s^{-1}), and ω_r is the rotor angular speed (rad s^{-1}). Equation (5.4) is typically expressed as a percentage. If the rotor current is zero, the motor does not produce a torque. In the absence of slip, operation of the motor is not possible. Electric machines that have rotors rotating at the same speed as the rotating magnetic field produced by the armature coils are referred to as synchronous machines. A typical synchronous motor consists of armature coils, installed at the stator, and field magnets at the rotor. A constant force is produced when two sets of magnets face each other at a fixed distance. Figure 5.22c shows a surface-mounted permanent magnet (SPM) synchronous motor scheme producing a torque. Instead of placing a rotor with SPM, a non-circular iron core also generates forces by position-dependent reluctance (Figure 5.22d). Figure 5.22e shows an interior permanent magnet (IPM) synchronous motor scheme.

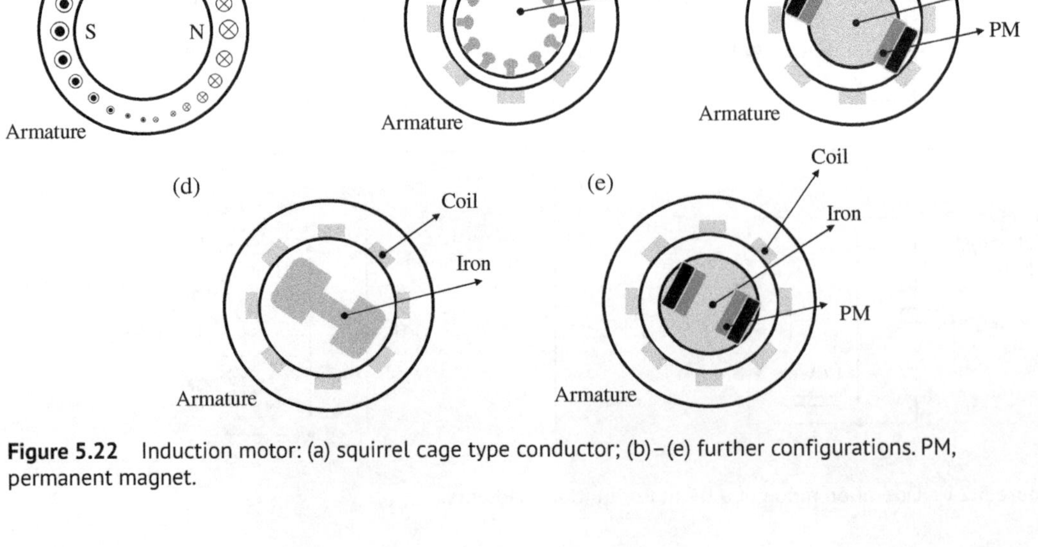

Figure 5.22 Induction motor: (a) squirrel cage type conductor; (b)–(e) further configurations. PM, permanent magnet.

5.2.1.5 Intelligent Control Integrating AI: Speed Regulation

In order to understand how it is possible to improve actuation by means of AI, it is important first to understand the analogy between a mechanical model and an electrical one. Concerning a DC electrical motor, the complete mechanical (Figure 5.23a) and electrical (Figure 5.23b) models provide the following equations describing the physical behavior of the DC motor. The mechanical model is described by:

$$k_{\text{me}} \cdot I - C_{\text{res}} - k_f \omega - J \cdot \frac{\mathrm{d}\omega}{\mathrm{d}t} = 0 \tag{5.46}$$

where $C_{\text{me}} = k_{\text{me}} \cdot I$ is the magnetoelectric torque caused by the current flowing in the rotor winding, C_{res} is the resistance torque of the mechanical part that must rotate, $k_f \cdot \omega$ is the torque due to the friction of moving parts, and $J \cdot \frac{\mathrm{d}\omega}{\mathrm{d}t}$ is the torque due to inertia (inductance which is null at steady state).

The circuit model of Figure 5.23b is described by:

$$V_g - RI - L \cdot \frac{\mathrm{d}i}{\mathrm{d}t} - k_e \cdot \omega = 0 \tag{5.47}$$

where V_g is the voltage source, R is the electrical resistance, and $L \cdot \frac{\mathrm{d}I}{\mathrm{d}t}$ is the voltage contribution due to electrical inductance (at steady state this term is null).

By considering the steady-state condition, the electrical current is provided by Eq. (5.47) as:

$$I = \frac{V_g}{R} - \frac{k_e}{R} \cdot \omega \tag{5.48}$$

The torque that drives the engine is then defined by:

$$C_{\text{me}} = \frac{k_{\text{me}} \cdot V_g}{R} - \frac{k_{\text{me}} \cdot k_e}{R} \cdot \omega \tag{5.49}$$

An advanced controlled system is provided by the scheme in Figure 5.24 with the following elements:

- DC engine controlled by a voltage signal V_g enabled by an actuation command.
- Sensors reading the working processing of the DC motor.
- Data acquisition module collecting sensor data.
- Database (DB) system.
- AI interface processing data sensors.
- Data protocol module (encoding the AI output structuring it in the used protocol).

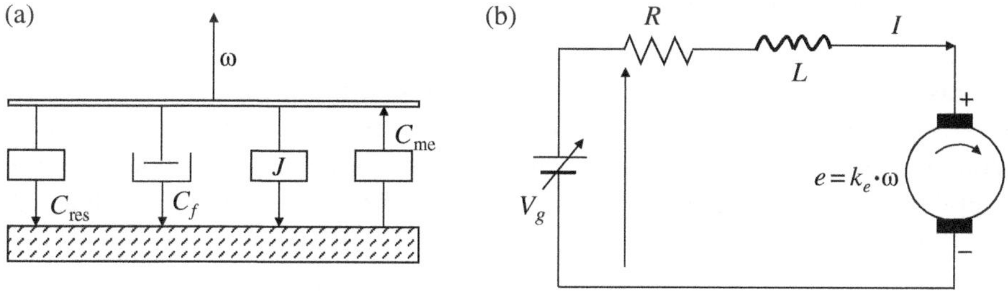

Figure 5.23 (a) Mechanical scheme and (b) electrical model of a DC motor.

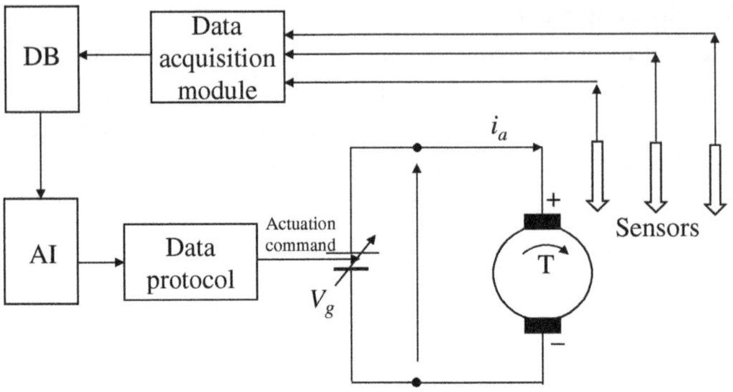

Figure 5.24 Advanced controlled system of a DC motor by AI algorithm implementation.

The systemic model of the controlled system is partially obtained by applying the Laplace transform able to convert the model equation from the frequency domain to the time domain. The transformed equations of the mechanical and electrical models are, respectively:

$$k_{\text{me}} \cdot I(s) - C_{\text{res}}(s) - k_f \cdot \omega(s) - s \cdot J \cdot \omega(s) = 0 \tag{5.50}$$

$$V_g(s) - R \cdot I(s) - s \cdot L \cdot I(s) - k_e \cdot \omega(s) = 0 \tag{5.51}$$

Using Eqs. (5.50) and (5.51), the frequency domain electrical current is obtained:

$$I(s) = \frac{V_g - k_e \cdot \omega(s)}{R + s \cdot L} \tag{5.52}$$

Using Eq. (5.50), the speed is obtained:

$$\omega(s) = \frac{k_{\text{me}}}{k_f + s \cdot J} \cdot I - \frac{C_{\text{res}}}{k_f + s \cdot J} \tag{5.53}$$

Equation (5.53) is modeled by the systemic circuit model in Figure 5.25 coupled at the input by the AI interface. The reference speed is controlled by the variable source V_g which is tuned by:

- Comparing the measured speed with the reference speed as for the basic feedback system.
- Setting the V_g by considering speed prediction (prediction of the speed variation by means of AI calculation).
- Setting the V_g by considering defect prediction (by analyzing processing anomalies).
- Simultaneously comparing data of sensors monitoring the processing (AI classification of correct and incorrect processing by the simultaneous analysis of different sensor data).

The AI interface can be implemented by a flexible programming language, such as Python, to decode the AI outputs into a programmable logic controller PLC data protocol.

The armature resistance control of the DC motor is modeled by the circuit in Figure 5.26a indicating the connection of a shunt motor implementing the armature resistance control method: in this approach, a variable resistor R_1 is added in the armature circuit and is changed by the output of the AI algorithm. The theoretical trend of the motor speed control versus the armature current is illustrated in Figure 5.26b: the effect of the series resistor is to decrease the speed, and the AI control unit can tune the armature current by properly setting the resistance value.

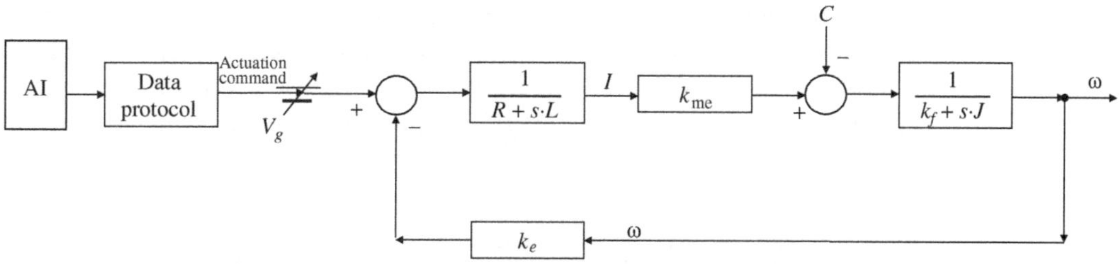

Figure 5.25 Systemic model of the DC motor feedback controlled by the AI model.

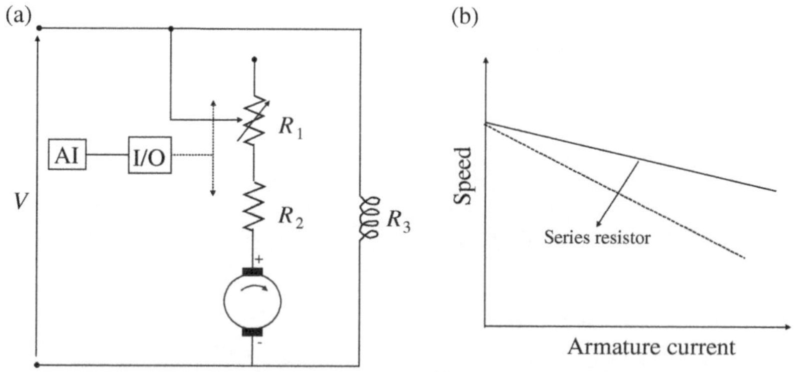

Figure 5.26 (a) Shunt motor and AI controlling electrical current. (b) Theoretical trend of the motor speed versus the armature current by adding the effect of the series resistor.

In the model of speed control of the DC series motor by the armature resistance control method in Figure 5.27a, control is accomplished by varying the armature circuit resistance. The variable resistance reduces the applied voltage to the armature, thus decreasing the motor speed. The theoretical speed–current characteristic of a series motor is illustrated in Figure 5.27b: when the value of variable resistance R_1 is increased, the motor decreases its speed. The AI algorithm can control the R_1 setting during the processing time according to sensor data processing.

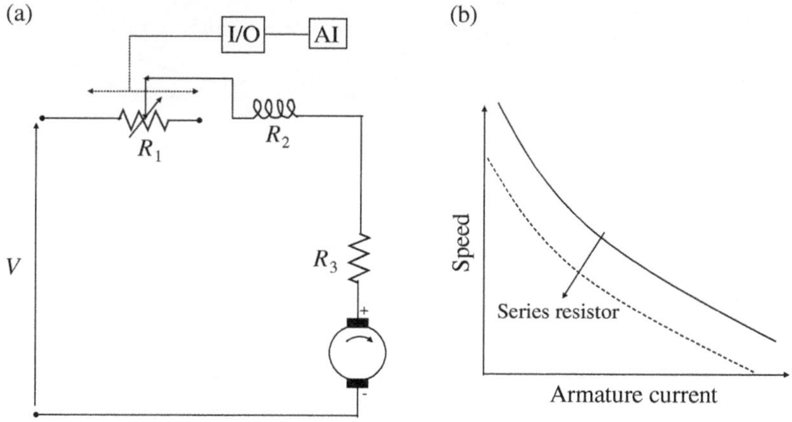

Figure 5.27 (a) Series motor and AI controlling resistance. (b) Theoretical trend of the motor speed versus the armature current by adding the effect of the series resistor.

Concerning the field flux control method of the DC motor, flux is produced by the field current. Speed control by this method is performed by controlling the field current. In the shunt motor, modeled by the circuit in Figure 5.28a, the variable resistor R_1 (shunt field regulator) is connected in series with the shunt field windings. The shunt current is provided by the following equation:

$$I_{SH} = \frac{V}{R_1 + R_3} \tag{5.54}$$

The electrical connection of R_1 reduces the field current, thus decreasing the field flux. The flux decrease increases the motor speed. The theoretical relationship between motor speed and torque is illustrated in Figure 5.28b.

In the model of the DC series motor in Figure 5.29a, a variable resistance R_1 is connected in parallel with the series field windings: part of the main current flows through the variable resistance R_1, thus reducing the current flowing through the field winding. The reduction in field current decreases the amount of flux and, consequently, the speed of the motor increases (see theoretical behavior in Figure 5.29b). Also, in this case, the AI algorithms drive the resistor values and the electrical current.

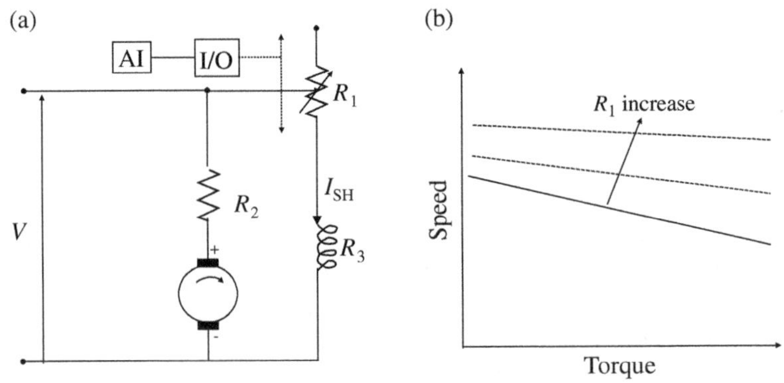

Figure 5.28 (a) DC shunt motor: modeling of the field flux control method. (b) Theoretical trend of the motor speed versus the torque by increasing the variable resistance R_1.

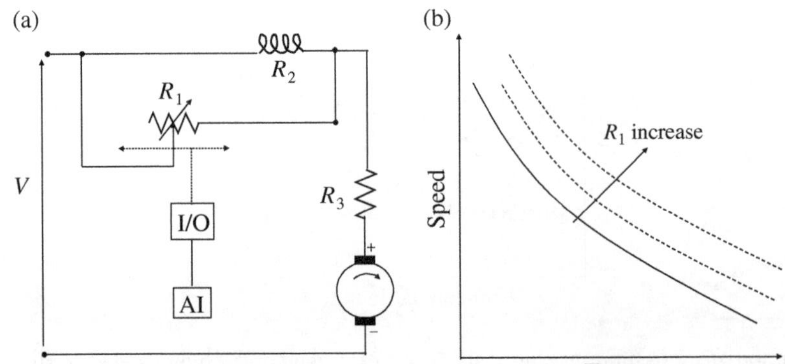

Figure 5.29 (a) DC series motor modeling. (b) Theoretical trend of the motor speed versus the torque by increasing the variable resistance R_1.

5.2.2 Improvement of Collaborative Exoskeletons by Auto-Adaptive Solutions Implementing Artificial Intelligence

Exoskeletons operate by supporting human activities such as for rehabilitation applications [20]. Human-driven motion reinforcement approaches are applied to support human physical activity, adapting the robotic arm trajectories required for a production line. The auto-adaptive motion control of trajectories is shown in Figure 5.30, composed of the following elements:

- Robot enabling motion trajectories q.
- Human forces supported by exoskeleton.
- Sensors (force sensors, torque sensors, position sensors, and others).
- Kinematic robot controller driven by the AI kernel.
- AI level constituted by a calculation kernel, a training module, a testing module, and a DB; all elements necessary for data processing.

The AI kernel activates the trajectory adaptation, generating movements according to the desired motion. In order to accelerate the AI calculation, the training and the testing modules are interfaced with another temporary DB behaving as data buffering. In order to preserve memory allocation, the buffer DB must be canceled after each operation, and after the data migration (data migration in the main dataset). All data are stored into the main DB collecting historical data useful for training. The trajectory adaptions occur with the following possible conditions:

- The sensors detect a strong human force, by decreasing the exoskeleton force (deceleration for a safety mode operation).
- The human trajectory is wrong (the AI automatically adjusts the trajectory, by comparing the position data of the position sensor with the correct positions allowed in the exoskeleton program; the AI also predicts the next human trajectories by analyzing the working movement thus anticipating the force to apply).
- The AI learns about the primary movements (a first training is performed by setting the correct movements performed by a human trainer; other correct movements are automatically saved in the training dataset during worker operation when a sensor detects a correct movement according to the true production operation).

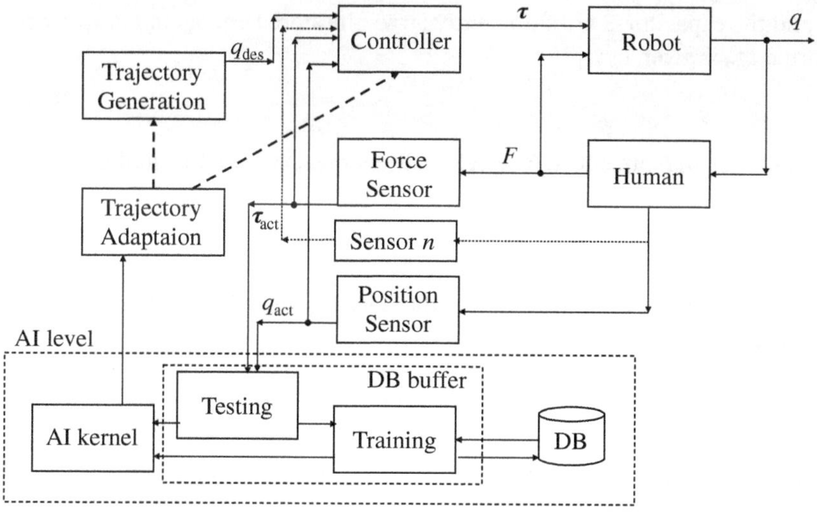

Figure 5.30 Adaptive-control diagram to automatically adapt to worker's desired motion by adapting trajectories in self-learning modality (act, actual; des, desired; F, force; q, motion trajectory; and τ, torque).

5.3 Human–Robot Self-Learning Collaboration in Industrial Applications and Electronic Aspects

Robots, specially CoBots, have grown exponentially in Industry 4.0 scenarios involving the use of sensors to facilitate collaborations with humans. There are many types of CoBots requiring supervision to guarantee that their actions do not hurt anyone or damage the robot. CoBots usually are characterized by a rounded design and integrate pressure sensors detecting collisions in order to stop the robot in damage risk conditions, thus recalculating a new path that avoids collisions. Electronics in CoBots can be complex. Different elements constitute the whole electronic and mechatronic system. Switches and the related control play an important role in the CoBot design. In this scenario, an AI controller is appropriate to automatize the switching control, by optimizing the auto-adaptive collaborative function.

5.3.1 DC–DC Converter

DC–DC converters are suitable for power regulation of a robot. DC–DC converters are power electronic devices that make it possible to connect DC robotic networks. These converters are generally controllable, so that the output voltages can be regulated or kept constant, even if there are variations in the supply voltage or load. Figure 5.31a illustrates a step-down converter circuit with electrical elements R and L. When the switch S is turned on, the diode is turned off, and the output voltage is represented by $e_{out} = E_S$, so that the output current rises depending on the value of the time constant L/R. In the period when the switch S is off, the stored energy of inductance L is released to show that the output current decays. In the steady state, the average value of the output voltage E_{out} is given by:

$$e_{out} = \frac{t_{on}}{t_{on} + t_{off}} E_S = \frac{t_{on}}{T} E_S = \alpha E_S \tag{5.55}$$

where T is the chopping period and α is the duty factor. Figure 5.31b shows the theoretical trend of the output current and voltage. Figure 5.31c shows a step-up converter to get a higher DC voltage than the source DC voltage. By turning on the switch S, the current flows through L and S to store energy in L. When the switch is turned off, the stored energy produces a current flow through L and D by the voltage $e_{out}-E_S>0$ induced in L, raising the voltage of the capacitance C. Assuming that both the reactor L and the capacitor C are sufficiently large, the stored energy in L, expressed in joules, in the time period t_{on}, is given by:

$$W_{L,\text{stored}} = E_S I_S t_{on} \tag{5.56}$$

where I_S is the average value of the current i_S. The released energy, expressed in joules, is provided by:

$$W_{L,\text{release}} = \left(E_{out} - E_S\right) I_S t_{off} \tag{5.57}$$

In the steady state, the stored energy becomes equal to the releasing energy from the reactor L, thus providing the following relationship:

$$E_{out} = \frac{T}{t_{off}} E_S = \frac{1}{1-\alpha} E_S \tag{5.58}$$

Figure 5.31d shows the theoretical trend of the electrical current and voltages of the circuit applying switching.

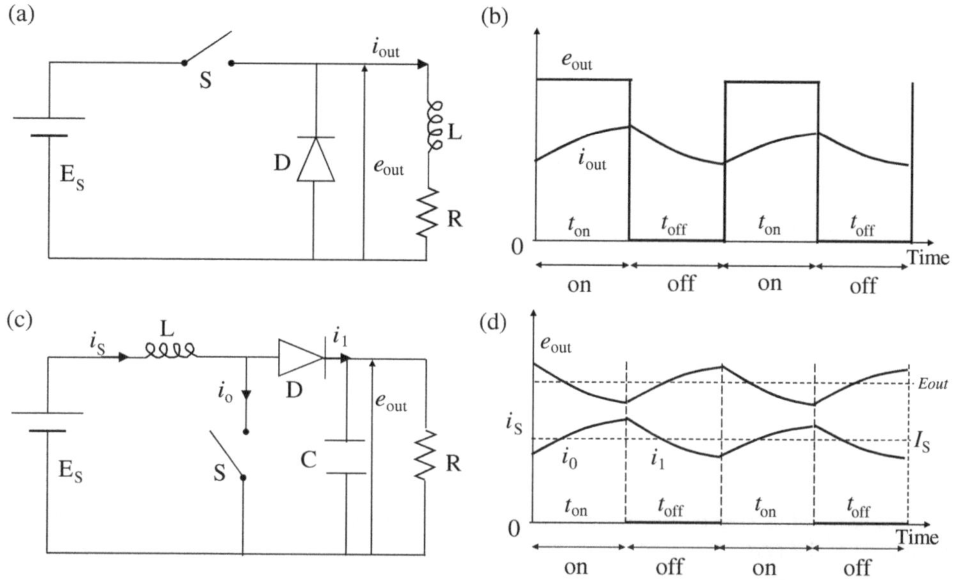

Figure 5.31 (a) Step down chopper circuit. (b) Voltage and current. (c) Step up chopper circuit. (d) Output and voltage current.

5.3.2 Voltage Source Inverter

Voltage-source inverters (VSIs) can reverse the voltage applied to the load, simply by acting on a switch. Typical applications are the reversal of the direction of rotation of robotic electric motors, or the inversion of the direction of movement of the linear actuators. The switching symbol is equivalent to the insulated-gate bipolar transistor (IGBT) circuit or to a metal-oxide-semiconductor field-effect transistor (MOSFET) (Figure 5.32a). Figure 5.32b shows the voltage source inverter principle where the switching control is enabled by the AI controller unit commutating the output voltage as depicted in Figure 5.32c, indicating the switch conditions.

Figure 5.33 represents the equivalent circuit of a three-phase VSI made by six switches to constitute three legs linking the motor M for each phase.

For the *n*-channel MOSFET the substrate semiconductor material is *p*-type, and the source and drain electrodes are *n*-type (Figure 5.34a). Moreover, biasing the gate terminal positive attracts electrons within the *p*-type semiconductor substrate under the gate region toward it. The free electrons within the *p*-type substrate generate a conductive channel effectively changing the *p*-type substrate into a *n*-type material allowing channel current to flow in the region near the electrodes. The reverse is also true for the *p*-channel MOSFET (*n*-type substrate generating a *p*-type conductive channel), where a negative gate potential generates holes under the gate region, attracted to the electrons on the outer side of the metal gate electrode. The theoretical *n*-channel MOSFET I_D–V_{DS} characteristic is illustrated in Figure 5.34b indicating the high signal switching function working in the saturation region. For the MOSFET, the drain current operating in the saturation region is defined as:

$$I_D = k\left(V_{GS} - V_{TH}\right)^2 \tag{5.59}$$

where k is a conduction parameter and V_{TH} is the characteristic threshold voltage.

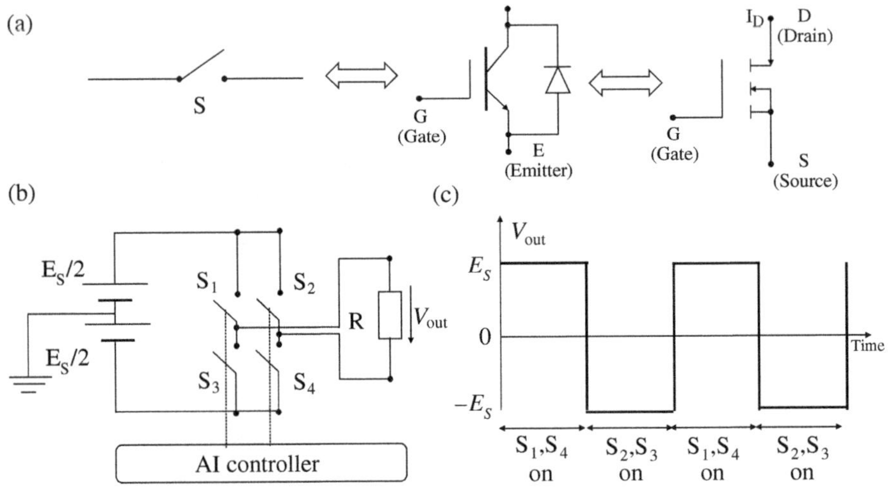

Figure 5.32 (a) Example of IGBT and of *n*-channel MOSFET switch equivalence; (b) voltage source inverter circuit; and (c) signals of the reversal voltage switching.

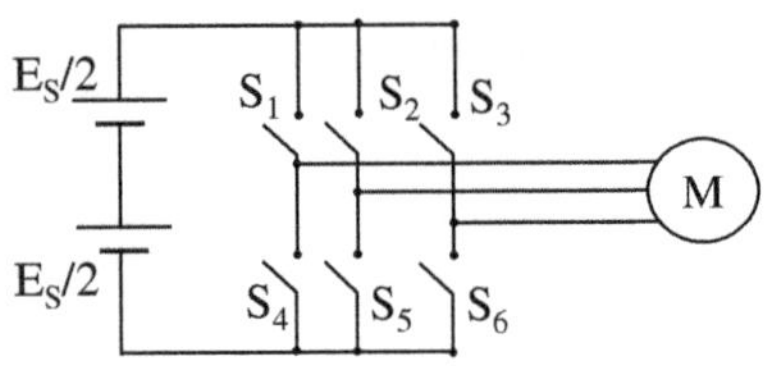

Figure 5.33 Three-phase VSI.

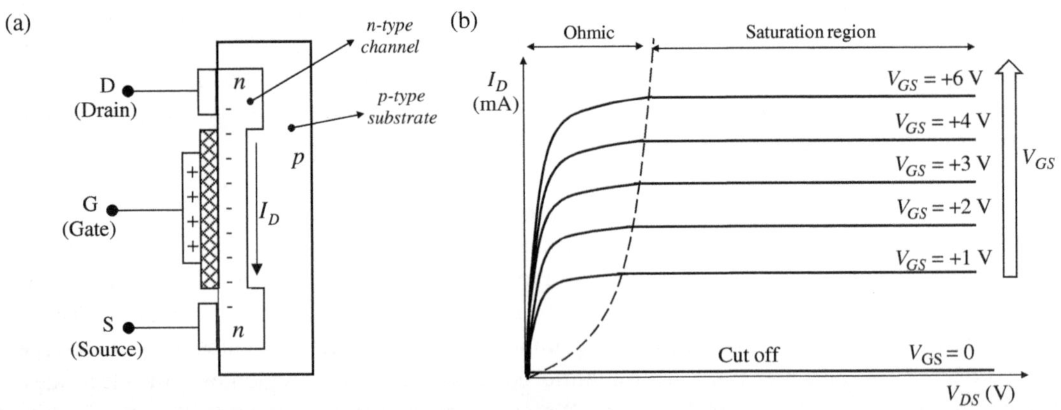

Figure 5.34 (a) Scheme representing the basic principle of electrical current generated into a *n*-channel MOSFET device. (b) Theoretical trend of the related I_D–V_{DS} characteristic.

Other elements behaving as a switch for the voltage source converter are the thyristor (or silicon controlled rectifier [SCR]) and the gate turn off (GTO) devices. Figure 5.35a shows the symbolic equivalence of the thyristor. The SCR is current driven and has a three terminal current controlled switching device with four layers. The device is turned on by providing a gate current I_G, when the device is forward biased, i.e. the anode is at a higher electric potential than the cathode. The GTO

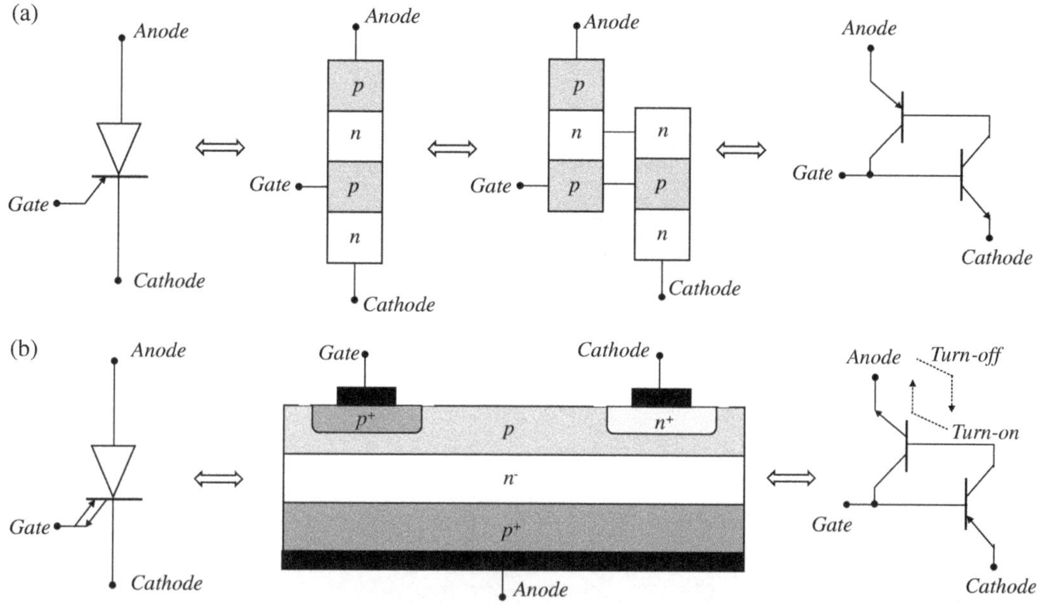

Figure 5.35 (a) SCR configuration and equivalences. (b) GTO symbols and circuit equivalence.

modeling of Figure 5.35b is an alternative to implement a switch by adopting *n*- and *p*-doped semiconductor materials. The GTO is then a controlled diode to be switched on and off, thus activating or blocking the passage of current, by acting directly on the gate electrode. It therefore overcomes the limitation that the current flowing, once triggered, can only be interrupted by external causes forcing electrical current elimination.

For SCR switching applications, the gate current does not have to be given continuously as depicted in Figure 5.36a and b, representing SCR circuit implementation and related processed signals, respectively.

The *I–V* characteristics of the SCR and GTO are illustrated in Figure 5.37a and b, respectively: the *I–V* characteristics are typical for switching conditions.

The pulse width modulation (PWM) technique is applied to improve the output signal by maximizing the fundamental frequency (optimization of the power supply quality). Figure 5.38a represents the normal switching output for a VSI, while Figure 5.38b illustrates the effect of the PWM modulation as a frequency variable voltage output, and Figure 5.38c illustrates an example of PWM modulation changing the output average voltage. One way to control the speed of a motor is

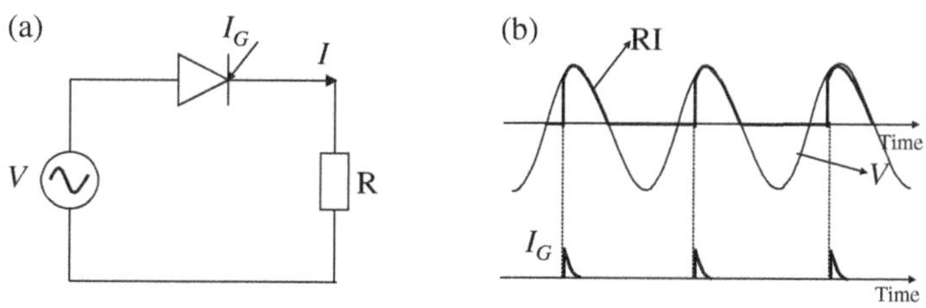

Figure 5.36 On-state caused by gate current: (a) equivalent circuit implementing SCR; (b) signal processing.

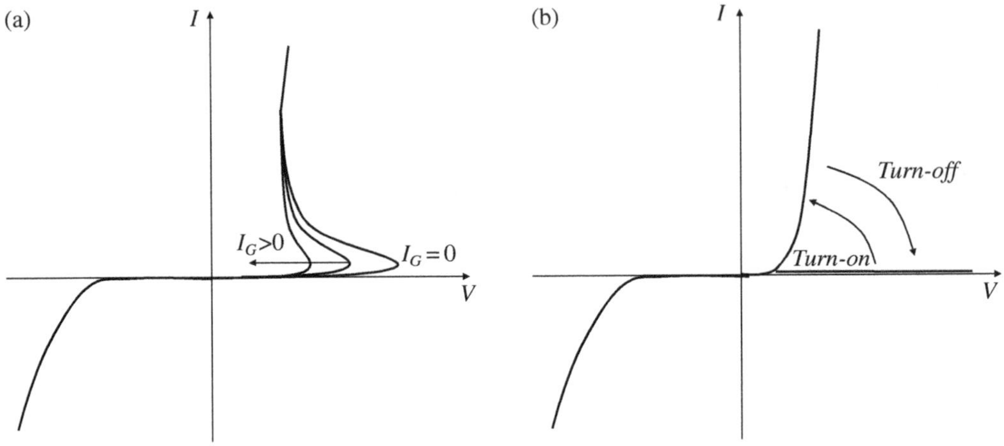

Figure 5.37 (a) SCR and (b) GTO $I-V$ characteristics.

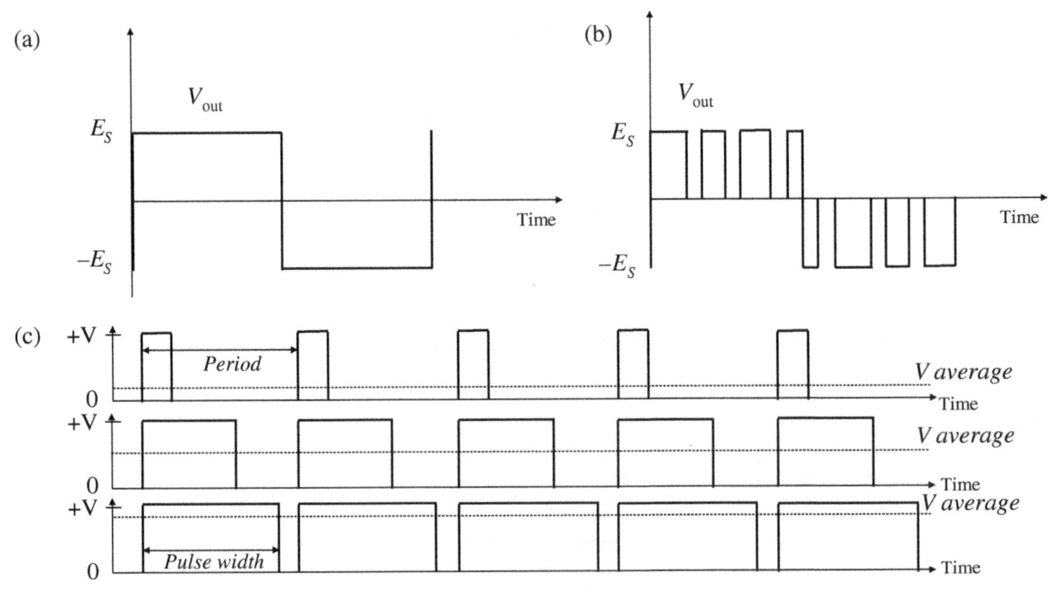

Figure 5.38 (a) Normal switching configuration; (b) PWM signal modulation concept; and (c) example of PWM modulation changing average output voltage.

to regulate the voltage across its terminals, and this is achieved by adopting the PWM technique: PWM of speed control drives the motor by a series of "ON-OFF" pulses and varying the signal duty cycle (fraction of time that the output voltage is "ON" compared with when it is "OFF"). The PWM is functional to change the duty cycle percentage as shown in Figure 5.39.

To control the speed of the motor, the controller changes by PWM technique the input voltage: the controller switches the full input voltage quickly by producing an effect equivalent to that of applying a fixed lower voltage. The AI controller is proper to predict the best motor speed velocity by changing the duty cycle during the production process.

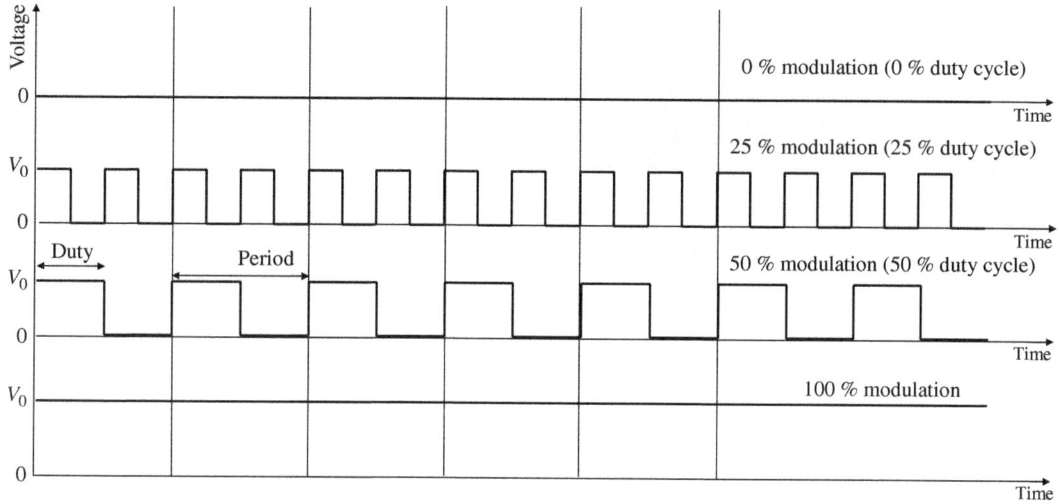

Figure 5.39 Example of PWM signal modulation.

5.3.3 Current-Source Inverter

A current-source inverter (CSI) converts the input DC into AC, and its output can be three-phase or single phase. An ideal current source is when current is constant and is independent of voltage. Figure 5.40a illustrates the basic principle of the CSI: a DC current source I_d is provided by a current generator generating as output a rectangular current wave form (Figure 5.40b).

Figure 5.41 shows the structure of a typical three-phase CSI converter: an inductor in the continuous connection allows the DC to be kept relatively constant, the voltage variations are balanced by $L_d di_d/d_t$ but only for small di_d/d_t, therefore, it effectively maintains a DC constant for a given duty point. Furthermore, due to the effect of the inductor L_d, the CSI circuit is not sensitive to short-circuit faults and switching errors in the circuit. The DC is switched between the upper group (S_1, S_2, S_3) and the lower group (S_4, S_5, S_6) of semiconductors. The fundamental assumption in the CSI circuit is that the current path exists only in one direction, only from the continuous connection; under this assumption a possible sequence switching is a square wave.

The AI controller can activate the switching condition and the motor control according to the AI outputs.

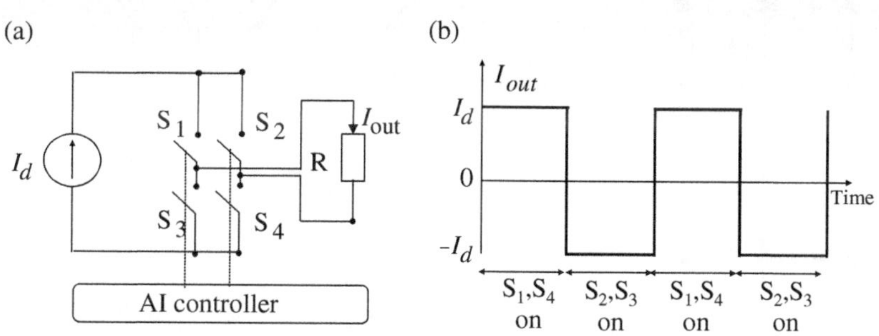

Figure 5.40 (a) Current-source inverter circuit; (b) signals of the reversal current switching.

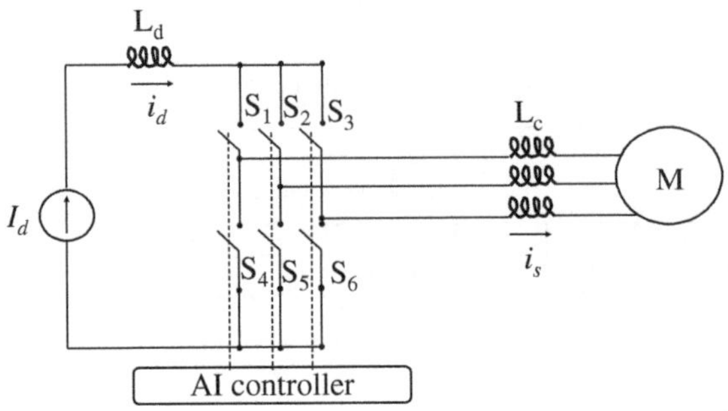

Figure 5.41 Scheme of a three-phase CSI.

5.3.4 DC Voltage Source

A way to implement DC voltage source converters is by means of diodes or switching devices configured as in Figure 5.42a (uncontrolled converter configuration), where the symbol of the switching device is expressed by a thyristor symbol. When a three-phase power supply is provided, and a diode bridge is constructed, the obtained wave form is shown in Figure 5.42b: the highest electric potential of the three phases appears as the terminal electric potential v_p, while the lowest potential appears as v_n (the output voltage is $v_{out} = v_p - v_n$).

Figure 5.43a shows the controlled converter layout using thyristors. In the diagram of the output wave form in Figure 5.43b, the delay denoted by the firing angle α (rad), which can be controlled to set the DC output voltage magnitude, is observed.

5.3.5 Capacitor and Reactor Effects on Signal Control

The inverters produce demanded voltage using the DC voltage given from the converters, generating a constant DC voltage or a constant DC current. Concerning the basic configuration in

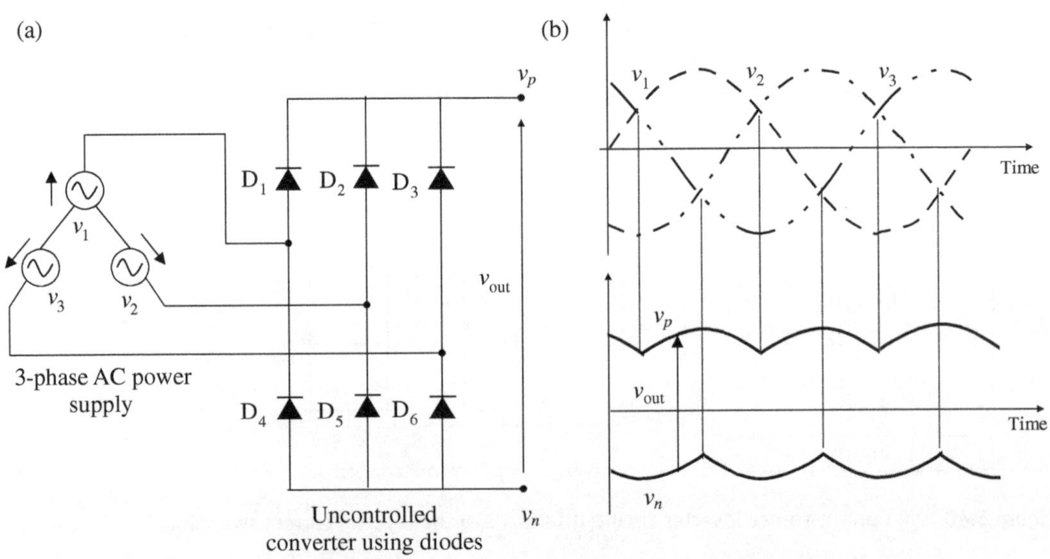

Figure 5.42 (a) Uncontrolled converter configuration; (b) signal processing of the converter.

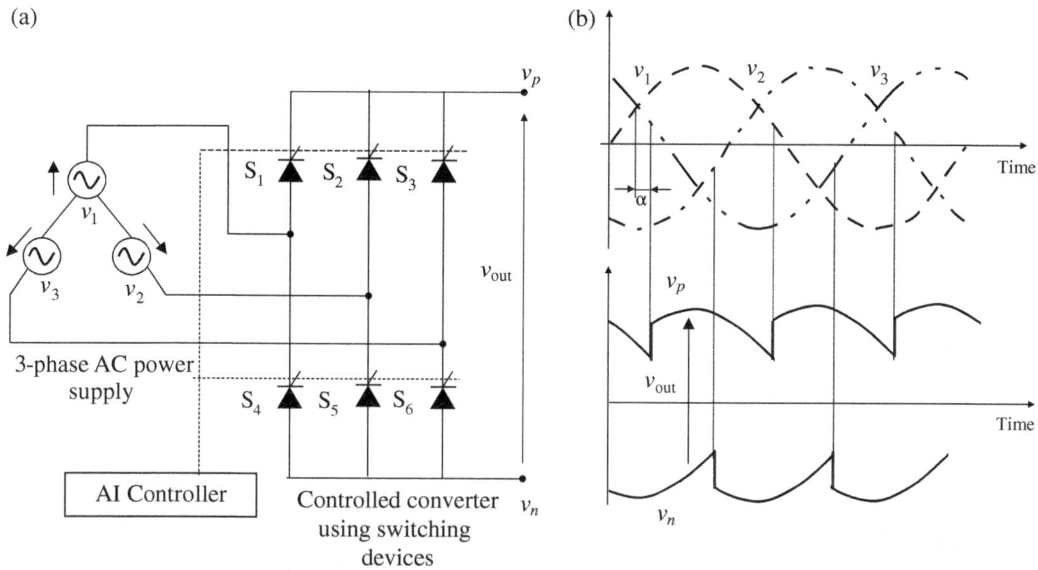

Figure 5.43 (a) Controlled converter; (b) signal processing of the converter.

Figure 5.44a related to a simple half-wave rectifier circuit, the output voltage v_D changes between 0 and the maximum value, moreover, the DC current i_D is characterized by a significant ripple component. The AI controller is appropriate to switch the diode component by a direct electrical link, by making on the load the same profile of the input voltage enabling the half-wave voltage profile. Capacitors are used for the smoothing of the output voltage in the time domain for the half wave rectifier (see circuit in Figure 5.44b): if the smoothing capacitor C is connected in parallel with the output, the diode is reverse biased during the period that the charged capacitor voltage is larger than the source voltage, when the capacitor supplies currents to the load. When the source voltage is larger than the capacitor voltage, the current flows from the voltage source to the load while the capacitor is charged, resulting in a smother waveform of the output voltage. The trend of voltage decay over time is provided by a time constant depending on the capacitor value C and on the load resistance R. In the case of larger capacity, the output voltage is strongly affected by the discharge process to the load. Accordingly, larger capacitor value is required to obtain a constant DC voltage. By considering the inductor in the configuration in Figure 5.44c, the DC current is smoothed due to the reactor voltage component v_L. In the case of a full-wave rectifier circuit (Figure 5.44d), the DC current contains significantly smoothed ripples. Larger inductance values of the inductor potentially produce a constant value of current.

A VSI and CSI are sketched in Figure 5.45 in a different configuration: capacitors are used in the DC link of the VSI to keep a constant DC voltage, while reactors are used to have a constant DC current in the DC link of the CSI.

5.3.6 Human-Robot System and Learning Approaches

Supervised systems are important for self-learning systems, including the possibility to apply neural networks supporting robotic manipulator control [21] and adaptive DC motor drive systems [22]. Traditional proportional integral derivative (PID) controllers [23–25] are commonly adopted to estimate the error value of a control variable as the difference between a desired

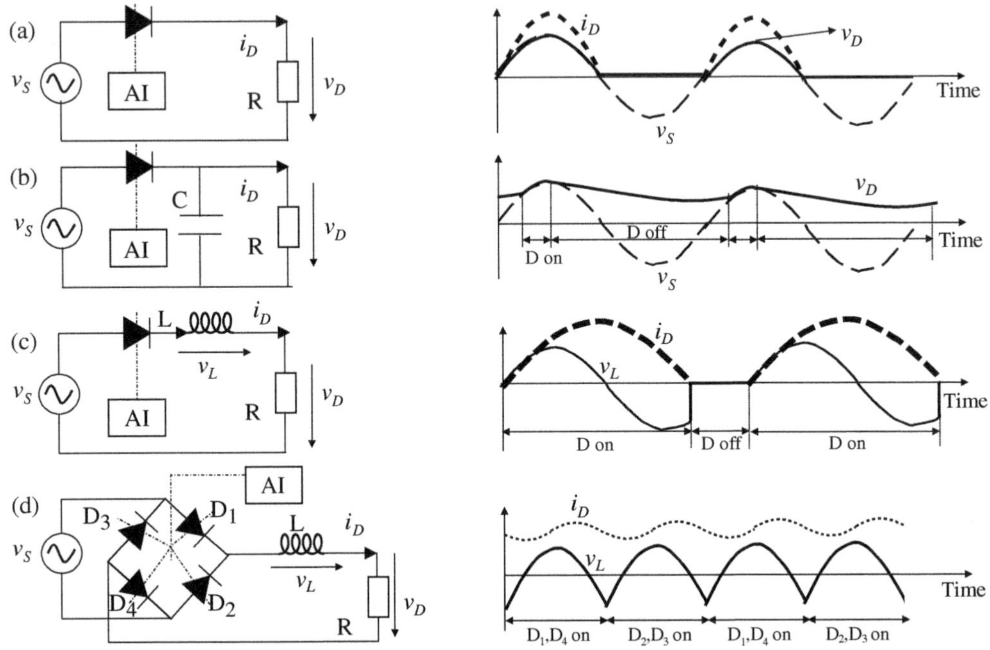

Figure 5.44 (a) Half-wave rectifier circuit basic scheme; (b) half-wave rectifier scheme with a capacitor; (c) half-wave rectifier scheme with an inductor; and (d) full-wave rectifier circuit layout.

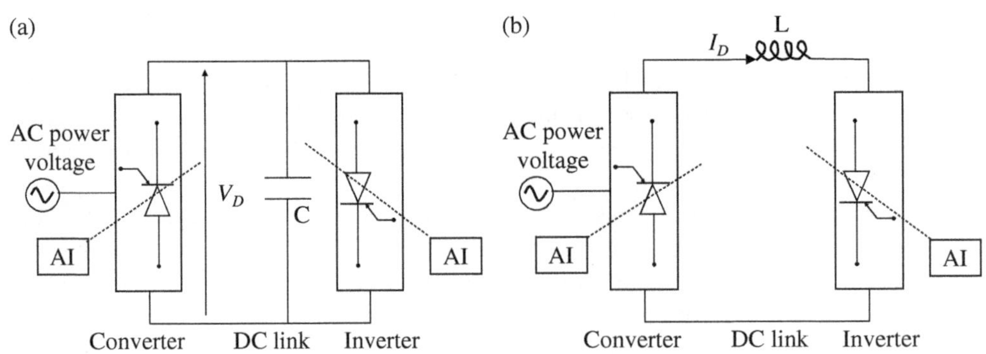

Figure 5.45 (a) Voltage-source inverter; (b) current-source inverter.

setpoint and a measured process variable, permitting feedback correction to be applied based on proportional (P), integral (I), and derivative (D) terms. The three terms of the controller cover the present error (P), past error (I) through low frequency compensation, and predicted error (D) through high frequency compensation. The time derivative term improves transient response or stability, but it does not necessarily always recover stability in practice. When there exists a dead time in signal transport, the derivative operation can cause stability degradation. Therefore, it is switched off in some cases, ending up with a PI controller. In any case the term D provides more information indicating if the error is increasing or decreasing, thus anticipating what will happen next, so it is a regulator predicting the error trend. This term makes the adjustment system even faster.

An advanced control system of robotic motors is provided by the feedback loop system in Figure 5.46a, where the PID controller and the AI Learning module are combined supporting together the error $e(t)$ estimation for the best setting of the control variable $X(t)$. The advanced system part able to control the motor compares the PID error with the AI one, thus improving and accelerating the best choice of the output variable actuating the motor (Figure 5.46b). Concerning PID, all three components P, I, and D are used simultaneously or independently, and the error signal $e(t)$ is the control input, while the control output $u_{\mathrm{PID}}(t)$ is defined as:

$$u_{\mathrm{PID}}(t) = K_p\left\{e(t) + \frac{1}{\tau_i}\int_0^t e(t)\,\mathrm{d}t + \tau_d \frac{\mathrm{d}e(t)}{\mathrm{d}t}\right\} \tag{5.60}$$

The function $u_{\mathrm{PID}}(t)$ in the form of a Laplace transformation is defined as:

$$U_{\mathrm{PID}}(s) = K_p\left(1 + \frac{1}{\tau_i s} + \tau_d s\right)E(s) \tag{5.61}$$

while the transfer function of the PID system is then defined as:

$$\frac{U_{\mathrm{PID}}(s)}{E(s)} = K_p\left(1 + \frac{1}{\tau_i s} + \tau_d s\right) \tag{5.62}$$

The $X(t)$ can denote a trajectory of a robot arm. By defining the PID gains with K_P the proportional gain, K_I the integral gain, and K_D the derivative gain as:

$$K_I = \frac{K_P}{\tau_i} \tag{5.63}$$

$$K_D = K_P \tau_d \tag{5.64}$$

the AI learning network will tune all the gain parameters. The gain tuning can minimize the following trajectory error:

$$e(t) = X(t)X_r(t) \tag{5.65}$$

Artificial neural networks (ANNs) are good candidates for the PID gain settings [26–28]. Deep learning (DL) techniques can totally replace PID control. In any case, the combination of a self-learning network and the PID system is a powerful tool to quickly decrease the trajectory errors. The advanced system in Figure 5.46a and b is characterized by the following further specifications improving the error minimization:

- Adding of other input data collected into a specific DB system.
- Integration into an information system processing other data of the same production line, which can be correlated with the variable $X(t)$.

Other data input are images, different data sensors typical for production monitoring (temperature, humidity, positioning, etc.), obstacle detection sensor data (laser or ultrasound sensor data), and other digital information which are the output of other AI algorithms.

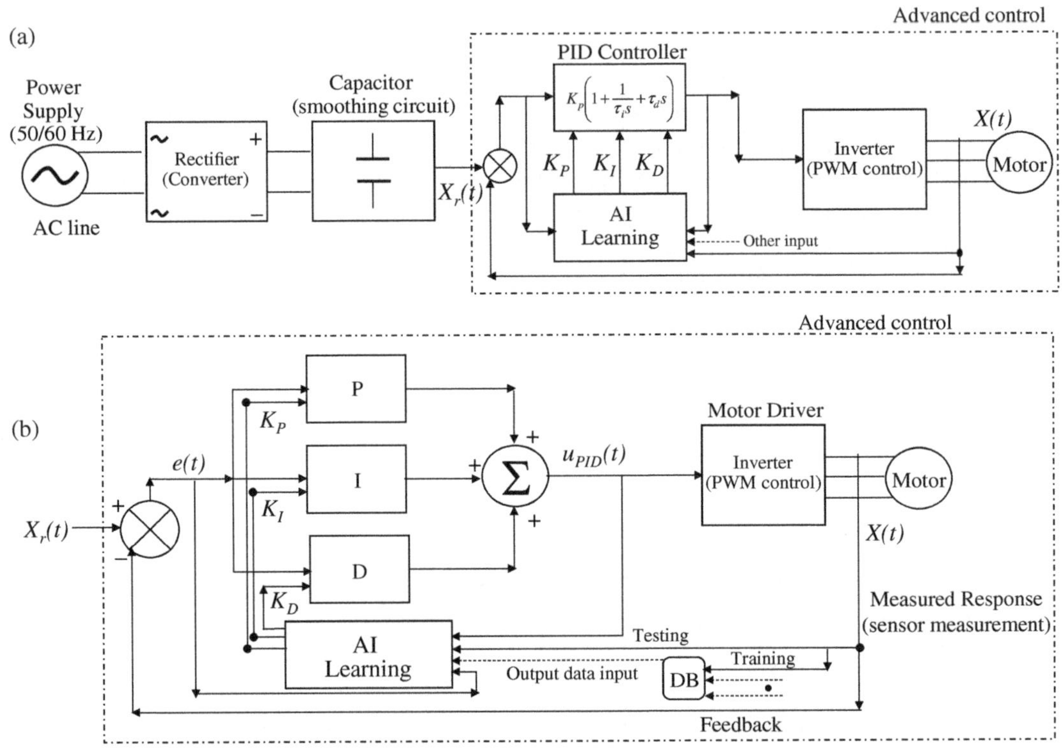

Figure 5.46 (a) Main scheme of an advanced robotic control by combining a PID controller and AI learning; (b) details of the advanced control part.

The proposed example is applied to the following case which is useful to explain how the control happens:

- A robot shoots a target with a bow and arrows.
- The target center is indicated by the reference $X_r(t)$ where for a fixed point $X_r(t) = \mathrm{O}(x_1, y_1)$.
- The image processing algorithm detects the target (for example indicated by a red colored target).
- The robot shoots the first arrow and measures $X(t)$ (point $X_r(t) = \mathrm{P}(x_i, y_i)$, where i indicates the attempt number) as a Euclidean distance between two points where the first point is the arrow position and the second point is the target, defined as:

$$d_i = \sqrt{(x_i - x_1)^2 + (y_i - y_1)^2} \tag{5.66}$$

- According to the error measurement the robot calibrates the force and the spatial orientation trying to shoot the target again.
- The procedure is repeated until the error is minimized and the target is centered within a tolerance range.

Figure 5.47 illustrates the theoretical case, indicating the adjustment of the trajectory improved by an advanced control system. The example is more complex if it is required to minimize the error in 3D space, where it is important to also consider spatial position data for the training dataset.

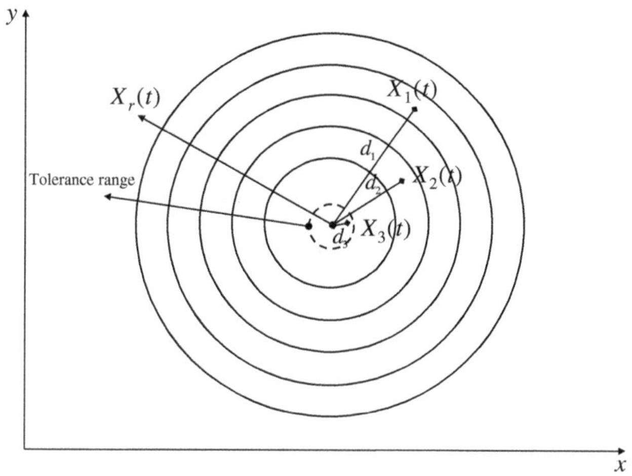

Figure 5.47 Example of an AI controlled system adjusting an arrow trajectory. X_r represents the target point and $X_1, X_2,$ and X_3 are the successive attempts of the robot adjusting trajectory. The Euclidean distance is within the tolerance range for the third attempt X_3.

5.3.6.1 Example of PID Implementation of Self-Adapting Gains

A possibility to implement a PID circuit is through resistances, capacitors, and operational amplifiers. The circuit in Figure 5.48a is characterized by the impedances Z_i and Z_r defined by the following relationships:

$$\frac{1}{Z_i} = \frac{1}{Z_{R_i}} + \frac{1}{Z_C} = \frac{1 + sCR_i}{R_i} \tag{5.67}$$

$$Z_r = R_r + \frac{1}{sC} = \frac{R_r sC + 1}{sC} \tag{5.68}$$

$$V_o(s) = -\frac{Z_r}{Z_i} V_i(s) = -\left(\frac{R_r sC + 1}{sC} \cdot \frac{1 + sCR_i}{R_i}\right) V_i(s) \tag{5.69}$$

The same circuit is summarized by the block diagrams in Figure 5.48b and c. The PID transfer function is then defined by:

$$\begin{aligned} G(s) &= \frac{V_o(s)}{V_i(s)} = -\frac{Z_r}{Z_i} = -\left(K_P + \frac{K_I}{s} + K_D s\right) = -\left(\frac{R_r sC + 1}{sC} \cdot \frac{1 + sCR_i}{R_i}\right) = \\ &= -\left(\frac{R_r sC + R_i R_r s^2 C^2 + 1 + R_i sC}{sCR_i}\right) = -\left(\left(\frac{R_i sC}{sCR_i} + \frac{R_r sC}{sCR_i}\right) + \frac{1}{sCR_i} + sCR_r\right) \end{aligned} \tag{5.70}$$

where

$$K_P = 1 + \frac{R_r}{R_i} \tag{5.71}$$

$$K_I = \frac{1}{R_i C} \tag{5.72}$$

$$K_D = R_r C \tag{5.73}$$

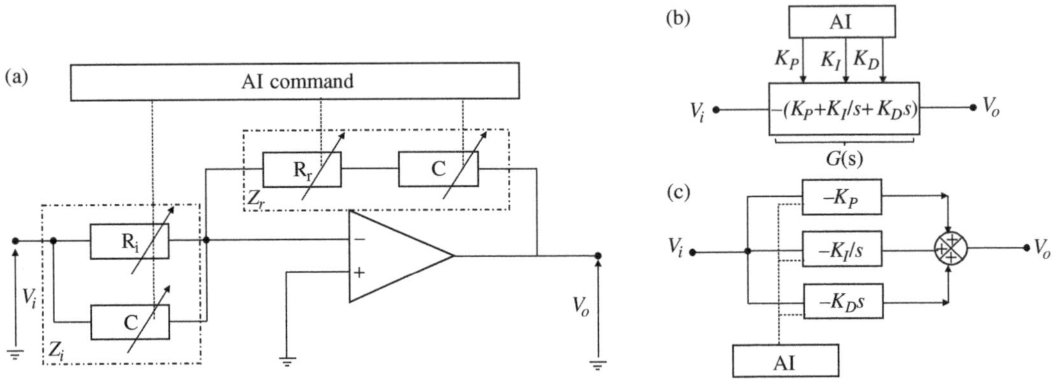

Figure 5.48 (a) PID implementation circuit layout tuned by AI commands; (b) block model of the PID circuit; and (c) alternative model representation.

The AI controller can change the resistances R_i and R_r and the capacitance C thus tuning the gains K_P, K_D, and K_I.

5.3.7 Unsupervised Learning Approaches

Image processing is also adopted for robots implementing unsupervised algorithms for automatic object selection [29]. In the application in Figure 5.49a a clustering-based technique is shown as an unsupervised approach to select an object from a production line. The object selection process is structured as follows:

- Different objects are placed on a first workstation.
- A camera detects an image containing all the objects to analyze for selection.
- The acquired image is pre-processed by neglecting noise due to imperfect lighting or other external disturbances.
- The features of each object are extracted by a CNN algorithm (data are collected into a DB useful for constructing a training dataset concerning all the objects to recognize including recognition of different object orientations).
- The clustering algorithm is applied to define object classes having the same features.
- Following the determined classes, a first robotic arm moves the recognized objects from the first workstation to a second one (the first clustering approach can allow the selection of similar objects in a wrong manner).
- Another image is acquired in the second workstation.
- The whole image processing procedure, shown in Figure 5.49b, is again applied to reject similar objects not included in the classes to select.
- A second robotic arm selects the recognized objects appertaining to the desired class.
- All the recognized objects are placed on a third workstation containing the final true objects.

The two image processing stages support the wrong object identification due to possible overlapping conditions. Each robotic arm continues its process selection, by optimizing during the time of object selection. The same image processing procedure can be adopted for defect recognition: the image processing recognizes defective pieces to reject them during the processing. The DB system,

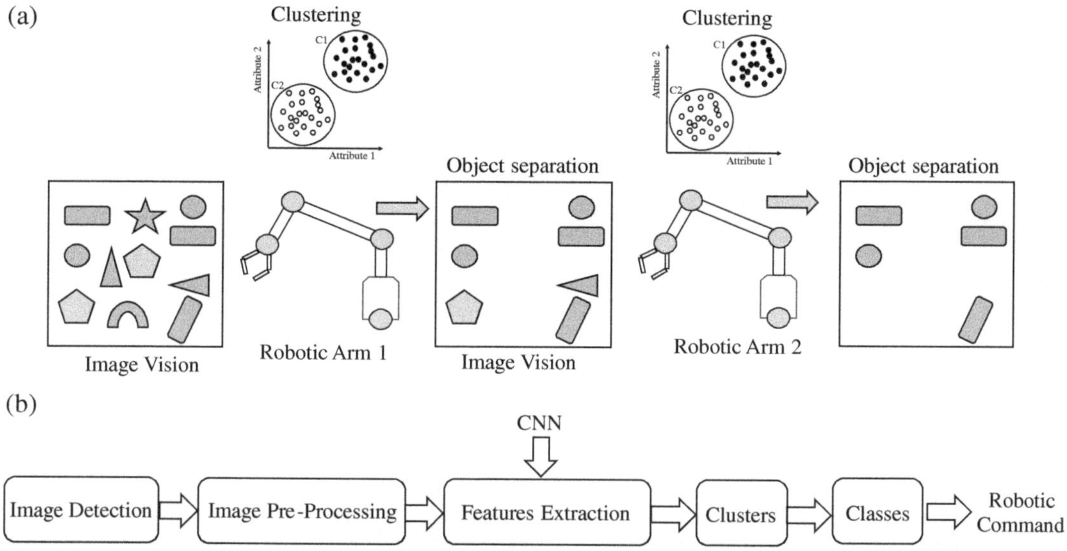

Figure 5.49 (a) Unsupervised process by selecting the object inline in two steps. (b) Image processing workflow applied at each selection stage.

containing the training dataset, continuously collects the object features by optimizing the clustering efficiency. The unsupervised approach is usually combined with supervised algorithms to increase object recognition efficiency.

5.3.8 Soft Robotics for Intelligent Collaborative Robotics

Soft robotics involves important industrial applications involving the manipulation of fragile objects. Four-dimensional printing and machine learning techniques provide new strategies to fabricate soft actuators and soft sensors [30]. Soft robots have promising characteristics derived from adapting to the working environment, such as:

- flexibility
- deformability
- energy-absorbing characteristics

Manipulation of soft robots may be difficult. The actuation and control of soft robots cannot be addressed by traditional methods. Novel actuation is required to realize the complex motion of these innovative configurations. Soft robot actuation can be enabled via electroactive polymers (EAPs), cable actuation, shape memory alloy (SMA) actuation and fluidic actuation. Control problems for soft robots involve disturbance management. Figure 5.50a shows a classical feedback control structure characterized by the following transfer function:

$$Y(s)=G(s)[U(s)+D(s)] \tag{5.74}$$

The control function is defined as:

$$U(s)=G_e(s)[G_f(s)R(s)-H(s)Y(s)] \tag{5.75}$$

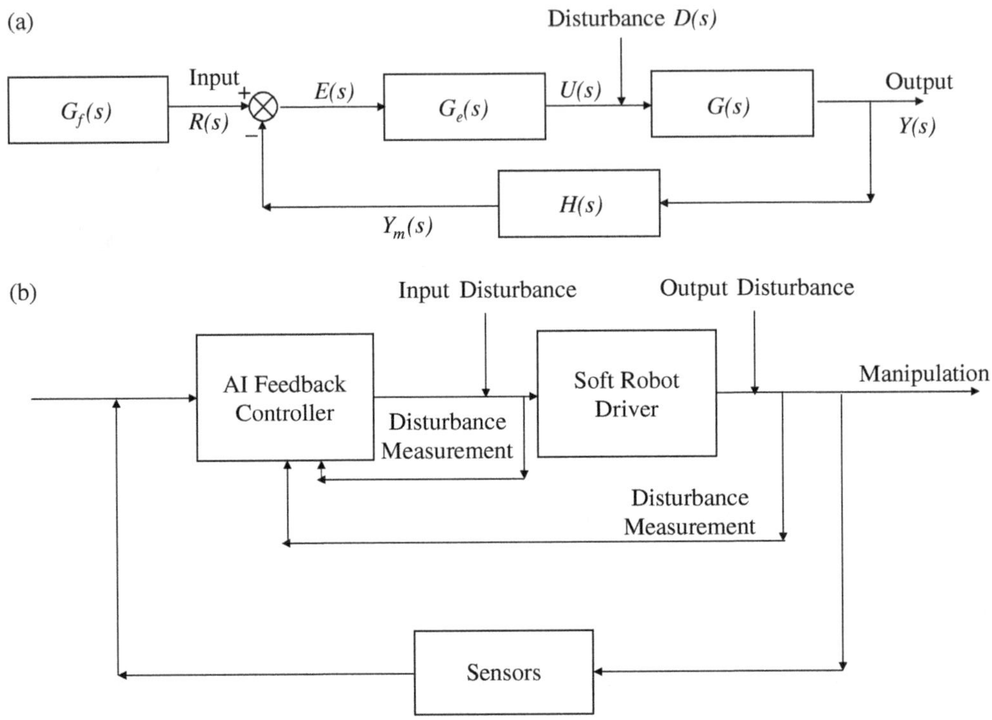

Figure 5.50 (a) Feedback loop systems. (b) Feedback system including AI feedback control and input/output disturbances.

where

$$Y_m(s) = H(s)Y(s) \tag{5.76}$$

The whole system is high sensitivity to the disturbances which can be classified as input and output disturbances. The feedback control system in Figure 5.50b integrates the AI module controlling disturbances. The AI processes as input signals, the input signal of the system, both the disturbances, and the output.

5.4 Robotics in Additive Manufacturing

AM is an important topic in industrial research. AM is a key issue of the Industry 4.0 scenario [31, 32]. Different techniques are applied in this scenario, involving the AI application [33, 34].

5.4.1 Additive Manufacturing in Industrial Production and Spray Technique

Smart material implementations are innovative approaches to direct research to new production approaches oriented towards AM. In particular, 4D printing techniques can be applied to polymeric materials [35, 36]. 4D printing takes into account time as the fourth dimension. 4D printing is an intelligent and programmable technique to apply to polymeric materials [34], and is able to change shape, properties or perform specific actions (such as self-assembly over time), following an external stimulus such as heat, vibrations, gravity, magnetism, or electricity. The output of this technique is the realization of self-assembling elements reacting with an external stimulus. An innovative approach to add materials is based on the pulsed spray technique [37]. This technique can be adopted for the fabrication of microcircuits and micro devices based on nanoparticles (NPs). In particular, the pulsed spray technique is applied in the literature [37] for nanodiamond (ND) layer deposition by using dichloroethane as the solvent for ND dispersion. The evaporation of the solvent allows the deposition of ND layers on a silicon substrate with a temperature of 120 °C to accelerate the solvent evaporation process (the boiling temperature of dichloroethane is 84 °C). The nozzle is controlled by the time duration of the spray pulse (parameter t_s defined in the transfer function $P(s)$) and the process is a function of solvent evaporation (t_e) and of the spray period ($T = t_s + t_e$). Another key parameter to control is the solution flux (transfer function $F(s)$). The fundamental principle of ND spray deposition is based on the atomizer driven: the standard control is achieved by a spray pulse of $t_s = 15$ ms and $t_e = 2$ s (time related to the solvent evaporation period). The AI engine is learned by sensor data characterizing the morphological deposition by means of images detected by a fixed microscope. Moreover, as for conductive surfaces, the data to read for the flux regulation is the electrical current measured on the conductive nanolayer. All data are processed by the AI algorithm adjusting in quasi real time the pulse timing and the flux (pressure variation able to tune ND flux capacity) according to the high-quality additive procedure. Figure 5.51 summarizes the control of both flux regulation and pulse regulation.

5.4.2 Artificial Intelligence Applications in Additive Manufacturing

Machine learning techniques are a new issue in AM. Table 5.4 lists some possible applications.

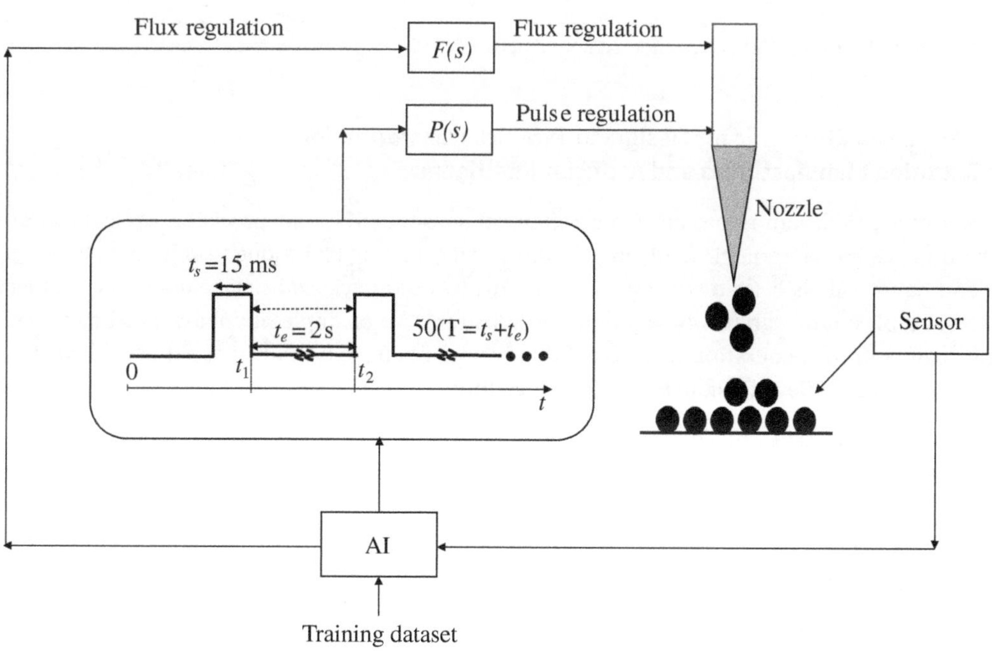

Figure 5.51 Pulsed spray technique in smart additive manufacturing controlled by an AI engine.

Table 5.4 AI facilities applied in additive manufacturing.

AI facility	Description	Reference
k-nearest neighbors, neural network, naïve Bayes, SVM, and decision tree	Classification of 3D laser scanner measurement points of additive manufactured parts	[38]
Bayesian predictive analytics and Augmented Layer-wise Spatial Log Gaussian Cox Process	Porosity detection	[39]
SVM	PID process control for fused deposition modeling	[40]
Markov decision process	Image-guided process control for metal AM application. In-situ AM control problem	[41]
Bayes Classifier, clustering algorithms	Selective laser melting process optimization	[42]
SVM	Defect detection by image classification (process monitoring in AM)	[43]
SVM, CNN	Anomalies detection and classification in AM	[44]
CNN	Anomaly detection and classification in a laser powder bed fusion AM process	[45]
Bayesian classifier	Powder-bed AM quality inspection. Fusion quality and defect detection	[46]
Least square SVM	Build precision	[47]
Genetic algorithm	Process parameter optimization	[48]
Neural network	Build-time prediction	[49]

AM, additive manufacturing; CNN, convolutional neural network; SVM, support vector machine.

5.4.3 Advanced Electronic for Design-to-Product Transformation: Laser Texturing Manufacturing and Artificial Intelligence

Nanoelectronics [50], nano-opto-electro-mechanical sensors [51], and precision optics [52] are important topics for technological advances. Another interesting technique usable for AM on the micro and nano scale is laser texturing [53, 54]. This technique can fabricate small and compact electrical and plasmonic circuits on a polymeric substrate. An example of an advanced nanoelectronic circuit is in the generation of gold conductive nanopatterns. Gold nanopattern fabrication on chitosan film is achieved by the following procedure:

- Preparation of chitosan film (polymeric substrate): 75–85% deacetyled chitosan polymer is dissolved in acetic acid (1 wt% in distilled water); chitosan is prepared using different polymer solutions (typically 0.5 and 1 wt%); chitosan polymeric films are deposited on a glass substrate by the drop casting approach; the concentrations used produce chitosan hydrogels capable of absorbing the gold (Au) precursor ($HAuCl_4$).
- Preparation of the gold precursor: the precursor used is a chloroauric acid salt ($M_w(HAuCl_4) = 339.5\,g\,mol^{-1}$) which is dissolved in distilled water ([Au] = 0.01 M); the precursor is mixed into the chitosan polymeric film by immersion tuning the precursor concentration in the chitosan substrate material.

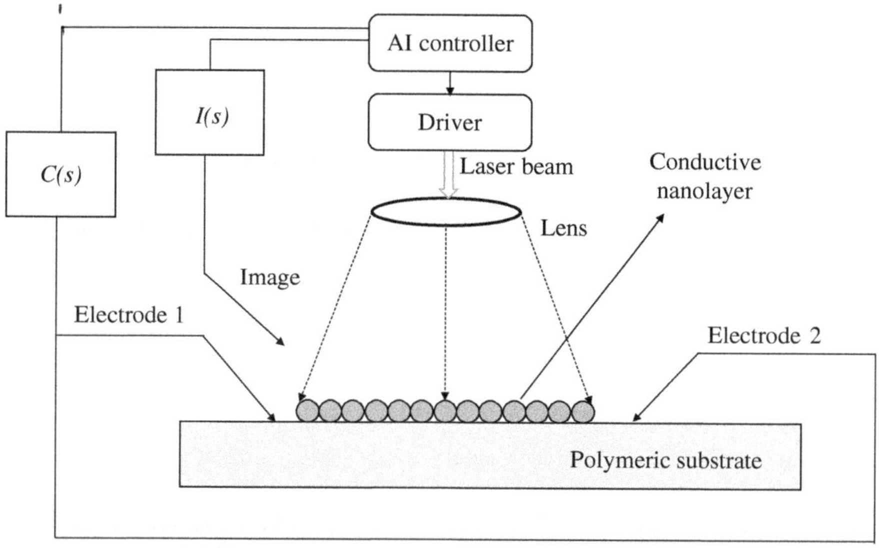

Figure 5.52 Laser texturing technique controlled by AI.

- Nanopattern generation: the gold nanopattern is generated by a laser light source using a pulsed Nd:YAG laser operating at 10 Hz with 4–6 ns pulse duration and with a working wavelength of 355 nm at room temperature.

In Figure 5.52 shows the block scheme of the laser texturing technique controlled by AI reading as input the image *C*(*s*) and the electrical current *I*(*s*).

References

1. Karabegović, I. and Husak, E. (2018). The fourth industrial revolution and the role of industrial robots: with a focus on China. *Journal of Engineering and Architecture* 6 (1): 68–75.
2. Vaidya, S., Prashant, A., and Bhosle, S. (2018). Industry 4.0 – a glimpse. *Procedia Manufacturing* 20 (1): 233–238.
3. Lichiardopol, S., Van De Wouw, N., and Nijmeijer, H. (2009). Control scheme for human-robot co-manipulation of uncertain, time-varying loads. *Proceedings of American Control Conference ACC'09*, St. Louis, USA (10–12 June 2009). Piscataway, NJ: IEEE.
4. Nikolaidis, S., Lasota, P., Ramakrishnan, R., and Shah, J. (2015). Improved human-robot team performance through cross-training, an approach inspired by human team training practices. *International Journal of Robotics Research* 34 (14): 1711–1730.
5. Rahman, S.M., Wang, Y., Walker, I.D., Mears, L., Pak, R., and Remy, S. (2016). Trust-based compliant robot-human handovers of payloads in collaborative assembly in flexible manufacturing. *Proceedings of the 2016 IEEE International Conference on Automation Science and Engineering (CASE)*, Fort Worth, USA, (21–25 August 2016). Piscataway, NJ: IEEE.
6. Johannsmeier, L. and Haddadin, S. (2017). A hierarchical human-robot interaction-planning framework for task allocation in collaborative industrial assembly processes. *IEEE Robotics and Automation Letters* 2 (1): 41–48.

7 Gabler, V., Stahl, T., Huber, G. et al. (2017). A game theoretic approach for adaptive action selection in close proximity human robot collaboration. *Proceedings of IEEE International Conference on Robotics and Automation ICRA*, Marina Bay Sands, Singapore (29 May –3 June 2017). IEEE.

8 Maeda, G., Maloo, A., Ewerton, M. et al. (2016). Anticipative interaction primitives for human-robot collaboration. *Proceedings of AAAI Fall Symposium Series*, Arlington, USA (17–19 November 2016). AAAI.

9 Wongphati, M., Osawa, H., and Imai, M. (2015). Gestures for manually controlling a helping hand robot. *International Journal of Social Robotics* 7 (5): 731–742.

10 Lenz, C., Rickert, M., Panin, G., and Knoll, A. (2009). Constraint task-based control in industrial settings. *Proceedings of IEEE/RSJ International Conference on Intelligent Robots and Systems*, St. Louis, USA (11–15 October 2009). Piscataway, NJ: IEEE.

11 Makrini, I. E., Merckaert, K., Lefeber, D., and Vanderborght, B. (2017). Design of a collaborative architecture for human-robot assembly tasks. *Proceedings of IEEE/RSJ International Conference on Intelligent Robots and Systems (IROS)*, Vancouver, Canada (24–28 September 2017). IEEE.

12 Guerin, K.R., Riedel, S.D., Bohren, J., and Hager, G. D. (2014). Adjutant: A framework for flexible human-machine collaborative systems. *Proceedings of 2014 IEEE/RSJ International Conference on Intelligent Robots and Systems IROS 2014*, Chicago, USA (14–18 September 2014). Piscataway, NJ: IEEE.

13 Cherubini, A., Passama, R., Fraisse, P., and Crosnier, A. (2015). A unified multimodal control framework for human-robot interaction. *Robotics and Autonomous Systems* 70 (2015): 106–115.

14 Matheson, E., Minto, R., Zampieri, E.G.G. et al. (2019). Human–robot collaboration in manufacturing applications: a review. *Robotics* 8 (100): 1–25.

15 Robla-Gómez, S., Becerra, V.M., LLATA, J.R. et al. (2017). Working together: a review on safe human-robot collaboration in industrial environments. *IEEE Access* 5 (1): 26754–26773.

16 Pauker, F., Ayatollahi, I., and Kittl, B. (2014). OPC UA for machine tending industrial robots. *Proceedings of the Second International Conference on Advances in Mechanical and Robotics Engineering AMRE 2014*. Zurich, Switzerland (25–26 October 2014). Institute of Research Engineers and Doctors.

17 Bances, E., Schneider, U., Siegert, J. et al. (2020). Exoskeletons towards Industrie 4.0: benefits and challenges of the IoT communication architecture. *Procedia Manufacturing* 42 (1): 49–56.

18 Wang, Y., Cheng, H., Hou, L., and c2AIDER (2019). A cognitive cloud exoskeleton system and its applications. *Cognitive Computation and Systems* 1 (2): 33–39.

19 Grosu, V., Guerrero, C.R., Brackx, B. et al. (2015). Instrumenting complex exoskeletons for improved human-robot interaction. *IEEE Instrumentation & Measurement Magazine* 2015: 1–7.

20 Riener, R., Lünenburger, L., and Colombo, G. (2006). Human-centered robotics applied to gait training and assessment. *Journal of Rehabilitation Research and Development* 43 (5): 679–694.

21 Jiang, Y., Yang, C., Na, J. et al. (2017). A brief review of neural networks based learning and control and their applications for robots. *Complexity* 217 (1895897): 1–14.

22 Zouari, F., Ben Saad, K., and Benrejeb, M. (2012). Adaptive internal model control of a DC motor drive system using dynamic neural network. *Journal of Software Engineering and Applications* 5 (1): 168–189.

23 Sariyildiz, E., Yu, H., and Ohnishi, K. (2015). A practical tuning method for the robust PID controller with velocity feed-back. *Machines* 3 (3): 1–15.

24 Tao, Y., Zheng, J., Linc, Y. et al. (2015). Fuzzy PID control method of deburring industrial robots. *Journal of Intelligent Fuzzy Systems* 29 (6): 2447–2455.

25 Premkumar, K., Kandasamy, P., Vishnu Priya, M. et al. (2019). Grey wolf optimized PID voltage and power factor controlled AC to DC system. *International Journal of Innovative Technology and Exploring Engineering* 9 (2): 5215–5220.

26 Elbelady, S.A., Fawaz, H.E., and Aziza, A.M.A. (2016). Online self tuning PID control using neural network for tracking control of a pneumatic cylinder using pulse width modulation piloted digital valves. *International Journal of Mechanical and Mechatronics Engineering* 16 (3): 123–136.

27 Hernández-Alvarado, R., García-Valdovinos, L.G., Salgado-Jiménez, T. et al. (2016). Neural network-based self-tuning PID control for underwater vehicles. *Sensors* 16 (1429): 1–18.

28 Cheon, K., Kim, J., Hamadache, M., and Lee, D. (2015). On replacing PID controller with deep learning controller for DC motor system. *Journal of Automation and Control Engineering* 3 (6): 452–456.

29 Guerin, J., Thiery, S., Nyiri, E., and Gibaru, O. (2018). Unsupervised robotic sorting: towards autonomous decision making robots. *International Journal of Artificial Intelligence and Applications* 9 (2): 81–98.

30 Zolfagharian, A., Kaynak, A., and Kouzani, A. (2020). Closed-loop 4D-printed soft robots. *Materials and Design* 188 (108411): 1–16.

31 Mehrpouya, M., Dehghanghadikolaei, A., Fotovvati, B. et al. (2019). The potential of additive manufacturing in the smart factory industrial 4.0: a review. *Applied Sciences* 9 (18): 1–34.

32 Dilberoglua, U.M., Gharehpapagha, B., Yamana, U., and Dolen, M. (2017). The role of additive manufacturing in the era of Industry 4.0. *Procedia Manufacturing* 11 (1): 545–554.

33 Wang, Y.B., Zheng, P., Peng, T. et al. (2020). Smart additive manufacturing: current artificial intelligence-enabled methods and future perspectives. *Science China Technological Sciences* 63 (9): 1600–1611.

34 Razvi, S. S., Feng, S., Narayanan, A. et al. (2019). A review of machine learning applications in additive manufacturing. *Proceedings of ASME 2019 International Design Engineering Technical Conferences and Computers and Information in Engineering Conference*. Anaheim, USA (18–21 August 2019). NIST.

35 González-Henríquez, C.M., Sarabia-Vallejos, M.A., and Rodriguez-Hernandez, J. (2019). Polymers for additive manufacturing and 4D-printing: materials, methodologies, and biomedical applications. *Progress in Polymer Science* 94 (1): 57–116.

36 Shie, M.Y., Shen, Y.F., Astuti, S.D. et al. (2019). Review of polymeric materials in 4D printing biomedical applications. *Polymers* 11 (11): 1–17.

37 Cicala, G., Massaro, A., Velardi, L. et al. (2014). Self-assembled pillar-like structures in nanodiamond layers by pulsed spray technique. *Applied Materials & Interfaces* 6 (23): 21101–21109.

38 Samie Tootooni, M., Dsouza, A., Donovan, R. et al. (2017). Classifying the dimensional variation in additive manufactured parts from laser-scanned three-dimensional point cloud data using machine learning approaches. *Journal of Manufacturing Science and Engineering* 139 (9): 1–14.

39 Liu, J.P., Liu, C., Bai, Y. et al. (2018). Layer-wise spatial modeling of porosity in additive manufacturing. *IISE Transaction* 51 (2): 109–123.

40 Liu, C., Roberson, D., and Kong, Z. (2017). Textural analysis-based online closed-loop quality control for additive manufacturing processes. *Proceedings of 2017 Industrial and Systems Engineering Conference*. Blacksburg, USA (20–23 May 2017). IISE.

41 Yao, B., Imani, F., and Yang, H. (2018). Markov decision process for image-guided additive manufacturing. *IEEE Robotics and Automation Letters* 3 (4): 2792–2798.

42 Uhlmann, E., Pontes, R.P., Laghmouchi, A., and Bergmann, A. (2017). Intelligent pattern recognition of a SLM machine process and sensor data. *Procedia CIRP* 62 (1): 464–469.

43 Gobert, C., Reutzel, E.W., Petrich, J. et al. (2018). Application of supervised machine learning for defect detection during metallic powder bed fusion additive manufacturing using high resolution imaging. *Additive Manufacturing* 21 (1): 517–528.

44 Zhang, Y., Hong, G.S., Ye, D. et al. (2018). Extraction and evaluation of melt pool, plume and spatter information for powder-bed fusion AM process monitoring. *Materials & Design* 156 (1): 458–469.

45 Scime, L. and Beuth, J. (2018). A multi-scale convolutional neural network for autonomous anomaly detection and classification in a laser powder bed fusion additive manufacturing process. *Additive Manufacturing* 24 (1): 273–286.

46 Aminzadeh, M. and Kurfess, T.R. (2018). Online quality inspection using Bayesian classification in powder-bed additive manufacturing from high-resolution visual camera images. *Journal of Intelligent Manufacturing* 30 (11): 1–19.

47 Lu, Z.L., Li, D.C., Lu, B.H. et al. (2010). The prediction of the building precision in the laser engineered net shaping process using advanced networks. *Optics and Lasers in Engineering* 48 (5): 519–525.

48 Fathi, A. and Mozaffari, A. (2014). Vector optimization of laser solid freeform fabrication system using a hierarchical mutable smart bee-fuzzy inference system and hybrid NSGA-II/self-organizing map. *Journal of Intelligent Manufacturing* 25 (4): 775–795.

49 Di Angelo, L. and Di Stefano, P. (2011). A neural network-based build time estimator for layer manufactured objects. *The International Journal of Advanced Manufacturing Technology* 57 (1–4): 215–224.

50 Bingheng, L., Hongbo, L., and Hongzhong, L. (2018). Additive manufacturing frontier: 3D printing electronics. *Opto-Electronic. Advances* 1 (1): 1–10.

51 Zobenica, Ž., van der Heijden, R.W., Petruzzella, M. et al. (2017). Integrated nano-opto-electro-mechanical sensor for spectrometry and nanometrology. *Nature Communications* 8 (2216): 1–8.

52 Zolfaghari, A., Chen, T., and Yi, A.Y. (2019). Additive manufacturing of precision optics at micro and nanoscale. *International Journal of Extreme Manufacturing* 1 (1): 1–18.

53 Spano, F., Castellano, A., Massaro, A. et al. (2012). Wettability control by laser texturing process generating localized gold nanoparticles on polymeric thin films. *Journal of Nanoscience and Nanotechnology* 12 (6): 4820–4824.

54 Spano, F., Massaro, A., Blasi, L. et al. (2012). In situ formation and size control of gold nanoparticles into chitosan for nanocomposite surfaces with tailored wettability. *Langmuir* 28 (8): 3911–3917.

6

Advanced Optoelectronic and Micro-/Nanosensors

This chapter proposes innovative and scientific solutions for nanocomposite and microelectromechanical system (MEMS) and nanoelectromechanical system (NEMS) sensors, by discussing the approaches to follow for the characterization of nanotechnology-based sensors from the perspective of producing innovative devices. The characterization concerns the use of specific machines and laboratory facilities. This chapter includes a section on possible experimental setups useful for nanocomposite sensor testing and provides more fundamental physical concepts for micro- and nanosensor prototyping. The main goal is to analyze some methods for theoretical characterization, the design and the fabrication of advanced optoelectronic and micro-/nanosensors useful in the research and development (R&D) industry. Starting with a discussion of measurement methodologies for the morphological characterization of thin films and nanoparticles (NPs), the study moves to (i) nanocomposite tip probes for optical enhancement, (ii) edge plasmonic resonance detecting liquids or dielectric contrasts identifying materials, (iii) nanocomposite pillars for liquids and loss detection in robotic finger integration, (iv) high sensitivity nanocomposite tactile pressure sensors detecting object shape profiles, small notches and pressure forces in general, and (v) thin film materials realizing GHz pressure sensors, and optical biomedical antennas based on rectangular micro apertures. The nanocomposite sensors, due to the small size of the fillers of the polymeric matrix, are typically implemented for optoelectronic sensors integrated into advanced robotic systems as robotic fingers. This chapter also discusses the modeling and synthesis of silica nanospheres behaving as nanoprobes for biomedical imaging, and how it is possible to characterize and to improve electrical properties of diamond material integrated in chip circuits based on silicon substrates. The analytical, numerical and measurement analyzed approaches show the basic characterization of the innovative devices to follow for industries involving the production of innovative prototype sensors, and artificial intelligence (AI) supporting detection discriminations in more complex sensor networks. The attention is then focused on approaches suitable to define methodologies and facilities for the characterization of sensors implemented by nanotechnology. The analyzed topics together with rapid prototyping processes allow understanding of how the micro- and nano-sensing technologies can be preliminarily studied and characterized to find innovative solutions for the production of industrial sensors. The possibility to fabricate nanoprobes, plasmonic resonators, and nanocomposite materials (NMs) orients the research towards the production of a new class of high sensitivity sensors which can be applied in both indoor and outdoor environments. All the cases analyzed propose a full overview of different micro-/nanosensors involving ultrasound, optical and radio frequency (RF) detection systems. Possible implementations of multiple micro-/nanosensors can be applied as

Electronics in Advanced Research Industries: Industry 4.0 to Industry 5.0 Advances, First Edition. Alessandro Massaro.

check points in different parts of the production line monitoring different processing stages, where a high sensitivity is required. The proposed examples are directed towards a new prototypal production of innovative sensors, and the use of new high sensitivity detection systems in an industrial environment providing a new concept of distributed small signal data processing, where decision support systems (DSSs) and AI algorithms play an important role in signal processing and the prediction of working conditions. All the discussed technologies are related to technology readiness levels (TRLs) ranging from 1 to 4.

6.1 Nanotechnology Laboratories in Industries

Some examples of micro-/nanosensors are discussed to show approaches suitable for industries which would start the production of innovative optoelectronic sensors. Starting with the description of basic facilities for the fabrication and characterization of sensors, some examples are provided of optoelectronic design methodologies, supporting the understanding of the research approach to follow for the realization of innovative sensor prototypes.

6.1.1 Facilities for Micro-/Nanosensor Fabrication and Characterization

MEMSs are technologies fully integrable due to the small dimensions compatible with microelectronic circuits, very low electrical power consumption, and very low costs. Furthermore, a MEMS can be interfaced with the microcontrollers. The evolution of MEMSs is so rapid that the current state of the art has moved from the micro dimension to the nano dimension with the introduction of NEMSs. Industries producing MEMSs and NEMSs must have facilities for design, reverse engineering (RE), and device characterization. The physical characterization is fundamental in order to provide guidelines for the design and optimization of new devices on the micro- and nanoscale. Some basic experimental facilities for micro- and nanosensor characterization and fabrication are the following:

- Chemical processes and lithography in a clean room (MEMS device development and optimization of the etching and optical lithography processes by means of a mask aligner).
- Growth of materials by RF sputtering; optimization of growth processes of materials used for MEMS manufacturing on commercial silicon (Si) and silicon dioxide (SiO_2) substrates.
- Polarization measurements of optical filters, photonic crystals, and optical fiber selecting electromagnetic field (EMF) polarization;
- Measurement setups for spectroscopy.
- Measurement setups for antenna and waveguides measuring the irradiated EMF.
- Measurements of in-plane source coupling of optical waveguides (experimental setups for the *in-plane* coupling of light into integrated optical guides).
- Measurements of source coupling and resonances of passive NEMSs.
- Measurements of scattering parameters and of complex impedances of MEMSs, by vector network analyzer (VNA).
- Equipment for vibration force measurements.
- Measurements of pressure and in general of mechanical forces for robotics.
- Characterization approaches of NMs by X-ray photoelectron spectroscopy (XPS), atomic force microscopy (AFM), Fourier transform infrared (FTIR) spectroscopy, transmission electron

microscopy (TEM), scanning electron microscopy (SEM), ultraviolet (UV), visible, and electrical conductivity measurements for sensors, determining protocols for the characterization of NMs.
- Electronic measurements of optoelectronic signals.
- Viscoelastic characterization of polymeric materials, and mechanical response measurement using a dynamical mechanical analyzer (DMA).
- Clean room for synthesis of NPs.
- Characterization of thin films by AFM using different methodologies such as Kelvin probe microscopy (KPM), scanning capacitance microscopy (SCM), and electric-magnetic force microscopy (EMFM), confocal microscopy, UV photoemission measurements, and Z-potential measurement.
- material growth by the microwave plasma enhanced chemical vapor deposition (MWPECVD) approach.
- Inductively coupled plasma etching (ICoP) for photonic crystal fabrication.

RF sputtering is a technique generally referred to as spray activated by RF, and is used for the deposition of insulating materials since in this case it is possible to disperse the charge that remains accumulated inside the target. Specifically, the sputtering is a process where particles of materials are expelled from a solid target material due to bombardment of the target by other energetic particles, using ion gases. This physical effect happens when the kinetic energy of the incident particles overcomes conventional thermal energies ($>>1\,eV$). This process is commonly adopted for thin film deposition and material etching. The sputtering method is performed by a direct voltage (direct current sputtering with a common voltage fixed at 3–5 kV) or by an alternating voltage (RF sputtering with a common frequency of 14 MHz). The network analyzing devices are classified into two categories:

- SNA (scalar network analyzer): analyzers reading only signal amplitudes.
- VNA (vector network analyzer): analyzers reading signal amplitudes and signal phases.

To characterize a network, the network analyzer sends a signal with an amplitude and a phase determined by the device under test and measures the reflected signals. VNAs commonly measure scattering parameters, since reflection and transmission of electrical networks are easier to measure at high frequencies. The possible measurements of a VNA are:

Return loss (RL), indicating the loss of signal strength resulting from the reflection caused by a discontinuity in a transmission line, defined as:

$$RL = 10 \cdot \log\left(\frac{P_i}{P_r}\right) \tag{6.1}$$

where P_i is the input power and P_r is the reflected one.

- scattering parameters S_{ij}.
- Standing wave ratio (SWR), defined as the ratio between the voltage maximum and minimum values of a stationary wave pattern; VSWR is a special value of SWR defined as $|V_{max}|/|V_{min}|$ (ratio between maximum and minimum voltage).
- Network gain.

The XPS technique characterizes the surface of a sample from a compositional point of view, both qualitatively and quantitatively, up to a depth of 10 nm, and detects the chemical bonds formed between the different elements present in the spectrum (oxidation, sulfides–sulfates, etc.). XPS is mainly used in:

- Dust analysis and the related chemical bonds.
- Surface treatment analysis.
- Analysis of coatings (<1 μm).

AFM is typically used for the study of atomic dimension surfaces of various types of compounds: thin or thick films of ceramic materials, amorphous materials, glass, synthetic or biological membranes, metals, polymers, semiconductors, etc. In AFM, the tip (a few μm) moves over the surface of a sample along the three Cartesian axes, by means of a movement induced by a piezoelectric mechanism. A servo control system (feedback) allows the tip to acquire information by means of the interaction force between the sample surface and the tip. Based on how the tip interacts with the surface of the sample, AFM is adopted in:

- Repulsive or contact mode (if the tip actually touches the sample, i.e. the distance between tip and sample is smaller than the average size of an atomic radius).
- Attractive or non-contact mode (if the tip does not touch it).
- Tapping mode (if the tip explores the sample by a discontinuous contact, due to the oscillatory movements continuously changing from the contact condition to the non-contact one; the frequency which the tip moves while scanning the sample surface varies between 50 kHz and 500 kHz).

In the non-contact mode (with distances between tip and sample >1 nm) the images are produced by van der Waals forces, or by electrostatic and magnetic forces. In contact mode, forces of ionic repulsion prevail. In addition to these forces is the friction force that acts between tip and surface, lateral forces and lateral deflections. By controlling the piezoelectric system, it is possible to vary:

- The scan size, which acts on the resolution of the images and represents the observation field of the sample.
- The scan angle, which permits the tip angle scanning the sample surface to be checked (it is important to reduce the associated artifact formation).
- The scan rate, defining the number of lines which are scanned per second (by reducing speed scan a better quality of image is obtained, but the interaction time is increased between tip and sample, thus increasing the risk of particle displacement).
- The vertical limit (Z limit), which fixes the vertical variation of the scan (to obtain high spatial resolutions, it is necessary to reduce this parameter as much as possible).

An AFM approach useful for surface electrical distribution is AFM-SCM, which consists of a conductive metallic probe tip modeling a capacitance sensor measuring surface electrical distribution. Figure 6.1a shows the AFM-SCM approach using UV excitation. Considering a diamond surface, the tip detects the surface electrical current distributed on the surface on the grain boundaries (Figure 6.1b). The basic principle of the electrical current reading is shown in Figure 6.1c,d, where it is possible to note the electrical current signal I assuming positive and the negative values: different electrical current directions are observed when the AFM tip is located above the diamond grain or on its boundary, this is due to the variation of capacitance ($CdV/dt = I$ thus $dI \propto dC$) as electrical current reversal which occurs when the current changes direction. The effect of current reversal is associated with voltage reversal. Figure 6.1e shows an example of the three-dimensional (3D) morphology of diamond film growth on a Si substrate, and Figure 6.1f shows an example of AFM

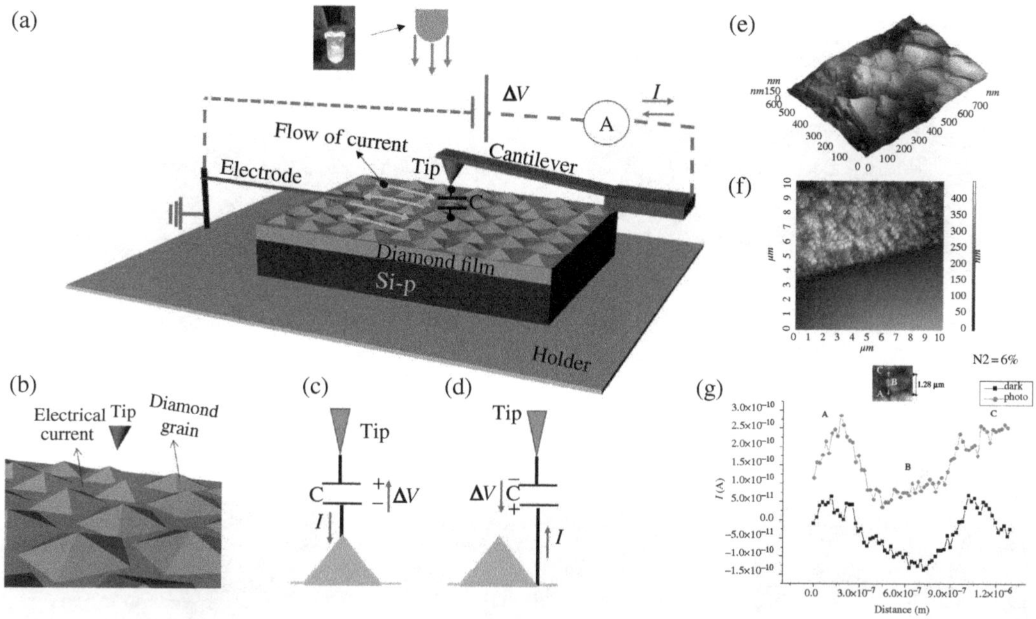

Figure 6.1 (a) Scheme of the AFM-SCM circuital approach. (b) Sketch of the electrical current distribution in valley boundaries. Electrical current *I*-direction for the AFM tip located above (c) and on the boundary (d) of the diamond grain, respectively. (e) Three-dimensional reconstruction of a diamond surface film. (f) Two-dimensional topography AFM image of a boundary region of ND deposition on Si substrate. (g) Electric current distribution on a diamond film surface with LED excitation and without the applied light source.

topographic images of nanodiamonds (NDs) sprayed on a Si surface. An example of the measured electrical current distribution on the diamond film surface with and without the effect of UV excitation is shown in Figure 6.1g: the electrical current density **J** is mainly observed on the grain boundaries representing the valleys of the diamond film surface. By means of UV excitation, the effect of the valley distribution is enhanced adding a photocurrent contribution. An example of NDs deposited on a laboratory glass is shown in Figure 6.2a.

FTIR spectroscopy is a useful technique for the analysis of organic compounds to identify chemical compounds present in the sample, through the determination of absorption spectra in the infrared (IR) band, characterized by absorption peaks.

The TEM approach is essentially based on the phenomenon of electronic diffraction. The generated electrons are accelerated and interact with a sample. In TEM there are two types of source electrons, a tungsten filament or lanthanum hexaboride, or a field emission gun (FEG), capable of extracting and accelerating electrons without the need to apply current. There are also an anode and a cathode to generate an electric field that accelerates electrons to energies between 80 keV and 300 keV. The accelerated electrons are directed along the microscope column, they are then deflected and focused by some electromagnetic lenses, interact with the sample, and are finally projected onto a fluorescent screen or detected by a charge-coupled device (CCD) sensor detecting the spatial distribution following the interaction. To obtain images of the sample in direct space, the microscope is focused on the foreground image of the objective lens. In this way, what is displayed on the screen or with the CCD is a two-dimensional (2D) projection of the 3D structure of the sample. The intensities revealed after the interaction are correlated to the morphology, thickness of the material passed through by the electrons, and the type of material (due to the

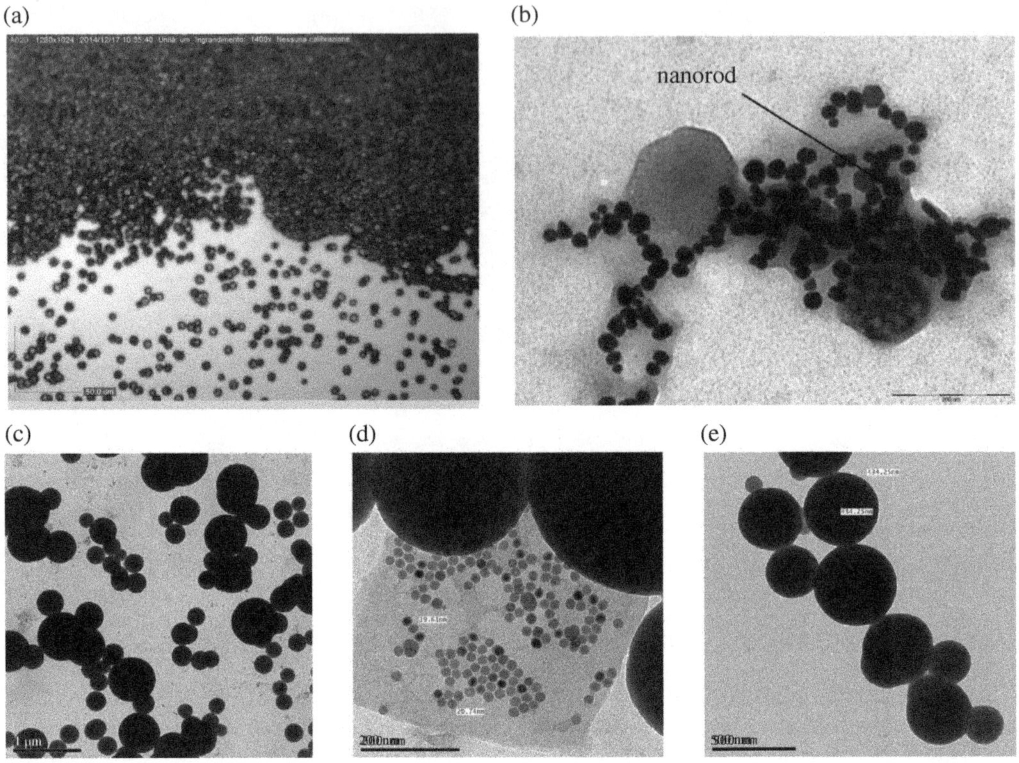

Figure 6.2 (a) Microscope image of NDs deposited on a glass layer. TEM images: (b) GNs extracted from a polydimethylsiloxane-gold (PDMS-Au) sample; (c) silica microparticles; (d) silica NPs; and (e) silica submicrometer particles.

differences in atomic numbers and charge densities). The interaction between electrons and matter is essentially a Coulomb interaction: there is an attraction due to the nuclear charge, and a repulsion due to electron cloud, according to the atomic scattering factor:

$$f_e(\theta) = \frac{me^2\lambda_e}{2h^2} \frac{\left[Z - f_\lambda(\theta)\right]}{\sin^2(\theta)} \tag{6.2}$$

where m is the mass, e is the charge, λ is the electron wavelength, h is the Planck constant, Z is the atomic number, θ is the scattering angle, and $f\lambda$ (θ) is the atomic scattering factor for X-rays. Since the electrons are accelerated to energies of the order of 10^2 keV, the following relativistic equation is used to calculate their wavelength:

$$\lambda_e = \left[2m_0E\left(1 + \frac{eE}{2m_0c^2}\right)\right]^{-1/2} \tag{6.3}$$

where m_0 is the electron rest mass (mass of a stationary electron). The TEM image is provided by the amplitude **A(k)**, which consider the presence of different atoms and the phase difference between incoming and outgoing waves, defined as:

$$\mathbf{A}(\mathbf{k}) = \sum_j f_j \exp\left[2\pi i(\mathbf{k} - \mathbf{k_0}) \cdot \mathbf{r_j}\right] \tag{6.4}$$

where f_j is the atomic diffusion amplitude of the atom in position $\mathbf{r_j}$ (in the exponential term there is the phase difference between the incoming wave and outgoing wave), and $\mathbf{k}$ is the wave vector. Examples of TEM images are shown in Figure 6.2b (gold nanoparticles [GNs] extracted from a polymeric matrix), and in Figure 6.2c–e (silica microparticles and silica NPs). TEM images can be analyzed by image processing tools enhancing the 3D morphology (Figure 6.3a), or surface characteristics as for functionalized NPs (Figure 6.3b).

In the SEM technique, an electron beam strikes the sample to be observed, where the beam is not fixed but is scanned. A scanning electron microscope has two main parts: the column containing the electronic optics and detectors, and the console which includes the control panel, the display and the processor. The electron beam is typically generated by thermionic emission from a tungsten filament or lanthanum hexaboride, or by emission of a field; the electrons thus produced are then accelerated by an electric field generated between the cathode (the filament) and the anode, at a potential difference that generally varies between 500 V and 30 kV, reaching energies between 500 eV and 30 keV. The beam is then focused using lenses. There are three main modes of operation: secondary electron imaging, backscattered electrons, and X-ray microanalysis. From secondary electrons information generally limited to surface morphology is obtained. From backscattered electrons, information is obtained from a greater depth and is also linked to the differences in the composition of the sample. X-rays are useful for determining the atomic composition of the sample.

Concerning the fabrication process of a thin film on a substrate, the MWPECVD technique is adopted to deposit thin diamond layers by plasma application. Figure 6.4 shows a MWPECVD reactor.

In order to start the production of innovative devices it is important to initially equip laboratories with basic facilities. A basic laboratory for nanocomposite fabrication and NP synthesis is equipped with the following fundamental machines suitable for rapid prototyping processes:

- mixer machin;
- analytical balance
- microscope
- fume cupboard
- laboratory sinks

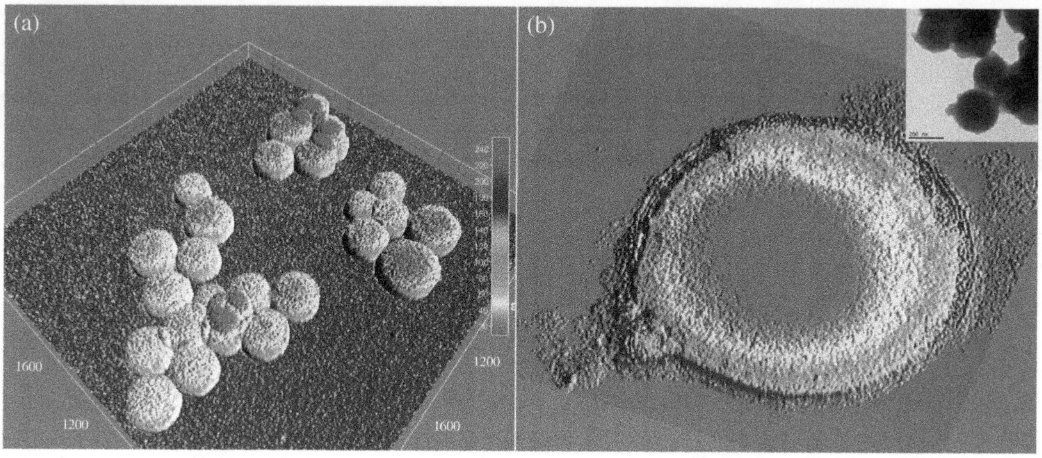

Figure 6.3 Post-processed TEM images: (a) 3D image processing of silica NPs; (b) 3D image processing of functionalized silica NPs.

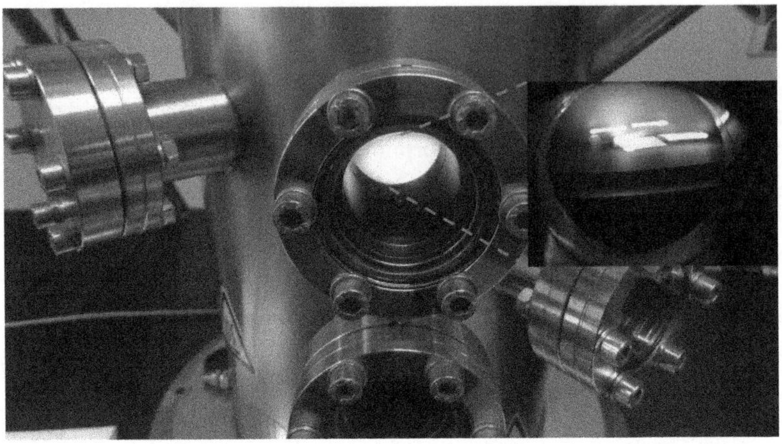

Figure 6.4 MWPECVD reactor and plasma generated during diamond film growth.

6.2 Micro- and Nanosensors as Preliminary Prototypes for Industry Research

Micro- and nanosensor fabrication represents an important innovation in industry production. The preliminary study and characterization of basic sensors is the first step to follow for rapid prototyping in advanced manufacturing and for patent development. In this section, elements are provided useful for the approach to follow for the characterization of physical and sensing aspects, important to define the micro- and nanosensor specifications. The discussed topics and examples are addressed for the production of advanced devices as innovative products, for the development of technologies for nanomaterial characterization and production, and for modeling and design of nanoscale devices. The analyzed methodologies are useful to understand the approaches to follow and to implement for enabling technologies in Industry 5.0.

6.2.1 Nanocomposite Optoelectronic Sensors and Optoelectronic Circuits for Pressure Sensors

The proposed examples are indicative to understand the complexity of the characterization of micro- and nanosensors thus highlighting how the scientific approach is fundamental in order to ideate new prototype sensors based on nanotechnology. In this section, optical fiber sensors, near field metallic probes, and nanocomposite pressure sensors are discussed. The proposed methodology is applied for rapid prototyping and for RE, concerning topics of fundamental research in nanotechnology and nanosensors. AI algorithms strengthen the detection analysis and the implementation of the proposed sensors in more complex robotic systems. The sensor characterization helps the reader to understand how physical basic principles can be implemented in nanoscale sensors, by showing the fundamental modeling useful for a first prototype.

6.2.1.1 Optical Fiber Nanocomposite Tip

NMs are used for many applications including permittivity sensors [1], tactile robotic sensors [2], and security detection systems [3]. A particular kind of polymer used for nanocomposite implementations is polydimethylsiloxane (PDMS). PDMS is known to be a good stabilizing and reducing agent of gold precursor (as chloroauric acid [$HAuCl_4$]) leading to the immediate generation of GNs by the chemical reduction process occurring at room temperature [4–7]. After chemical reduction, the GNs are embedded in the PDMS polymeric matrix as nanofillers by constituting the PDMS-Au NM. This material is easy to apply directly on an optical fiber core as sketched in Figure 6.5.

It is important to note that the relative contributions of absorption and scattering to the overall extinction spectra (extinction = scattering + absorption) is a function of the nanofiller size; for example, in [8] the authors showed, by classical electrodynamic calculations, that the extinction spectra of small GNs, with diameters less than about 50 nm, are dominated by absorption, whereas scattering becomes increasingly important for larger diameters. Since the specific optical properties of the metal nanostructures, i.e. absorption, scattering, and local electric field, are size dependent, size control of the GNs dictates the final application. It is therefore of considerable interest to investigate methods by which the radiation field spectra and local electric field generated by an optical fiber source are enhanced, and to study the enhancement as a function of the nanostructure size or of the filling factor of the nanofillers. Starting with the Brendel–Bormann (BB) method [9] which describes the experimental data of the complex permittivity of metals over a wide spectral range, the analytical approach extends the model to the design of an efficient PDMS-Au material able to amplify a light signal coming from an optical fiber. The proposed approach takes into account spherical GNs at different concentrations in the PDMS polymer showing the possibility of tuning the PDMS-Au material directly by varying the gold precursor concentration in order to calculate the best light scattering efficiency of GNs. The PDMS-Au material enhances the light coming from an optical fiber at a defined working wavelength. A light source excites the nanocomposite tip made by GNs. The light is scattered inside the material by GNs, consequently, some working wavelengths are absorbed, and others are enhanced due to the conservation of energy. By means of the model and design, it is possible to choose a narrow light source working at the wavelength λ_0 corresponding exactly to the amplified wavelength. Starting with the condition found in the analytical approach, it is possible to model by an appropriately designed finite element method (FEM) tool the PDMS-Au tip, by proving the possibility to enhance the light as for active materials. In order to fabricate the optoelectronic sensor, it is important to:

- Model by the BB approach the PDMS-Au NM by evaluating the scattering efficiency conditions.
- Model by FEM, under the best working condition of scattering a PDMS-Au unit cell representing the optical fiber tip.

In order to estimate the PDMS-Au scattering efficiency, the effective dielectric function must be estimated. The effective dielectric function ε_{eff} for spherical metallic particles having dielectric function ε_m embedded in a medium (polymer film) ε_s is defined as [10]:

$$\varepsilon_{\text{eff}} = \varepsilon_s \frac{\varepsilon_m(1+2\phi)+2\varepsilon_s(1-\phi)}{\varepsilon_m(1-\phi)+\varepsilon_s(2+\phi)} \tag{6.5}$$

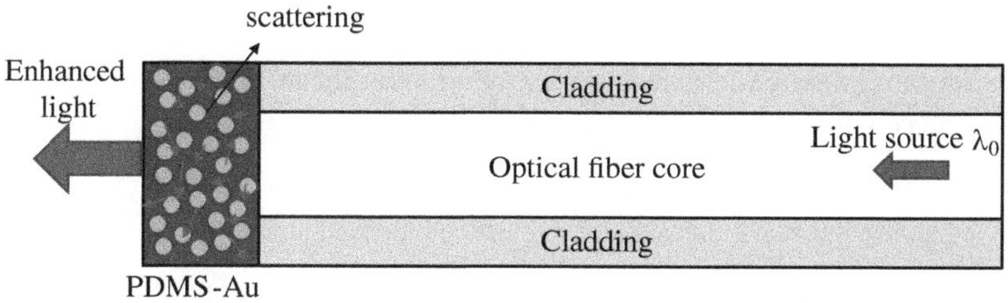

Figure 6.5 Enhanced light of an optical fiber end embedded in a PDMS-Au tip of NM: schematic sketch of the basic principle representing the light scattering in a PDMS-Au tip and the light enhancement due to the active behavior.

where $\phi = 0, \ldots, 1$ indicates the metal concentration as volume filling factor of the GNs. The polymer of choice was PDMS ($\varepsilon_s = 1.96$) since it is suitable for the direct generation of GNs in the polymeric matrix in the presence of suitable precursors. The complex dielectric function of gold is expressed in the following form:

$$\varepsilon_m(\omega) = \varepsilon'_m(\omega) + i\varepsilon''_m(\omega) \tag{6.6}$$

According to the BB model, Eq. (6.6) becomes

$$\varepsilon_m(\omega) = 1 - \frac{\Omega_p^2}{\omega(\omega - i\Gamma_0)} + \sum_{j=1}^{k} \chi_j(\omega) \tag{6.7}$$

where k is the number of BB oscillators used to interpret the intraband part of the spectrum, $\Omega_p = (f_0\omega_p)^{1/2}$ is the plasma frequency associated with intraband transition with oscillator strength f_0 and damping constant Γ_0, and with the infinite number of oscillators (part which replaces the Lorentz oscillators) given by:

$$\chi_j(\omega) = \frac{1}{\sqrt{2\pi}\sigma_j} \int_{-\infty}^{\infty} \exp\left[\frac{-(t-\omega_j)^2}{2\sigma_j^2}\right] \times \frac{f_j\omega_p^2}{(t^2-\omega^2)+i\omega\Gamma_j} \mathrm{d}t \tag{6.8}$$

A high concentration of GNs in a PDMS polymeric matrix increases the effective refractive index of the NM by absorbing a large quantity of electromagnetic energy especially at optical frequencies. When a small spherical metallic NP is irradiated by light, the oscillating electric field causes the conduction electrons to oscillate coherently. The electric field $\mathbf{E}_{\mathbf{out}}$ is approximated for small particles (dipole plasmonic resonance) as:

$$\mathbf{E}_{\mathbf{out}} = \mathbf{A}_1 - g_d A_1\left[\frac{\mathbf{x}}{r^3} - \frac{3x}{r^5}(x\mathbf{x} + y\mathbf{y} + z\mathbf{z})\right] \tag{6.9}$$

where **x**, **y**, and **z** are the usual unit vectors, $\mathbf{A_1}$ is the propagating wave amplitude, and g_d is the normalized sphere polarizability:

$$g_d = \frac{\varepsilon_m - \varepsilon_{\text{eff}}}{\varepsilon_m + 2\varepsilon_{\text{eff}}} \tag{6.10}$$

where r is the radius of the sphere. The resonance conditions are estimated by calculating the extinction efficiency:

$$Q_{\text{ext.}} = 4u\, Im(g_d) \tag{6.11}$$

with $u = 2\pi(\varepsilon_{\text{eff}})^{1/2}/\lambda_0$. In order to estimate the scattering contribution to the efficiency, the following scattering efficiency is also considered:

$$Q_{\text{scatt.}} = \frac{8}{3}u^4|g_d|^2 \tag{6.12}$$

Figure 6.6a–c shows the theoretical scattering efficiency versus different filling factors (concentration ϕ) and different GN radius values. Two main peaks are observed. Peaks shift versus higher

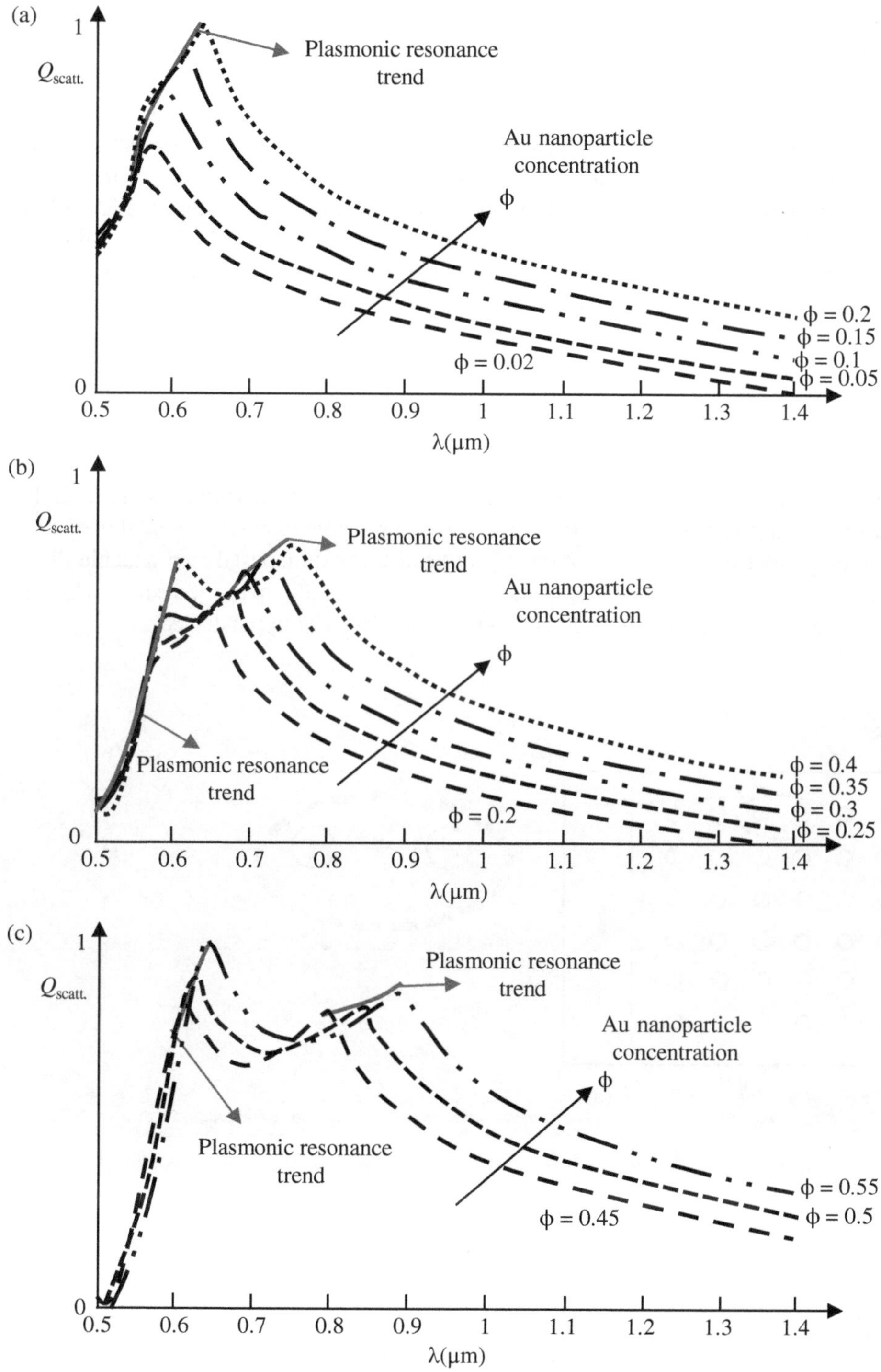

Figure 6.6 PDMS-Au scattering efficiency versus the working wavelength for different gold micro-/nanoparticle concentration ϕ in PDMS material. GN radius: (a) 25 nm; (b) 0.5 nm; and (c) 180 nm.

wavelengths are always observed by increasing ϕ. Consequently, by tuning the GN concentration, it is possible to define the working wavelength of a source coupled to a fiber with the end embedded in PDMS-Au and behaving as an optical fiber amplifier with active tip.

The FEM tool is adopted for the design of the nanocomposite tip thus supporting rapid prototyping processes. The described analytical model does not consider the layout and the displacement of GNs in the PDMS polymeric matrix. Moreover, a GN monodispersion condition is assumed. An accurate modeling representing the light scattering versus the geometry is necessary. Figure 6.7a shows the FEM unit cell model [11] of the PDMS-Au. A light source excites the PDMS-Au unit cell, and numerical input and output probes estimate the input and output time-domain electric fields, respectively. The unit cell concept is fundamental for the design phase in rapid prototyping. The output probe is placed in the external air region. Figure 6.7b indicates the geometrical parameters L and D used to represent the distance and the diameter of the GNs, respectively. Figure 6.7c summarizes the L and D parameters used for the simulations in Figure 6.7d, indicating the gain of the PDMS-Au versus the scale factor: assuming a working wavelength of 0.9 μm (corresponding to the second peak of Figure 6.6c), a reference value of a scale factor is fixed, and, successively, all the geometrical unit cell dimensions are scaled. In the case analyzed, the scale factor 0.9 provides the best gain condition. Analysis of Figure 6.7 explains how it is possible to create a model useful for the rapid prototyping process: the concept of the 2D unit cell simulation based on a scalable simulation approach is important for preliminary design of a nanocomposite tip probe system, which is the fundamental part of a more complex optical fiber sensor (first prototypal device).

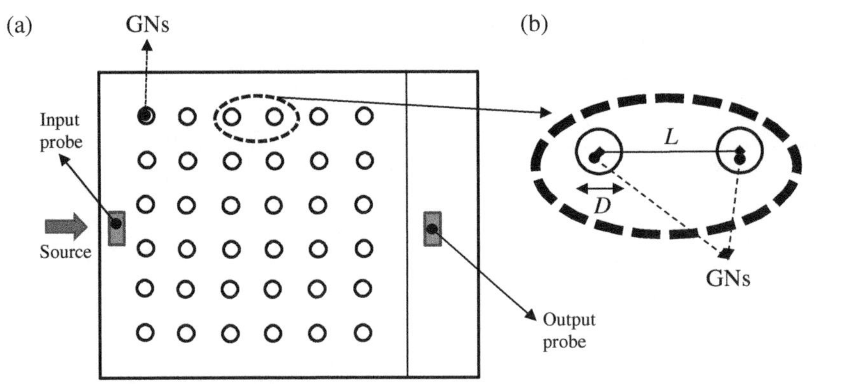

(c)

Scale factor	L (μm)	D (μm)
0.6	0.718	0.24
0.7	0.837	0.28
0.8	0.95	0.32
0.9	1.07	0.36
1 (Ref.)	1.2	0.4
1.1	1.31	0.44
1.2	1.43	0.48

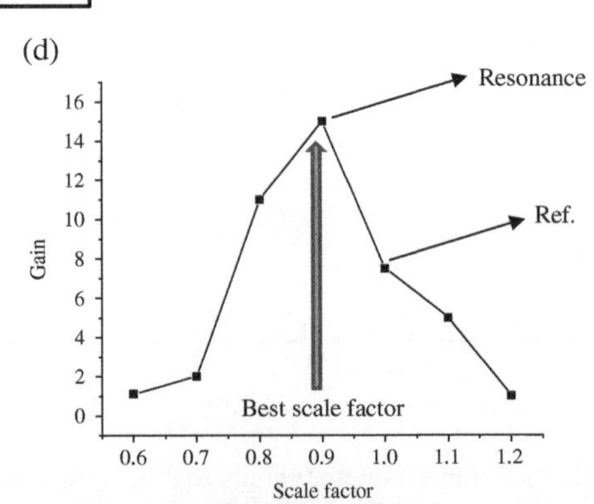

Figure 6.7 (a) Unit cell: modeling of PDMS with monodisperse GNs. (b) Zoomed image indicating the GN layout. (c) Scale factor and L and D values. (d) Optical gain due to resonance.

Figure 6.8 (a) shows the total electric field propagation inside a pure PDMS material (without GNs) at $\lambda = 0.9\,\mu m$: a small numerical aperture (NA) is checked. The NA is an important specification for the design of optical fiber probes as endoscopic systems where it is important to establish this parameter. If GNs are considered inside the PDMS unit cell (Figure 6.8b), a strong light scattering is observed, a gain factor of about 5 at the output air region (see input and output time domain total electric fields in Figure 6.8c), and a higher NA. The design criteria are summarized by the following two main points:

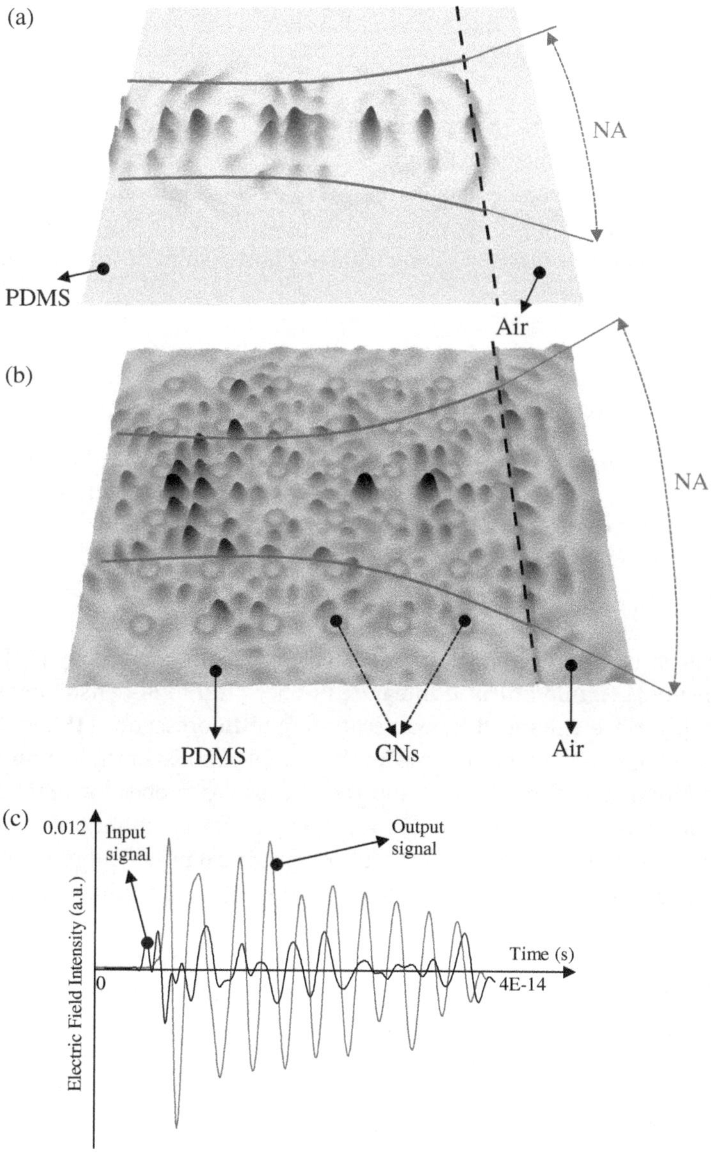

Figure 6.8 FEM simulations: (a) light propagating in the PDMS material; (b) light propagating in the PDMS material with metallic NP inclusions (light scattered in the PDMS-Au by the GNs); (c) input and output time domain signals checking the output optical gain (light enhancement).

- tuning of the GN concentration by evaluating the working wavelength λ_0 corresponding to the high scattering efficiency.
- Starting from the GN size found in the analytical (BB) approach, model the PDMS-Au by FEM in order to calculate the best layout providing a high gain factor.

The theoretical model of light scattering in PDMS-Au NM is adopted to formulate criteria for the FEM modeling, thus completing the analysis of the scattering effect, and proving the behavior of PDMS-Au as an active material characterized by a gain. The PDMS-Au is used for a fiber end tip able to enhance the light at output as for fully integrated optical probes. This enhancement is suitable for many applications including optical fiber probe systems, robotic optical sensors, and medical endoscopy [12]. Both analytical and FEM modeling are used in order to define design criteria based on the tuning of GN concentration and on the control of the monodispersion in the polymeric matrix.

Monodispersion in a polymeric matrix is a strong hypothesis to understand the main physical effects of metallic NPs characterized by an average size (average diameter dimension). In order to understand the real behavior of the light crossing the nanocomposite material, it is fundamental to first measure the output light, by adopting a simple experimental setup suitable for a basic check of the optical gain. Light enhancement at the output of the optical fiber is also suitable for sensing applications such as:

- Detecting object shape.
- Pollution detection in air or in liquids.
- Color detection by considering the different optical abosrbance of the colored objects.

The detection system is improved by an AI engine supporting data processing for the detection (classifying colors or pollution typology), and optimizing automatized actions, such as the sensor trajectories and movements to perform to scan the object.

6.2.2 Plasmonic Probes

Concentration of optical energy is used in many applications, such as near-field microscopy [13], single-molecule biochemical sensing [14], miniaturized integrated optics [15], high-density data storage [16], nonlinear optics [17], surface enhanced Raman scattering or fluorescence [18], and implanted wireless systems [19]. Metallic wedges are suitable for these applications for their ability to focus the energy in the proximity of the discontinuity, and they behave as probes for optical detection systems. The optical source is a surface plasmonic wave which can be enhanced by a micro-/nanograting placed directly on the probe [20]. An example of an optical near field probe is shown in Figure 6.9a. The simultaneous transverse resonance diffraction (STRD) method allows the near field singularity to be calculated by distinguishing the near field from the irregular field and consequently the noise from the signal. Figure 6.9b shows an example of transmission line modeling suitable for the calculation of resonance conditions. The micro-/nanoscale of the edge antennas in Figure 6.9c,d requires an accurate modeling especially for near field applications. The traditional numerical approaches such as FEM and the finite difference time domain (FDTD) technique require high computational cost for the evaluation of the radiated field in the proximity of a dielectric or metallic discontinuity. The new approach of multipole expansion of Green's function (MEG) based on the novel principle of STRD [21–24] is suitable for near field characterization. STRD/MEG is applied for the near field evaluation of a 3D dielectric corner by showing a good convergent solution with a low computational cost and with a reduction of the computational time

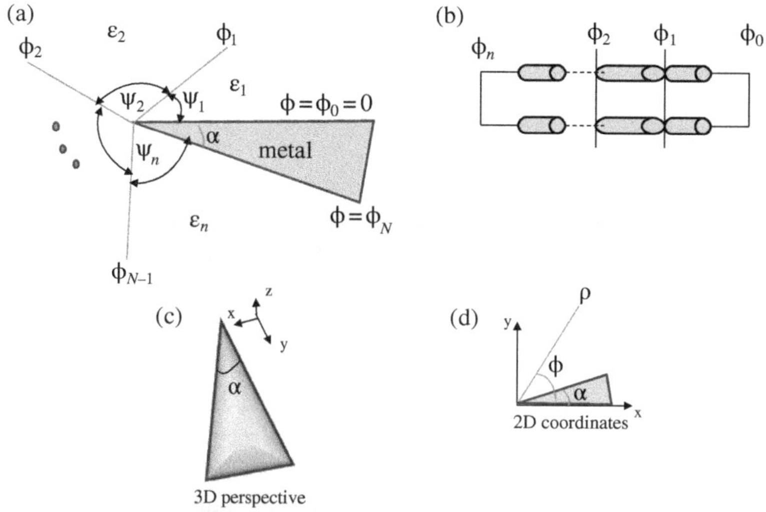

Figure 6.9 (a) Metallic wedge in dielectric materials. (b) Transmission line circuit for a generic metallic wedge profile. A perfectly conducting wedge infinite in the z-direction. (c) Three-dimensional view. (d) Geometry in the transverse plane also showing the cylindrical coordinate system.

of 12- to 15-fold compared with traditional numerical methods. STRD/MEG calculates the near field radiation patterns of metallic wedges. Analytical tools are typically used for a primary design. For a correct first design it is important to:

- Describe accurately the STRD/MEG model of a generic metallic wedge by introducing the equivalent transmission line circuit.
- Calculate the radiation pattern of different kinds of wedges (radiation characterization).
- Apply the STRD/MEG to a particular layout of the probe system able to detect a variation of dielectric contrast as for permittivity sensors.

The radiated electric field at an external point P is expressed as:

$$\mathbf{E}(P)=\frac{1}{j\omega 4\pi\varepsilon_c}\nabla_P\times\nabla_P\times\left(\int_V \mathbf{J}_i G(\mathbf{r},\mathbf{r}')\mathrm{d}V_Q+\int_S(\mathbf{H}\times\hat{n})G(\mathbf{r},\mathbf{r}')\mathrm{d}S_Q\right)- \\ -\frac{1}{4\pi}\nabla_P\times\left(\int_S(\hat{n}\times\mathbf{E})G(\mathbf{r},\mathbf{r}')\mathrm{d}S_Q\right)-\frac{\mathbf{J}_i}{j\omega\varepsilon_c} \tag{6.13}$$

where in spherical coordinates the wedge point is at $Q = Q(r',\theta',\varphi')$ (Q is identified by a vector $\mathbf{r}'$), and the field is evaluated at $P = P(r,\theta,\varphi)$ (P is identified by a vector $\mathbf{r}$). The permittivity ε_c indicates the cladding permittivity, where the cladding encloses the wedge. In spherical coordinates, the Green function satisfies the Helmholtz scalar equation and the multipole expansion of Green's function is given by:

$$G(\mathbf{r},\mathbf{r}')=\sum_{l,m} g_l(r,r')Y_{l,m}^*(\theta',\varphi')Y_{l,m}(\theta,\varphi) \tag{6.14}$$

By setting equal to zero the z-derivatives in Maxwell's equation in cylindrical coordinates the three components of the transverse electric (TE) field TE_z are obtained:

$$E_\rho = -j\omega\mu\frac{1}{\rho}\frac{\partial\Phi}{\partial\phi}$$
$$E_\phi = j\omega\mu\frac{\partial\Phi}{\partial\rho}$$
$$H_z = k^2\Phi \tag{6.15}$$

The MEG approach takes into account the radiation conditions especially for the near field calculation, but an accurate evaluation of the potential Φ is requested. Assuming the potential Φ as

$$\Phi(\rho,\phi) = \frac{1}{k_0^2}\rho^\nu\left[A_i\sin(\nu\phi) - B_i\cos(\nu\phi)\right] \tag{6.16}$$

where $k_0 = 2\pi/\lambda_0$ is the free space wavenumber and λ_0 is the optical working wavelength, the near field is characterized by the singularity ν which is calculated by means of a circuital approach. If the metallic wedge is embedded in N dielectric materials, as indicated in Figure 6.9a, the circuital approach takes into account the EMF propagating along the cascade of transmission lines in Figure 6.9b. Each transmission line is characterized by an electric length $\delta_i = \nu\psi_i$ (which is a function of the angle ψ_i related to the dielectric material ε_i), and by a characteristic impedance ζ_i. The simultaneous resonance of the transmission lines provides the singularity ν of the EMF. The resonance condition is performed by annulling the total impedance of the circuit with respect to a reference. If the conductor is embedded in a single dielectric material $\varepsilon_1 = \varepsilon_c = n^2$, as reported in Figure 6.9b,c, the equivalent transmission line circuit of the metallic edge is a single transmission line enclosed in two short circuits. By considering for simplicity the TE-polarized (H_z, E_ρ, $E\varphi$) field in Eq. (6.15), the equivalent voltages and currents related to the scheme in Figure 6.9b are defined as:

$$\begin{pmatrix} V_{i+1} \\ I_{i+1} \end{pmatrix} = \begin{pmatrix} \cos(\nu\phi_i) & \zeta_i\sin(\nu\phi_i) \\ \zeta_i^{-1}\sin(\nu\phi_i) & \cos(\nu\phi_i) \end{pmatrix}\begin{pmatrix} V_i \\ I_i \end{pmatrix} \tag{6.17}$$

where $\zeta_i = 1/\varepsilon$ is the normalized characteristic impedance. By introducing the load impedance $Z_i = V_i/I_i$ and $T_i = \tan(\nu\psi_i)$, and by using the standard formula for the impedance transformation along a transmission line:

$$Z_i = \zeta_i\frac{Z_{i-1} + j\zeta_i T_i}{\zeta_i + jZ_{i-1}T_i} \quad i = 1,2,\ldots,N \tag{6.18}$$

it is possible to calculate the simultaneous resonance condition of the total impedance $Z_1 + Z_2 = 0$ in the case of $N = 1$ (the case in Figure 6.9c,d), where Z_1 and Z_2 are the impedances calculated starting from a reference plane. In our case the resonance condition provides the following singularity parameter

$$\nu = \frac{\pi}{\Psi_1} \tag{6.19}$$

where $\psi_i = 2\pi - \alpha$.

As application of the STRD approach implemented in MEG theory (STRD/MEG), the working wavelength is fixed to $\lambda_0 = 1$ μm (indium gallium arsenic phosphide [InGaAsP] laser source) and is implemented Eq. (6.16). With respect to the evaluation of the total electric field reported in Eq. (6.15), the following relationships are adopted, which are obtained by the construction in

Figure 6.10a,b. In particular, the cylindrical coordinates are correlated with the Cartesian coordinates by:

$$\begin{cases} x = \rho\cos\phi \\ y = \rho\sin\phi \\ z = z \end{cases} \tag{6.20}$$

with

$$\rho = \left(x^2 + y^2\right)^{1/2}, \tan\phi = y/x \tag{6.21}$$

while the spherical coordinates are correlated with the Cartesian coordinates by:

$$\begin{cases} x = r\sin\theta\cos\varphi \\ y = r\sin\theta\sin\varphi \\ z = r\cos\varphi \end{cases} \tag{6.22}$$

$$\begin{cases} r = \left(x^2 + y^2 + z^2\right)^{1/2} \\ \tan\theta = \left(x^2 + y^2\right)^{1/2} / z \\ \tan\varphi = y/x \end{cases} \tag{6.23}$$

When the surface plasmonic resonance occurs (the surface plasmonic wave excites the wedge probe), and part of the incident light is absorbed, the propagation constant k becomes:

$$k = k_d + k_r \tag{6.24}$$

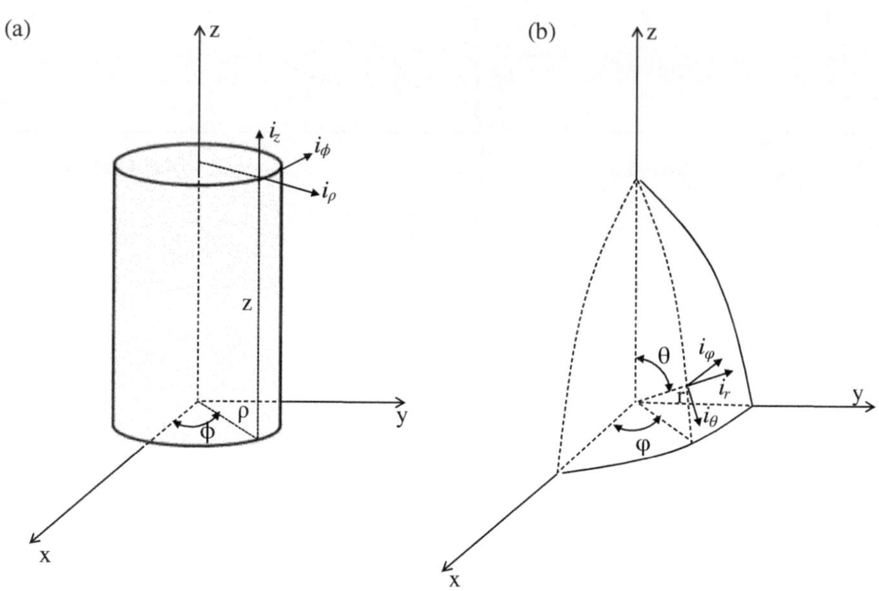

Figure 6.10 (a) Cylindrical coordinate system. (b) Spherical coordinate system.

where k_d is the dispersion relation of the surface plasmonic wave of a metal–air interface in a one-sharp-boundary model and k_r is the perturbation to k_d due to the physical volume of the metal defined by the thickness (in the proposed case it is considered an infinitely thin metallic probe). The propagation constant is a function of the metallic permittivity. The complex permittivity function of a metallic film is expressed in the following form:

$$\varepsilon_m(\omega) = a(\omega) + ib(\omega) = \varepsilon'_m(\omega) + i\varepsilon''_m(\omega) \tag{6.25}$$

and can be evaluated by the BB approach. Figure 6.11 shows the dielectric permittivity theoretical trend for different kinds of metal such as gold (Figure 6.11a), silver (Figure 6.11b), platinum (Figure 6.11c) and titanium (Figure 6.11d), which can be used as plasmonic probes.

The infinitely thin metal conductor ($\alpha = 0$) represents a critical case regarding the convergence of the near field solution; therefore it is useful to fix the minimum value of the order l performing a closed solution for all the geometrical configurations with $\alpha \neq 0$. The TE_z assumption is due to the hypothesis of TE_z optical source excitation, and the total electric field is calculated by Eq. (6.13) by supposing the impressed currents $\mathbf{J_i} = 0$. It is noted that the solution with $l = 1$ (l is the order of the expansion of Eq. (6.14) and $m = -l, \ldots, +l$) is weakly accurate, while the solution with $l = 3$ is convergent. For this purpose, the choice of the order for the conductor edges must be at least $l = 3$.

Figure 6.12 shows a 90° wedge near field theoretical trend ($\alpha = 90°$).

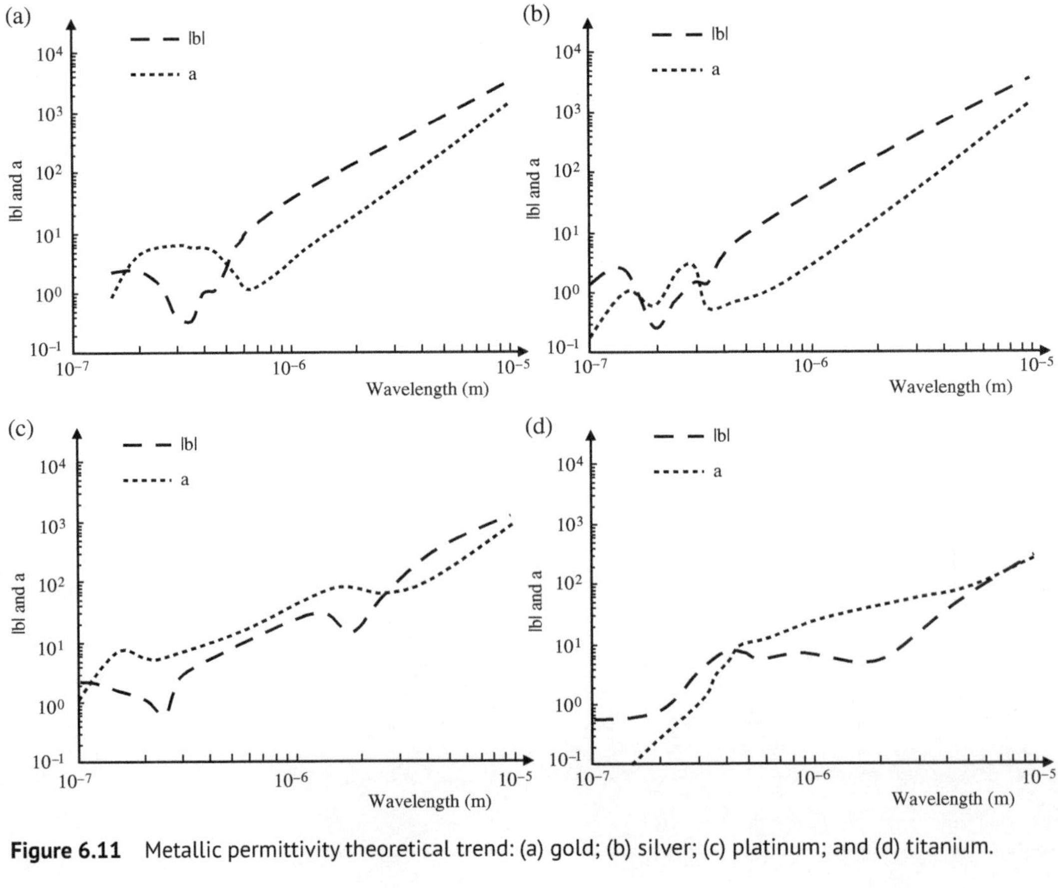

Figure 6.11 Metallic permittivity theoretical trend: (a) gold; (b) silver; (c) platinum; and (d) titanium.

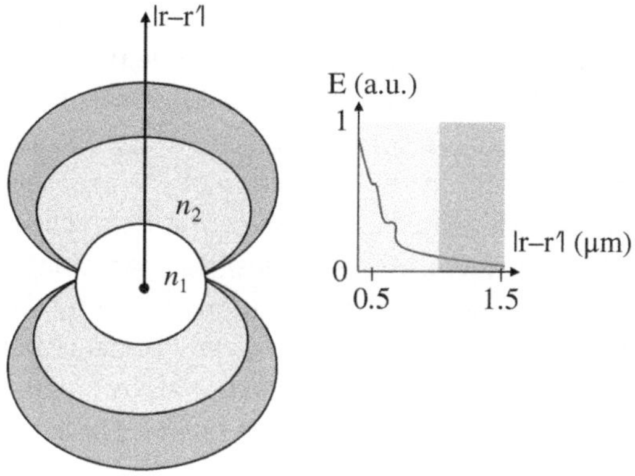

Figure 6.12 STRD theoretical near field radiation pattern for a gold metallic wedge with $\alpha = \pi/2$ working at $\lambda_0 = 1\,\mu m$. The order is fixed to $l = 3$.

In order to focus the study on a permittivity detection system, it is supposed that the receiver probe is placed at a distance $|\mathbf{r}-\mathbf{r}'|$, which depends on the working wavelength and on the near field region. In this example, the solution of Eq. (6.14) is considered for materials characterized by different dielectric contrasts $dn = n_2 - n_1$. Figure 6.13 proves that the STRD/MEG approach also allows to distinguish the near field amplitudes for low dielectric contrasts directly by observing the near field amplitude. Observing Figure 6.13, it is possible to conclude that the receiver probe must be located at a distance between 0.3 μm and 0.8 μm (the near field region where it is possible to distinguish the variation of the electric field amplitude), thus defining the main sensor characteristics.

The STRD/MEG theoretical model is necessary to characterize the near field generated by a generic metallic edge. The STRD technique based on MEG is particularly suitable for the design and rigorous analysis of wireless nanosystems. The analytical approach is a step necessary to define a specific design tool oriented on rapid prototyping and microsensor characterization. This

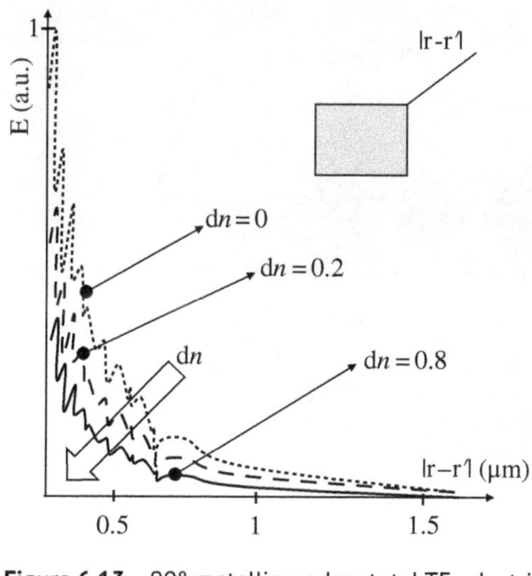

Figure 6.13 90° metallic wedge: total TE_z electric field theoretical trend versus $|\mathbf{r}-\mathbf{r}'|$ expressed in micrometers and for different variations dn of the refractive index. The order of the MEG expansion is $l = 3$. Inset: schematic plot of the distance $|\mathbf{r}-\mathbf{r}'|$.

model efficiently evaluates the near field distribution because of the interaction of optical sources, such as surface plasmonic waves, with perfectly conducting wedges. In order to show the pertinence of the tool, the proposed approach is applied to a 90° metallic probe for the detection of low dielectric contrasts but can be applied for different edge layouts. According to the material permittivity detection, the metallic wedge behaves as a near field sensor: the estimation of the variation of the refractive index permits the detection of materials with related dielectric permittivity scheduled into a matrix (matrix of physical properties). In manufacturing processes, it is particularly important to realize high resolution wedges in order to preserve the specifications found by the theoretical model. The probe characterizations and the near field measurements are fundamental for device fabrication optimization. An AI algorithm can pre-classify different materials by reading as input data all the materials stored into a table, thus training the classification model. The same algorithm can identify the material by reading the digital value of the refractive index contrast dn allowing the material classification. The embedded system of sensor and AI software can be applied in a robotic sensing application for the automatic detection of materials. Experimental and numerical results can contribute to enrich the training dataset of the model to adopt for the discrimination of materials.

Another application of the enhancement of the near field emitted light is shown in Figure 6.14, where an optical waveguide is characterized by two metallic planes embedding a chitosan slab guiding the light generated by an external laser source. The top metallic plane can be fabricated by the laser texturing technique. The device in Figure 6.14 is appropriate for biomedical applications for skin melanoma detection [25]. The wavelengths useful for the detection of melanoma are $\lambda = 1390$ nm and $\lambda = 1458$ nm. The waveguide behaves as an optical antenna sensor measuring the relative variation of the refractive index $\Delta n/n$ of the skin expressed in percentage: for $\Delta n/n < 10\%$ no risks are detected, for $10\% < \Delta n/n \leq 20\%$ a risk is detected, and the condition $\Delta n/n > 20\%$

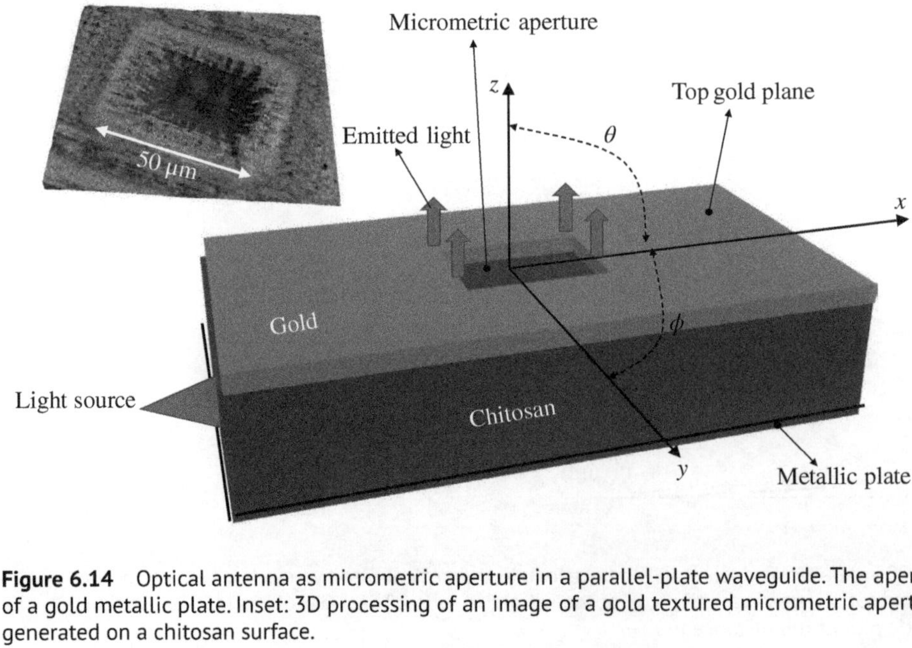

Figure 6.14 Optical antenna as micrometric aperture in a parallel-plate waveguide. The aperture is on top of a gold metallic plate. Inset: 3D processing of an image of a gold textured micrometric aperture generated on a chitosan surface.

indicates a melanoma. A neural network can be trained to classify automatically the $\Delta n/n$ parameter automatizing skin melanoma detection.

6.2.3 Nanocomposite Pressure Sensor

Pressure sensors made by NMs represent an important topic in nanotechnology. The basic scheme of the pressure sensor is illustrated in Figure 6.15a: a tapered optical fiber is embedded into a NM constituted by a polymer matrix containing spherical GNs. The pressure sensor works as follows:

- An input light signal is transmitted by the input of the optical fiber.
- The light is coupled inside the NM in the corresponding tapered region.
- A pressure is applied on the region above the tapered region increasing the light coupling in the NM.

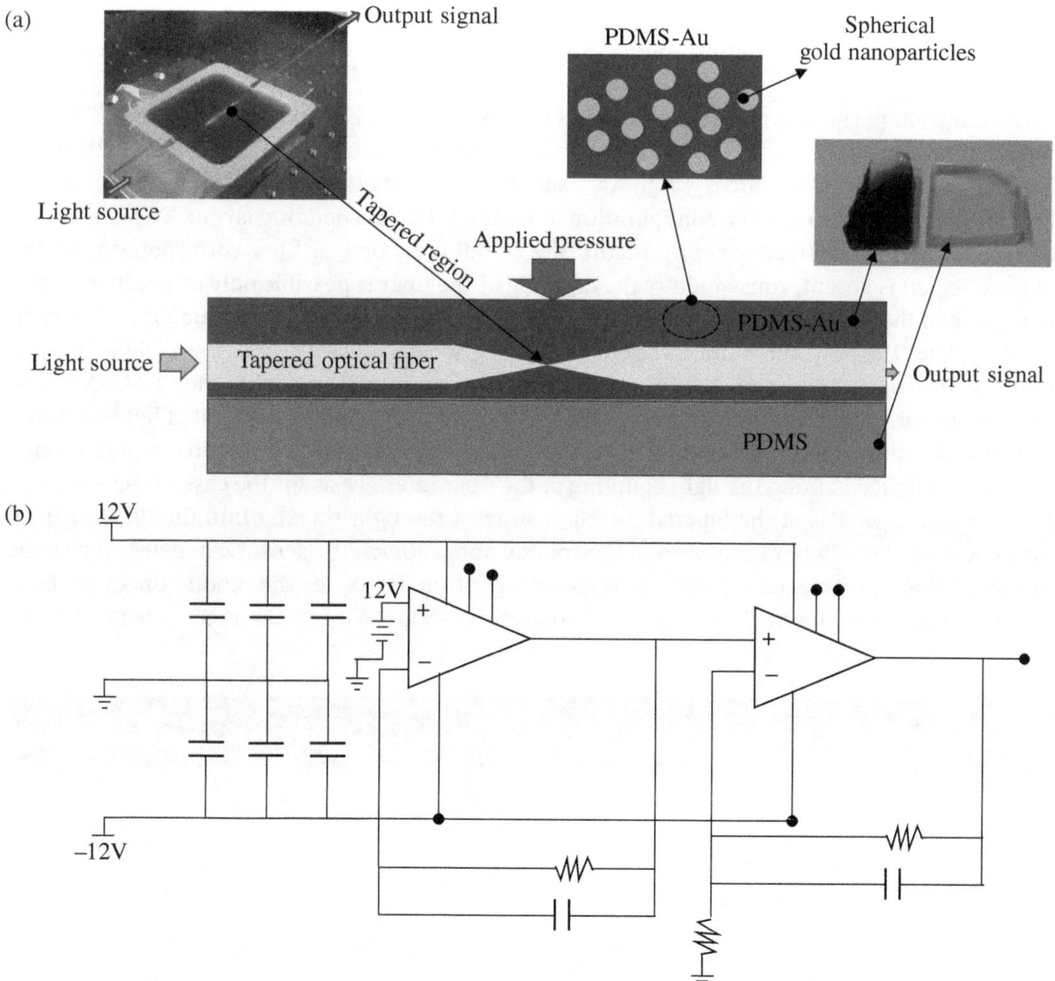

Figure 6.15 Basic scheme of pressure sensor (longitudinal section): (a) tapered optical fiber embedded into PDMS-Au NM (insets: photo of the prototypal device, and basic layout of the NM, and photos of PDMS and PDMS-Au samples); (b) basic circuit layout enhancing optical fiber output signal, made by resistors and capacitors.

- After the application of pressure, the light intensity at the sensor output decreases, and it is detected by a photodetector reading a voltage signal.

Figure 6.15a shows a pressure nanocomposite sensor made by a tapered fiber embedded in a PDMS-Au material [2, 26]: when a pressure is applied the effective dielectric permittivity of the PDMS-Au material changes gradually according to Eq. (6.5), where the concentration of gold nanofillers is ϕ. When a pressure is applied, the effective permittivity index ε_{eff} assumes a maximum value at the tapered interface, and a minimum value at the interface between PDMS-Au and air; the output light intensity decreases due to the enhanced light coupling into the NM. The light intensity at the output of the fiber is transformed into a voltage signal V_{out} by a photodetector, and then gained by the circuit amplifier in Figure 6.15b increasing the signal/noise ratio. The output of the electronic circuit amplifier is the input of a signal analyzer reading voltage intensity: the electronic circuit is connected to the oscilloscope with a coaxial cable used for the connection between the photodiode and the electronic circuit and for connection with the oscilloscope. The double stage layout in Figure 6.15 enables the signal amplification due to the applied pressure. For control of the output a threshold comparator is adopted at the end of the two amplifier stages [2].

The nanocomposite pressure sensor can detect a few grams as applied pressure forces [26]. The senor is implemented in totally optical robotic fingers detecting surface profiles [2], small notches [27], and shear forces [28]. An example of a robotic application is illustrated in Figure 6.16a,b representing the configuration of Figure 6.15a in a bending layout. The fabrication of a bending tapered fiber is very difficult: the cladding of optical fiber corresponding to the tapered region is absent, consequently the tapering of the fiber is possible only by gradually heating up when the fiber is curved, and then by cooling the deposited PDMS-Au material. The cooling completes the polymerization process. The final viscoelastic property of the PDMS-Au tip is tuned by the concentration of GNs. Figure 6.17a shows another 3D model of the PDMS-Au pressure sensor for robotic tactile implementation and Figure 6.17b shows the related light coupling principle describing pressure sensing: the GNs are displaced near the fiber tapered region when a pressure is applied gaining the light coupling at the fiber interface as for the case of the sensor in Figure 6.15a. The GNs at the tapered interface support the light transfer into the PDMS-Au tip thus decreasing the light output signal. For robotic applications, the sensor can detect the shape profile of objects or surfaces. A data processing algorithm associates the robotic finger position along a vertical direction with the measured voltage signal due to the shear forces detected [2, 28].

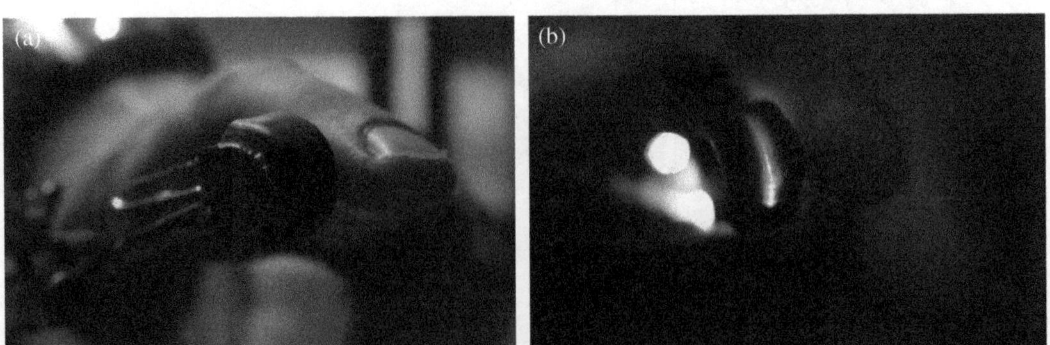

Figure 6.16 (a) PDMS-Au robotic finger (tactile pressure sensor). (b) Light pressure sensor (light coupling into the PDMS-Au material).

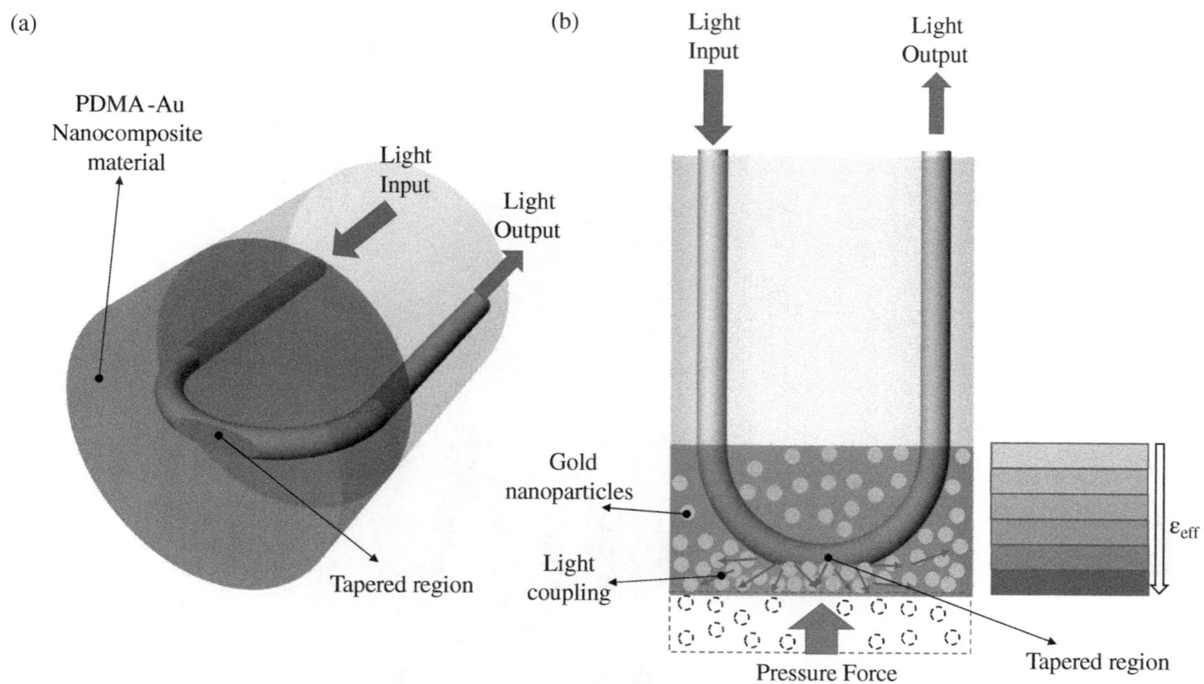

Figure 6.17 (a) PDMS-Au tip of the optical pressure sensor for robotic finger configuration. (b) Light coupling basic principle after the application of a pressure force on the tip.

The tactile sensor in Figure 6.17a is adopted for the detection of small notches [27]. A sensing approach for robotic application is the scanning rotation modeled in Figure 6.18: the sensor touches a surface (region I) and the output signal is decreased for the light coupling, then it is automatically moved by rotating its position into the small notch to detect (region II), and consequently the output signal increases (because the pressure intensity decreases) until the tip is moved outside the notch (region III) strongly touching the external surface again. The data post processing favors the automatic recognition of the notch: after a calibration of the sensor (sampling, scanning velocity, etc.), a neural network is appropriate to recognize the notch size by means of a classification approach, matching output signals with position along the z-direction as for the V_{out} trend in Figure 6.18. Notch recognition is useful for internal detection of pipeline cracks or in application environments where it is difficult to perform a measurement.

Table 6.1 and Table 6.2 list some of the main properties of PDMS and PDMS-Au, respectively.

Another application of PDMS-Au NM is in security systems. Figure 6.19a shows the basic scheme of the application of a sensor following the configuration in Figure 6.15a of a tapered fiber

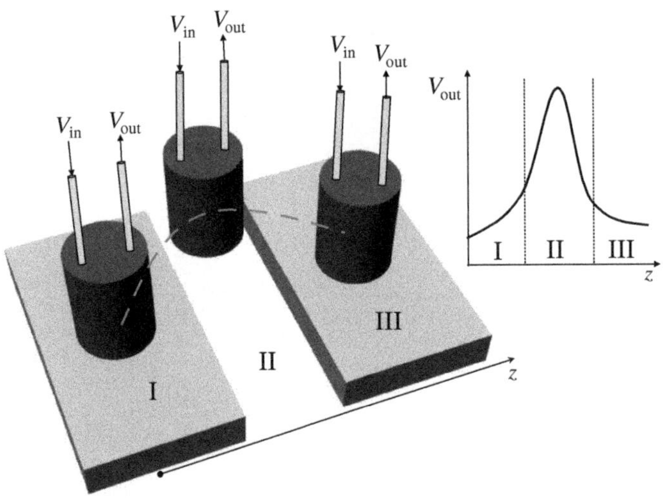

Figure 6.18 Small notch sensing approach and signal detection principle and V_{out} trend versus the sensor position.

Table 6.1 Main properties of the PDMS polymer.

PDMS properties	Value
Density	$970\,kg\,m^{-3}$
Young's modulus (E_{PDMS})	500×10^3 Pa
Poisson's ratio of PDMS (ν_{PDMS})	0.5
Relative permittivity (ε_{PDMS})	2.25
Polymerization	PDMS is obtained by a classical chemical procedure (a curing agent with a 1:10 weight ratio of base polymer)

Table 6.2 Main properties of the PDMS-Au material.

PDMS-Au properties	Value
Elastomeric reactivity measured by a dynamic mechanical analyzer	0.6 s for applied pressures ranging from 10 to 100 g (equivalent to ranging from 0.098 to 0.98 N)
Typical Au GN concentration	10% by weight using a chloroauric acid salt as a gold precursor in water $[M_w(HAuCl_4)] = 339.785\,g\,mol^{-1}$ $[HAuCl_4] = 0.01\,M$
Average friction coefficient at velocity $v = 0.005\,m\,s^{-1}$ (layout in Figure 6.17a)	0.454
Average friction coefficient at velocity $v = 0.05\,m\,s^{-1}$ (layout in Figure 6.17a)	1.140
Optical sensitivity (pressure sensitivity)	5 g
Notch size detection	Up to 1 mm
Tip diameter (layout in Figure 6.17a)	1.5 cm

coupling light with nanocomposite cladding. When a pressure is applied on a floating floor, the PDMS-Au cladding changes its effective permittivity thus increasing the light coupling inside it due to its deformation (Figure 6.19b). AI classification is able to discriminate between the applied pressures reading the light intensity at the output of the tapered optical fiber, thus distinguishing the passage of an animal or of a human body, or of a mobile object moving on the floor. The classifier is learned associating the classes of applied pressures to the measured output light intensities.

6.2.4 Nanocomposite Sensor for Liquid Detection Systems and Fluid Loss Systems

Sensory information for feeling materials and determining their physical properties is provided by different kinds of sensors. Presently, many researchers are attempting to apply detection ability to intelligent robotic systems detecting contact forces, vibrations, textures, temperatures, and liquids. Of particular scientific interest is a new class of material oriented towards liquid detection able to increase the sensing effect: this class, named NMs, represents metallic/organic nanostructures added in polymer matrices. NMs strongly absorb the light becoming technologically interesting, since the light absorbed by the nanostructure is locally released as energy. This class of NMs is suitable for dielectric permittivity sensors for the detection of materials. NMs are also implemented to

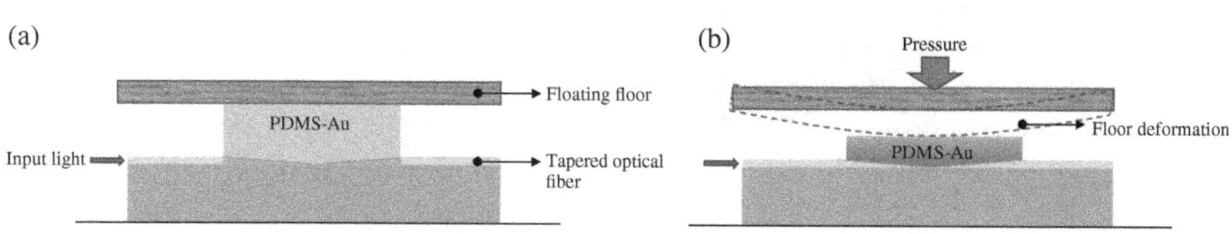

Figure 6.19 PDMS-Au used for surveillance security systems. (a) Schematic configuration of a floating floor without the application of pressure. (b) Deformation scheme after the application of pressure on the floor top.

obtain new sensors for liquid detection [1] exhibiting enhancement of optical radiation and of light coupling. The detection of fluid losses is important in industries, especially when a lot of machines are involved in the production. The proposed sensor is an example of how a sensitive liquid sensor can be achieved with NMs, starting with the basic physical principles. Another important aspect is the possibility to design a system based on multiple sensors for advanced fluid loss detection systems.

6.2.4.1 Nanocomposite Sensor for Liquid Detection Systems Based on a Pillar-Type Layout

In this section, the optical transmittivity response of a millimeter pillar-type sensor made of gold micro-/nanoparticles in PDMS material is experimentally characterized. The choice of the millimeter pillar-type layout is studied in order to collect the light in a multimode receiver optical fiber better. After a discussion of basic principles concerning physical aspects of the proposed sensor, the PDMS/PDMS-Au pillar-type layout is optically characterized, and the variation of the transmittivity response due to the detection of different liquids, such as ethanol and oil, placed on the PDMS-Au sensor is measured. The proposed sensor is apposite for industrial applications such as oil spill and in a robotics finger layout. The pillar-type sensor is illustrated in Figure 6.20a,b and shown in the photograph in Figure 6.21. The device is used specifically for the detection of liquid materials. Two measurement approaches are used. The first one (Figure 6.20a) takes into account the guided modes and the radiation properties of the layout: the light source is guided inside the bulk and is radiated outside by means of pillars. The second one (Figure 6.20b) considers the light directly coupled with the pillar and does not exploit the guided modes. The proposed sensor is mainly based on the detection of the radiated optical field enhanced by gold micro-/nanoparticles in PDMS polymeric film. As the PDMS is able to generate, by chemical reduction, GNs starting from gold precursors, it is easy to fabricate the PDMS-Au sensor. The PDMS elastomer is obtained by mixing a curing agent of 1:10 weight ratio to base monomer. PDMS polymer is known to be a good stabilizing/reducing agent of gold precursor leading to the immediate generation of GNs by the chemical reduction process occurring at room temperature. The effect of the chemical

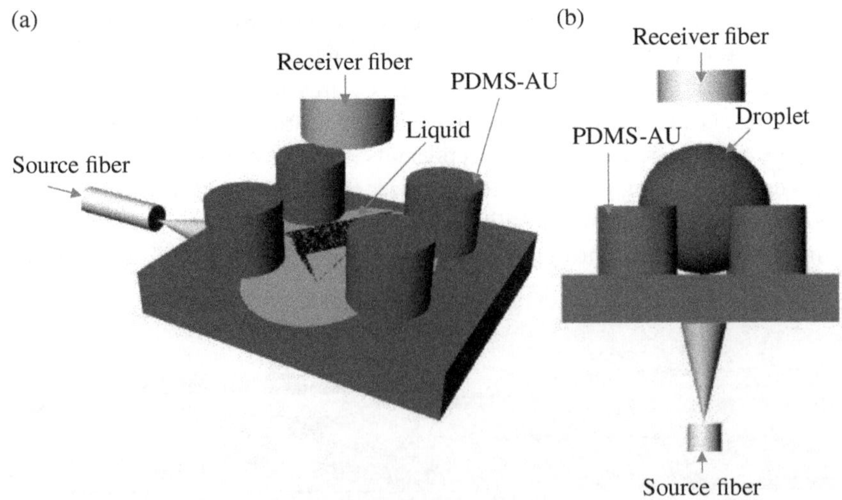

Figure 6.20 PDMS-Au 2×2 matrix pillar-type sensor layout for liquid detection. (a) In-plane scheme exciting in plane the PDMS-Au bulk. (b) Vertical scheme exciting the bulk from the bottom. In both cases the receiver fiber is placed above the pillar structure.

reduction is to generate the violet color of the sample: this color indicates the chemical reduction process which happens by mixing the liquid PDMS for about 10 minutes with the gold precursor prepared in water. The PDMS-Au device is characterized by a good EMF confinement due to the high refractive index because of the micro-/nanometallic inclusions.

For the prototype in Figure 6.21, a very high GN concentration is obtained in order to increase the effective refractive index of the PDMS, and to support the EM coupling. In the specific case, a chloroauric acid salt is used as the gold precursor in water with a concentration of about 10 wt%. A rigid silicon mask with millimeter holes is immersed in PDMS-Au liquid in order to achieve the pillar layout. Two days at room temperature is enough to easily remove the mask and to obtain solid PDMS-Au pillars without the heating process generally used to accelerate the solidification process. The AFM image of the cross section of the gold NM (Figure 6.22a) provides information on the dimensions of

Figure 6.21 PDMS-Au 2×2 matrix pillar-type layout. The height of a single pillar is 1.5 mm, the diameter is 1 mm, and the length of the side of the square layout is 2.5 mm.

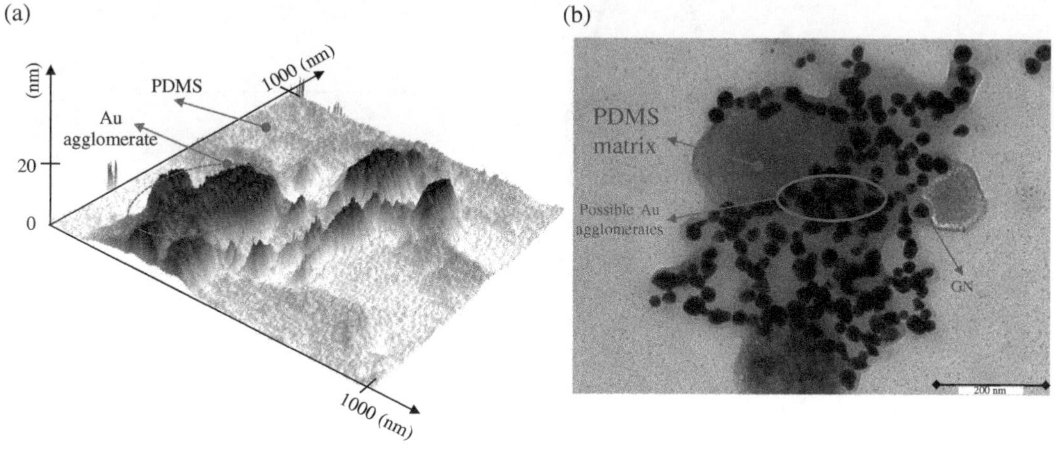

Figure 6.22 (a) AFM image: topography indicating gold fillers in a PDMS matrix. (b) TEM image of GNs embedded in the PDMS polymeric matrix.

the gold fillers inside the PDMS-Au NM used for the sensor. A Park System AFM instrument (XE-100) is used in true non-contact mode to acquire the AFM images. The images are acquired in air using an anti-vibration table (Table Stable TS-150) and an acoustic enclosure. Single-beam silicon cantilevers tips (PPP-NCHR-10) are adopted for data acquisition (of about less than 10 nm nominal radius and with 42 N m^{-1} elastic force constant for high sensitivity). The resonance frequency is around 280 kHz, and the scan rate is between 0.2 Hz and 1.0 Hz. The polydispersive PDMS-Au material contains micro/nano gold particles as confirmed by the TEM image in Figure 6.22b.

The micro/nano fillers are dispersed and agglomerated inside the polymeric matrix: the mono-dispersion of gold fillers is very difficult to control, and for high gold concentration the probability to find agglomerate clusters increases. The TEM image is recorded by a JEOL Jem 1011 microscope operating at an accelerating voltage of 100 kV. TEM samples are prepared by dropping a dilute solution of NPs in distilled water and acetone on carbon-coated copper grids (Formvar/Carbon 300 Mesh Cu). In order to separate the GNs from the polymeric matrix, a droplet of gold precursor is deposited on fresh PDMS and a curing agent sample. After a few minutes more NPs are generated by the chemical reduction and remain suspended until reduction is complete. Afterwards, these GNs are easily removed from the polymer and merged in distiller water to be analyzed by the TEM technique. The presence of gold, deducted by Figure 6.22, is also confirmed by Figure 6.23 which shows the experimental spectrum of the gold nanocomposite sample used for the sensor prototype: pure PDMS does not show any absorbance in the visible region, whereas PDMS-Au has an absorbance centered around $\lambda = 1190$ nm due to the average effect of GNs (see the trend in Figure 6.23). The inset in Figure 6.23 shows the radiated field of a single PDMS/PDMS-Au pillar. Concerning the optical properties of the PDMS-Au material, the effective refractive index of the

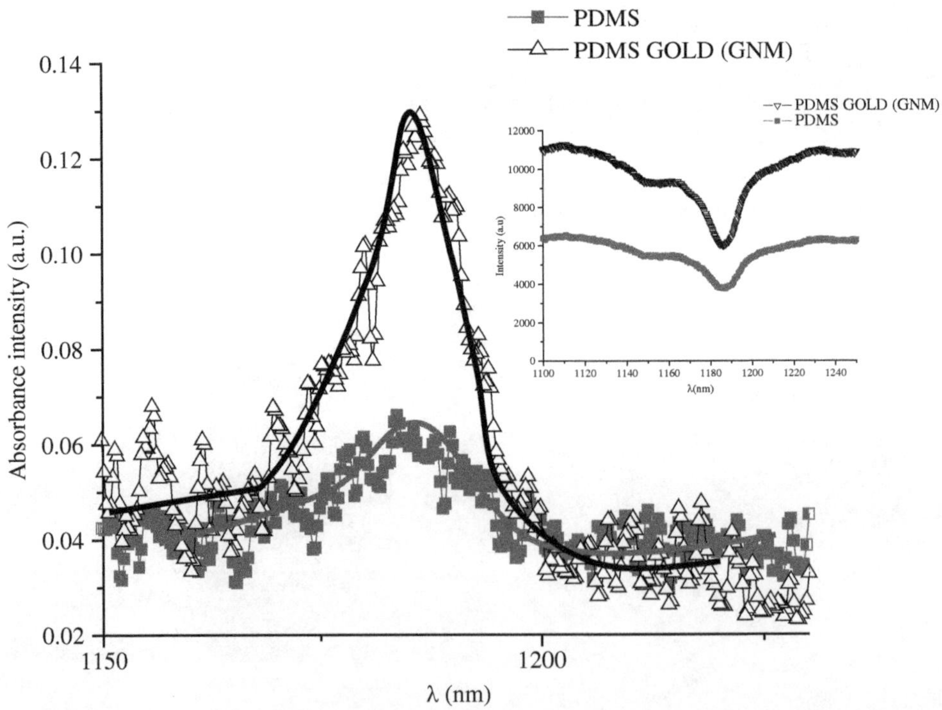

Figure 6.23 Absorbance trend of a PDMS/PDMS-Au bulk type sample. Inset: radiated optical intensity of a single PDMS/PDMS-Au pillar.

PDMS-Au waveguide is approximately estimated assuming spherical GNs in PDMS material. In this case the exact effective dielectric function $\varepsilon_{\mathrm{eff}}$ is difficult to model due to the different sizes and geometries of gold micro-/nanoparticles. In order to analyze the difference between the optical behavior of the PDMS and PDMS-Au structures it is assumed, as an approximation, that spherical gold particles have a dielectric function ε_m which varies with the optical working wavelengths. The spherical gold particles embedded in a medium ε_s (dielectric permittivity of the PDMS) define the effective permittivity shown in Eq. (6.5).

The EMF is radiated in the air space above the pillars (in the region closest to the pillars). The radiated electric field $\mathbf{E}(x,y,z)$ (radiation pattern which defines the optical transmittivity) is defined by:

$$\mathbf{E}(x,y,z)=\int_S\left(\left(j\omega\mu_0\mathbf{H}_{\mathrm{tan}}\right)G+\left(\mathbf{E}_{\mathrm{tan}}\times\nabla G\right)+\left(\mathbf{E}_{\mathrm{normal}}\nabla G\right)\right)\mathrm{d}S \tag{6.26}$$

where S represents the radiation boundary spherical surface, j is the imaginary unit, $\mathbf{H}_{\mathrm{tan}}$ is the component of the magnetic field that is tangential to the surface, $\mathbf{E}_{\mathrm{normal}}$ is the component of the electric field that is orthogonal to the surface, and G is the free space Green's function.

Figure 6.24 shows a simulated EMF confinement for a PDMS-Au 2×2 matrix pillar-type sensor. In the simulation, ethanol liquid behaves as a cladding material able to absorb EMF energy: the electric field radiated by the pillar decreases due to the energy lost in the liquid material. The pillar layout allows the light to be directed as in T optical junctions. This enables light collection in the region above the pillar as radiated EM field. Due to the light scattering of gold micro-/nanoparticles in PDMS, the intensity of the radiated light increases by facilitating the detection. In accordance with the optical collecting properties, and with the proper dimensions of the pillars (the pillar diameter is of the same order of the optical fiber probe receiving the radiated optical signal), it is

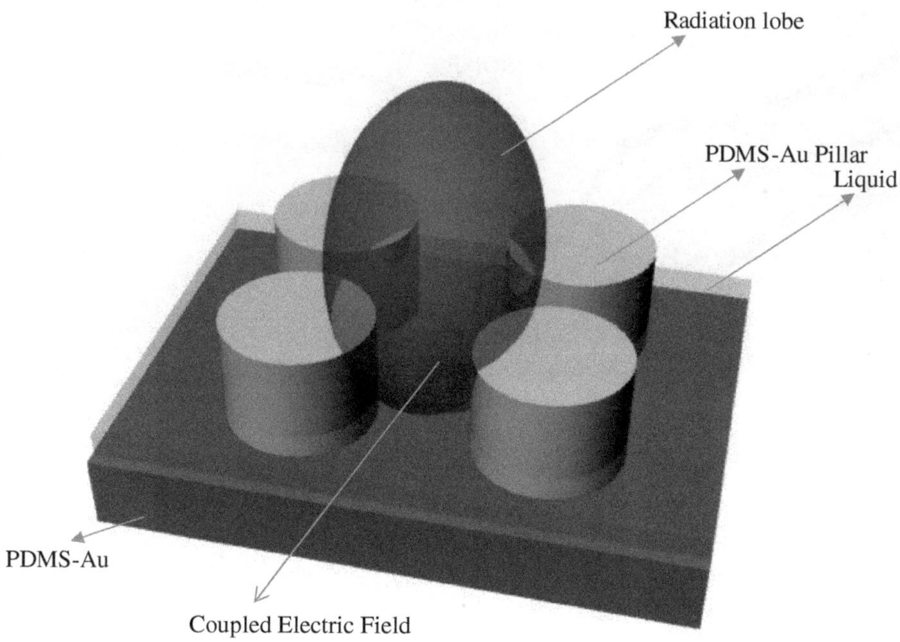

Figure 6.24 Total electric field confinement: sketch of a simulation for a PDMS-Au 2×2 matrix pillar-type sensor with ethanol liquid.

possible to characterize the single pillar layout simply by placing on the pillar an optical fiber behaving as a receiver probe. A layout useful for the detection of liquids is the pillar matrix layout. In particular, the 2×2 square layout configuration is the basic layout used to trap and to detect liquid materials: part of the EMF is coupled with the liquid which behaves as a waveguide cladding by changing the intensity of the radiated field and, consequently, by detecting the variation of the permittivity. The intensity of the optical transmittivity decreases by increasing the permittivity value of the cladding (liquids to detect). Due to the symmetry, the cylindrical geometry facilitates the measurements. For the sensor in Figure 6.24, the variation of the radiated transmittivity intensity is measured by putting on the sensor prototype two liquids, such as ethanol and oil, which have different refractive index values.

The optical intensity is measured by an optical multichannel analyzer (OMA) connected to a multimode fiber; a broad lamp source MIKROPACK halogen light source HL-2000-FHSA having a wide spectrum ranging from 200 to 2000 nm is used as light source. From the measurements performed an optical absorbance around $\lambda = 1190$ nm is observed confirming the spectra in Figure 6.23. The variation of the transmitted intensity a is function of the refractive index permittivity and, as expected, a lower optical intensity is found in Figure 6.25 when the oil is detected: oil has a higher

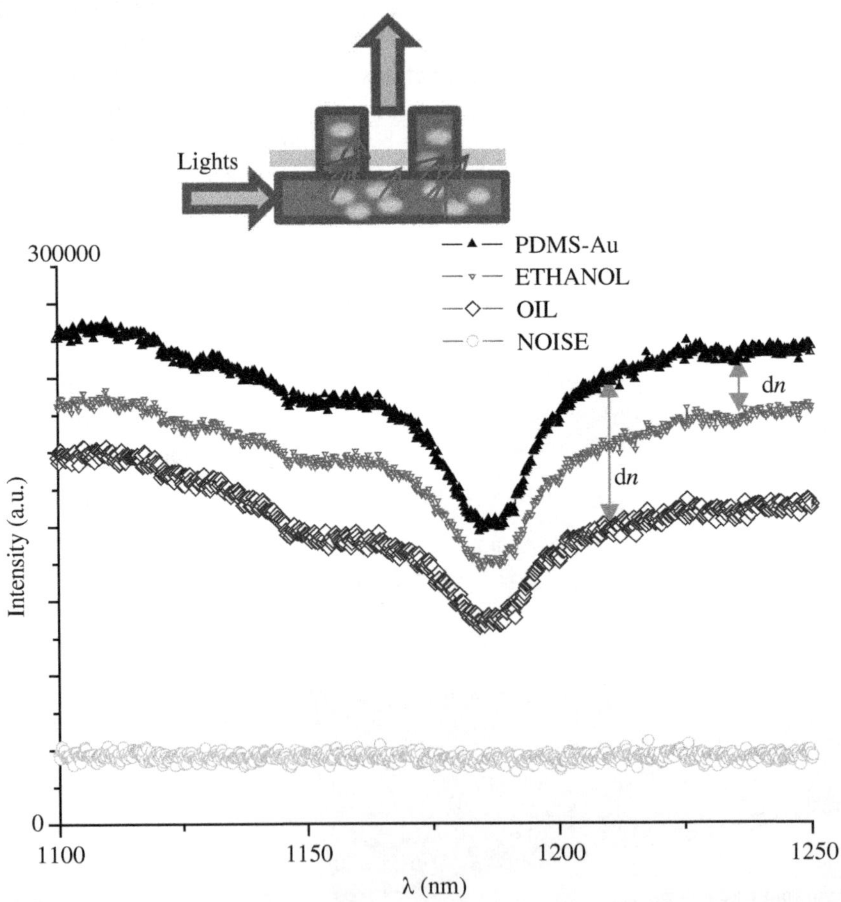

Figure 6.25 2×2 PDMS-Au matrix: variation of the transmitted intensity by considering ethanol and oil as liquids to detect. Inset: schematic diagram of a PDMS pillar and light propagation as in a T optical junction (light propagation and scattering in PDMS-Au pillars).

refractive index compared with ethanol (n_{oil} [$\lambda = 590$ nm] = 1.468 and $n_{ethanol}$ [$\lambda = 590$ nm] = 1.3571), so, more energy is confined inside the oil cladding. The detection principle is in the variation of the transmittivity intensity: this variation is associated with the variation of the dielectric contrast ($dn = n_{liquid}$-n_{air}). An intelligent system can correlate the optical intensity variations to a well-defined liquid material database, where an initial acquisition of the transmittivity intensity (calibration process) is necessary in order to distinguish the relative permittivity. It is noted that oil and ethanol transmittivities are measured in separate time steps (no mixed solutions are considered). Due to the impermeability, the proposed sensor prototype of PDMS can be reused at different times (a useful feature for robotics applications).

The second experimentation scheme, shown in Figure 6.26, considers a small oil droplet placed on the top spherical surface, and the effect of an optical ray coupled with the whole pillar waveguide: part of the light is backscattered depending on the oil concentration, and some of the light passes through the liquid supported by the enhancement property of the PDMS-Au, and is reflected by the upper surface of the liquid. The intensity of the reflected light is a function of the oil concentration. The oil concentrations, expressed in wt%, correspond to the oil droplets used during the experimentation (Table 6.3). Figure 6.27 shows the spectra measured by the OMA: as expected the transmitted optical intensity decreases when the oil concentration increases until saturation (10–15 oil droplets). This saturation is also a function of the geometry of the sensor.

The analyzed sensor can be easily implemented in robotic finger systems. By using small laser diode sources and small photodiode detectors converting the optical signals into electric signals, the robotic sensor requires small spaces assuring extremely fast detection responses. A possible schematic representation is shown in Figure 6.28, where the sensor is coupled by optical fibers and the variation of the signal is processed by an intelligent system providing commands about the

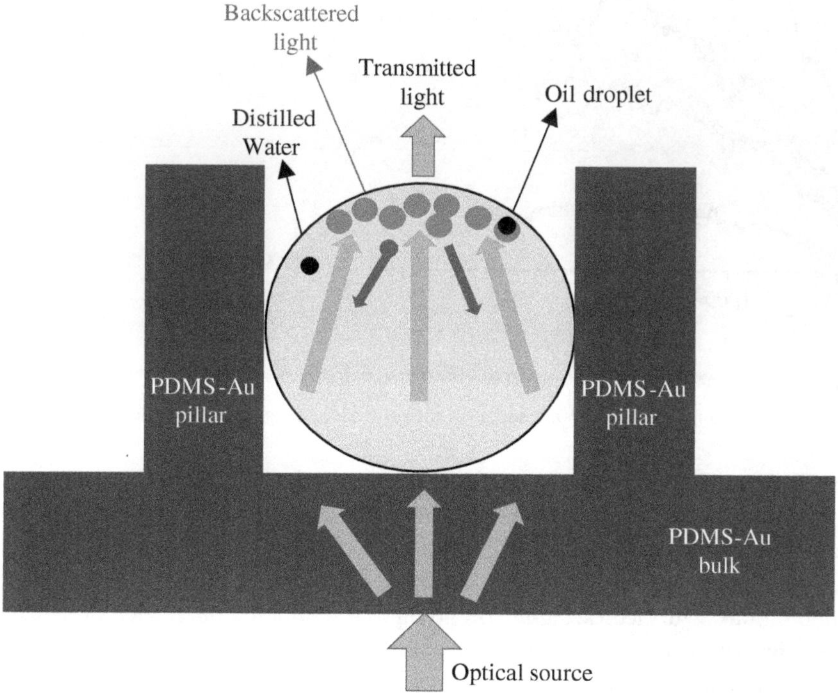

Figure 6.26 Basic principle of light scattering.

Table 6.3 Oil concentrations associated with the oil droplets.

Droplet	Oil concentration (wt%)
0	0 (distilled water)
1	0.45
2	0.9
3	1.35
4	1.8
5	2.25
8	3.6
10	3.5
15	6.75

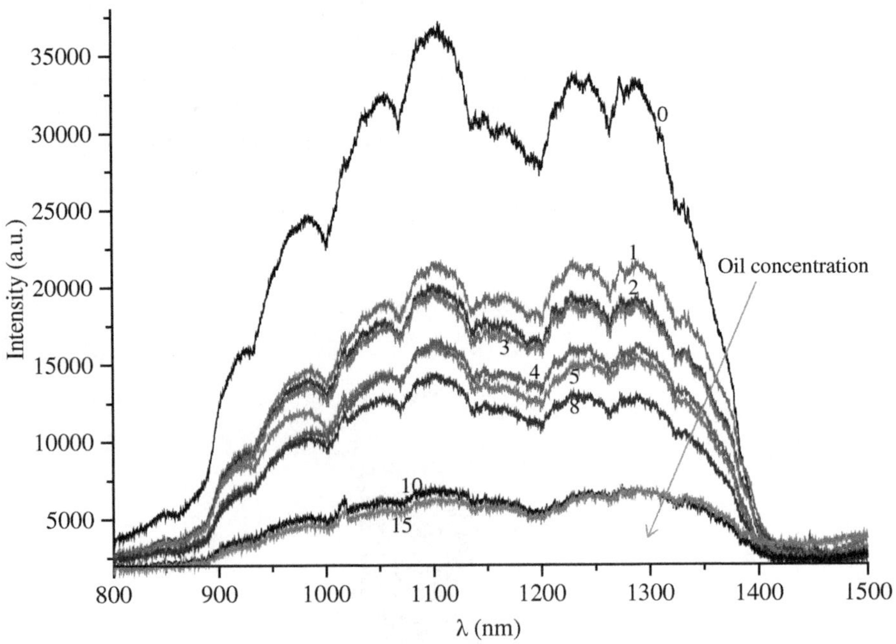

Figure 6.27 Sensor optical spectra: transmitted optical intensities for different oil concentrations. The numbers shown on the spectra indicate the number of oil droplets deposited on the sensor.

movement of the robotic finger by following defined procedures. Analyzing different simulations, it is observed that an optical fiber can also detect optical intensity variations if it is embedded in the PDMS-Au material as for integrated systems. The liquids are collected by means of PDMS borders arranged along the boundaries of the sensor. A fiber bundle is coupled to an optical coupler connected to the small photodiode. The electric signal, outgoing to the photodiode, is then processed by an electronic chip able to control the mechanical movements of the finger, by following the procedure related to a defined application. The detected data are ready to be post-processed by a logic network comparing different values of a database. By means of an accurate database

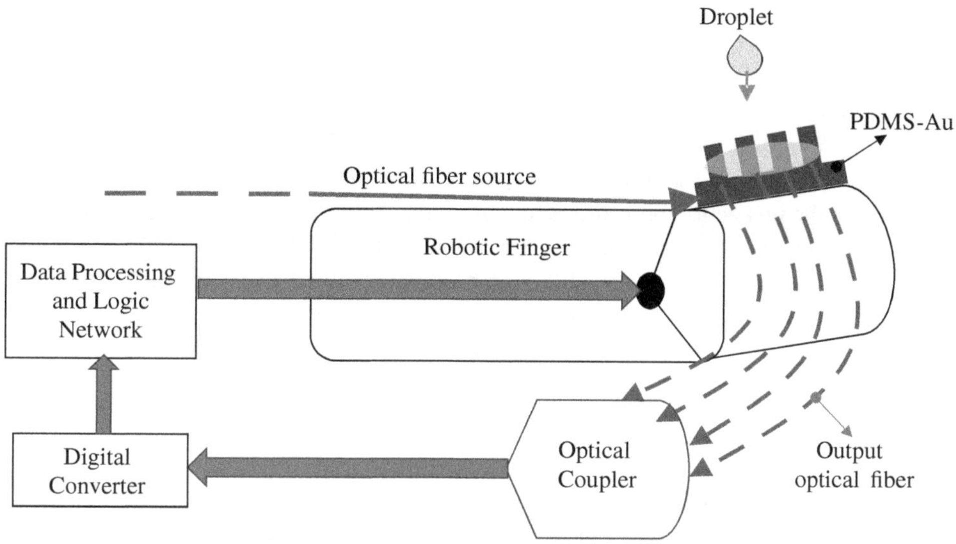

Figure 6.28 Implementation of the sensor in a robotic finger.

implemented in an intelligent system, it is possible to detect many liquid typologies. As the proposed sensor is sensitive to low variations in permittivity, it can also be used to verify the quality of ethanol in automotive fuels.

The proposed technology, also discussed in [1], is suitable for all industry applications involving oil loss leakage detection, and for environment applications of oil spill. Many research projects focus on the development of accurate systems for monitoring oil content in water, particularly in the neighborhood of port authorities and facilities because some commercial vessels clean their machinery rooms by releasing bilge waters containing a mixture of fuel, cleaning solvents, etc.

6.2.4.2 Micro- and Nanosensors in the Monitoring of Production Processes: Leakage Monitoring

Micro- and nanosensors are smart sensors oriented toward the monitoring of production processes detecting oil leaks, by combining electro-optical in "line-type" and "point-type" detection systems. Other detection systems are implemented by transmitting a wireless pressure or acoustic signal to a unit capable of processing anomalous signals and defining the position of the leak. Detection can be carried out at several points in the monitoring system by a single probe which has its movement appropriately controlled: in this way the problem is solved of space for installation, and of the insertion of multiple sensors along the line or in the machinery to be monitored. Other approaches are performed by image vision techniques. Among the algorithms supporting data analysis and oriented toward predictive maintenance of machinery, mechanical systems, and production lines, artificial neural networks (ANNs) play an important role. An ANN is useful in the risk assessment of breakage, by processing data representative of punctual or distributed oil leakage. The neural network algorithms, intended as a DSS, therefore process more data, even from multiple sensors, in order to support the interpretation of the reading of the leak signals: this is an important aspect, especially if multiple sensors must control complex fluidic, mechanical and pneumatic networks whose detection process is designed through a workflow simulation. The design of the control systems is one of the strengths for the evaluation of the different application contexts and industrial areas of leak detection systems. The workflow computing environment offers in this case the possibility of designing dynamic multi-user detection systems helping to plan the sensor layout to implement, thus allowing the control of leakage monitoring. The monitoring system is integrable with other sensors controlling temperatures, pressures, volumes, flow rates, etc. By means of a multisensor system based on smart devices, it is possible to design, simulate, implement, and test various optimized loss control systems through the development of different supervision logics, automatically controlling the production lines. A dynamic simulation of the circuits provides useful information about the inspection procedures that are carried out in static conditions. Moreover, numerical tools simulate the effects of fluid dynamic losses, by formulating models able to parameterize the different leakage conditions in different production scenarios. The leak simulation serves to optimize the displacement of the inspection check points, by predicting the load loss, the distributed evaluation of the lost volumes in different points of the circuit, and the variation of other parameters such as temperature, and to model other forms of leaks, and other anomalies. The models also help to configure the PLC systems managing sensors.

6.2.5 Examples of Digital MEMS/NEMS Sensors: Technological Aspects and Applications

MEMS and NEMS technologies are advanced topics for industry research. In this section, some examples of thin film sensors and their uses in biomedical imaging are given. The approaches and the methodologies are fully analyzed in the following chapters about rapid prototyping and RE applied to micro- and nanostructures.

6.2.5.1 Thin Film MEMS

MEMSs are a particular category of smart sensors made by thin film layers. Microwave MEMS filters and pressure sensors are an interesting group in this category. Consider a layout example consisting of seven, six and four circular air holes arranged in a transmission line as illustrated in Figure 6.29a. Figure 6.29b shows a six-hole prototype. The holes are etched in the upper molybdenum (Mo) layer (Mo layer 1) and in the aluminum nitride (AlN) layer, and are characterized by a diameter of 2 mm and a period of 3 mm [29]. The width of the periodic structure is 4 mm. The thickness of Mo layer 1, the AlN layer and Mo layer 2 are 70, 500, and 100 nm, respectively. The cross section of the MEMS sensor is illustrated in Figure 6.29b, where it is possible to observe the etched parts and the basic principle of the pressure sensing. The transmittivity responses are characterized in a frequency range between 1 GHz and 36 GHz by means of a network analyzer measuring the S_{12} scattering parameter of the chip, by finding possible depths of the frequency response due to device resonance following the application of a pressure force: the effect of pressure on the chip is shown in the frequency response (Figure 6.29d indicating clearer depths, by increasing applied pressure forces). The first structure in Figure 6.29a presents an evident S_{12} depth around 11.19 GHz for pressure forces ranging from 2 to 20 g, and a bandgap around 40 GHz due to the central cavity [29]. The presence of the bandgap allows the devices to be used as "microwave photonic band gap" waveguides at high frequencies. The pressure sensing behavior is shown in Figure 6.29d.

The fabrication process of the Mo/AlnN/Mo devices is similar to the ring MEMS (RMEMS) process [30]; it is developed by means of bromograph photolithography. The MEMS is connected to the network analyzer at input and to the output port by a properly dimensioned tapered waveguide able to transfer the maximum power between the two ports.

Another MEMS device is the RMEMS antenna shown in Figure 6.30a,b; the Mo/AlN/Mo freestanding structure is discussed in [30–32]. The RMEMS behaves as a piezoelectric antenna: the piezoelectric frequencies resonance may change the ring diameter and in general the geometrical

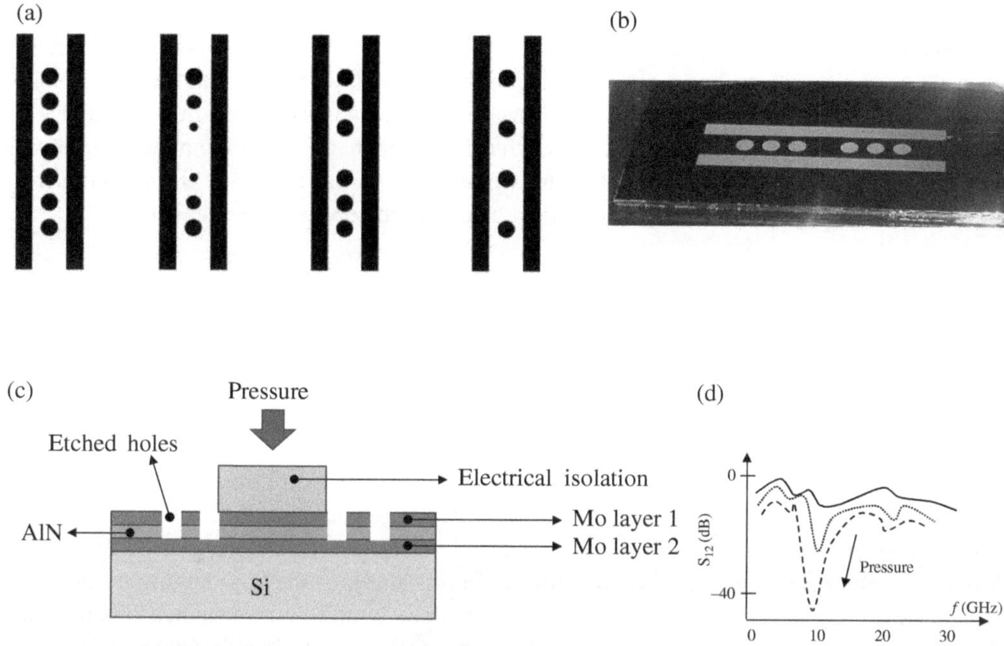

Figure 6.29 Microwave MEMS pressure sensor: (a) masks used for photolithography; (b) MEMS prototype on silicon substrate; (c) cross section sketch of the pressure sensor device; and (d) theoretical trend of the S_{12} frequency response versus the frequency when pressure forces are applied on the central region of the device.

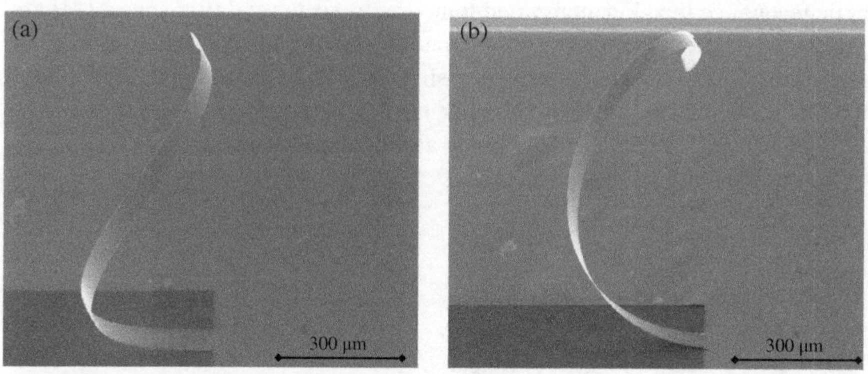

Figure 6.30 (a) Ring MEMS and (b) zooming of the base of the antenna.

configuration as ring deformations. The geometrical deformation of a transmitter ring modulates the phase or the amplitude of a transmitted GHz/THz signal: the modulation is due to the piezoelectric effect of the AlN piezoelectric layer and is performed by applying a resonance control signal found at lower frequencies. Concerning radiation aspects, according to theory, a wire antenna where length l is usually $\lambda/50 < l \leq \lambda/10$ behaves as an infinitesimal dipole (small dipole). For this purpose, by considering diameters between 100 μm and 800 μm and lengths between 250 μm and 3 mm, the RMEMS structure behaves as a small dipole or a dipole ranging from GHz to THz frequency range. In order to design the RMEMS it is important to estimate the scattering parameter at the input port of the ring structure (S_{11} parameter) which defines the reflectivity at the input port: a low reflectivity indicates a good match with a source and a high transferred power. Focusing on ring diameters of 100 and 140 μm, a good performance of the antenna at working frequencies between 2.6 THz and 3.6 THz is found. Concerning measurement approaches, the antenna must be placed into an anechoic room where a horn receiver antenna measures the transmitted signal (S_{12} parameter). The anechoic chamber is composed of the electromagnetic absorbing materials shown in Figure 6.31, where the pyramidal objects have dimensions as a function of the working frequency (by increasing the working frequencies the object dimensions must decrease).

6.2.5.2 Nanoprobes for Medical Imaging

Nanoprobes are used in different application fields including biomedical ones. NPs are potentially adopted as nanoprobes in bio-imaging. In order to design a biomedical imaging system with NPs able to enhance imaging of lymph nodes located in the mediastinum (a division of the thoracic cavity), it is necessary to approach the study as an interdisciplinary analysis, merging knowledge of numerical engineering modeling with knowledge of ultrasound facilities and clinical aspects. The medium- and long-term follow-up of aggressive non-Hodgkin's lymphomas with mediastinal involvement at diagnosis is crucial in lymphoma patient management and the evaluation of nodal status can be decisive in early identification of mediastinal relapse, considering the difficulty to differentiate between fibrotic tissue and active tumor. In this analysis, the bio-imaging study must be oriented toward the detection of abnormal and anomalous lymph nodes by silica NPs used as ultrasound imaging contrast agent. The study of lymph node activity is an important issue in oncology. Different techniques observing size, shape, nodal border, echogenic hilus, echogenicity, calcification, intranodal necrosis, ancillary features, vascular pattern, and vascular resistance are used in order to detect abnormal lymph nodes [33]. Different imaging modalities such as positron emission tomography, single-photon emission computed tomography, bioluminescence, optical fluorescence, magnetic resonance imaging, computed tomography scans, and ultrasound [34] are used for this kind of detection. Nanotechnology, implementing proper functionalized NPs [35], potentially helps to interpret clinical images better, decreasing diagnostic errors. In particular, after intravenous administration, NPs of ferumoxtran-10 can be used as contrast agent for the detection of lymph flow disturbances or disturbances of nodal architecture by metastases leading to

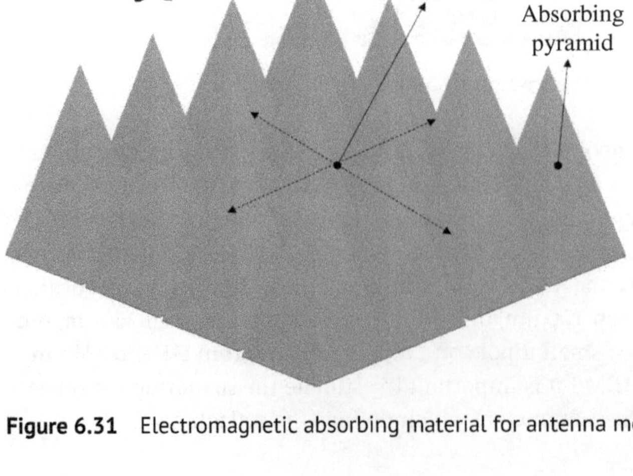

Figure 6.31 Electromagnetic absorbing material for antenna measurements.

abnormal accumulation patterns [36]. Due to magnetic properties, properly encapsulated or functionalized silica NPs [37, 38] substitute the use of iron oxide NPs as contrast agent [39]. As for many NPs, production and exposure of NPs less than 100 nm in diameter may pose unknown risks [40]. To consider this toxicity risk, silica NPs with a diameter up to 100 nm may be a possible topic to study. The goal of the use of NPs for bioimaging is to apply engineered numerical modeling used for minimally/non-invasive melanoma wireless detection systems based on micro antenna probe implementation [41], [42] to ultrasound techniques applied with silica NPs as contrast agents [43]. The numerical model provides 2D maps indicating risk levels and working conditions of the electromagnetic waves and probes used. In addition, ultrasound facilities can localize the possible NP accumulation into an anomalous lymph node. The image enhancement is provided by the following considerations: silica NPs go from the lymph vessel into the anomalous lymph node (see the model in Figure 6.32). Assuming proper functionalization, the silica NPs accumulate inside the mediastinum lymph node. An ultrasound wave excited by an external probe, before arriving in the mediastinum cavity, travels inside the human tissues as indicated in the multilayered model in Figure 6.33a. The ultrasound wave propagation is modeled by the circuit in Figure 6.33b representing all the electrical impedances modeling the different human tissues. The detection can be performed by an external probe able to detect the acoustic impedance variations due to the contrast of silica NPs: the ultrasound signal is backscattered and enhanced by the silica NPs providing the imaging of the mediastinum lymph node. In the model an ultrasound plane wave is considered to excite the unit cell model in Figure 6.34 made by a polydispersion of silica NPs embedded in an agarose gel (the use of agarose gel is useful for a preliminary "in vitro" experimentation stage oriented toward the comparison with numerical results). Concerning the micro-/

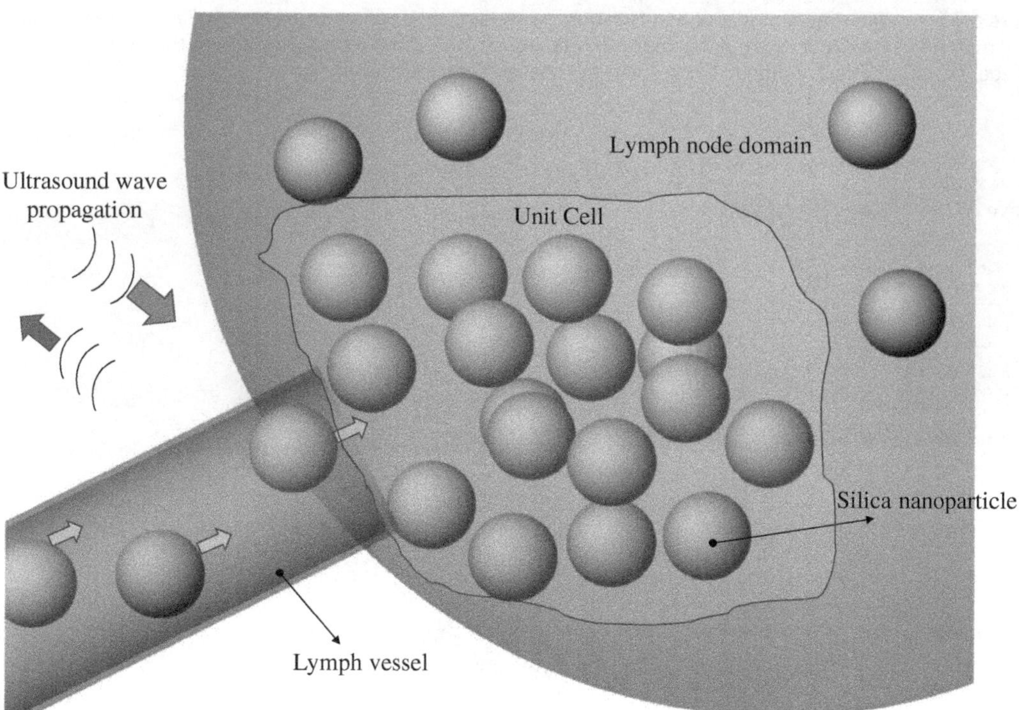

Figure 6.32 Three-dimensional model of silica NPs embedded in a lymph node.

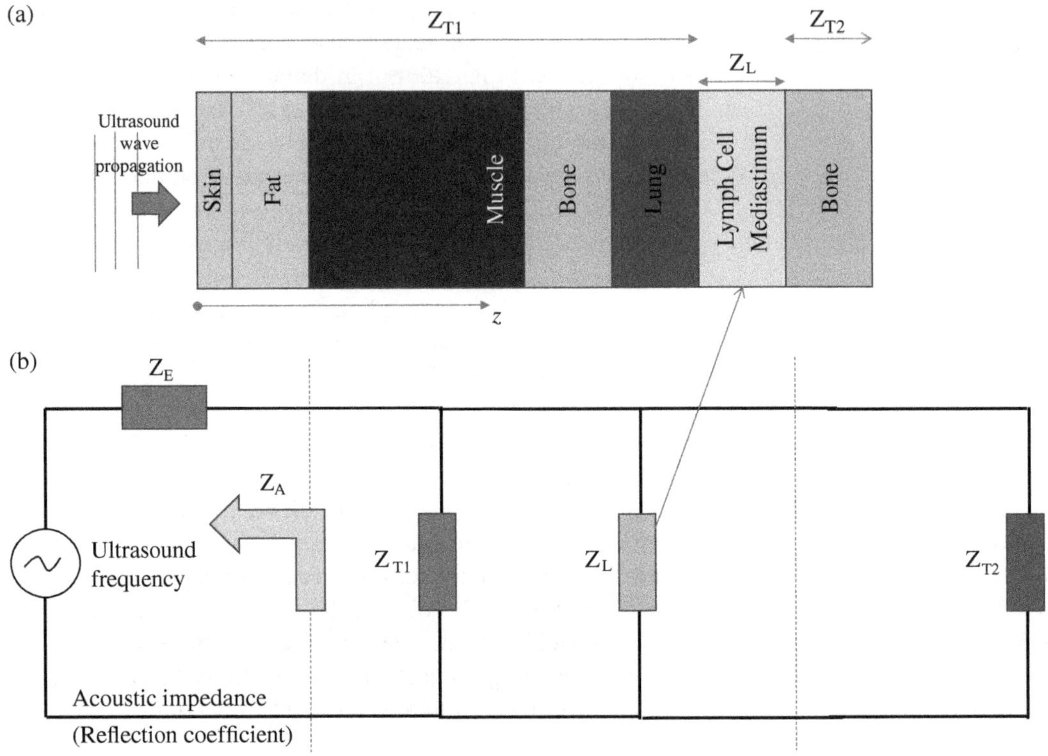

Figure 6.33 (a) Multilayer model of an ultrasound wave propagated in human tissues. The final target is the lymph node located in the mediastinum. (b) Equivalent circuit modeling considering the effective impedances related to the human tissues and the acoustic impedance.

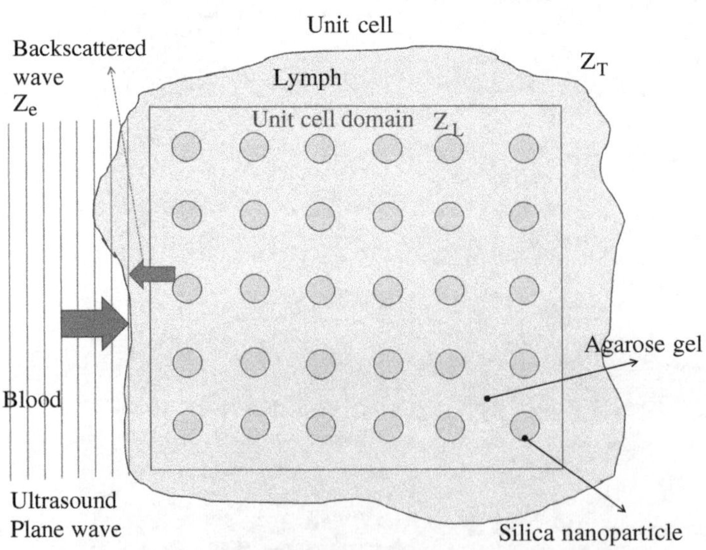

Figure 6.34 Unit cell model of in the mediastinum lymph node.

nanoscale, the unit cell model is useful in order to characterize the electromagnetic behavior (frequency resonance and reflectivity) of a high-density region of Si NPs which is assumed located in the mediastinum layer indicated in Figure 6.33a. This unit cell is repeated many times in order to obtain a macroscopic effect for a larger scale. The results are useful for testing on real human/animal tissues. In fact, by using numerical results it is possible to tune optimally the input power and the frequency of the ultrasound probe to obtain high resolution ultrasound images. The results are also useful to design innovative antennas for mediastinum lymph nodes, and to simultaneously compare other images obtained by other techniques, finding a possible correlation and thus drastically reducing the diagnostic error. The next step is to apply *in vitro* the proposed research, by observing the silica NPs electromagnetic behavior in liquids simulating human tissues. Other steps before application in the human body include the application of silica NPs in rats and observing the toxicity, and thus having important information about silica NP mobility in the human body.

The NPs interact with the EMF/ultrasound wave coming from the external ultrasonic source. The interaction is defined by means of the wave reflectivity (see backscattered wave in Figure 6.34) due to the presence of NPs, which behave as the contrast agent for imaging. The reflected EMF/ultrasound wave can be detected by means of an ultrasonic probe able to map the NP dislocation. The EMF/ultrasound characterization is performed by means of time or frequency domain numerical tools, able to provide a mapping of NP concentration and, consequently, to indicate a strong signal reflectivity due to anomalous conditions.

The electromagnetic and ultrasound characterization of silica NPs is performed in the analyzed case by means of a dedicated locally conformal finite-difference time-domain (FDTD) scheme [44]. Such a procedure allows accurate modeling of local electrical and geometrical properties of the complex shaped NPs, so providing a clear advantage over the use of conventional FDTD algorithms based on staircasing approaches or unstructured and stretched space lattices and, therefore, potentially suffering from significant numerical dispersion and/or instability. By using the developed procedure, the preoperative discrimination between benign and cancerous tissues and, in particular, the early prediction of potential malignancy can be carried out conveniently by evaluating the scattering response of NPs as a function of the relevant density, size, as well as of the dielectric properties featured, at ultrasound or high frequencies, by the biological tissue being analyzed. Figure 6.35a,b shows the theoretical normalized power spectrum of the wave signal reflected/

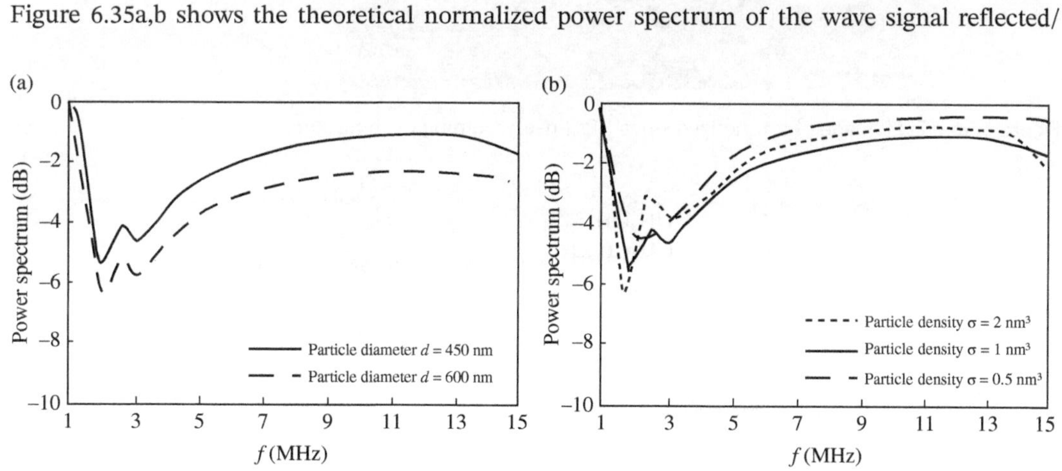

Figure 6.35 (a) Theoretical normalized power spectrum of the diffracted wave signal by silica NPs as a function of the particle diameter *d* when assuming a volumetric density $\sigma = 1.0\,\text{mm}^{-3}$. (b) Theoretical normalized power spectrum of the diffracted wave signal by silica NPs as a function of the particle density δ when assuming a diameter $d = 600\,\text{nm}$.

diffracted by a distribution of NPs as a function of the relevant diameter and density. As observed, the spectral response of the NPs strongly depends on their average interspacing, and low-frequency resonances are triggered as the particle density increases. On the other hand, the diffracted power level tends to become smaller as the particle diameter decreases. By using the proposed modeling methodology, it is possible to retrieve, in a fully non-invasive way, useful data regarding the cancer-related risk level in different biological tissues. The same approach can be adopted for NPs with smaller dimensions up to 100 nm according to the toxicity risk.

NPs have emerged as promising cellular probes for diagnostic imaging, since their surface is easy to conjugate with specific antibodies or ligands for selective recognition and binding to the target cells. Moreover, they are a suitable candidate because of their biocompatibility, ease of bioactive molecule encapsulation, and simplicity of formation. The silica NPs in Figure 6.36 can be synthetized by different approaches. Figure 6.37 shows a possible scheme for functionalized silica NPs using mercaptopropyltrimethoxysilane (MPTS) as the precursor. Another possible scheme for NP

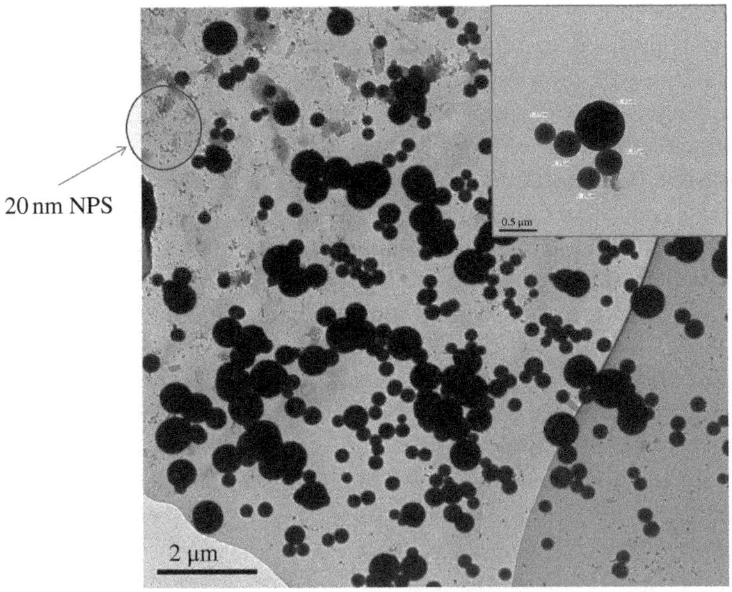

Figure 6.36 TEM image of synthetized silica NPs (inset: zooming of some silica NPs).

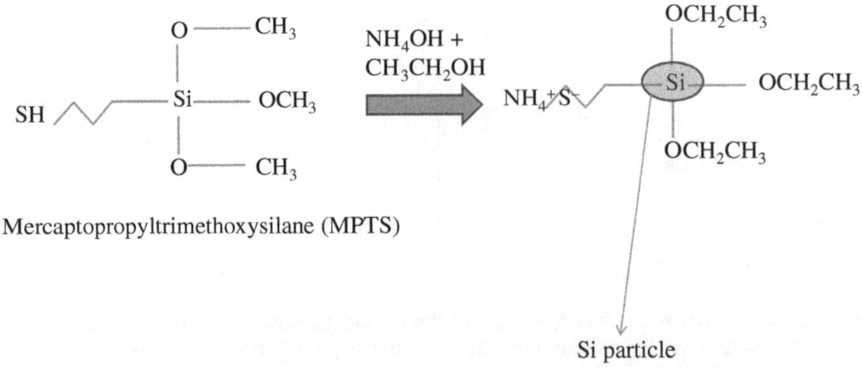

Figure 6.37 Example of functionalized silica NP synthesis by MPTS.

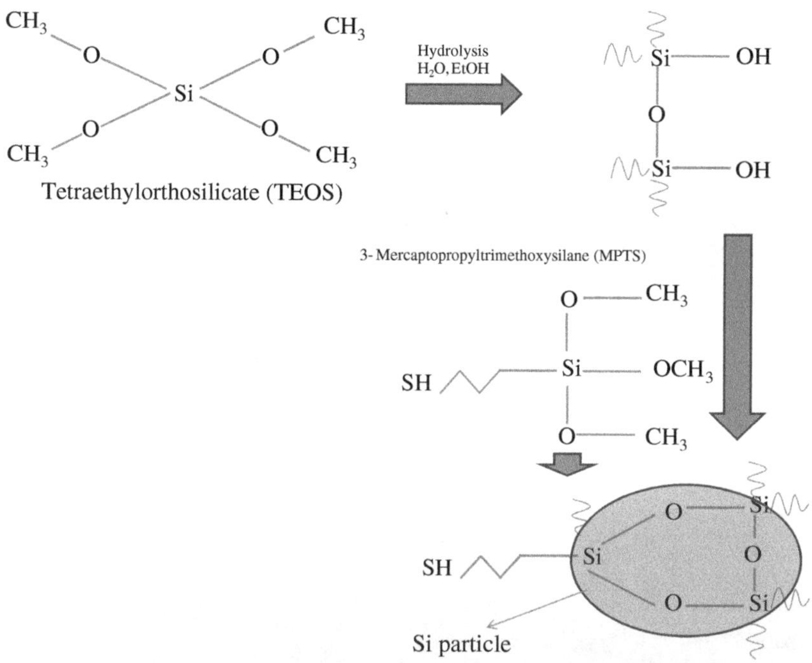

Figure 6.38 Example of functionalized silica NP synthesis by TEOS.

synthesis is shown in Figure 6.38, where tetraethylorthosilicate (TEOS) is used as the precursor. Another alternative approach for the synthesis of silica NPs is discussed in [45].

6.2.5.3 Diamond Thin Film Devices: Sensing Improvements

Diamond material is suitable for many applications involving sensing and mechanical properties. Diamond exhibits exceptionally good mechanical and heat resistance. Concerning electrical conductivity applications, diamond can be implemented as an electrical insulation material, or as a waveguide for antennas. A first step is to characterize the diamond material into integrated devices. By growing diamond film or diamond assembled grain on *p*-doped silicon (*p*-Si) substrate, it is possible to characterize the electrical insulation properties. As shown in Figure 6.39, the surface density of current **J** calculated on the *p*-Si interface is characterized by low current values corresponding to the diamond grain region, thus proving a decrease in current on diamond of about one order compared with the **J** value of the *p*-Si material.

In order to improve electrical current flowing also on the diamond material, it is possible to dope the diamond with N_2. As illustrated in Figure 6.40, in order to improve electrical conductivity a high N_2 concentration is required (for low concentration there are no significance improvements). The diamond film layout on substrates changes the radiation behavior of the whole device behaving as a patch antenna, thus providing the desired radiation pattern. An alternative way to improve electrical current on the diamond is to implement the graphite component. The electrical current is changed by depositing a layer of NDs on the substrate by means of the pulsed spray technique. Figure 6.41a,b shows the images of a standard substrate and of the ND layer sprayed on the same substrate, respectively. Figure 6.42 shows the measured electrical current–voltage (*I*–*V*) characteristic of a sample with a deposition of NDs on the surface. The ND deposition significantly enhances the surface electrical conductivity due to the rectifying action [46]. An assembly of NDs in a

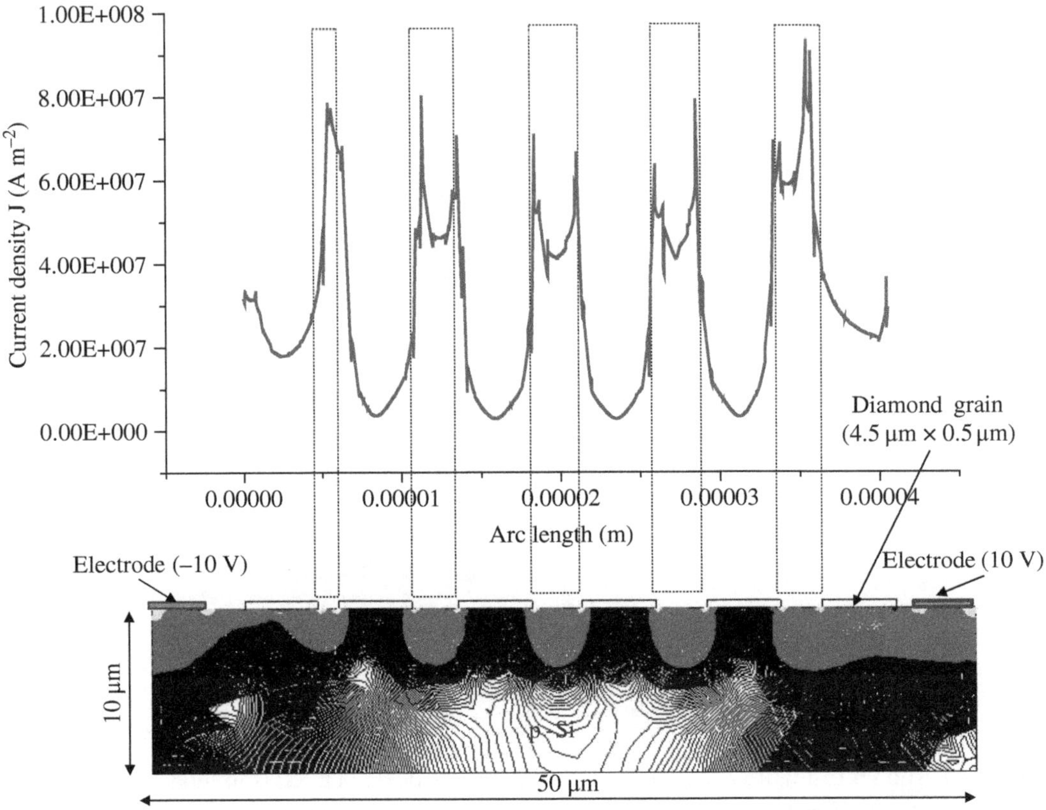

Figure 6.39 Current density **J** calculation characterizing insulation behavior of diamond deposited or growth on a *p*-Si substrate. The model takes into account the voltage excitation by two electrodes.

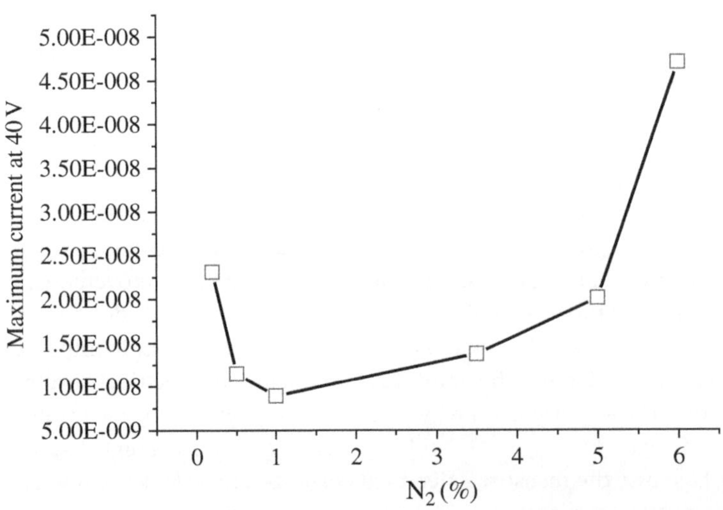

Figure 6.40 Measured current of diamond film growth on a silicon substrate.

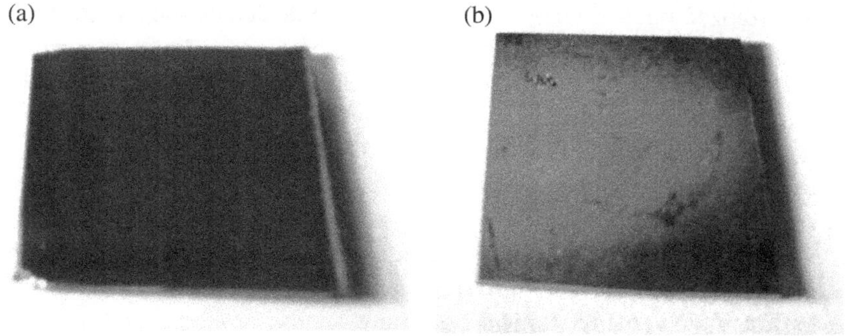

Figure 6.41 (a) A piece of commercial substrate. (b) Example of a ND layer deposited on the same substrate.

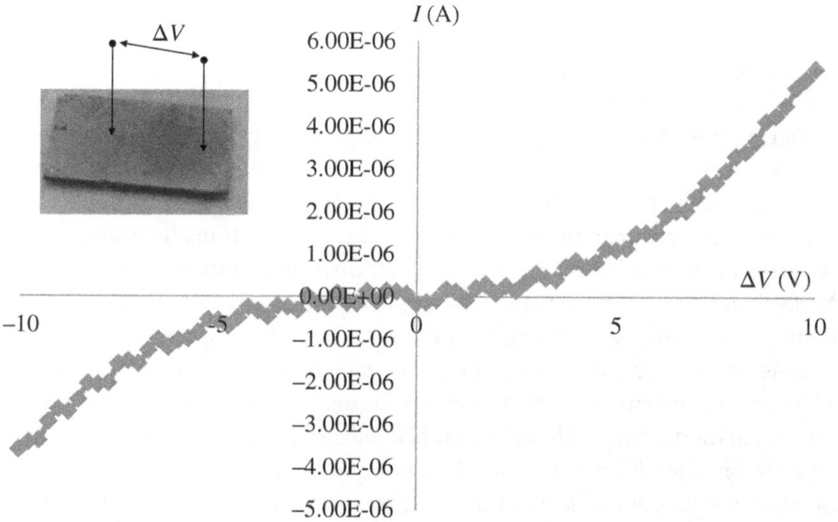

Figure 6.42 *I*–*V* characteristic of a ND sprayed layer deposited on a silicon substrate. Inset: example of a ND sprayed sample.

pillar-type structure can locally increase the voltage but decreases the electrical current flowing along the pillar structure [47]. The proposed examples highlight the importance of characterizing the device for all physical aspects, including materials used, layouts, and technology used. Device characterization is important to improve sensing properties.

6.3 Multisensor Systems and Big Data Synchronization of Micro-/Nanoprobes

Micro- and nano sensors based on innovative technologies can be implemented in matrices containing multiple sensing areas or on the same substrates as for micro antenna arrays. Otherwise, small sensors can be allocated in different parts of a production system. The integration in the same chip of multiple sensors, or the layout design concerning the allocation of sensors in strategic production check points, represents the engineering of the monitoring system. Possible advances are in the ideation of synchronized micro-/nanoprobes detecting small signals such as EMFs to control electromagnetic compatibility analysis, or small vibrations by means of highly sensitive accelerometers, or highly sensitive pressure sensors. Another application of small sensors is the array configuration as for multiple antenna patches realized on the same substrate and optimizing the radiation transmittivity. In innovative gas sensors, the realization of micro or nano assembled particles is appropriate to trap and detect small quantities of gas particles, also reading small gas losses which can degenerate in big gas leakages: the continuous monitoring of small gas losses is useful for the prediction of pipeline damage or machine breakdowns. Electrospun polymeric fibers are good candidates for gas detection in industrial and environmental applications [48]. Concerning oil leakage detection, NMs are implemented in optical detection systems in robotic hands [1]. The main research issue is to design and fabricate innovative sensors enhancing sensing properties for predictive maintenance of machines by means of extracting very useful signals in a noisy environment. In order to correctly process data sensors, big data systems are designed collecting all historical signals of the sensors allocated in the industrial indoor or outdoor environment thus powering AI training datasets and, consequently, the learning approach. The synchronization of the data transmission is the primary specification for the information system design, avoiding signal overlapping. The possibility to define crons or to consider electronic devices allowing time delays or temporaneous data buffering are solutions for data collecting, especially when a lot of sensors transmit simultaneously to a receiver system. The DSSs implementing AI algorithms are important to predict the function of the production machines continuously providing small signals containing important information. Multisensor network systems are also able to transduce signals into electrical ones. Piezoelectric and thermoelectric materials can transform vibrations or temperature gradients into a low voltage signal. By considering for example a production line characterized by a lot of machines vibrating continuously, and parts of them with high and low temperatures, it is possible to install piezoelectric and thermoelectric materials generating energy for sensor consumption or for local lighting. Furthermore, in multisensor systems, AI algorithms are executed in a DSS to extract useful information from small signals embedded and hidden in noises. The cross-correlation analysis of different signals coming from the same production line environment, and the possibility to analyze historical data collected into a big data system, facilitate the prediction and the evaluation of the machine working conditions. The availability of distributed small multisensor systems also allows to correlate information, thus facilitating the interpretation of results. The malfunction of a precise part of a machine is predicted by extracting features from the small signals and by correlating the detected data. By considering for example a network of small accelerometers distributed in different parts of the same production line, it is potentially possible to predict failures by correlating all the detected signals and the predicted ones in the time domain and in the frequency domain by applying the Fourier transform. The extraction of vibration frequencies from a noisy signal represents a challenge for AI application: the big data system in this case permits a learning model to be constructed that is able to classify and identify possible features of normal or anomalous machine processing conditions.

References

1 Massaro, A., Spano, F., Cingolani, R., and Athanassiou, A. (2011). Experimental optical characterization and polymeric layouts of gold PDMS nanocomposite sensor for liquid detection. *IEEE Journal of Sensors* 11 (9): 1780–1786.

2 Massaro, A., Spano, F., Ekuakille, A.L. et al. (2011). Design and characterization of nanocomposite pressure sensor implemented in tactile robotic system. *IEEE Transactions on Instrumentation and Measurement* 60 (8): 2967–2975.

3 Massaro, A., Spano, F., Cingolani, R., and Athanassiou, A. (2011). Innovative optical nanocomposite sensor for security systems. *Journal of Engineering Technology* 1 (2): 30–33.

4 Zhang, Q., Xu, J.J., Liu, Y., and Cuen, H.Y. (2008). In-situ synthesis of poly(dimethylsiloxane)-gold nanoparticles composite films and its application in microfluidic system. *Lab on a Chip* 8 (2): 352–357.

5 Hoppe, C.E., Ridriguez-Abreu, C., Lazzari, M. et al. (2008). One-pot preparation of gold-elastomer nanocomposites using PDMS- graft-PEO copolymer micelles as nanoreactors. *Physics Status Solidi (a)* 205 (6): 1455–1459.

6 Goyal, A., Kumar, A., Patra, P.K. et al. (2009). In situ synthesis of metal nanoparticle embedded free standing multifunctional PDMS films. *Macromolecular Rapid Communications* 30 (13): 1116–1122.

7 Devi, P., Mahmoud, A.Y., Badilescu, S. et al. (2010). Synthesis and surface modification of poly(dimethylsiloxane)-gold nanocomposite films for biosensing applications. In: IEEE Proceedings of the International Conference on Biosciences, Cancun, Mexico (7–13 March 2010). Piscataway, NJ: IEEE.

8 Jain, P.K., Lee, K.S., El-Sayed, I.H., and El-Sayed, M.A. (2006). Calculated absorption and scattering properties of gold nanoparticles of different size, shape, and composition: applications in biological imaging and biomedicine. *The Journal of Physical Chemistry B* 110 (14): 7238–7248.

9 Rakic, A., Djurisic, A.B., Elazar, J.M., and Majewski, M.L. (1998). Optical properties of metallic films for vertical-cavity optoelectronic devices. *Applied Optics* 37 (22): 5271–5283.

10 Chen, M. and Horn, R.G. (2007). Refractive index of sparse layers of adsorbed gold nanoparticles. *Journal of Colloid and Interface Science* 315 (2): 814–817.

11 Massaro, A., Errico, V., Stomeo, T. et al. (2008). 3D FEM modeling and fabrication of circular photonic crystal microcavity. *IEEE Journal of Lightwave Technology* 26 (16): 2960–2968.

12 Massaro, A., Giannoccaro, N.I., Spedicato, L. et al. (2014). Characterization of an innovative like-eye sensor for feature detection and robot sensing. *Measurement* 55 (1): 153–167.

13 Hoshino, K., Gopal, A., and Zhang, X. (2009). Near-field scanning nanophotonic microscopy –breaking the diffraction limit using integrated nano light-emitting probe tip. *IEEE Journal of Selected Topics of Quantum Electronics* 5 (5): 1393–1399.

14 Wenger, J. and Rigneault, H. (2010). Photonic methods to enhance fluorescence correlation spectroscopy and single molecule fluorescence detection. *International Journal of Molecular Sciences* 11 (1): 206–222.

15 Issa, N.A. and Guckenberger, R. (2007). Optical nanofocusing on tapered metallic waveguides. *Plasmonics* 2 (1): 31–37.

16 Shaw, G.A., Trethewey, J.S., Johnson, A.D. et al. (2005). Thermomechanical high-density data storage in a metallic material via the shape-memory effect. *Advanced Materials* 17 (9): 1123–1127.

17 Ichimura, T., Hayazawa, N., Hashimoto, M. et al. (2004). Application of tip-enhanced microscopy for nonlinear Raman spectroscopy. *Applied Physics Letters* 84 (10): 1768.

18 Huh, Y.S., Chung, A.J., and Erickson, D. (2009). Surface enhanced Raman spectroscopy and its application to molecular and cellular analysis. *Microfluidics and Nanofluidics* 6 (285): 285–297.

19 Simons, R.N., Miranda, F.A., and Wilson, J.D. (2006). Wearable wireless telemetry system for implantable Bio-MEMS sensor. In: Proceedings of the 28th IEEE EMBS Conference, New York, USA (30 August –3 September 2006). Piscataway, NJ: IEEE.

20 Ropers, C., Neacsu, C.C., Elsaesser, T. et al. (2007). Grating-coupling of surface plasmons onto metallic tips: a nanoconfined light source. *Nano* 7 (9): 2784–2788.

21 Massaro, A., Pierantoni, L., Cingolani, R., and Rozzi, T. (2009). A new analytical model of diffraction by 3D-dielectric corners. *IEEE Transactions on Antennas and Propagation* 57 (8): 2323–2330.

22 Massaro, A., Caratelli, D., Yarovoy, A., and Cingolani, R. (2011). Analytical approaches for nano-plasmonic and micro-millimetric antennas. *IET Microwaves, Antennas and Propagation* 5 (3): 349–356.

23 Massaro, A. and Cingolani, R. (2010). STRD and near field modeling of metallic wedges for optical detection systems. *IEEE Journal of Lightwave Technology* 28 (24): 3562–3568.

24 Massaro, A. and Cingolani, R. (2011). MEG modeling and measurements of optical antenna sensors as micro rectangular apertures. *IEEE Lightwave of Technology* 29 (12): 1835–1841.

25 Massaro, A., Spano, F., and Athanassiou, A. (2012). Modeling and innovative technology of optical 3D antenna sensors as micro-rectangular apertures. *Optical and Quantum Electronics* 44 (4/6): 213–218.

26 Massaro, A., Spano, F., Cazzato, P. et al. (2013). Robot tactile sensing: gold nanocomposites as highly sensitive real-time optical pressure sensors. *IEEE -RAS- Robotics and Automation Magazine* 20 (2): 82–90.

27 Giannoccaro, N.I., Massaro, A., Spedicato, L., and Ekuakille, A.L. (2015). Detection analysis of small notches damages using a new tactile optical sensor. *IEEE Transaction on Mechatronics* 20 (1): 313–320.

28 Massaro, A., Troia, M., Spano, F., and Carbone, G. (2013). Friction in totally optical robotic finger oriented on shear force characterization. *IEEE Journal of Sensors* 13 (2): 548–555.

29 Massaro, A., Venanzoni, G., Farina, M. et al. (2011). Experimental pressure sensing and technology of piezoelectric microwave/RF MEMS filters. *International Journal of Microwave and Wireless Technologies* 3 (5): 587–593.

30 Massaro, A., De Guido, S., Ingrosso, I. et al. (2011). Free standing piezoelectric rings for high efficiency energy harvesting at low frequency. *Applied Physics Letters* 98 (5): 1–3.

31 Massaro, A., Cingolani, R., and Passaseo, A. (2011). 3D FEM modelling and technology of piezoelectric ring MEMS antenna. *Progress In Electromagnetics Research C* 23 (1): 123–135.

32 Massaro, A., Ingrosso, I., Giordano, C. et al. (2009). Micromachining Mo/ALN/Mo piezoelectric ring resonators. In: IEEE Proceedings of EuMA, Rome, Italy (29 September–1 October 2009). Piscataway, NJ: IEEE.

33 Ahuja, A. and Ying, M. (2003). Sonography of neck lymph nodes. Part II: abnormal lymph nodes. *Clinical Radiology* 58 (5): 359–366.

34 Massoud, T.F. and Gambhir, S.S. (2003). Molecular imaging in living subjects: seeing fundamental biological processes in a new light. *Genes & Development* 17 (5): 545–580.

35 Liu, Y., Miyoshy, H., and Nakamura, M. (2007). Nanomedicine for drug delivery and imaging: a promising avenue for cancer therapy and diagnosis using targeted functional nanoparticles. *International Journal of Cancer* 120 (12): 2527–2537.

36 Torabi, M., Aquino, S.L., and Harisinghani, M.G. (2004). Current concepts in lymph node imaging. *The Journal of Nuclear Medicine* 45 (9): 1509–1518.

37 Murcia, M.J. and Naumann, C.A. (2005). Biofunctionalization of fluorescent nanoparticles. In: Nanotechnologies for the Life Sciences .Vol. 1. Biofunctionalization of Nanomaterials (ed. C.S.S.R. Kumar), 1–39. Weinheim: Wiley-VCH.

38 Tsuchimochi, M., Hayama, K., Toyama, M. et al. (2013). Dual-modality imaging with 99mTc and fluorescent indocyanine green using surface-modified silica nanoparticles for biopsy of the sentinel lymph node: an animal study. *EJNMMI Research* 3 (1): 33.

39 Varna, M., Ratajczak, P., Ferreira, I. et al. (2012). in vivo distribution of inorganic nanoparticles in preclinical models. *Journal of Biomaterials and Nanobiotechnology* 3 (2): 269–279.

40 Choi., J. and Wang, N.S. (2011). Nanoparticles in biomedical applications and their safety concerns. In: Biomedical Engineering – From Theory to Applications. Ch. 13. InTech House.

41 Caratelli, D., Massaro, A., Cingolani, R., and Yarovoy, A. (2012). Accurate time-domain modeling of reconfigurable antenna sensors for non-invasive melanoma skin cancer detection. *IEEE Sensors Journal* 12 (3): 635–643.

42 Caratelli, D., Yarovoy, A., Massaro, A., and Ekuakille, A.L. (2012). Design and full-wave analysis of piezoelectric micro-needle antenna sensors for enhanced near-field detection of skin cancer. *Progress in Electromagnetism PIERS* 125 (1): 391–413.

43 Malvindi, M.A., Greco, A., Conversano, F. et al. (2011). Magnetic/ silica nanocomposites as dual-mode contrast agents for combined magnetic resonance imaging and ultrasonography. *Advanced Functional Materials* 21 (13): 2548–2555.

44 Caratelli, D., Yarovoy, A., Massaro, A. et al. (2011). Locally conformal FDTD modeling of MEMS-based antenna sensors for melanoma detection. In: Proceedings of IEEE MeMeA International Symposium on Medical Measurements and Applications 2011, Bari, Italy (30–31 May 2011). Piscataway, NJ: IEEE.

45 Johnston, A.P.R., Battersby, B.J., Lawrie, G.A., and Trau, M. (2005). Porous functionalised silica particles: a potential platform for biomolecular screening. *Chemical Communications* 7 (1): 848–850.

46 Cicala, G., Massaro, A., Velardi, L. et al. (2015). Enhancement of surface electrical current on silicon via nanodiamond particles deposited by pulsed spray technique. *Physics Status Solidi (a)* 212 (12): 2862–2868.

47 Cicala, G., Massaro, A., Velardi, L., and Senesi, G.S. (2016). Morphological and electrical characterization of pillar-like structures in nanodiamond layers. *IEEE Transactions on Nanotechnology* 15 (6): 890–895.

48 Genovese, M.A., Colusso, E., Colombo, M. et al. (2017). Acidochromic fibrous polymer composites for rapid gas detection. *Journal of Materials Chemistry A* 5 (1): 339–348.

7

Image Vision Advances

Image vision techniques are applied mainly to detect defects in manufacturing quality processes. In this chapter different solutions are discussed oriented toward image segmentation, image classification, and image clustering for *in-line* production processes. Moreover, important solutions are provided for digitalized information and data monitoring production lines, discussing some image processing architectures and industry applications. The hardware and software solutions are combined allowing big data analytics, augmented reality (AR), and predictive maintenance of machines supporting defect monitoring and quality improvement. Different solutions of sensors and implementations are matched with different scenarios in manufacturing, and in general in production processes, where autodetection of defects is fundamental for a self-adaptive production setting. Image vision approaches are applied in different industry sectors, also involving inspection services. Some image processing approaches are discussed concerning food industries. This chapter shows pseudocodes that help to understand image processing implementation in smart systems. The discussion is mainly focused on different application fields in industries, adopting artificial intelligence (AI) solutions to improve product defect recognition. To this end, the combined use of discrete Fourier transform, K-Means and long short-term memory (LSTM) neural network algorithms increases the performance of the image processing techniques detecting defects. Image segmentation methods represent an alternative approach to extract defect information to combine with AI classification results. The watershed image segmentation and active contour snake techniques are indicative for extracting the contours and measuring possible variations of the object shapes representing defectivities. Furthermore, the LSTM networks are applied to check and enhance defectivities estimated by other image processing approaches such as infrared thermography (IRT), K-means, and morphological image segmentation. The combined used of different image vision techniques is an innovative methodology to classify defects, by automatizing the quality check of the production process and improving the quality check accuracy. AI algorithms are also suitable to predict machine failures, thus behaving as a risk prediction tool to be adopted in predictive maintenance procedures. The real challenge in the next era of Industry 5.0 is to integrate AI data processing *in-line* with the production network, enabling a real time risk prediction, and self-adaptive corrective actions in pick and place systems. Concerning production monitoring, image vision techniques are appropriate for implementation in AR systems, contributing to the optimization of the production layouts and quality checkpoints based on augmented information. Examples of electronic implementations by means of logic ports show how AI can be applied specifically to a matrix of pixels constituting the image for analysis. Moreover, this chapter introduces theoretical formulas facilitating the understanding of the application of image processing techniques for specific case studies. The proposed applications highlight the methodology to

Electronics in Advanced Research Industries: Industry 4.0 to Industry 5.0 Advances, First Edition. Alessandro Massaro.

follow to apply advanced image vision algorithms, by suggesting possible procedures for the quality check, mainly based on the simultaneous use of several techniques.

7.1 Defect Classification by Artificial Intelligence and Data Processor Units

Defect classification is potentially performed and improved in industry production processes by AI algorithms, combined with image vision techniques. In this section, different examples are provided proving the efficiency of these tools matched with the quality ISO 9001:2015 standard. Specifically, image vision techniques are discussed concerning tire production and welding checks. Furthermore, a theoretical approach is introduced matching pixel detection systems with AI classification and prediction, by showing possible logic circuit implementations. The AI improvement addresses new production with a self-adaptive approach.

7.1.1 Artificial Intelligence Algorithms and Automatism for Defect Classification: Case Study of Tire Production

The setup in Figure 7.1a is suitable for the quality check of tire assembly monitoring the wheel alignment and possible defects. In this setup a laser profilometer is coupled with a detector verifying the alignment, a rotary encoder is controlled by a direct current motor able to rotate the tire along the cylindrical axis, a motor is connected by RS232 standard, and a camera is used for other images to further check the alignment and defects [1, 2]. The setup is connected to the local network by an enterprise service bus (ESB) system enabling an Ethernet network. An image vision system detects the profile generating a three-dimensional (3D) reconstruction of the image. The 3D reconstruction is then converted to a linearized two-dimensional (2D) image (TIFF image), thus decreasing the computational cost required for defect analysis. The K-means image processing of the linearized image in Figure 7.1b enhances the defect region to analyze. Another approach used for the evaluation of defects is discrete Fourier transform (DFT). DFT is one of the main techniques used to characterize the spatial morphology of an image and infer the presence of image patterns in the spatial domain. In particular, DFT has been used for defect characterization [3]. The DFT of an image of size N × M is defined as:

$$F(u,v)=\sum_{l=0}^{N-1}\sum_{m=0}^{M-1} f(l,m)e^{-2\pi i\left(\frac{lu}{N}+\frac{mv}{M}\right)} \tag{7.1}$$

where $f(l,m)$ is the intensity function, (u,v) is the frequency domain, i is the imaginary unit, and the exponential functions

$$e^{-2\pi i\left(\frac{lu}{N}+\frac{mv}{M}\right)} \tag{7.2}$$

are orthogonal basic decomposing the bidimensional image (sinusoidal composed functions). The Fourier function is decomposed into a real and an imaginary part as:

$$F(u,v)=F_{\mathrm{Real}}(u,v)+iF_{\mathrm{Imag.}}(u,v) \tag{7.3}$$

The amplitude spectrum is given by:

$$|F(u,v)|=\sqrt{F^2_{\mathrm{Real}}(u,v)+F^2_{\mathrm{Imag.}}(u,v)} \tag{7.4}$$

The DFT reconstruction of the detected defect is shown in Figure 7.1c: the fringe dispersion near the center of the defect is representative of the defect. In Figure 7.1d the detected defect is confirmed by an artificial neural network (ANN) due to the material stretching: the LSTM network confirms the defect found by the K-means algorithm.

The ANN is an alternative approach for defect check, typically requiring different images for the training model. When there are not enough images to create an efficient training model, it is possible to use a subset of the same image as the training dataset. This methodology allows to deduce the regions where the model fails, to predict the next slices of pixel, or classify a pixel configuration representing the defect, by considering the recurrent morphological features (features extracted from frames) using the LSTM fully connected architecture. The use of the memory capabilities of the LSTM network allows to infer which frame of pixel enhances a defect since its spatial features are not consistent with the features of the whole image: in the case of tire defects, the defect features are automatically identified as a region of the tire where the pixel configuration is classified as different compared with the other homogeneous pixel subsets. For simplicity, a defect is identified as the subset region where the maximum of the mean square error function falls outside one (indicating the higher error condition). Figure 7.2 shows the architecture based on a neural network detecting defects. In this model it is possible to distinguish:

- The input image containing the training and the testing datasets.
- The frame decomposition stage enabling feature map extraction for each image subset.
- The convolution stage able to classify a detected defect.

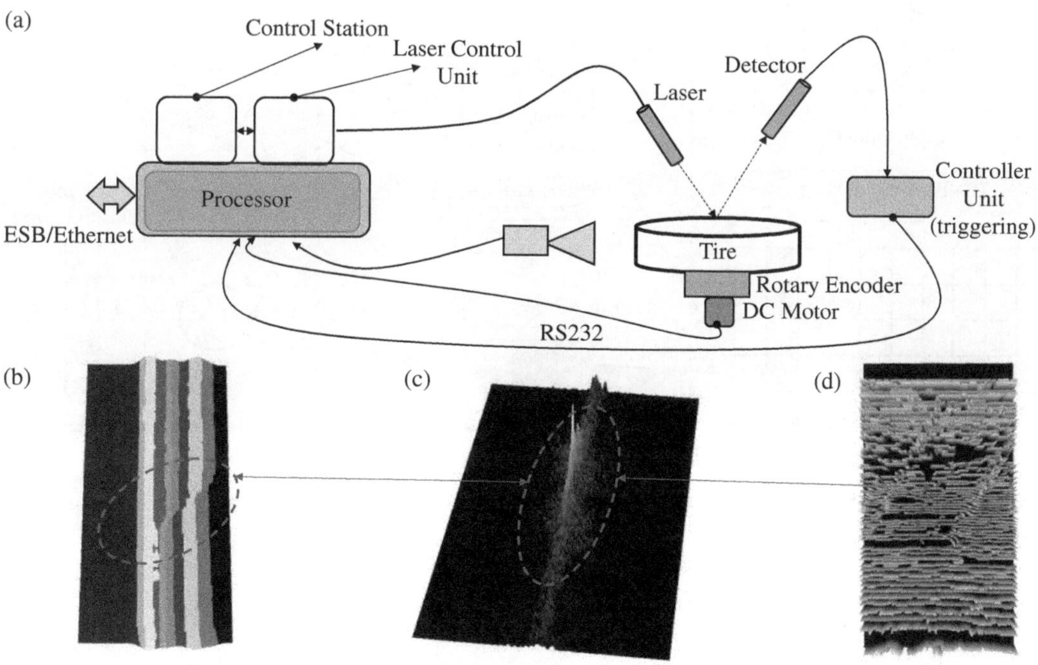

Figure 7.1 (a) Setup for defect monitoring of tire assembly. (b) K-means algorithm enhancing the defect profile by a 3D representation (K = 10 is the cluster number). (c) 3D DFT analysis of the defective area. (d) LSTM image processing of the defect region reconstructed in 3D.

The neural network image processing represents only a module of the approach illustrated in Figure 7.3. Concerning for example tire assembly. the method is divided into different stages such as:

- A first stage representing the first quality check controlling the positioning of the tire assembly into the workstation, by analyzing the tire/rim alignment (sidewall control).
- An advanced quality check stage made combining different AI algorithms, DFT, and the K-means clustering algorithm enhancing defect classes.
- A long-term quality check based on defect prediction and on the use of other classification tools, by analyzing historical production process data collected into a big data system.

The complete flowchart in Figure 7.3 can be implemented in business process modeling (BPM) software improving quality control.

7.1.2 Welding Classification and Nondestructive Testing Suitable for the Quality Check

Defects in welding processing are classified as external and internal defects. The following classes are listed for external defects:

- **Lack of penetration.** This defect is found when the fusion at the top presents some more or less intermittent discontinuities. The lack of penetration may occur in the area of overlap or in areas where "spot welding" is carried out to hold the joint together. In some cases, it is caused by relevant differences in the composition of the material of the two edges to merge.

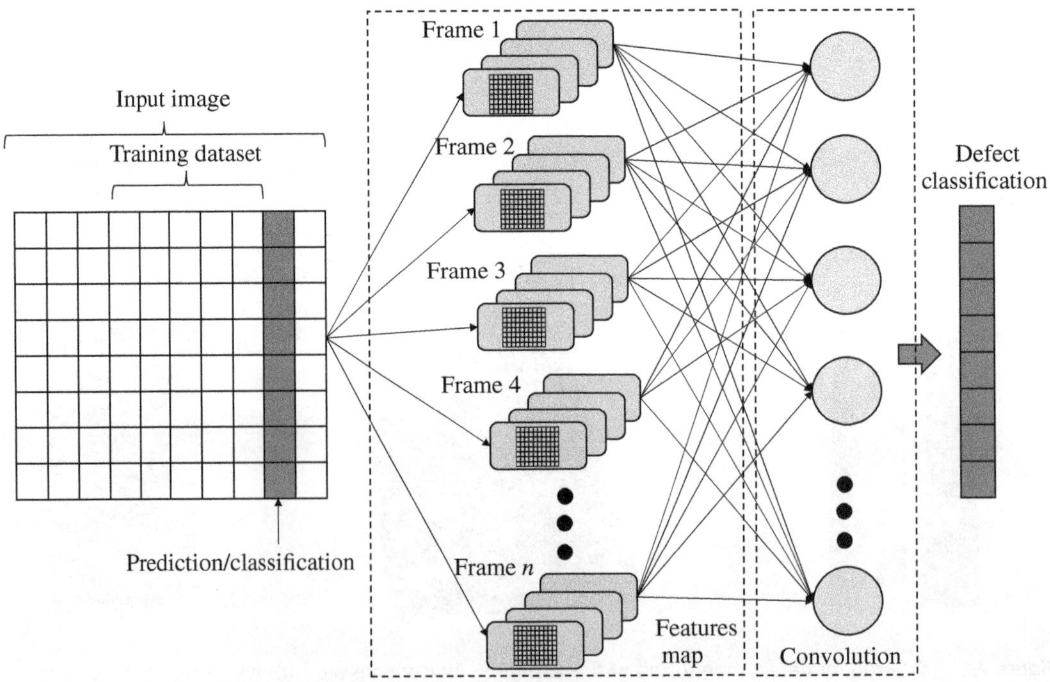

Figure 7.2 Architecture of the stack of the neural network used for defect detection.

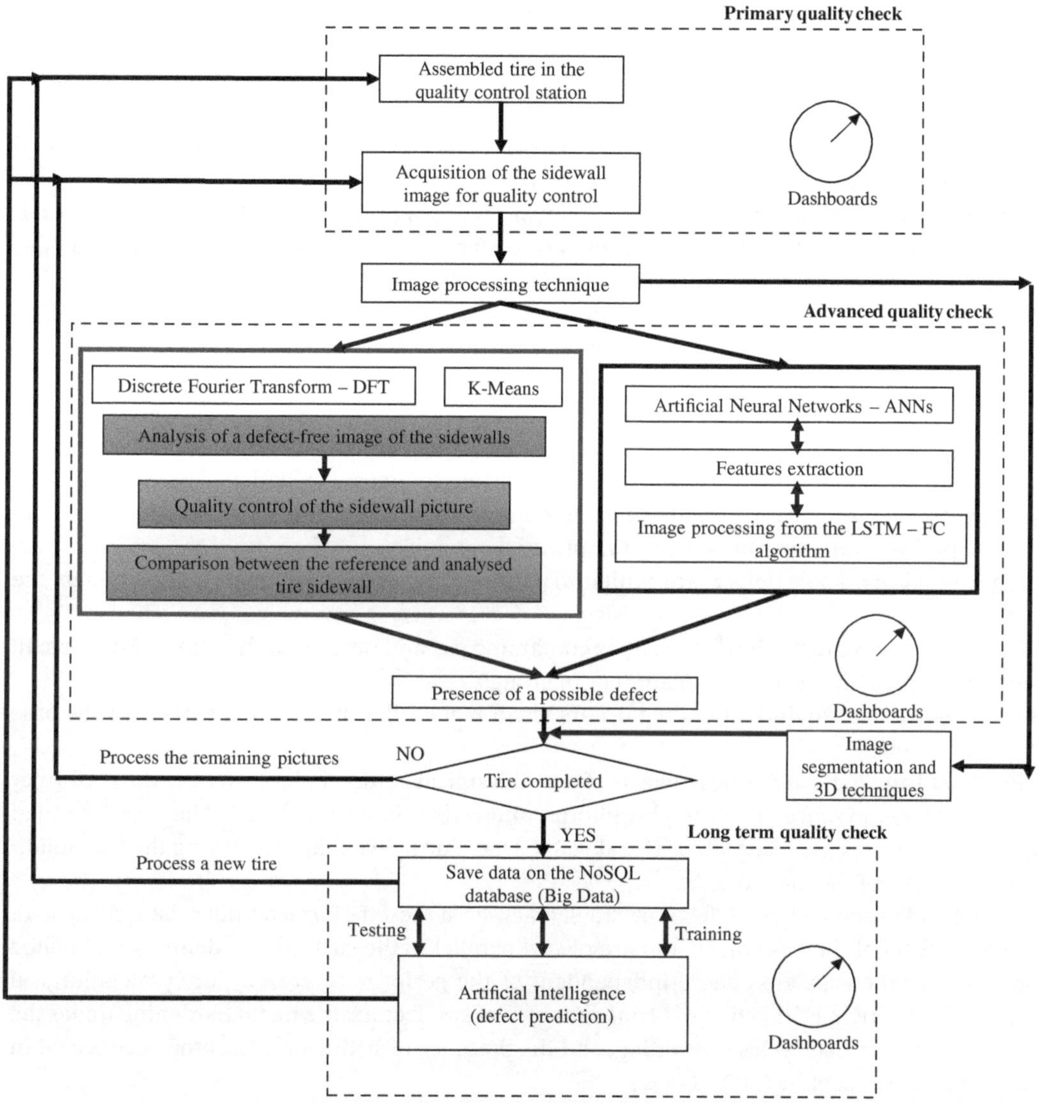

Figure 7.3 Flowchart of a full approach for tire quality check and defect monitoring by means of image processing techniques. DFT, K-Means, ANN, and LSTM are applied for the advanced quality check. Other image segmentation and 3D reconstruction techniques are adopted to confirm the image processing results furthermore enhancing defects.

- **Excess penetration.** This defect may occur if the soldering iron uses excessive current or operates between edges that are too far apart. The defect takes on the appearance of a drop hanging irregularly. The features of this defect are craterizations, formation of inclusions and porosities, and dark coloring.
- **Excessive convexity.** The regulations prescribe a limit to the external thickness of the weld. In addition to an unnecessary waste of filler material, excessive welding is not always synonymous with robustness, but may on the contrary cause a weakening of the coupling.
- **Irregular mesh (poor surface appearance).** The mesh is irregular with the width being variable.

- **Marginal incisions.** These defects are the small channels generated at the connection between the joint over-thickness and the surface of the piece. The marginal (or lateral) incisions involve a decrease in the resistant section of the joint, and therefore create a point of possible breakage trigger.
- **Excessive coloring (discoloration).** The excessive coloring of the area is due to the leakage of chromium carbides from the stainless-steel matrix.
- **Spatters and starting scratches.** Sprays are warm drops projected outside the boundaries of the joint and anchored firmly to the workpiece surface due to too high current or an excessive arc length. Spits are traces of the arc ignition projected onto the metal surface around the joint. Both must be removed mechanically. On some types of carbon steel, the incandescent drop causes the formation of small hardened areas often accompanied by micro cracks.

The following classes are listed for internal defects:

- **Lack of penetration.** The lack of penetration occurs in internal welding in the case of joints with "X" preparation to be made with multiple passes. To prevent such deficiencies, it is necessary to grind the bead before the next pass to remove the slag and make it smooth and homogeneous and free from imperfections and impurity.
- **Slag inclusions.** These defects are similar to points or elongated. They mainly occur where the welder loses control of the bath by the slag.
- **Blowholes (porosity).** They are cavities containing gas and occur mainly in the form of small spheres (pores). They occur in isolation or in groups.
- **Bonding (lack of fusion).** Bonding occurs when metal filler drips on the surface of the base metal not melted.
- **Hot cracking.** These are critical defects of a metallurgical nature, independent of the performer and due to the presence of sulfur, phosphorus, and carbon in the weld pool. They are formed at high temperatures (between 1400 °C and 900 °C), i.e. during welding and during the first solidification stages of the melted zone.
- **Underbead cracks.** These defects are small cracks that they are formed under the weld bead in the first layers of the base metal and are placed parallel to the surface that delimits the melted area. The cold cracks are defects independent of the performer but of a purely metallurgical nature. Their origin is linked to the concomitance of two factors: the metal hardening under the action of the thermal cycles of welding, and the presence of hydrogen. Cracking occurs cold in the last stages of cooling (300–200 °C).
- **Fish eyes.** This is a class of defect that affects the melted area and that on the fracture sections looks like a clear patch (1–2 mm in diameter) with a microscopic dark spot in the center.

The main visible defects are:

- Irregular welding shapes and layouts.
- Too big or too small welds.
- Missing welding.
- Irregular patterns (linking electrical connections).
- Agglomerates (too much deposited material).
- Dirty particles on the welds.
- Excess or less deposited welding materials.
- Shifting welding.
- Component lift.
- Damaging welding.

Nondestructive testing (NDT) refers to a set of non-invasive physical tests aimed at investigating the possible presence of defects in a piece without having to destroy all or part of it. NDT methodologies include, in reference to UNI EN 473, the following:

- penetrating liquids checks
- radiographic and gammagraphic checks
- ultrasound checks
- magnetoscopic checks
- eddy currents
- eddy-current testing
- vision test
- acoustic emission
- thermography

Concerning quality improvement, NDT, if used in appropriate points of the manufacturing cycle, permits defects to be identified in the different production stages. Control methodologies are classified as superficial, subsuperficial, or volumetric. Volumetric methodologies are classified as methodologies for transmission and methodologies for reflection. The advantages of transmission methodologies are lower signal attenuation, which must cross the thickness of the object only once, however they require that both surfaces of the object are accessible, while reflection methodologies allow only one surface of the object to be accessible. Surface control methodologies obviously require accessibility to the surface on which the checks are carried out.

7.1.2.1 Watershed Image Segmentation and Automatic Welding Defect Classification

Image vision is typically used for the detection and classification of surface defects of pin through hole (PTH) and surface mount technology (SMT) boards. A typical approach is based on the image segmentation technique. The watershed method [1] is a segmentation approach based on the transformation of the grayscale image by applying morphology operators. By defining the function representing an image as $v(x,y)$, and the spatial domain $(x,y) \in \Re$ [2], the morphological gradient of the image is defined as:

$$\delta_I v = D - E = (v \oplus I) - (v \odot I) \tag{7.5}$$

where D and E indicate the elementary dilation and erosion of v, respectively, by transforming the element I. The operator used for the morphological transformation is the Laplacian:

$$\Delta_I v = (v \oplus I) - 2v + (v \odot I) \tag{7.6}$$

This operator discriminates the areas of minima intensity ($\Delta_I v > 0$), the areas of maxima intensity ($\Delta_I v < 0$), and edge parts ($\Delta_I v = 0$). The image segmentation processing of the watershed approach is configured as a landscape immersed in a lake, where holes are the local minima, the catchment basin is filled by water, and dams (watersheds) are built linking points of water levels coming from different basins (Figure 7.4a). The segmentation image processing by the Laplacian is stopped when the water level reaches the highest peak in the landscape. The landscape region is then segmented into basins separated by dams indicating edges of the image and the segmented extracted regions. The gray level image is viewed as a topographic relief, where the gray level of a pixel is considered to be its height in the relief. Assuming that a drop of water falls on a topographical relief, it flows along a path to eventually reach a local minimum. The watershed of the analyzed relief corresponds to the limits of the adjacent water catchment areas of the drops. The image pixel clusters embedded by watershed contours are the extracted welding shapes. By selecting the dimension of a specific class of welding, it is possible to define the dimensions of defects by area ranges expressed in pixel^2 or mm^2 (an example of defect classification is shown in Figure 7.4b). The example used for the test can select a range of four areas related to defects having areas below x_1, between x_2 and x_3, between x_4 and x_5, and greater than x_6.

The steps followed by the watershed approach are:

1. Transformation of the source color image into a grayscale image.
2. Selection in the image of pixels that have an intensity value less than a threshold and transformation of the image into an array of 0 and 1 (each pixel with an intensity value greater than or equal to the threshold value is converted to 0 and the others to 1, thus structuring an image matrix into a binary matrix).
3. Removal of all connected components that have less than a minimum number of pixels from the binary image.
4. Filling operation of the extracted holes (a hole is a set of background pixels that cannot be reached by filling the background from the edge of the image).
5. Suppressing the lighter structures of the surrounding environment which are connected to the edge of the image.
6. Matrix reversal to calculate the transformation of the Euclidean distance of the binary image.
7. Forcing the non-object pixels to the $-\infty$ value.
8. Eliminating local maxima.

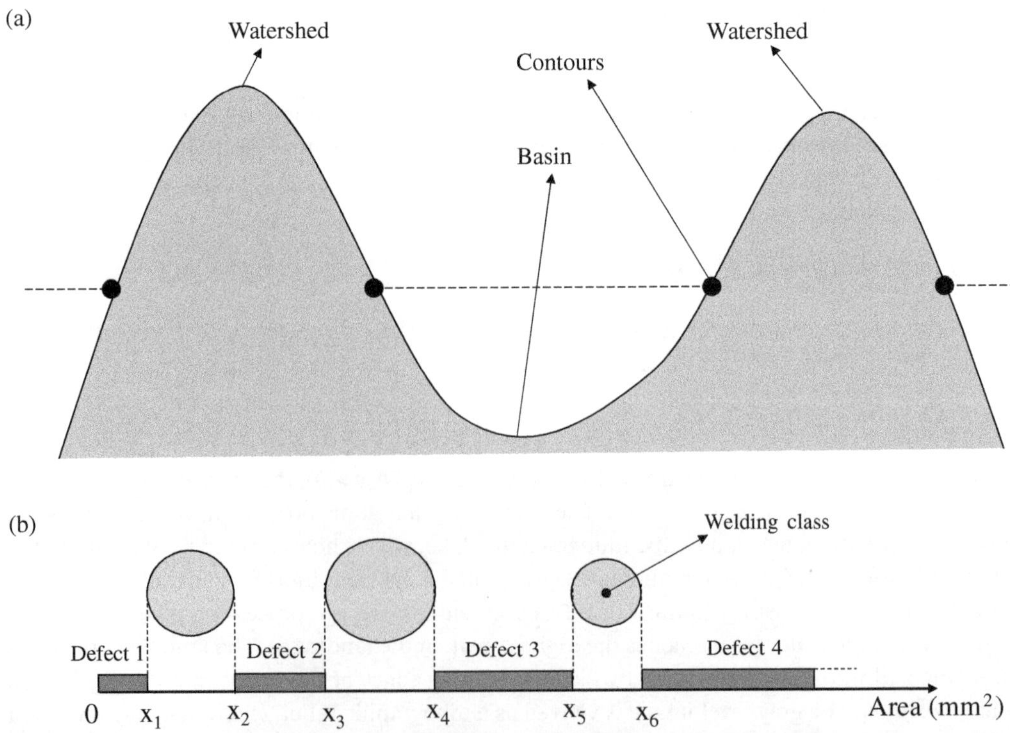

Figure 7.4 (a) Watershed and geometrical analogy of local minima and local maxima. The contour is defined by the level of the imaginary water filling the basin. (b) Defect range classification.

9. Calculating the label matrix identifying the watershed regions (elements labeled 0 do not belong to a single watershed region, elements labeled 1 belong to the first watershed region, elements labeled 2 belong to the second watershed region, and so on).

The watershed approach is integrated as ImageJ library (open source): in particular this tool uses java language to estimate the center of each object and execute the morphological erode operation: by the calculation of the distances from the object center points to the edges of the same objects, the tool evaluates and extracts the contours by considering defined thresholds for the contour resolution. An example of a pseudocode code implementing watershed segmentation (*Analyze Particles function*) is:

```
1. Opening the image to process;
2. Binary image conversion;
3. Thresholding setting (the default is the black and white filter);
4. Call of the library implementing the Watershed segmentation;
5. Setting the library with the Region of Interest (ROI);
6. Estimation of ROI coordinates to crop the processed image surroundings defects (extracted contours).
```

The logic condition defining defect classes is described by the following pseudocode taking into account the range defined in Figure 7.4b:

1 "For" cycle defining defect ranges and related coordinates (diameters considered as defects);

2. "If" logic condition of $area<=x_1$, $x_2=<area<=x_3$, $x_4=<area<=x_5$, $area>=x_6$;

3. The images with Rectangle as Region of Interest (ROI) coordinates, is stored inside a defined directory as a .jpg file.

Figure 7.5 shows a flowchart representing the model of the welding quality monitoring system by image segmentation. The model includes the AI and the big data for defect classification and quality prediction. The Unified Modeling Language (UML) sequence diagram in Figure 7.6 explains the timing of the different stages of the image vision model.

A way to adopt the image segmentation approach to measure an object is to detect a reference scalebar in the image, thus permitting the conversion of the number of pixels to cm or mm dimensions, and the areas from $pixels^2$ to mm^2 or cm^2.

7.1.3 Encoding and Decoding Circuits in Artificial Intelligence Data Processing

An AI logic can be implemented into an integrated circuit by means of encoding and decoding electronic units: the encoding process is able to convert into bit sequences the input information, and the decoding function is applied for final information conversion. Figure 7.7 shows an example of a theoretical integration of an AI algorithm into an information system: sensor data or the information extracted from an image (high or low pixel intensities) are converted into digital information representing a number which is encoded to be processed by the AI algorithm; the output of the AI algorithm is a sequence of bits which must be decoded in order reconstruct the value. The encoding process is useful to decrease the complexity of the network model allowing a better prediction accuracy.

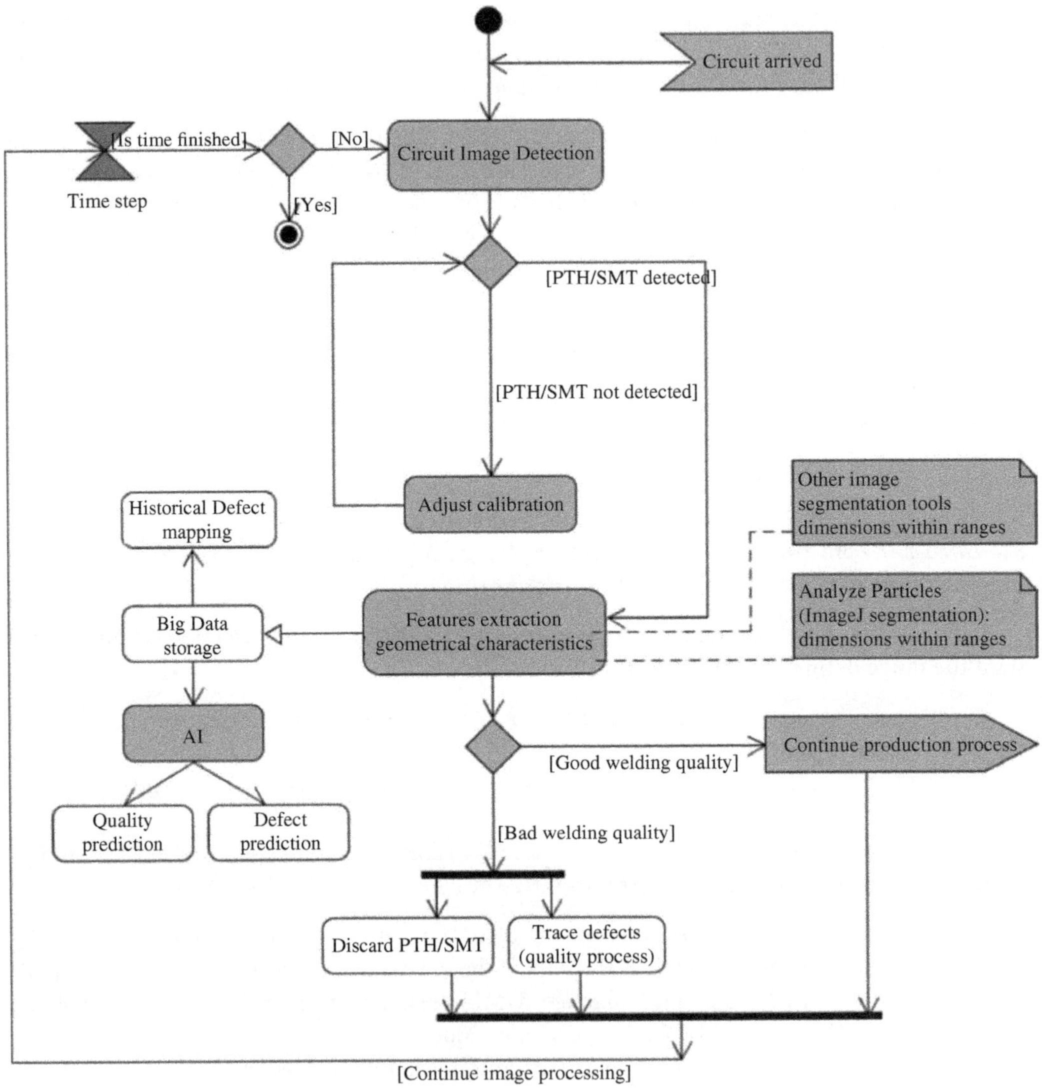

Figure 7.5 Flowchart modeling for welding defect check.

Consider for example seven sensor signals at the input of a production line, producing a low (bit 0) or a high value (bit 1), with eight encoded thresholds (T) fixed as a combination of all the bits, by implementing the electronic circuit in Figure 7.8a, which is simplified to the circuit in Figure 7.8b. The threshold values care defined in the proposed example as T = $O_0 \cdot 2^{O0}$ + $O_1 \cdot 2^{O1}$ + $O_2 \cdot 2^{O2}$.

The following equations are the output of the encoder defining the threshold value:

$$O_0 = I_1 + I_3 + I_5 + I_7 \tag{7.7}$$

$$O_1 = I_2 + I_3 + I_6 + I_7 \tag{7.8}$$

$$O_2 = I_4 + I_5 + I_6 + I_7 \tag{7.9}$$

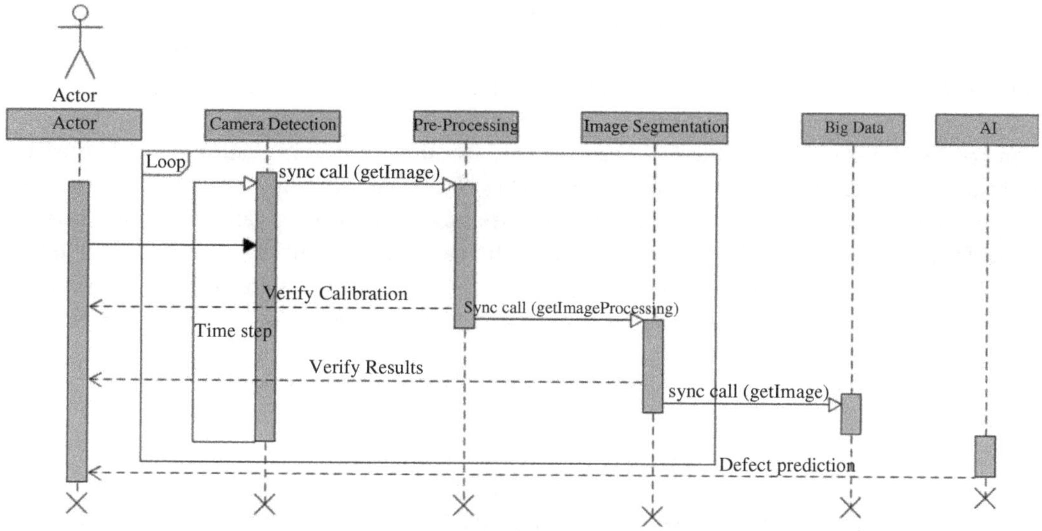

Figure 7.6 Sequence diagram of the image vision system based on image segmentation.

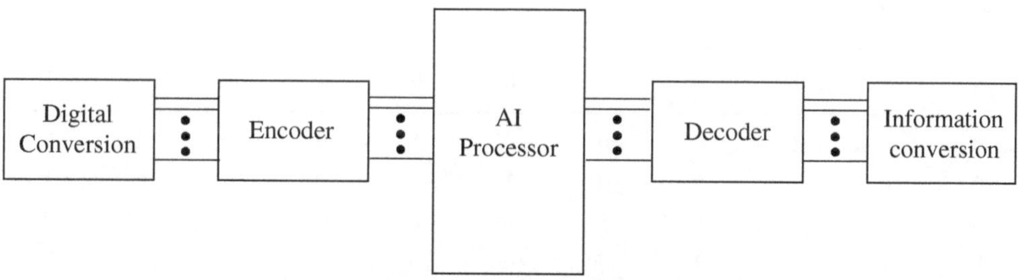

Figure 7.7 Encoding and decoding enabling AI algorithm processing.

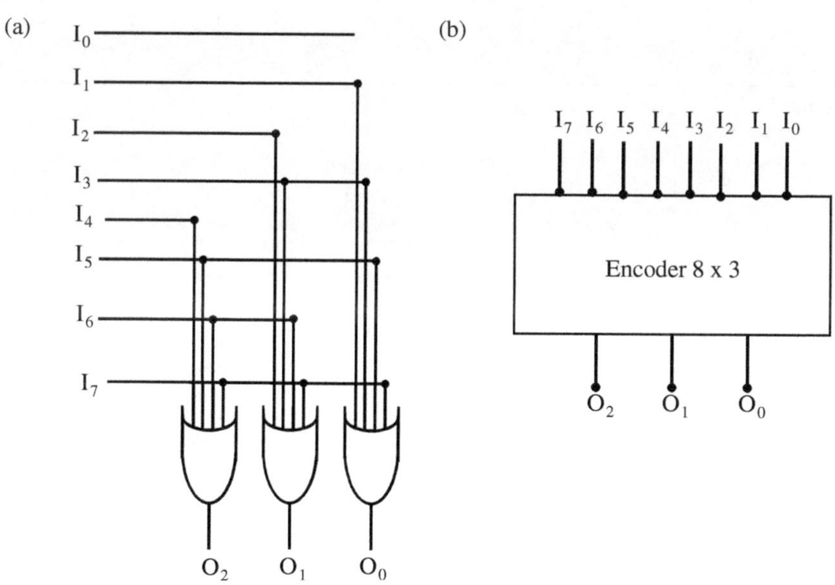

Figure 7.8 (a) Digital circuit configuration of an encoder. (b) Black box modeling of the encoder.

The encoder outputs of the example in Figure 7.8 are summarized in Table 7.1:

The AI engine can process the sequence of bits by a gradient boosted trees algorithm (GBTA) predicting the threshold value. The Konstanz Information Miner (KNIME) workflow in Figure 7.9 provides the results summarized in Table 7.2: a low threshold value is predicted by using the training and the testing dataset of Table 7.2. The predicted threshold values are estimated near the values of the samples used for the training, which are low threshold values for the analyzed case. The testing results are not known by the AI engine that assigns the nearest value making a certain mistake (calculation error). The prediction results change with the training and the testing datasets according to real conditions. This example proves that AI algorithms are applicable especially when different bits are considered at the input, by replacing a behavior similar to that of the encoder, but oriented to a learning process. The training process can self-learn how the combination of bit values generates a certain output as for a logic circuit and functions.

The example in Table 7.2 explains how it is possible to associate a sequence of bits for the learning process as a succession of pixel intensities of part of an image (vector line or matrix). The defectivity can be associated with a particular sequence of pixel intensities identified by the AI classification algorithms as for the encoder. The training of the AI algorithm is useful for self-adaptive defect recognition.

Table 7.1 Encoder 8×3 inputs and outputs.

I_7	I_6	I_5	I_4	I_3	I_2	I_1	I_0	O_2	O_1	O_0	T
0	0	0	0	0	0	0	1	0	0	0	0
0	0	0	0	0	0	1	0	0	0	1	1
0	0	0	0	0	1	0	0	0	1	0	2
0	0	0	0	1	0	0	0	0	1	1	3
0	0	0	1	0	0	0	0	1	0	0	4
0	0	1	0	0	0	0	0	1	0	1	5
0	1	0	0	0	0	0	0	1	1	0	6
1	0	0	0	0	0	0	0	1	1	1	7

T indicates the threshold value.

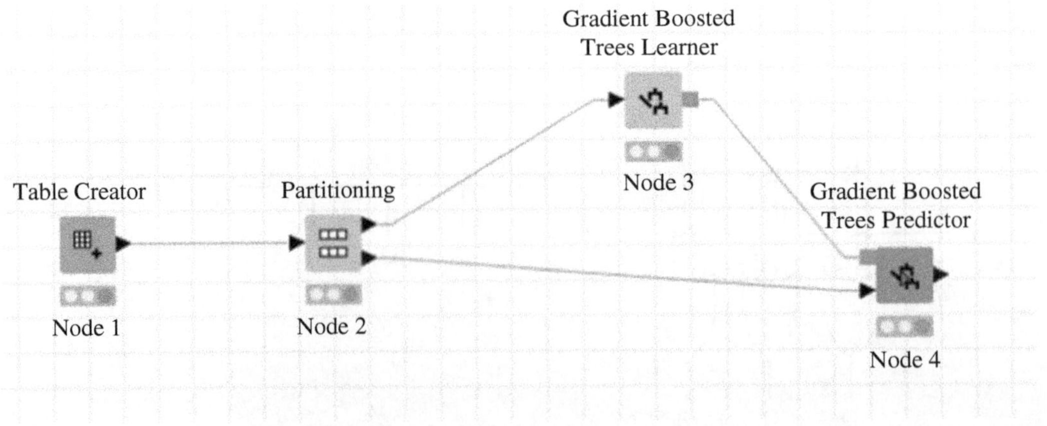

Figure 7.9 Algorithm running into an AI processor (AI engine).

Table 7.2 Data processed by the network model in Figure 7.9.

Table creator			Defined threshold	First partition (training)			Second partition (testing)			Predicted threshold
O_2	O_1	O_0	T	O_2	O_1	O_0	O_2	O_1	O_0	T
0	0	0	0	0	0	1	0	0	0	1
0	0	1	1	0	1	0	0	1	1	2
0	1	0	2	1	0	0				
0	1	1	3	1	0	1				
1	0	0	4	1	1	0				
1	0	1	5	1	1	1				
1	1	0	6							
1	1	1	7							

The decoder circuit models in Figure 7.10a,b convert the encoded information into the original one, where the inputs and the outputs of the decoder correspond to the same as in Table 7.1 with $X_2 = O_2$, $X_1 = O_1$, $X_0 = O_0$, and $I_i = Z_i$ with $i = 0, 1, 2, 3, \ldots, 7$.

By considering a database system collecting data, the digital converted signals are directly extracted to the database and processed by the model in Figure 7.9.

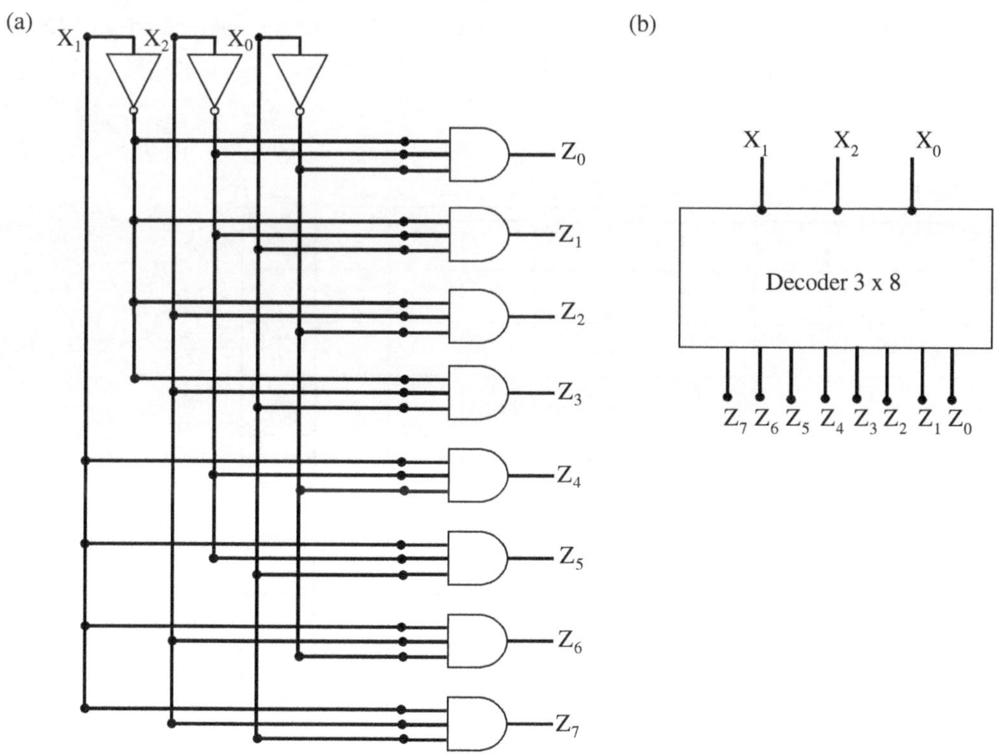

Figure 7.10 (a) Digital circuit configuration of a decoder. (b) Black box modeling of the decoder.

7.1.4 Electronic Logic Port Implementations: Pixel Matrix Logic Condition

A cluster of pixels having a high gray scale intensity represents a feature to extract. A decision support system (DSS) can be applied directly to the pixel matrix of a subset of the whole image. Supposing that the matrix frame is centered with respect to a feature to detect, the logic condition is represented by the simultaneous high intensities of the pixels indicated in Figure 7.11. This simultaneous condition is modeled by the circuit in Figure 7.12. Other conditions are applied to other subsets of pixels by properly setting logic OR conditions. A pseudocode implementing logic conditions to the whole image is:

```
IF in the frame1 Y1=1 or
IF in the frame2 Y2=1 or
then the piece is defected but can be adjusted;
IF Y1=1 and Y2= 1 then discard the piece;
IF Y1 =0 and Y2 =0 the piece is regular.
```

In order to automatize the image processing logic condition, it is important to calibrate the image vision setup centering the frame coordinates with respect to the critical points where defects are usually detected (frame position calibration).

7.2 Image Vision Architectures and Electronic Design

In this section, different examples and methodologies are discussed of image vision architectures applied to industry systems. The attention is focused on IRT, K-means, watershed and LSTM approaches, by introducing AR technology supporting AI processing.

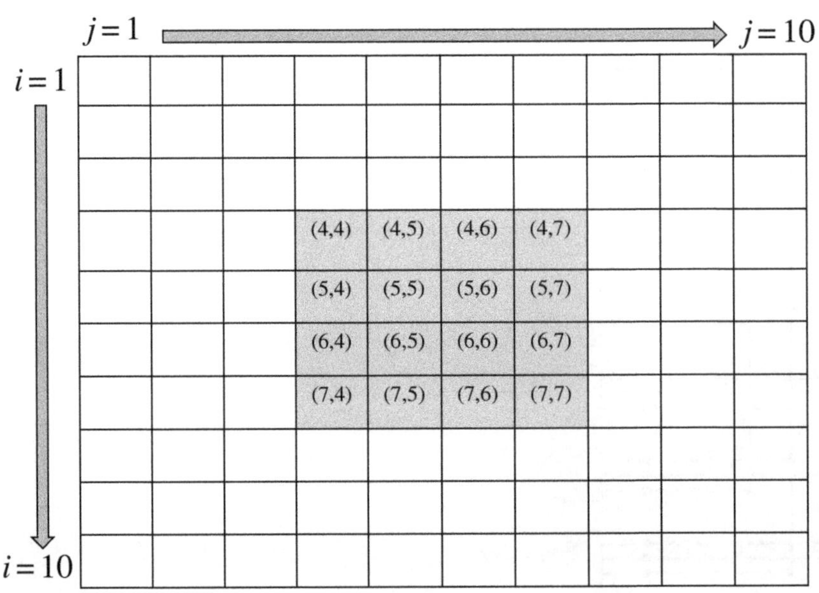

Figure 7.11 Pixel matrix of an image subset.

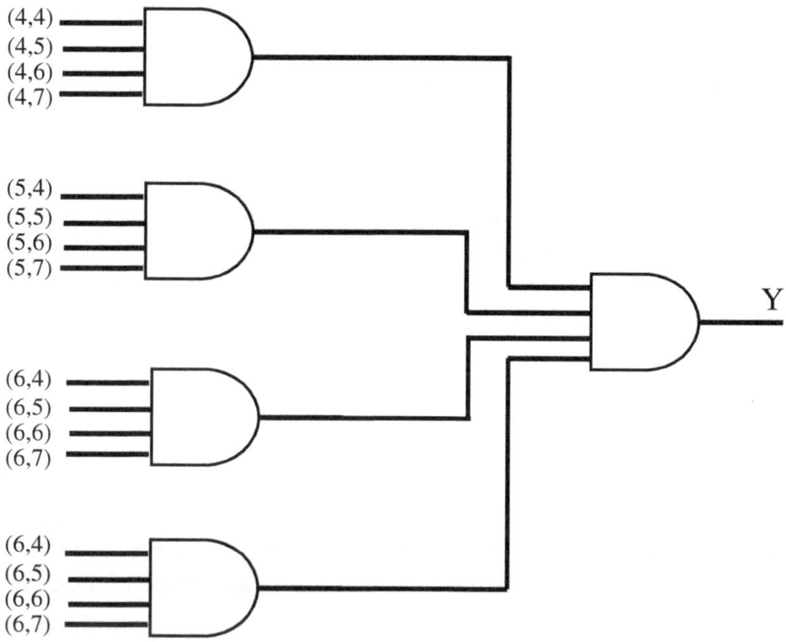

Figure 7.12 AND logic ports implementing the feature in Figure 7.11.

7.2.1 Infrared Thermography Monitoring Industrial Processes

Image vision techniques are improved by IRT tools integrated in automatic defect detection systems. In advanced image vision systems, IRT is combined with other image processing tools to gain information about defectivity. The image vison hardware must be compatible with the production environment. Different aspects need to be considered for correct installation of the infrared (IR) elements such as:

- Thermal resolution according to the defect resolution and tolerance to detect.
- No vibrating operating conditions (thermal camera allocation).
- No thermal reflection conditions (thermal camera allocation).
- Frame rate compatible with the production velocity.
- No environment characterized by air pollution.
- Correct distances according to the camera optics.

Possible improvements of the IRT are in the acquisition of 3D radiometric images measuring the local temperature of complex products. Mechatronics is useful in this case to construct intelligent systems able to detect and reconstruct a 3D radiometric image. The circuits to implement must convert the radiometric image, as a matrix of pixel intensities, into standard protocols traveling in the local area network (LAN). Concerning the simultaneous use of different image processing techniques, an application involving welding processing is discussed in the following.

7.2.1.1 Welding Image Vision Processing and Architecture Design: Radiometric Post Processing

Defect detection in the welding process is performed by different techniques [4]. The first approach is the acquisition of a radiometric image of the welded part by using an IR thermal camera

(Figure 7.13a). In order to acquire correct information, it is preferable to heat the welded region by a thermal gun, and then measure the radiation emitted by the heated region. By defining a calculus line along a central region of the thermal image, the temperature profile is calculated (expressed in °C). The anomalous regions are indicated by the presence of troughs or peaks in the temperature profile. Another approach to follow is to post-process by other image processing algorithms. By applying the K-means algorithm (Figure 7.13b), it is possible to select clusters of pixels enhancing the inhomogeneous regions (inhomogeneous clusters). Another supporting image processing technique is image segmentation by morphological operations such as erosion and dilation [5]. The erosion of the binary image i obtained with a structuring element e generates the new binary images $g = i \ominus e$, having eroded boundary regions. The dilation of the binary image i performed with a structuring element e produces the new binary images $g = i \oplus e$, where it is important to expand all the shapes and the boundaries embedded in the original image. Figure 7.13c shows the results of the combination of dilation and erosion processing of the radiometric image. Finally, a further technique highlighting defects uses convolutional neural networks (CNNs). Typically, CNNs require a dataset with a large number of pixels to train and test the model to apply (large number of images to process). In typical cases the images to train the model are insufficient, and welds are characterized by different features thus requiring advanced analyses to extract correct defect information. A strategy to follow is the use of the LSTM approach, as a variation of the CNN technique. The LSTM method is also applied on a single image: the input image is partitioned into a large number of subimages of small pixel size, and all the subimages represent the sequence adopted for the construction of the neural network model. In the approach used the equivalent time dimension is assumed by the horizontal dimension of the whole image. Feature extraction is based in this case on the estimation of non-recurrent features. In the analyzed case in Figure 7.13d a test dataset is used equal to the training data set, thus identifying the welded region where the mean square error is maximized. In all the image processing output in Figure 7.13 an anomalous region is detected characterized by a lower temperature compared with the homogeneous surrounding region. It is preferable to apply tall the approaches together to optimize the welding process check, and to predict the production quality. Figure 7.14 shows a flowchart modeling the welding process optimization.

Concerning application of the K-means algorithm, it is observed that the best cluster number is selected according to the image to process: if there are different clusters of pixel intensity values characterizing the images, it is preferable to select a cluster number that is not too high.

7.2.2 Electronic and Firmware for Inline Image Monitoring Systems: Hole Precision in Milling Quality Processes

In this subsection, an example is proposed of firmware implementation of an automatic image vision system to classify tolerances of holes generated in milling processes [6]. The architecture of the image processing module enabling an inline monitoring system involves the configuration of two crons where:

- The first cron enables the image acquisition (the acquisition code can be implemented in Python language through the use of the camera's software development kit [SDK]).
- The second cron can pre-process the image acquired in the first cronjob, and identify and analyze its features.

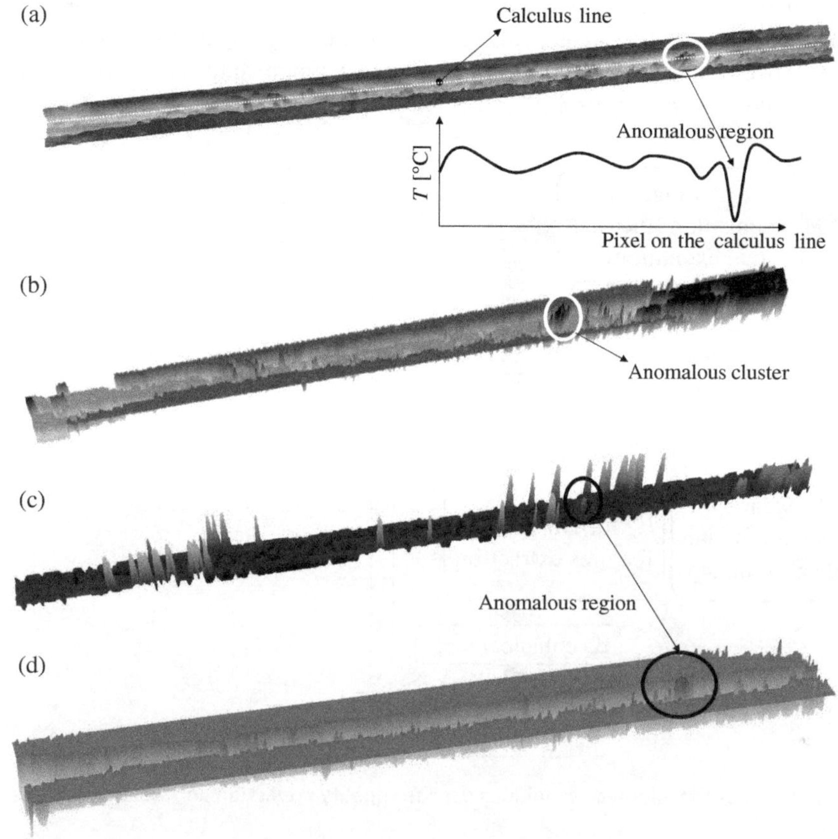

Figure 7.13 3D image processing enhancing welding defect. (a) Infrared thermogram of a welding part and related temperature profile estimated along the calculus line (the trough indicates the anomalous region). (b) K-means image processing of the post-processed radiometric image (where the cluster number is 4); (c) feature extraction of a labeled image (morphological segmentation of the radiometric image by erosion and dilation); (d) LSTM image classification of the radiometric image.

In order to acquire the image, the camera must be connected to the local network. One option is to establish a direct peer-to-peer connection between the camera and the host computer. In industrial networks, the cameras are connected to individual Gigabit Ethernet (GigE) network adapters or to a multiport adapter on the host computer. Alternatively, the connection is established between the cameras and the host computer via a network switch. Figure 7.15a shows a typical scheme implementing the two crons.

For an optimized image vision system able to estimate hole precision, the information network is shown by the block diagram in Figure 7.15b: the watershed approach is able to extract hole shape and a K-means algorithm supports the automatic hole classification, where the tolerances are defined by the centroid positions and by the Euclidean distances.

During the data processing, the image is converted into an 8-bit image (black and white format) allowing the threshold parameter to be applied for the recognition of details. Subsequently, each identified detail is within the image, and analyzed by detecting the size of the areas. The two images (the original and the images with areas), and the Excel file containing all measurements, are saved as output into a specific folder by using the timestamp to distinguish the various detected frames. The storage of the measurement is useful for the quality process mapping and analysis.

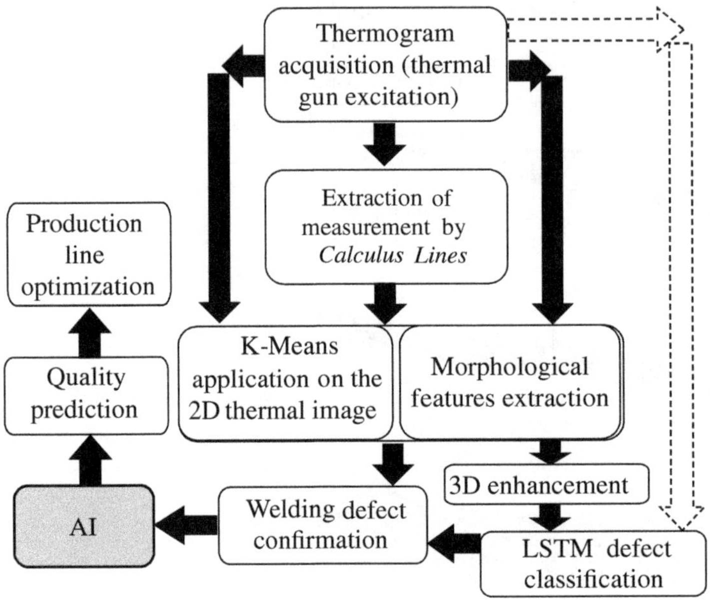

Figure 7.14 Architecture model of the adopted techniques used for quality prediction and welding process optimization.

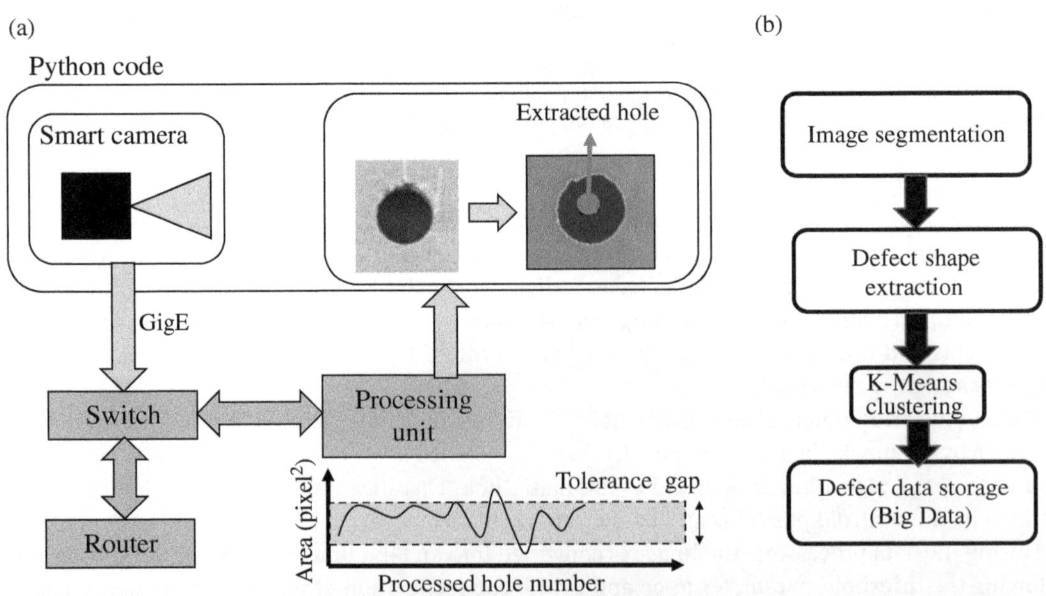

Figure 7.15 (a) Image vision architecture of a system checking hole precision. (b) Flowchart supporting quality procedure.

The application of the K-means algorithm is suitable to confirm the checked defects as outlier results. Figure 7.15b shows the flowchart optimizing defect check. In order to estimate the quality of the holes drilled during processing, the area inside the hole is calculated. In particular, the defect parameter, Δ/σ, is estimated which is defined as the deviation of the measured area of the hole A_{hole} from the average area $\bar{A}_{\text{hole}}$ and normalized to the standard deviation σ:

$$\frac{\Delta}{\sigma} = \frac{\left(A_{\text{hole}} - \bar{A}_{\text{hole}}\right)}{\sigma} \tag{7.10}$$

The algorithm to detect defects is improved by means of a statistical test applied to area values: the test is based on the evaluation of the probability $P(\Delta/\sigma)$ of the measured hole area A, and by the comparison with the mean value $\bar{A}$. The tolerance gap is defined by the parameter Δ/σ (tolerance gap): if the probability $P(\Delta/\sigma)$ is less than a specific percentage, it is possible to confirm that the hole is not accurate. The LSTM algorithm furthermore predicts the hole shape and consequently the defect thus accomplishing the predictive maintenance of the milling machine and properly setting the tool velocity in order to avoid defectivities.

7.2.3 Image Vision and Predictive Maintenance by Artificial Intelligence

Image vision techniques are typically improved by AI. In this subsection, some examples are provided of AI implementation in industrial applications.

7.2.3.1 Profilometer for Image Vision

Image processing by a 3D profilometer system is an innovative approach to check the damage risk for conveyor belts. When using a 3D system, the light source and the receiver (light sensor area) are independent of each other, so installation and adjustment on the site are very expensive in terms of both time and effort. For this reason, it is preferable to integrate into a unique system the 3D camera and the light source, which can facilitate the image detection calibration; therefore, it is necessary to implement an algorithm performing a linearization correction. In the case of the profilometer, the laser light source and the receiver are contained in a single body, avoiding loss of alignment. Concerning conveyor belt monitoring, the profiler can focus a laser light onto the surface to check. Specifically, the laser light is reflected from the target at different angles thus providing certain information about the profile of the surface to detect. From the angle between the direction of emission and that of reception, the position of the object is then calculated by means of mathematical triangulations. The laser source, the sensor camera, and the target form a triangle, hence the name of the technique used (laser triangulation). Using this technique, the sensor acquires a series of 3D profiles, creating a complete scan of the object. The sensor speed and exposure time required to measure the object are generally critical factors in profilometer applications. The main specifications of a profilometer are:

- Data points/profile.
- Linearity along Z (±%).
- Resolution along Z (mm).
- Resolution along X (mm).
- Repeatability along Z (μm).
- Safety distance (clearance distance, CD) (mm).
- Measuring range (mm).

- Field of view (FOV) (mm).
- Laser classes.
- Scan speed.
- LAN interface (GigE).
- Input (differential encoder, trigger, etc.).
- Output (2× digital output, RS-485 Serial [xxx kBaud], 1× mA analog output, etc.).
- Voltage (power).
- Protection Index (IP67).
- Working temperatures.
- Vibration resistance (range in Hz).
- Shock resistance.

In particular, the CD, FOV, and measuring range (MR) are very important concepts for understanding sensor configuration and results. The CD is the minimum distance for scanning an object from the sensor. Any object located at a shorter distance produces invalid data. The MR is the vertical distance at which the conveyor belt can be scanned and measured. The FOV is the width on the X axis along the measuring range. The profilometer is composed of the following components:

- Sensing area keeping light reflected from target surfaces.
- Laser emitter.
- Input and output connectors.
- Power connector for power supply.
- LAN connector for Ethernet network.
- Power indicator.
- Range indicator.
- Laser safety indicator.

The sensing area is a position sensitive detector (PSD) or a charge-coupled device (CCD). It is preferable to connect the sensor for remote control via wireless through an access point to be interfaced with devices such as encoders, photocells, or PLCs. By remote control an operator can access the configuration of the profilometer through a web interface or through software. The profilometer functional configuration is shown in Figure 7.16a. Figure 7.16b shows the basic principles involved in checking damage on a conveyor belt.

In this scenario, the user uses a graphical user interface (GUI) interfacing the profilometer and setting the necessary parameters for the control process. The detection system operates when the conveyor belt is completely empty in order to check possible cracks or damage. The profilometer collects information on the profile of the conveyor belt generating a map of the surface of the entire conveyor belt. The parameter setting is established according to the defect resolution and the technological limits. The computer vision module interacts directly with the vision system to process the images acquired by the profilometer. The computer vision module is ideally divided into two parts, in relation to the main purposes of the image processing algorithms: a data pre-processing module and a data processing one. Data pre-processing is performed by image filters avoiding noise and background errors. The extracted image is loaded into a database through a data management controller (DMC) which manages the exchange of data between the various modules of the system. The extracted features are then processed by post-processing software based on AI algorithms providing information about breakage risk and predicting possible lacerations. A LSTM

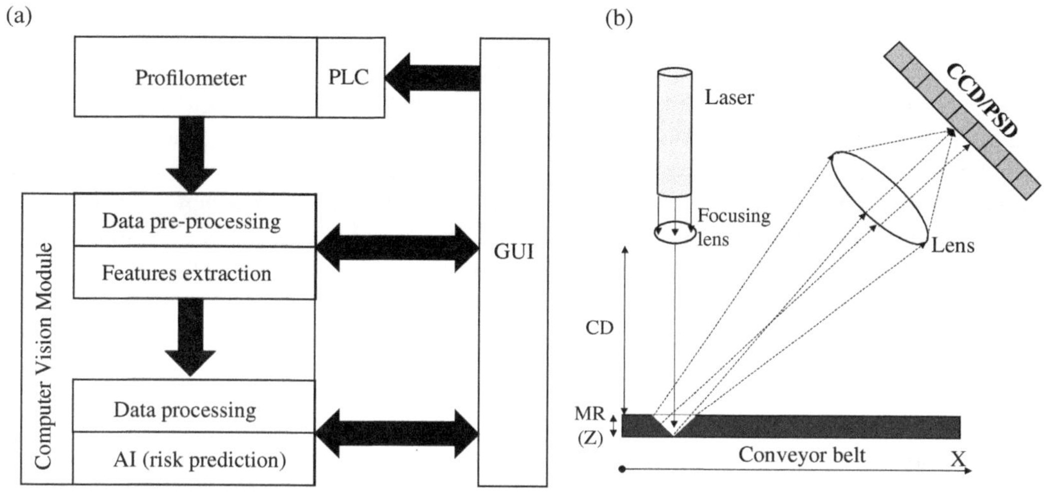

Figure 7.16 (a) Block diagram modeling the profilometer image vision system. (b) Basic principle of laser scanning detecting conveyor belt damage.

network can compare the scanned lines with the trained ones: the comparison of the succession of the pixel intensities of the scanned line with the trained successions of normal and anomalous lines provides the detection of a possible laceration with greater certainty. The operator, through the GUI, views all the image processing results.

7.2.3.2 In-Line 3D Image Vision AI System Integrating Profilometer and Image Processing

The production, preparation, and quality control by inspection techniques of printed circuit boards (PCBs) is a research topic [7–11]. An application of integration of image processing 3D and profilometer detection is the *in-line* integrated quality control for manufacturing. The system integrates image vision technologies and optical sensors which are be able to check the quality of the manufactured PCBs. Image vision techniques and laser signal processing generate the 3D model of the manufactured components. The manufactured component is compared with the 3D nominal geometry in order to control the correct position of a lot of miniaturized electrical components assembled and soldered on the same electronic board. In order to obtain a good performing result, AI algorithms and a database could support defect detection and classification. By using for example OpenCV libraries, it is possible to enable image vision procedures involving frame acquisition matching with the production velocity. The data are used to operate the actual-to-nominal comparison between the manufactured components and the nominal specification which these must comply with. The quality control is finalized from the adoption of a robotic system capable of picking the components classified as defective from the analysis. Those components must be fixed with the indications resulting from the AI classification. Alternatively, the components can be considered as scrap, depending on the choice given to the algorithm. The described system is applied at different points in the manufacturing process and not only at the end of the process. The AI is trained in order that the actual characteristics appear with the nominal characteristics the component must have at that precise process phase as for agile manufacturing. The development and testing of the first prototype provide the input to integrate more than one of these systems within different process phases and ensure 100% quality of the manufactured components, certified from such a systematic approach. Furthermore, each component is uniquely identified by a traceable system, e.g. QR code or barcode, which is given to the robot in order to pick the correct component from the production line. The instructions to repair the rejected pieces are saved in the database, assigning proper references to the component. As a consequence, the operator repairs the rejected components in a reduced amount of time. Otherwise, automatic machines designed for repairing are instructed with information to complete the repairing process. The information and suggestions are provided from the results given by the AI algorithms: the AI analyses the occurrence of non-compliant boards and supports the pick and place process in a synchronized way. A schematic workflow of the intelligent system is illustrated in Figure 7.17. The facilities of the production line are:

- Smart 3D laser profiling.
- CCD high resolution camera.
- 360° image vision system.
- Anthropomorphic robot.
- Picking and place arm for robot.
- Hardware and software connected to the system.
- AI algorithms predicting defects.
- Traceability software tool.

The architecture in Figure 7.17 can be applied for the reconstruction of 3D radiometric images for the quality check of other products characterized by temperature treatment. A complete image vision system is characterized by the fusion of the whole information extracted to the image processing of the classical image and of the radiometric one. Data fusion, together with the use of AI and big data fusion, represents the best approach to upgrade industry tools for verifying the conformity of products.

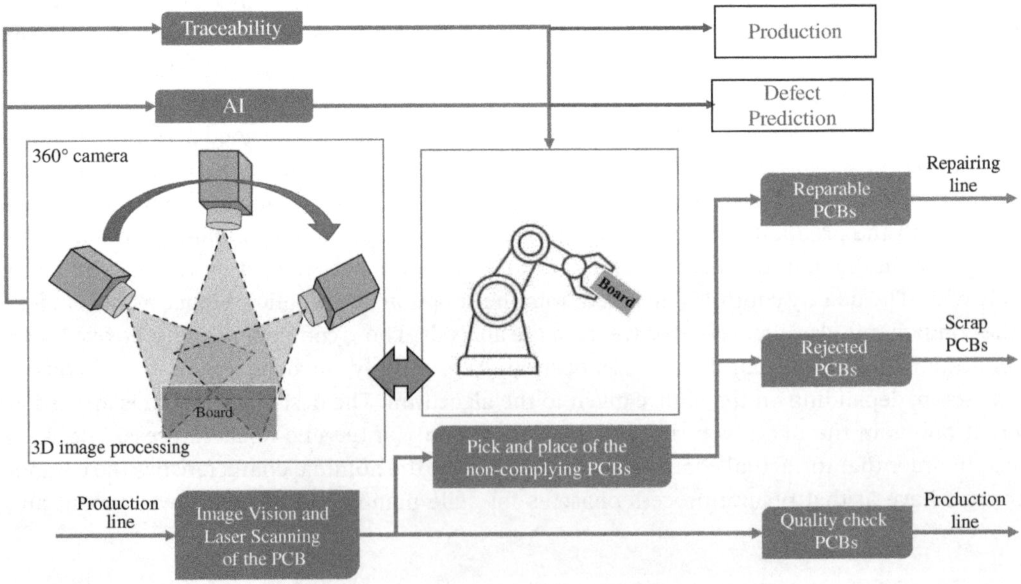

Figure 7.17 Architecture of detection system integrating AI and 3D technologies.

7.2.4 Augmented Reality Systems and Artificial Neural Networks: Image Vision Supporting Production Processes

High-resolution AR systems are useful to support production processes and to optimize production layouts and quality. Figure 7.18 shows an architecture merging AI and AR facilities. The AR visors are adopted to control production together with sensors allocated in the production line. The images acquired by the visors are processed in the cloud by AI algorithms detecting production anomalies and recognizing defects. All data are stored into a big data system controlled and managed by external specialists. All processed data are used to optimize production layouts and quality. The images detected during the production process are adopted for the training model of the self-learning AI algorithms.

The production models are designed and uploaded in the AR device. The augmented vision superimposes the model with the real environment suggesting the operation to be done. The information about quality check is stored and processed by a big data system optimizing the AI models classifying algorithms about automatic defect detection. The real issue is to use graphics processing units (GPUs) to configure the visor in order to locally process all data about AI: in this way the classification algorithms provide a "quasi" real-time detection and identification of product failures simultaneously suggesting the operations to correct the production. A possible upgrade of the technology is to adopt the same AR hardware technology to optimize the training dataset of the AI classification algorithms. In this way a self-adaptive recognition process can be enabled directly by means of workers viewing through AR visors. A typical problem of AI adoption is the creation of a stable and massive training dataset. The use of AR technology is a solution for acquiring the images, selected by the operator, to improve the training model. Different operators can detect images by the use of other AR visors, thus contributing to the data fusion. Data fusion includes the following data typology which can be visualized on the visor screen:

- 3D CAD files used for rendering vision reconstructing a virtual environment on the real one.
- Information about working procedures to follow (picking, assembling, etc.).
- Machine data.
- AI defect detection results.

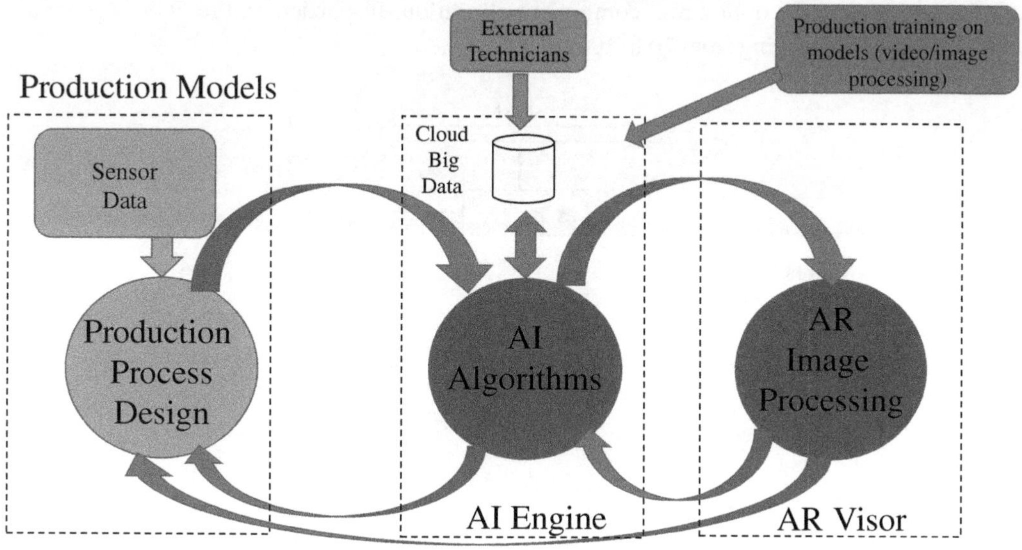

Figure 7.18 AR architecture improving production processes.

- intra-logistics information.
- Information about production traceability (integration with traceability technology).
- Measurements by image segmentation algorithms (reading on the image of a scalebar converting the pixels into a dimension to measure).

The external specialists, by visualizing all data, are able to suggest corrections and machine settings, thus interactively driving the workers also using voice and chat functions.

7.2.5 Infrared Thermography Circuit Design and Automated System

Different thermal cameras are suitable to control the production quality, concerning different production processes mainly involving the product heating. Figure 7.19 shows an example of a monitoring circuit preliminary design, suitable for the control of two production lines. The monitoring system is composed of the following elements:

- Different thermal GigE cameras or radiometric thermal sensors for *in-line* detection of the temperatures, and providing thermal maps of the temperature distribution; the thermal cameras must be distributed along the two production lines.
- Each production line is interconnected to an industrial mini personal computer (Mini PC) behaving as the central unit for the acquisition, registration, processing, and possible transmission of the radiometric map or other data; the outputs of the Mini PC are thermal maps, and pre-processed thermal images used for features extraction; pre-stored thermograms frames and raw images.
- A switch for connecting the Mini PC to the LAN network using Ethernet protocol, and for remote control and automatic data transmission.

USB technology has been adopted for the connections between the Mini PC and the thermal imaging camera as for webcam systems. In addition, to make the application to be implemented on the central unit stable and reliable, Python is preferred as the programming language. The acquisition of radiometric images involves the use of thermal sensors to be mounted on the top of the production lines and having a sufficient frame rate compared with the conveyor belt velocity. The distance between the measuring device and the object to be measured determines the model of the thermal imager, its optics, and component allocation. In particular, the FOV is generally checked for distances ranging from 70 to 100 cm.

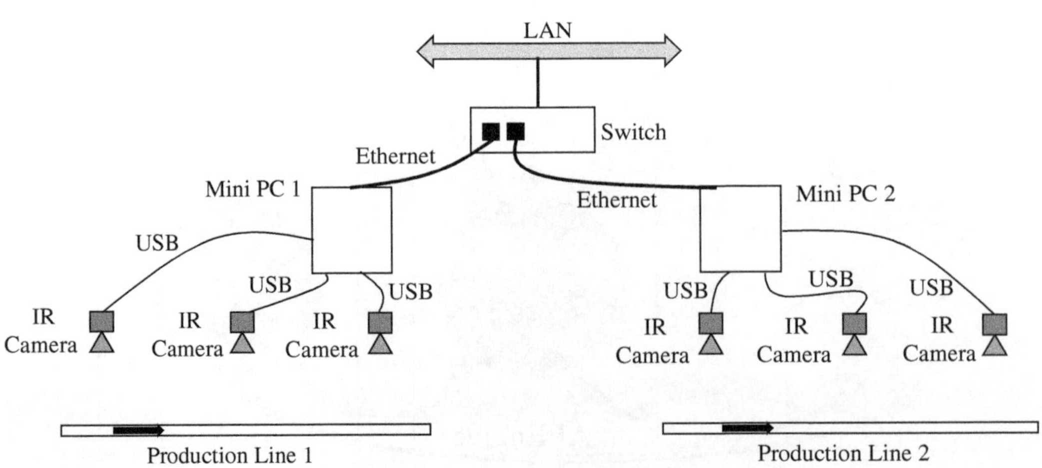

Figure 7.19 Infrastructure integrating infrared camera circuits.

Concerning automated systems, IR imaging supports the actuation for example by automatically discarding the defective pieces by ensuring the production efficiency. Table 7.3 lists some critical aspects concerning thermal hardware and software implementation.

The thermal camera is also used for machine predictive maintenance, including the monitoring of the conveyor belt rollers. The example in Figure 7.20 shows a thermal image proving that each roller is subjected to a different mechanical stress. The continuous check of the rollers could predict possible conveyor belt movement stop conditions. The 3D image processing can further improve the image processing by enhancing the defect reading. As shown in Figure 7.21a, a 2D radiometric image indicates a possible defect on the conveyor belt. The "intensity" of this defect is highlighted by 3D image processing in Figure 7.21b, where it is possible to better observe high radiometric pixel values compared with the pixel intensities of the surrounding regions.

Table 7.3 Critical aspects of infrared thermography implementation in industrial environments.

Critical aspects	Descriptions and mitigations
Interconnection between software and hardware	A mechatronic board is needed to improve control and implementation of the actuation system. The AI engine, detecting defects, must be correctly interconnected in order to adapt the operating conditions of the machine in real time. The best synchronization between software and hardware provides the best efficiency of the production machine based on the operating phases
Real time adaptation of production speed based on thermographic detections	The production speed may change according to the data collected by the thermal camera during the production phases. The AI algorithm correctly foresees failures in order to guarantee optimal production thus eliminating possible defects
Self-adaptive production based on the processing of historical production data	The historical data provide long-term information about possible failures (construction of the training model). The correct application of the predictive maintenance based on AI historical data processing reduces these production risks
Construction of a correct training dataset for the mechatronic board model to be implemented	The implementation and feedback control of the mechatronic board is performed in AI models, by means of a training dataset constructed considering the correct operations of the production lines. The setting of the parameters must be planned considering different working conditions in order to formulate a correct training dataset. Self-learning AI algorithms may optimize control and implementation mechanisms, but errors or incorrect data in the training dataset can be crucial for the production efficiency
Self-adaptive system	A good self-adaptation system can be performed by simultaneously analyzing several attributes (temperature, machine energy consumption, vibrations, etc.). For best analysis, multi-attribute AI data processing should be applied to a set of unrelated attributes (no errors due to redundancies). A low computational cost is required in order to provide a real time self-adaptation. For this, cloud computing can be considered (the risk in this case is the failure of online computing)
Correct choice of hardware and mechanical prototyping components	The electronic boards must check and move the mechanical components according to the designed spaces and specific functions. Compatibility between hardware, software and mechanical parts guarantee the quality of production. Incompatibilities may lead to defects and poor precision
Reliable real-time image vision techniques	The computational cost of image processing must follow the production speed. The adoption of a low-performance processor is inappropriate for real-time data processing. Integrated GPU boards are a possible solution for a decrease in the computational costs addressing the calculation in real time

Figure 7.20 Thermal image of conveyor belt rollers showing areas of anomalous heating.

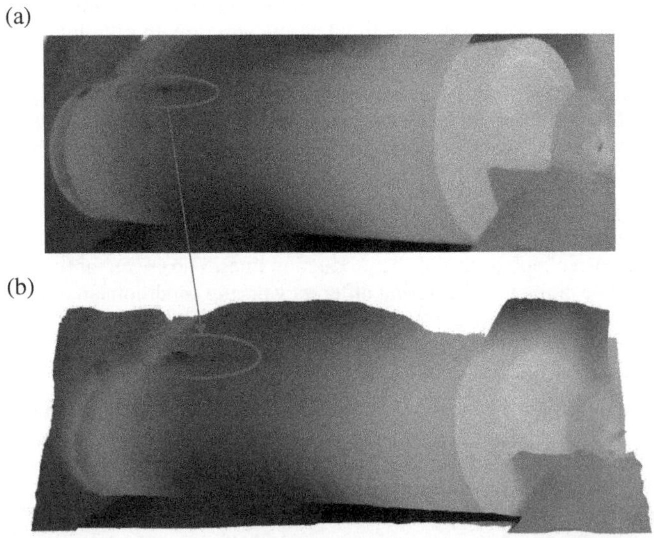

Figure 7.21 (a) Thermal image of conveyor belt rollers in gray scalebar. (b) 3D image reconstruction enhancing a carpet defect.

LSTM classification algorithms are suitable to classify the defect by means of a training dataset construction by considering a single frame rate (comparison between different lines of pixel intensities) or different images.

7.3 Image Segmentation and Image Clustering

Image segmentation and image clustering are especially important to detect defects. For this purpose, it is important to understand how a smart camera automatically detects images to be processed. In this section, examples are proposed facilitating the comprehension of firmware implementations. The examples are addressed on the image segmentation code setting.

7.3.1 Electronic and Firmware for In-Line Monitoring Systems: Camera Connection

In-line monitoring systems require cameras with a specific frame rate following the production velocity. In smart systems or decentralized ones, the camera is linked to microprocessors and processors. The camera systems linked to a processor unit are easily managed by configuring a Python script, for example by using OpenCV libraries. A typical firmware pseudocode useful for the USB *in-line* monitoring system, checking the correct installation of a camera or a thermal camera, is as follows:

```
    1. Setting image importing process and capture function;
    2. Storing state for capture from Pure Thermal Camera 1 or Smart
Camera with OpenCV;
    3. Capture from the recognized camera;
    4. Run loop for image capturing from IR sensor.
```

By considering radiometric measurements, the firmware begins with the definition of a class dedicated to connecting and viewing the video from a thermal imager. Two methods are implemented. The first class enables the connection of the thermal imager. This connection occurs first by capturing the video of the device *i*, and then checking whether this connection is open. The second method basically involves opening a window, reading the single frame, and viewing the image. The time, expressed in milliseconds, temporizes the next frame capture. A microprocessor is used as a central unit, connecting the different thermal cameras or simple cameras of the production line to control.

7.3.2 Image Segmentation and Clustering Techniques: Automated In-Line Monitoring Systems

Using the watershed approach and ImageJ libraries, the image segmentation can be implemented by means of two main classes:

- *Main.java*
- *Analyzer.java*

In the first class the main commands are added for acquiring the image from the predefined path and for managing the image. In the second class, however, all the commands are added related to image analysis and pre-processing, from pre-processing to analysis of details.

In particular, with the *Analyzer.java* class the image is converted into an 8 bit (black and white) image, so that the ImageJ tool applies the threshold for the recognition of details. Subsequently, each identified detail is checked within the image and analyzed by detecting the size of the segmented areas. The two images, the original one and images with extracted areas, are saved as outputs together with an Excel file containing the complete table of details. Each analyzed image is saved into a specific folder based on the timestamp to distinguish the various frames taken into consideration.

In image processing, the K-means clustering approach is applied directly to the image. The K-means clustering processes an input image and segments based on pixel clusters. The main parameters to set are:

- Cluster center tolerance: at each iteration the cluster centroid is updated; if cluster centers are under a tolerance value, it is fixed and the algorithm converges.
- Interpretation of the image as 3D or 2D.
- Enabling randomization seed of the initialization (pseudorandom values).
- Clustering number optimization and printing results.

A way to execute the K-means analysis is the use of the Python programming language and some open-source libraries that are commonly used, such as Pandas and Tensorflow. All libraries are imported first by the following commands:

```
import numpy as np            # For data management
import pandas as pd           # For data management
import matplotlib.pyplot as plt  # For data visualization
```

Then, the script launching the training on a pixel dataset is executed:

```
from sklearn import datasets     # To import the sample dataset
from sklearn.preprocessing import StandardScaler  # To transform
the dataset
from sklearn.cluster import KMeans    # To instantiate, train and
use model
from sklearn import metrics           # For Model Evaluation
from sklearn import cluster, datasets
import pandas.util.testing as tm
```

7.3.3 Circuit Timing In-Line Monitoring and Data Storage Systems

A multithread approach is typically used for sequential image processing as for *in-line* monitoring where conveyor belts transporting products are always in movement, or for the *in-line* check of electronic boards. Concerning circuit quality monitoring, an example of a pseudocode, highlighting the multithread processing, is as follows:

```
  1. Create the frame;
  2. Set of the box dimension of the java window (windows detecting
the object to process);
  3. Set of the box dimension of the camera window;
  4. Selection of the USB output (webcam/camera selection);
  5.Thread webcam aperture setting (streaming camera);
  6. Thread starting;
  7. Thread shots a photo every x seconds (automatic detection
process: thread management);
  8. Threads stop condition.
```

After the conclusion of the thread acquisition process, the image frame is stored automatically into a database system, or in a buffer memory directly linked to an AI engine estimating defects. The defect estimation is executed on an external server or locally by means of a high-performance smart processor.

A possible future application of quantum computing is the possibility to locally process the images in real time. This aspect is important especially for production processes characterized by very high production velocity. In any case, an important aspect to analyze is the matching between the firmware timing and hardware performance, and also the computational time of the quantum calculators. This gap is crucial for an effective integration of quantum computing technologies in *in-line* production stages. A possibility in Industry 5.0 is to adopt quantum computing for 3D images or laser scanner images containing a lot of data point measurements (data point cloud), thus increasing the accuracy of the defect analysis.

7.3.4 Image Segmentation in Product Quality Monitoring: Snake Contour Approach

Image segmentation is the process of partitioning an image into regions that are homogeneous for one or more features. In an image, each pixel is assigned to one of the different classes just defined. The segmentation process ends when the objects or regions of interest have been identified, according to the information that needs to be extracted from the image. The goal of segmentation is to simplify the image into something that is more meaningful and easier to analyze. Image segmentation is generally used to identify objects and boundaries in images, such as lines and curves. During the segmentation process, each pixel of an image is labeled so that pixels with the same label share certain features. The result of the image segmentation is both a set of segments covering the image, and a set of outlines extracted from the image itself. Each pixel that belongs to the same region is similar with regards to a characteristic property of interest, such as color, intensity, or texture. Adjacent regions can be significantly different with respect to any of these characteristics. Image segmentation techniques by means of an active contour snake approach [12] represent a way to enhance image feature extraction. This approach is dynamic: snakes are active contours converging during the time on shape edges embedding a cluster of pixels having a minimum energy (energy defined by the grayscale pixel intensity). The snake contour is defined by the function $c(s) = (x(s), y(s))$. The functional energy associated with the snake is defined as:

$$E_{\text{snake}} = \int_0^t E_{\text{snake}}(c(s))ds = \int_0^t \left[E_{\text{int}}(c(s)) + E_{\text{image}}(c(s)) + E_{\text{con}}(c(s))\right]ds \tag{7.11}$$

where E_{int} is the internal energy of the spline due to the bending enclosing contour to extract, E_{image} is the image forces, and E_{con} is the constraint forces acting on the features to extract. Figure 7.22 shows a theoretical example of an active contour snake application.

E_{int} represents the internal energy deformation of the snake, responsible for its extensibility and flexibility, and is defined as:

$$E_{\text{int}}(c(s)) = \alpha(s)\left|\frac{\partial c(s)}{\partial s}\right|^2 + \beta(s)\left|\frac{\partial^2 c(s)}{\partial s^2}\right|^2 \tag{7.12}$$

where the first term of the equation is the internal tension forces characterizing the elastic behavior of the snake, while the second term expresses rigidity and resistance to contour bending.

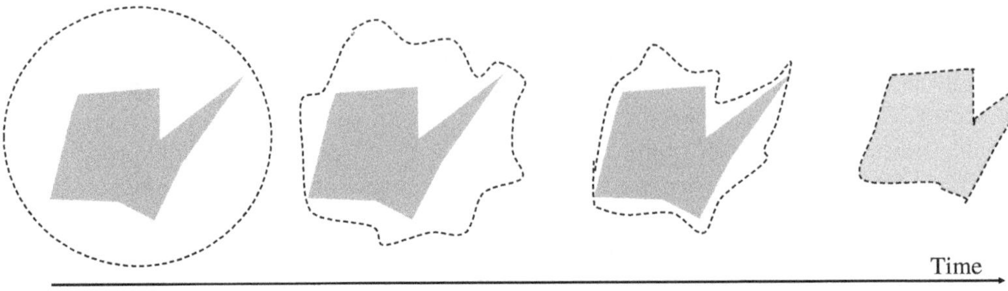

Figure 7.22 Ideal dynamic extraction of the active contour snake method.

The two terms are controlled, respectively, by the terms $\alpha(s)$ and $\beta(s)$, able to tune the lengthening and flexing of the curve enclosing the object to detect. The external energy $E_{\text{ext}} = E_{\text{image}} + E_{\text{conv}}$ is an external potential function $P(x, y)$, dependent on the image and by external constraints, taking into account the snake forces toward particular gradient characteristics: corresponding to a part of the image having high gradient values, E_{ext} is defined by the following potential:

$$P(x,y) = -\lambda \left| \nabla \left[G_\sigma(x,y) * I(x,y) \right] \right| \tag{7.13}$$

where the gradient is calculated on the convolution of the image $I(x, y)$ with a Gaussian smoothing filter $G_\sigma(x, y)$, whose variance σ controls the spatial extent of the minimum local energy, and the amplitude of the potential $P(x, y)$ is regulated by the coefficient λ. Therefore, snake segmentation follows the energy minimization principle. From the calculation of the variations, a contour (s) which minimizes the function $E_{\text{snake}}(s)$ satisfies the following Euler–Lagrange equation:

$$-\frac{\partial}{\partial s}\left(\alpha(s)\frac{\partial c}{\partial s}\right) + \frac{\partial^2}{\partial s^2}\left(\beta(s)\frac{\partial^2 c}{\partial s^2}\right) + \nabla E_{\text{ext}}\left(c(s)\right) = 0 \tag{7.14}$$

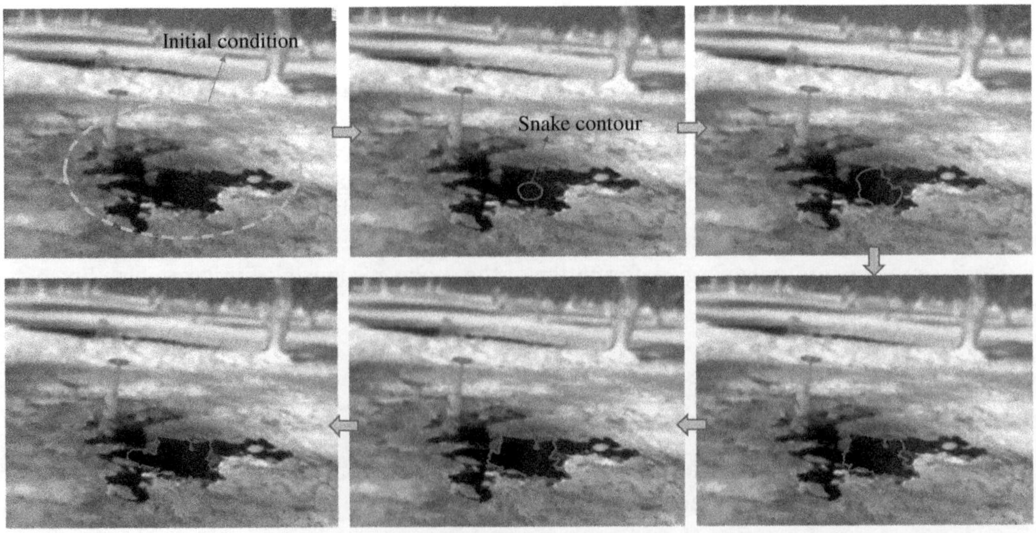

Figure 7.23 Snake contour plot applied to a radiometric image: dynamic convergence of the snake contour applied to a water leakage plume.

which expresses the balance between internal (snake) and external forces (of the image and external constraints). Generally, this partial differential equation is solved using numerical algorithms, making the snake evolve in different steps. The main problem with the snake approach is that the region to analyze must be near the object of interest (initialization condition of the algorithm). In order to check the quality of a product, the snake contour algorithm is applied with the following steps oriented toward checking production quality:

- The object is detected by a camera.
- The image area containing the object is centred (the object is in the center).
- The snake contour algorithm is initialized by choosing the initial point (starting point of the algorithm to execute).
- After the snake contour convergence, the image is filtered in order to extract the geometric shape (geometrical feature).
- The geometrical feature is compared with other profiles in order to check possible defects (as logic difference been the two compared shapes) or to prove the product quality.
- All data are stored into a database system for statistical analyses.

Snake contour segmentation applied to a water leakage plume is shown in Figure 7.23 indicating algorithm convergence for the initial time steps. The snake contour algorithm is applied in Figure 7.24, directly on the image of a processed piece: the dynamic convergence of the processes contour, and the setting of the gradient threshold and the initial condition permits to extract the contour of a hole of a processed metallic piece.

7.3.5 Advanced Image Clustering: K-Means Applied to Radiometric Images

Another approach to follow for advanced image processing is to apply the K-means algorithm directly to the radiometric image [13]. The clustered image of radiometric processing is provided

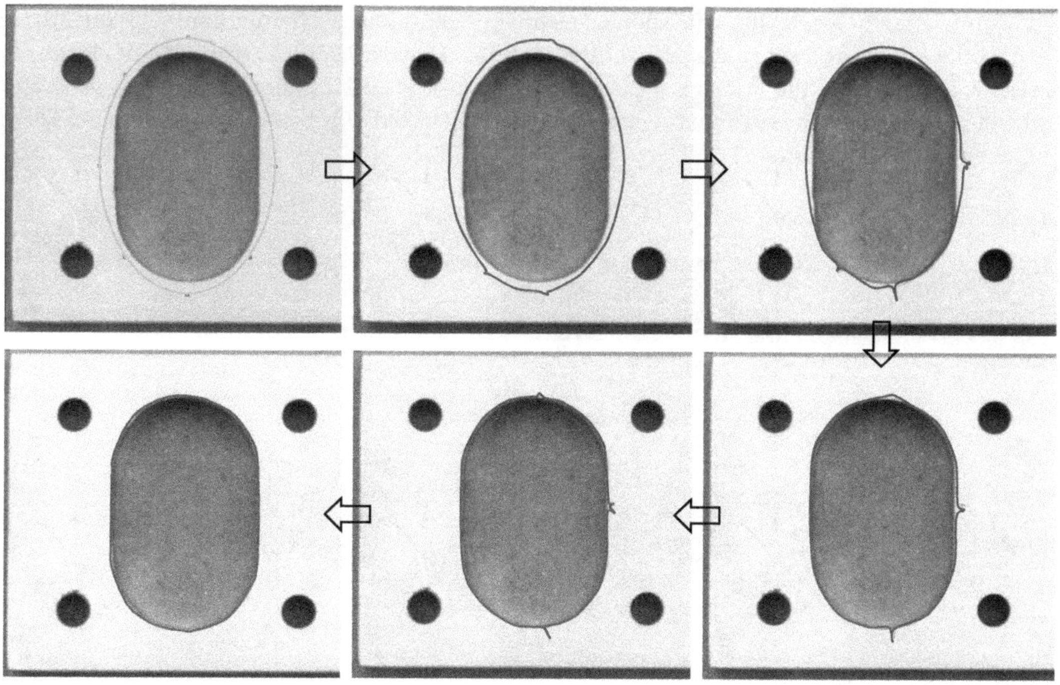

Figure 7.24 Snake contour plot applied to the image of a processed metallic piece (holes): dynamic convergence of the snake contour on the main hole.

by the following considerations. Consider the processed radiometric image as $I: \Omega \rightarrow \mathbb{R}$, $\Omega \subset \mathbb{R}^2$. A contour C of the image I can be represented by a Lipschitz function $\phi: \Omega \rightarrow \mathbb{R}$, known as a level set function, which assumes negative values in the internal region of the contour C and positive values in the region external to C:

$$\phi(x) = \begin{cases} <0 & \text{for } x \in \text{in } C \\ >0 & \text{for } x \in \text{out } C \end{cases} \tag{7.15}$$

The choice of the signs of the function ϕ is arbitrary. This function is characterized by opposite signs inside and outside the contour, so that this it is identified by the points where the level set function is equal to zero; the contour C is defined as:

$$C = \{x \in \Omega | \phi(x) = 0\} \tag{7.16}$$

The K-means algorithm can be applied before the snake contour algorithm to define the contour to extract better (clusters of pixels having the same grayscale intensities), or it is applied directly on the radiometric image [14]. The clustering algorithm follows these steps:

- Initialize cluster centroids $o_1, o_2, o_3, \ldots, o_k \in \mathbb{R}^n$ randomly.
- Repeat until convergence for every i and j set as follows:

$$\begin{aligned} c^{(i)} &= \arg\min_j \left\| x^{(i)} - o_j \right\|^2 \\ o_j &= \frac{\sum_{i=1}^{m} 1\{c^{(i)} = j\} x^{(i)}}{\sum_{i=1}^{m} 1\{c^{(i)} = j\}} \end{aligned} \tag{7.17}$$

where $x^{(i)}$ is a value of the training dataset and represents the image pixel having a specific intensity (radiometric value). The K-means application of the thermogram in Figure 7.23 is illustrated in Figure 7.25a: for low K values it is possible to highlight the leakage plume, and, subsequently, to extract the geometrical shape feature by filtering colors (Figure 7.25b). The number of clusters to select is a function of the number of pixel intensities of the entire radiometric image: generally, when the object with the surrounding environment is detected, it is preferable not to increase the number of clusters as much.

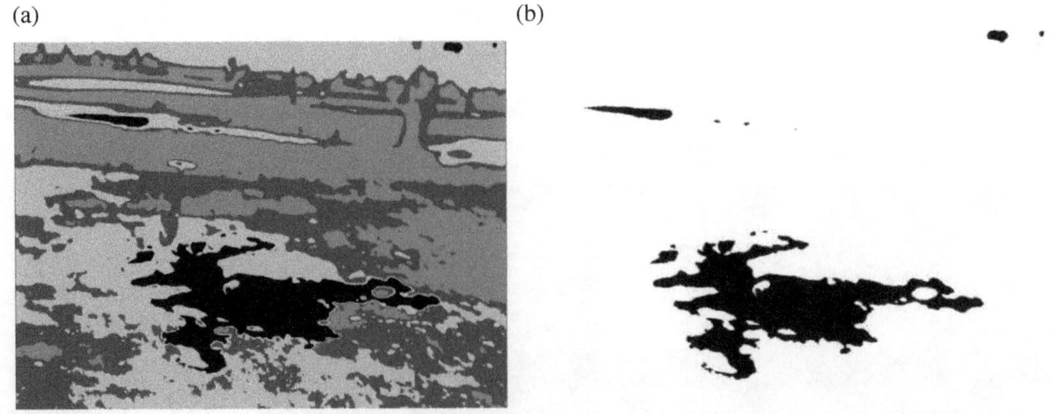

Figure 7.25 (a) K-means image processing calculated for a cluster number of 4. (b) Filtered image extracting the cluster enhanced in black (water leakage plume).

7.4 Image Segmentation for Food Defect Detection

Concerning pasta production, image processing tools are adopted for the detection of different defect classes, such as white or dark stains as spots distributed on the pasta surface, misshapen parts as deformed shapes, sticky parts, and cracks. The white spots are mainly due to the hydration process or the irregular granulometry of flour. The watershed approach extracts the contours to calculate the white parts measuring the areas in mm^2 enclosed by the calculated contours (Figure 7.26). The following summarizes the data processing steps for the detection of irregular areas (over a defined threshold):

- Filtering pre-image processing by setting properly the red, green, and blue (RGB) color threshold (this setting procedure is a function of different parameters such as luminosity, RGB color intensities, and image resolution).
- Shape extraction (approximation of elliptical shapes embedding detected white regions).
- Statistical evaluation of the defect populations.

The threshold limit in Figure 7.26 indicates the quality limit: all the estimated areas above the threshold are considered as an outlier (checked defect), and the percentage of defects define the pasta quality index. The quality index is also predicted by applying the ANN algorithm [15, 16] able to provide alerts useful to readjust production machine parameters as for dynamic self-adaptive processes.

Concerning the treatment and disinfection processes of vegetables, different approaches are adopted such as ozone treatment, electrolyzed water washing, and hydrogen peroxide (H_2O_2) application [17]. An alternative is to combine ozone simultaneously with hydrogen peroxide (H_2O_2/O_3). The pretreatment and the treatment processes could generate product defects. In order to check evident and visible defects, image filtering processing is generally applied setting a color threshold brightness and saturation. Figure 7.27a shows an example of a 3D reconstruction of vegetables enhancing defects and Figure 7.27b shows the extracted defect color features by setting

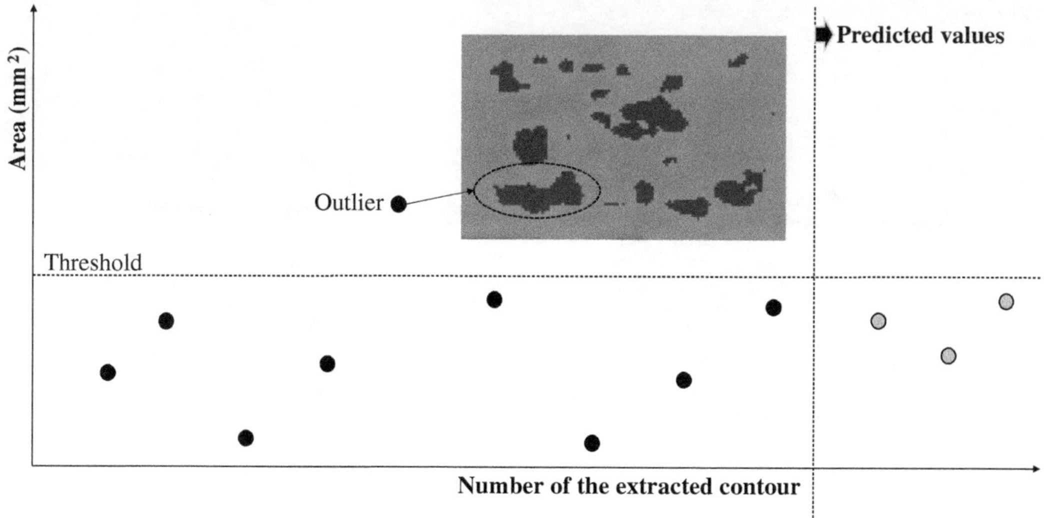

Figure 7.26 Theoretical areas of extracted contours versus the defect number. Inset: defect detected in pasta product.

the hue saturation brightness (HSB), filtering and adopting a 3D reconstruction. The whole image processing in the specific case study is summarized by the following steps:

- Image brightness and saturation filtering.
- Color filtering (HSB filtering).
- Contour extraction.
- Anomalous area calculation (damaged areas).
- Calculation of the index of area percentage of product damaged compared with total area of the detected object (index of quality).

Another application of image processing and segmentation is the real-time monitoring of fresh foods. For monitoring meat, a methodological approach to follow is based on checking the distributed temperatures of the whole product during the time. A fresh food cannot be placed for more time outside the refrigerator. For this purpose, IRT combined with K-means analysis highlighting the colder and hotter regions provides more information about the probability of product damage [18]. Figure 7.28a shows a radiometric image of a meat product and Figure 7.28b shows the applied K-means image processing enhancing different temperature areas.

The methodological approach to follow for meat monitoring is based on the following two main steps [18]:

- Measurement by IR thermal camera of the average minimum temperature of the whole product to control.
- Verification of the minimum temperature trend versus time.

The average minimum temperature estimated during the time indicates the product deterioration risk. In Figure 7.29 the alerting threshold is defined as the minimum temperature value beyond which the food begins to deteriorate. The food damage risk increases drastically versus the time defining a high risk region, which is a function of the fresh food to control. In food production lines, the temperature is also checked for the environment which potentially influences product deterioration. Other AI algorithms based on food defect classification can facilitate risk analysis. A fully automatic alerting system detecting damage is appropriate to increase product quality.

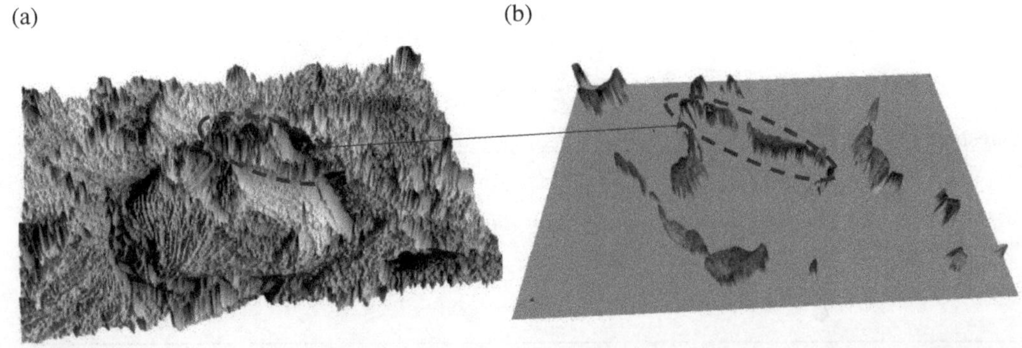

Figure 7.27 (a) 3D image reconstruction of vegetables with defects. (b) 3D image processing filtering defects.

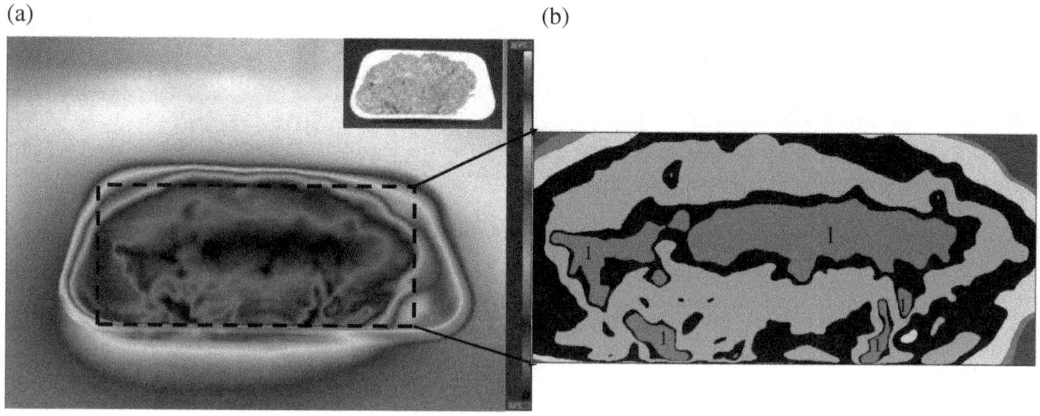

Figure 7.28 (a) Thermogram of a meat product (inset: original image of the product). (b) K-means (5 clusters) image clustering applied on the radiometric image (zone 1 indicates colder areas).

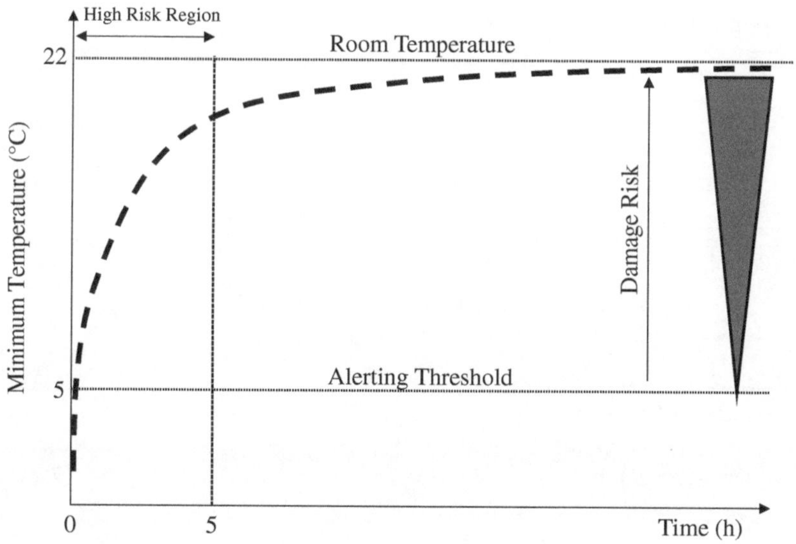

Figure 7.29 Theoretical trend of minimum temperature indicating risk regions.

7.5 Random Forest Pixel Classification

Image segmentation and pixel classification are research issues supporting difficult target identification. An algorithm suitable for this approach is the fast random forest (FRFo) algorithm implementing the random forest (RFo) classification technique [19]. Random forest is a machine learning (ML) technique executing an ensemble of decision trees to classify in the specific case the pixels of an image, by training the model with pixel features and by defining labeled cluster data as clusters of pixels. Decision trees are initially structured from randomly selected clusters of pixels, and the decision splits randomly the features at each node in the tree. The final classification is estimated as the average classification of all constructed trees. The RFo algorithm exhibits good accuracy, relatively low computational cost, and is based on the multiclass segmentation approach [20, 21]. A tool implanting this approach is discussed in [22, 23]. The image processing approach shown in Figure 7.30 is explained by the following steps: (i) the input image is processed to extract different image features constituting a cluster of pixels of the training dataset; (ii) the regions to discriminate are marked in the input image in order to define the classes to check in the same image; (iii) the RFo classifier is executed finding the requested classes in the same image. The training is performed by applying different pixel processing operations such as:

- Gaussian Blur providing the blurred version of the image.
- Sobel filtering calculating for each pixel the approximation of the gradient of the image intensity.
- Hessian filtering able to estimate for each pixel the Hessian matrix defined as

$$H = \begin{pmatrix} a & b \\ c & d \end{pmatrix} \tag{7.18}$$

and the related matrix parameters module, trace, determinant (Det), first eigenvalue Fe), second eigenvalue (Se), orientation (angle orientation for which the second derivative is maximal), gamma-normalized square eigenvalue difference (Gamma), and square of the gamma-normalized eigenvalue difference (Square gamma):

$$\text{Module} = \sqrt{a^2 + bc + d^2} \tag{7.19}$$

$$\text{Trace} = a + d \tag{7.20}$$

$$\text{Det} = \sqrt{a^2 + bc + d^2} \tag{7.21}$$

$$\text{Fe} = \frac{a+d}{2} + \sqrt{\frac{4b^2 + (a-d)^2}{2}} \tag{7.22}$$

$$\text{Se} = \frac{a+d}{2} - \sqrt{\frac{4b^2 + (a-d)^2}{2}} \tag{7.23}$$

$$\text{Orientation} = \frac{1}{2}\arccos\left(4b^2 + (a-d)^2\right) \tag{7.24}$$

$$\text{Gamma} = t^4 (a-d)^2 \left((a-d)^2 + 4b\right)^2 \quad \text{with } t = 1^{3/4} \tag{7.25}$$

$$\text{Square gamma} = t^2 \left((a-d)^2 + 4b\right)^2 \quad \text{with } t = 1^{3/4} \tag{7.26}$$

- Difference of gaussians.
- Projection processing (by rotation of the original kernel).
- Pixel value parameters (mean, variance, median, minimum, and maximum).
- Anisotropic diffusion filtering (reducing image noise trying to preserve sharp edges).
- Bilateral filtering (filtering preserving edges).
- Lipschitz filtering (noise-reduction preserving edges).
- Gabor filtering (linear filtering used for texture analysis).
- Derivative filtering calculating high order derivatives of the input image.
- Structure filtering calculating the eigenvalues of all the elements constituting the input image.
- Neighbors (shifting of the image in different directions by a certain number of pixels).

For colored images it is preferable to train the model by the following operations:

- Gaussian Blur performing *n* individual 3D convolutions with Gaussian kernel.
- Hessian filtering.
- Derivative filtering.
- Laplacian filtering.
- Structure filtering.
- Minimum, maximum, mean, variance, and median.

The pixel features identification is applied for the image in Figure 7.31a representing a micrometric rectangular gold aperture: a mask with rectangular aperture is applied on a chitosan surface containing a gold precursor, to generate by laser texturing a thin gold layer on the surface. The gold surfaces of the first prototypes are usually irregular and inhomogeneous, and ML is applied to optimize the quality check. By defining the three classes marked by ellipses in Figure 7.31a, it is possible to distinguish by RFo algorithm the main surface morphological parts by extracting the features of the regions with average gold content, with higher gold thickness and with only chitosan. Figure 7.31b shows these estimated classes. More classes are defined in order to classify more irregularities on the same images. Another application of the RFo algorithm is the identification of the same object characteristics contained in the same image. In the case of nanotechnology image processing a typical problem is to estimate the population of nanoparticles having the same size diameter. The first and unoptimized nanoparticle syntheses provide assembled nanoparticles difficult to classify from the image by calculating the diameter sizes. Some algorithms are applied to automatically calculate the different areas [24]. The example in Figure 7.32a is a typical case to automatically distinguish different classes of nanoparticle sizes by analyzing the same TEM image. For this case, three main classes are defined by choosing the particles enhanced as circles in Figure 7.32b, related to the three classes to estimate. Figure 7.32c and d illustrates a better extraction of the class indicated in the figure by the number 1 by distinguishing these particles in agglomerates.

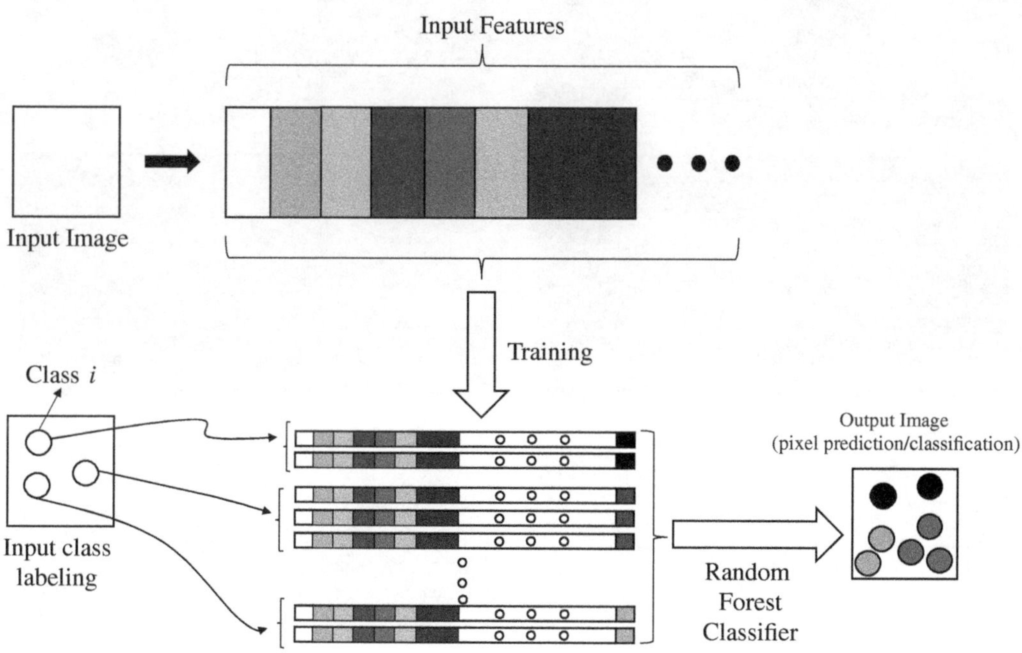

Figure 7.30 Image processing procedure based on pixel features training and RFo classifier.

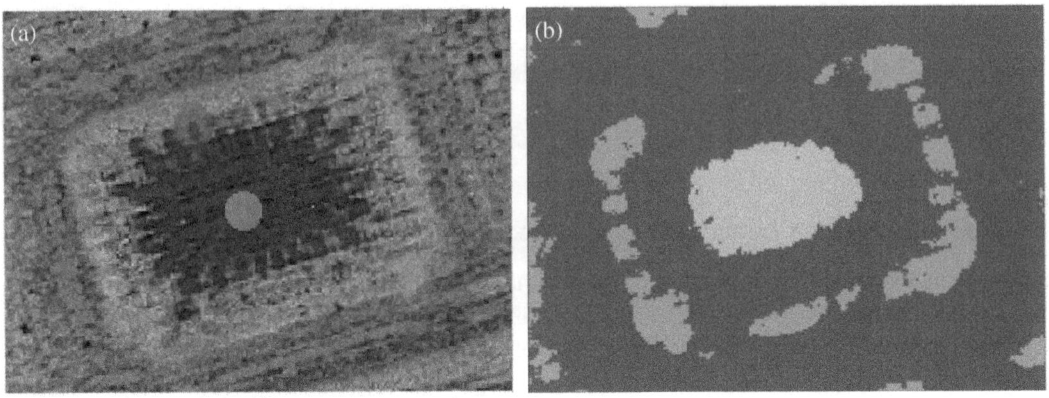

Figure 7.31 (a) Original image of a micrometric aperture with the definition of three classes marked by ellipses. (b) Pixel classification using the RFo classification technique (the training features selected are Gaussian blur, Sobel filter, Hessian filter, projection processing, and Difference of Gaussian).

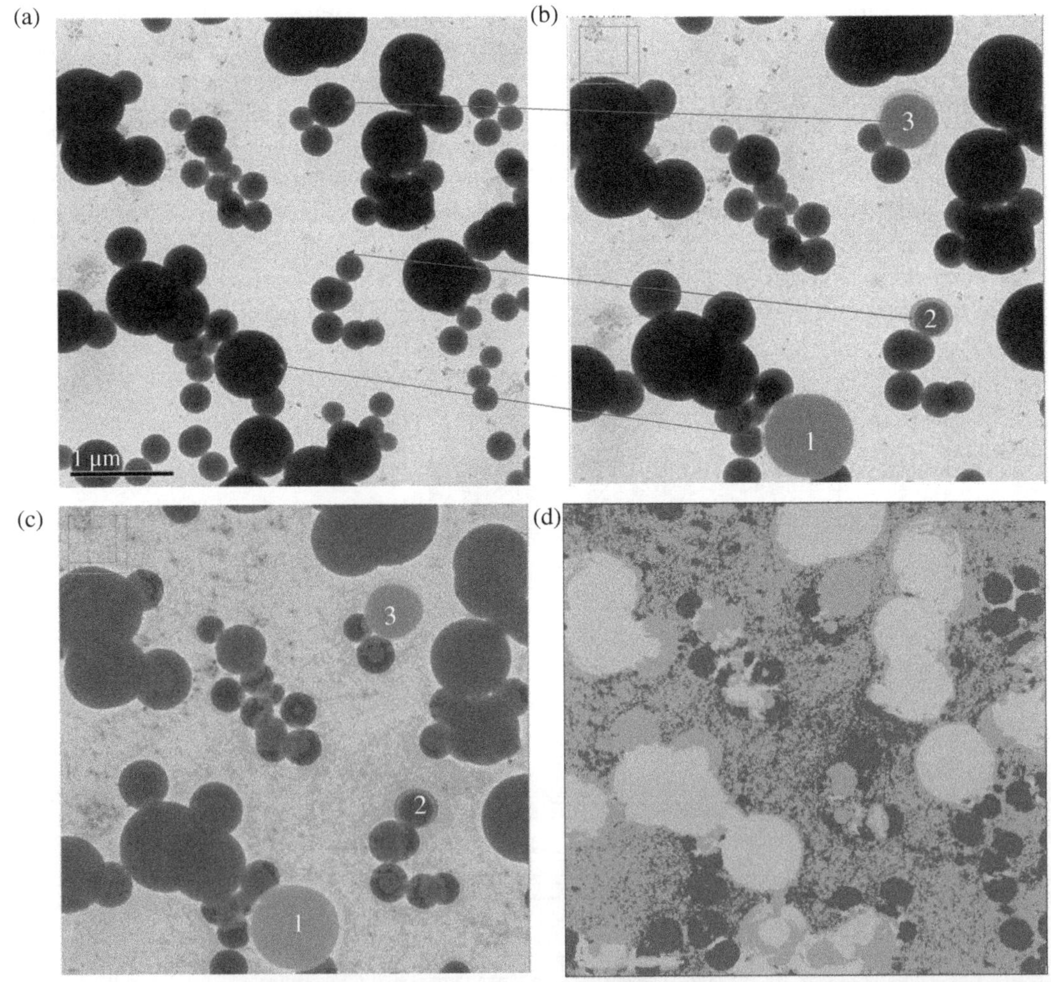

Figure 7.32 (a) TEM image of silica nanoparticles. (b) Class definition identifying the three main diameters of the nanoparticle. (c) Initial classification of the recognized classes. (d) Example of feature extraction.

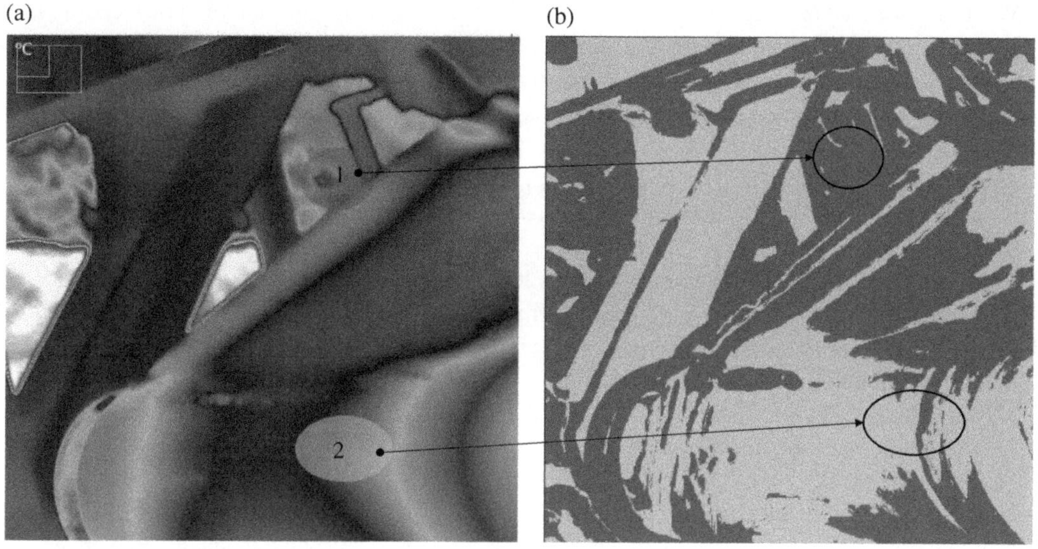

Figure 7.33 (a) Thermogram of a conveyor belt carpet. (b) Feature extraction of two main classes, where class 1 indicates the hottest region related to major stressed regions and class 2 the coldest areas.

The same RFo approach is applied directly on thermograms to find classes of information defined by radiometric pixels: in the case of the radiometric image in Figure 7.33a, it is processed by the RFo approach, by classifying in Figure 7.33b two main classes where class 1 is the hottest region and class 2 is a colder one. In this case the training model is constructed by processing some pixel areas of the same image.

References

1 Massaro, A., Dipierro, G., Cannella, E., and Galiano, A.M. (2020). Comparative analysis among Discrete Fourier Transform, K-Means and artificial neural networks image processing techniques oriented on quality control of assembled tires. *Information* 11 (5): 1–20.

2 Massaro, A., Manfredonia, I., Galiano, A., and Contuzzi, N. (2019). Inline image vision technique for tires Industry 4.0: quality and defect monitoring in tires assembly. In: *Proceedings of IEEE International Workshop on Metrology for Industry 4.0 and IoT*, Naples, Italy (4–6 June 2019). Piscataway, NJ: IEEE.

3 Chan, C.H. and Pang, G.K.H. (1999). Fabric defect detection by Fourier analysis. In: *Proceedings of the 34th IEEE–IAS Annual Meeting Conference Record Industry Applications Society*, Phoenix, USA(3–7 October 1999). Piscataway, NJ: IEEE.

4 Massaro, A., Panarese, A., Dipierro, G. et al. (2020). Infrared thermography and image processing applied on Weldings quality monitoring. In: *Proceedings of IEEE International Workshop on Metrology for Industry 4.0 and IoT*, Rome, Italy (3–5 June 2020). Piscataway, NJ: IEEE.

5 Beucher, S., Rivest, J.F., and Soille, P. (1993). Morphological gradients. *Journal of Electronic Imaging* 2 (4): 326–336.

6 Massaro, A., Panarese, A., Galiano, A. et al. (2020). Image processing segmentation applied on defect estimation in production processes. In: *Proceedings of IEEE International Workshop on Metrology for Industry 4.0 and IoT*, Rome, Italy (3–5 June 2020). Piscataway, NJ: IEEE.

7 Van Laarhoven, P.J.M. and Zijm, W.H.M. (1993). Production preparation and numerical control in PCB assembly. *International Journal of Flexible Manufacturing Systems* 5 (3): 187–207.

8 Tseng, T.L., Jothishankar, M.C., and Wu, T. (2004). Quality control problem in printed circuit board manufacturing – an extended rough set theory approach. *International Journal of Flexible Manufacturing Systems* 23 (1): 56–72.

9 Mar, N.S.S., Yarlagadda, P.K.D.V., and Fookes, C. (2011). Design and development of automatic visual inspection system for PCB manufacturing. *Robotics and Computer-Integrated Manufacturing* 27 (5): 949–962.

10 Chavan, R.R., Chavan, S.A., Dokhe, G.D. et al. (2016). Quality control of PCB using image processing. *International Journal of Computer Applications* 141 (5): 28–32.

11 Edinbarough, I., Balderas, R., and Bose, S. (2005). A vision and robot based on-line inspection monitoring system for electronic manufacturing. *Computers in Industry* 56 (8–9): 986–996.

12 Massaro, A., Lay-Ekuakille, A., Caratelli, D. et al. (2012). Optical performance evaluation of oil spill detection methods: thickness and extent. *IEEE Transactions on Instrumentation and Measurement* 61 (12): 3332–3339.

13 Massaro, A., Galiano, A., Meuli, G., and Massari, S.F. (2018). Overview and application of enabling technologies oriented on energy routing monitoring, on network installation and on predictive maintenance. *International Journal of Artificial Intelligence and Applications (IJAIA)* 9 (2): 1–20.

14 Massaro, A. and Galiano, A. (2020). Image processing and post-data mining processing for security in industrial applications: security in industry. In: *Handbook of Research on Intelligent Data Processing and Information Security Systems* (eds. S.M. Bilan and S.I. Al-Zoubi), 117–146. Hershey, PA: IGI Global.

15 Massaro, A. and Galiano, A. (2020). Re-engineering process in a food factory: an overview of technologies and approaches for the design of pasta production processes. *Production & Manufacturing Research* 8 (1): 80–100.

16 Massaro, A., Manfredonia, I., Galiano, A. et al. (2019). Sensing and quality monitoring facilities designed for pasta industry including traceability, image vision and predictive maintenance. In: *Proceedings of IEEE International Workshop on Metrology for Industry 4.0 and IoT*, Naples, Italy (4–6 June 2019). Piscataway, NJ: IEEE.

17 Massaro, A., Maritati, V., Savino, N. et al. (2020). Blockchain and technologies matching with the case of study of vegetables production. *Information Technology in Industry (ITII)* 9 (2): 1–8.

18 Massaro, A., Panarese, A., and Galiano, A. (2020). Infrared thermography applied on fresh Food monitoring in automated alerting systems. In: *Proceedings of IEEE International Workshop on Metrology for Industry 4.0 and IoT*, Rome, Italy (3–5 June 2020). Piscataway, NJ: IEEE.

19 Breiman, L. (2001). Random forests. *Machine Learning* 45 (1): 5–32.

20 Mahapatra, D. (2014). Analyzing training information from random forests for improved image segmentation. *IEEE Transactions on Image Processing* 23 (4): 1504–1512.

21 Belgiu, M. and Drăguţ, L. (2016). Random forest in remote sensing: a review of applications and future directions. *ISPRS Journal of Photogrammetry and Remote Sensing* 114: 24–31.

22 Arganda-Carreras, I., Kaynig, V., Rueden, C. et al. (2017). Trainable Weka segmentation: a machine learning tool for microscopy pixel classification. *Bioinformatics* 33 (15): 2424–2426.

23 Lormand, C., Zellmer, G.F., Németh, K. et al. (2018). Weka trainable segmentation plugin in ImageJ: a semi-automatic tool applied to crystal size distributions of microlites in volcanic rocks. *Microscopy and Microanalysis* 24 (6): 667–675.

24 Lay-Ekuakille, A., Spano, F., Mvemba, P.K. et al. (2019). Conductivity image characterization of gold nanoparticles based device through atomic force microscopy. In: *Proceedings of IEEE 2018 Nanotechnology for Instrumentation and Measurement (NANOfIM)*, Mexico City, Mexico (7–8 November 2019). Piscataway, NJ: IEEE.

8

Electronic and Reverse Engineering

This chapter provides important solutions about reverse engineering (RE) approaches to adopt in industry for production quality optimization. RE is an important topic for manufacturing industries oriented toward production upgrade and Industry 5.0 facilities. Specifically, some solutions are proposed for industrial applications with useful and innovative aspects of RE improved by automated systems, involving simulations, experimentation, and artificial intelligence (AI). RE approaches are mainly addressed to optimization processes involving rapid prototypes and production. The passage from Industry 4.0 to Industry 5.0 involves the introduction of a self-adaptive re-engineering of the device, by taking into account simulations, measurements, and data processing with the support of AI algorithms. The self-adaptive optimization is provided by the automatisms implemented to produce a specific component or device. To this end, a new aspect for optimized production is automated testing as a new concept of automatic redesign and re-engineering of a product. The discussion is mainly focused on RE in advanced manufacturing describing possible optimization approaches of devices constituted by micrometric structures, thin films, and nanocomposite materials (NMs). Advanced RE approaches are applied in fabrication processes for nanocomposite sensors, for the source connection of antennas, and in general for the realization of electronic boards. Other important aspects of RE methodologies are the measurement approaches for electrical characterization of thin film multilayered devices. Particular attention is paid to the analyses of RE tools as finite element method (FEM) simulators in the design and modeling of micrometric optical devices. RE methodologies are also addressed to microelectromechanical system (MEMS) technological process optimization including limit characterization, layout control, and experimentation. Other proposed examples include nanocomposite optical sensors and related engineered technologies, compact antennas, photonic crystals (PCs), and methodologies to implement electronic circuits. The proposed overview facilitates comprehension of how RE can intervene in electronic and optoelectronic devices. This chapter highlights the possibility to merge FEM techniques into advanced RE processes, In order to improve manufacturing processes, AI and experimentation approaches involve devices with millimetric and micrometric scale dimensions. The goal is to provide an industrial overview with specific case studies of possible RE approaches to help the reader to understand RE principles and the importance of RE in automatic production systems. RE is a concept which is extended to the optimization of materials, technology, measurement approaches, production machine setting, and specifications of the device to fabricate. Rapid prototyping and RE are closely interrelated and contribute together to product industrialization. The upgrade and automatisms of RE methods are topics of interest in the Industry 5.0 scenario.

Electronics in Advanced Research Industries: Industry 4.0 to Industry 5.0 Advances, First Edition. Alessandro Massaro.

8.1 Reverse Engineering Systems and Mechanical Precision

RE is defined as the process of duplicating an object in its features and dimensions through a physical analysis and measurements, obtaining the technical data required for processing. To simplify, RE is the process of creating a three-dimensional (3D) drawing obtained by measuring an object to be redesigned or modified, to determine its dimensions and tolerances when there are no specific data. The definition of tolerance values is a crucial aspect of the RE process: tolerances are assumed by a careful study of the parts and features of the object. RE is suitable:

- When a new physical model of a product is created, but there are no drawings and models.
- To fabricate different prototypes.
- When drawings are created, but the prototypes used in manufacturing have been modified or lost.
- When a new computer-aided design (CAD) system is introduced and all existing product data must be loaded into the new system.
- When an accident has destroyed all existing data.
- For a redesign process, as in the case of new product development.

In some cases, it may be easier and less expensive to extract a design from an existing product and modify it, rather than creating new ones. The models necessary for product development can be numerous and different both in size and in material. In most manufacturing sectors, manual changes are made during the pre-production phase, in order to improve the ergonomics, functionality, esthetics, and final quality of the product or part of it. Many designers still prefer to make changes on concrete models rather than working in a CAD environment. Therefore, the need arises for the designer to retrieve as much information as possible to adapt it to customer needs effectively and quickly, with the possibility of creating customized products. RE tools facilitate surface scanning: the ability to scan with times of up to 100 times less than traditional approaches with a consequent reduction of costs. RE facilitates the use of production systems in any sector, from mechanics to electronics, and from clothing to sports, creating a new approach for the redesign of parts for components that are defective or do not conform to the desired specifications.

Companies that produce foundry molds have the possibility of recovering a parametric 3D file from old models or molds not designed with CAD by obtaining a digital and more contemporary product. The use of scanners and 3D measurement systems allows new applications and models to be created: through the introduction of this new technology and 3D CAD programs, it is possible to generate a file ready to be printed using common 3D printers.

The tools commonly used in RE are:

- Probes verifying the surfaces to identify holes or other discrepancies with respect to the CAD model (touch probes).
- Photogrammetry (any object to be produced in rapid prototyping is detected and reconstructed by photographs).
- Laser scanner.
- Holography.
- Tomography (internal scans and quality control).

Data acquisition is the primary specification of RE processes. There are two main different data acquisition techniques: contact and non-contact. RE data processing mainly involves a 3D CAD reconstruction. The RE phases involve:

- Parametric redesign (optimization of design parameters, analysis of parameter sensitivity and tolerance, and construction of parametric tests).

- Adaptive redesign (search for new effects, new principles, subsystems to adapt, flow force analysis, and construction of adaptive tests).
- Original redesign (construction of the prototype test and application of new original concepts).

Redesign is therefore a fundamental step for CAD modeling of the components and must take into account both the parameterization of the constraint points and the optimization of the design drawing trying to avoid the generation of manufacturing defects. This application field is named reverse engineering for manufacturing (REFM). Data acquisition technologies must be applied following a specific methodology that is a function of the class of components to inspect and redesign. The CAD libraries, virtual machines, adaptive processes, and computer-aided manufacturing (CAM) agents are system specifications that are part of the manufacturing and automation solutions of Industry 4.0. RE procedures become indispensable when the original CAD designs are not available for a component, a component is to be reproduced starting from an existing one, or for component updating or upgrading, thus redesigning a component without any production defects, and optimizing the design and production process. The choice of sensors and methodologies to be applied is also the goal of industry research: the studies of the components to be reconstructed involve different techniques, depending on the available measuring instruments. A technique for measuring objects involves the use of coordinate measuring machines (CMMs): the so-called CMM allows the acquisition of surfaces by points through the use of contact probes. Another technique for measuring an object is the 3D scanning approach using non-contact optical scanners capable of reconstructing any real object in the form of a polygonal digital model and providing a cloud of points as output of the measurements. The most common output formats of 3D data from scanning are STL, OBJ, and PLY. Starting from these formats, RE accomplishes the reconstruction of CAD entities such as curves, surfaces, and solids.

The various procedures and techniques of RE for reconstructing a CAD model from a mesh are:

- RE for auto surfacing (automatic fitting of surfaces suitable for free-form objects).
- RE for entity extraction (mesh processing with functions for extracting mathematical entities from the mesh, where the CAD model can be reconstructed by carrying out a redesign of the drawing).
- Integrated or complete RE (integration of mesh processing tools and mesh modeling by creating curves, surfaces, and solids that interpolate the scan results).

The three main phases of RE are summarized in Table 8.1.

Table 8.1 RE phases.

Phase	Description
Phase 1 (scanning process)	The first phase is related to a scanning strategy producing 3D clouds of measured points, by defining object surfaces to interpolate. The measurements can be performed by contact scanners (contact probes touching object surfaces to detect), and by non-contact scanners by lasers, optical systems, and charge-coupled device sensors
Phase 2 (point processing)	The second phase concerns processes enabling the importation of the measured data, the decrease of the noise by filtering processed data, and the cleaning of insignificant data for final 3D object reconstruction.
Phase 3 (application of the geometric model)	In the final RE phase the 3D computer-aided design model is generated in file formats such as IGES, VDA, STL, DXF, OBJ, VRML, and ISO G Code

3D, three-dimensional.

All the RE approaches are tailored to specific systems which are a function of the case study of mechanical precision. The RE techniques must be improved by allowing the measurement of hidden points especially for objects with a complex 3D shape. The scanning program must be optimized in order to detect the most significant points by decreasing the computational cost. The AI can facilitate the planning of a scanning program by means of automatic defect classification and recognition following wear analysis: the scanning can be refined for the region where the small defect is usually detected, by providing more details about the defect and suggestions about the 3D object redesign.

8.1.1 Reverse Engineering Platform: Tools, Approaches, and Facilities

RE techniques are typically implemented in workstations designed for the specific product. The workstation is composed of hardware and software RE tools. A fully integrated RE platform takes into account the integration of the following tools:

- A software module calculating raw material requirements, generating the production plan of the production lines, and providing data to the data mining engine for the prediction of production costs.
- Dashboards for monitoring production progress, providing efficiency indicators useful for evaluating the prototype system.
- Other facilities (electricity consumption, detection and monitoring, traceability tools, etc.).

The architecture in Figure 8.1 describes an example of a full RE platform showing the integration of AI. The platform is structured in the following phases:

- Application of the scanning techniques.
- Feature extraction of the scanned object (AI defect check and classification).
- 3D CAD modeling (reconstruction of the object by RE CAD interpolation techniques able to construct a surface from the detected data points).
- Redesign (according to the detected defects).

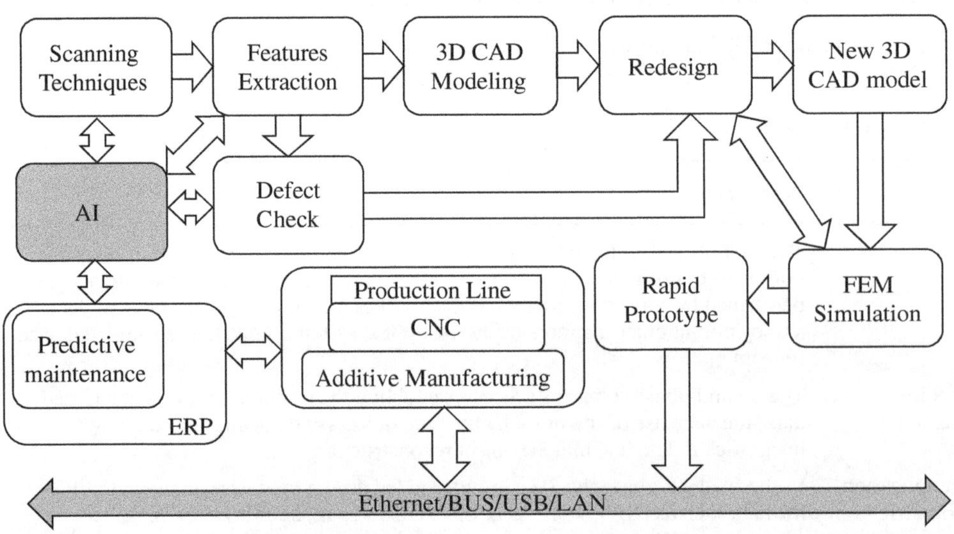

Figure 8.1 Example of a full RE platform.

- Construction of a new 3D model to simulate.
- Simulation by FEM.
- Rapid prototype (the final 3D CAD model is optimized for production).
- Object fabrication (the modeled object is fabricated by computer numerical control [CNC] machines or by additive manufacturing tools including 3D printing).
- Enterprise resource planning (ERP) data processing (mapping of all the production phases and improving predictive maintenance).

All the phases are managed by tools interconnected with the local network.

Pre-assembly and assembly must be applied to pieces of high mechanical precision. The machining precision tolerance and the corresponding functionality is assessed in order to highlight the technological limits in the implementation of the prototype (definition of the validity ranges).

During the design phase, the dimensional and geometric tolerances related to the assembly are carefully quantified. Therefore, the effect of assembly on wear is experimentally evaluated in order to re-engineer the assembly procedures, by redefining at the same time tolerances of the components. The main specifications of a complete RE platform are:

- Precision measuring machines such as laser scanners and CMMs.
- Processing techniques for surface analysis and surface reconstruction.
- Specifications and characteristics of the CNC machinery.
- Digitization of production processes.

The RE platform includes other facilities and approaches upgrading the redesign technique such as:

- Measurement and classification of piece defects.
- Predictive maintenance of CNC machinery.
- Optimization of the working parameters of the CNC machinery.
- Monitoring and prediction of production quality.
- Monitoring of production progress through Key Performance Indicators (KPIs) and calculation of raw material requirements for predicting production costs and waste materials.

An advanced approach to follow for the RE is the Design for "X" [1]. Design for X (DFX) is essentially composed of four design parameters which must be taken into account for correct work execution: Design for Manufacturing (DFM), Design for Assembly (DFA), Design for Disassembly (DFD), and Design for Environment (DFE). DFX is a design method that aims to reduce costs and working times, but it is also a way to pay attention to unnecessary waste and care for the environment. DFM is a design methodology focused on satisfying the subsequent work phases for the realization of a project. DFA is the design methodology that aims to facilitate the subsequent assembly phases of the components, in addition, it favors the decrease of costs and times, improving the product reliability of the components, by supporting their composition and their positioning. By this approach, it is essential to reduce the number of parts to assemble, avoiding accidental joints, and adapting all the parts according to the machinery to be used and the orientation and connection of the components. Another important aspect of the method is to reduce movements and passages to assemble the product, by increasing the elements of symmetry, by avoiding shapes overlapping between the parts, and by facilitating handling and coupling of the various components. DFD is the design methodology oriented toward the reduction of the elements that make up the final product. Reduction of the number of components implies several advantages; in fact, the eliminated part will never be designed, overhauled, tested, subject to breakage or deterioration or obsolescence. A component is removed if:

- During use of the product, the component commercially changes or is withdrawn from the market.
- The component is made from different materials.
- For a correct operation, the component must be functionally separated from the other components.

DFE is the design methodology to ensure greater compatibility with the environment: it considers the different production processes such as the extraction of raw materials and their processing, the production of the model, the use and transport of the product and its disposal without negatively affecting the environment with harmful emissions (low environmental impact). Figure 8.2 shows a model of the DFX approach highlighting the main processes involved in RE production controlled by AI.

RE facilities are important for production optimization. Flexible production oriented toward Industry 5.0 production must follow precise guidelines. Table 8.2 lists some facilities regarding a modern production line integrating RE facilities, integrating a control room and an advanced workstation in the production of nanocomposite sensors.

Figure 8.3 shows an example of the architecture of a flexible production platform integrating Industry 5.0 facilities applied to a nanocomposite sensor to be integrated into a robotic system. The architecture highlights the phases of creation, design, development, prototyping and testing of the sensor necessary for product realization, which can be performed by a CAD/computer-aided engineering (CAE) workstation. The use of CAD/CAE tools allows an integration of the design, production and testing activities, according to RE approaches. In this framework, the design phase must be related to production, by taking into account the technological limits of the available technology and materials. In this sense, choosing to use AM techniques certainly speeds up production compared with traditional production techniques, given the possibility of creating even complex sensors in microscale. The data acquisition of a used sensor is then compared with the original design model (nominal model) to check possible variations. The workstation upgrades the production of the new sensor, with a view to designing and building self-adaptive production lines, by combining the workstation's functionality with data processing units in a unique control room. The control room is equipped with hardware, and software AI algorithms mainly suitable for

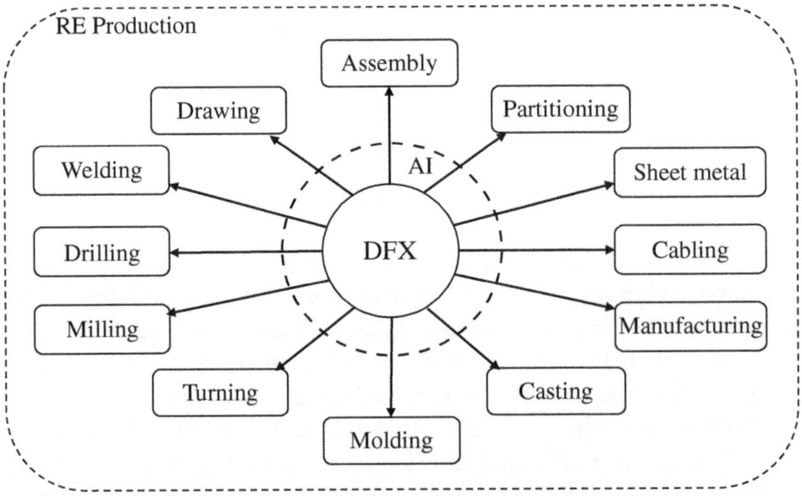

Figure 8.2 DFX model defining RE advanced production.

Table 8.2 Example of workstation characteristics implementing RE facilities for nanocomposite sensor production.

Production facilities	Description
Digital industrial system	Testing of digital production machines enabling Industry 4.0 facilities. The digital systems provide digital data for the RE of the components, and for the analysis of defects (improvement of quality)
CAD and CAE workstation, 3D printer unit	CAD and CAE are facilities for the RE of the prototype components and for defect classification. Following the wear analysis of the manufactured product, the RE facilities are useful to analyze the nanocomposite sensor behavior exposed to external forces. Specifically, AI algorithms, such as deep convolutional neutral networks [2], are applied for the defect classification of each individual prototype component, and for the predictive maintenance of the assemblies. The CAD and CAE techniques reproduce the individual pieces analyzed in the 3D drawing model, by generating data about design variations. The 3D printing units reproduce the new designs of the individual sensor to be re-engineered. The re-engineering of the innovative sensor is useful for planning the industrialization phase of the prototype. The workstation also stores further data provided by laser scanner and touch probe systems in the case of complex packages or measurements of assembled parts
3D FEM simulator	FEM is an example of a tool useful to simulate the physical and mechanical responses of a nanocomposite sensor prototype. The simulation data are crossed with that of RE in order to optimize the design and construction of the individual components. The simulator also provides useful elements for the definition of critical issues, and for the definition of the technical risks relating to the product industrialization. The industrialization analysis takes into account the product lifecycle [3], and the choice of the most resistant reusable materials. The simulations also provide information about the effective forces executed on each component by simulating wear resistance and electromagnetic exposure. By RE measurements, it is also possible to analyze the data to predict maintenance actions through AI algorithms. The simulations therefore: • Characterize the behavior of the sensor prototype. • Analyze the characteristics of the materials regarding impact and wear resistance (creation of simulation models capable of replicating the conditions of use of the analyzed sensor). • Provide important information for RE and the component re-engineering. • Simulate all the physical effects acting on the component
Process simulator	A process simulator tool activates the testing simulation to be executed for the parameter setting, for the methods and for the timings of tests to be performed. The simulator is used for the formulation of production processes of the new assembled sensing system
Mechatronic interfaces for testing	Mechatronic interfaces are applied to test devices. The tests are fundamental for the redesign of the nanocomposite sensor prototype. The main functionalities of the testing interfaces are: • Automatic setting of the testing parameters on the basis of data experimentally collected (digital values of sensors such as temperatures, humidity, vibrations, positions, and contours extracted by image vision techniques [4]). • Automatic adjustment of the amplitudes of the applied forces simulating the real working condition of the sensor (working environment scenario). • Other types of implementations and control tools based on the test benches

(Continued)

Table 8.2 (Continued)

Production facilities	Description
AI engine	An AI engine is implemented in the control room, enabling: • RE clustering and classification of defects. • Component classification [5]. • Predictive maintenance of the sensor prototype. • Prediction of component wear by crossing information from FEM simulators, and image processing outputs. • A decision support system for the formulation of a new layout design

AI, artificial intelligence; CAD, computer-aided design; CAE, computer aided engineering; 3D, three-dimensional; FEM, finite element method; RE, reverse engineering.

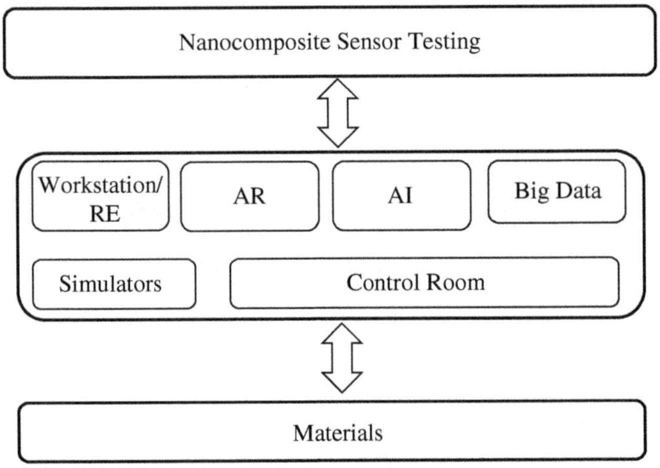

Figure 8.3 Example of a flexible production line oriented toward Industry 5.0: RE platform applied to a nanocomposite sensor in robotics.

predictive maintenance, defect and failure predictions. This innovation generates production automatisms that are part of the logic of advancement of the skilled technologies of Industry 5.0. Of particular interest is the control and automation of the testing process. Concerning the production of advanced gas nanocomposite sensors, the tests can be managed by a mechatronic interface controlled by an AI engine, able to automatize all test according to partial test results, thus optimizing the test plan. If more gases are considered as input of the sensor, the mechatronic interface has the function to tune the input percentage of each gas, analyzing the sensitivity and capability to discriminate the gas typology. Furthermore, the big data together with the AI engine are useful to provide the capability to discriminate and automatically recognize the gas composition. The tests are automatized by the mechatronic interface able to define tolerances and limits considering critical input conditions generating sensor breakage or damage (high input pressures, high input voltages, high temperature, high vibrations, etc.). Augmented reality (AR) visors can be adopted supporting laboratory tests, by addressing the worker to execute defined testing procedures. The fully automated testing process is an innovative aspect for the RE approach oriented toward sensor optimization and quality check, involving analysis of reusability, durability and wear check.

One of the goals of an innovative production is to develop specific RE techniques automatizing the process of redesign and quality check, by automatically correcting sensor defects and by providing a good sensing precision. Concerning metallic sensors or components, the metallic part can be processed by means of the redesigned 3D CAD using typical methods such as:

- Forming (for example extrusion, profiling, and rolling).
- Cutting.
- Thermal treatments.

8.2 Working Processing and Adaptation

Different approaches and techniques are adopted to adapt intelligently the different processes involved for an optimized production. Process simulation integrating AI algorithms as decision trees (DTs) represents a new concept of advanced RE oriented toward self-adaption of the production. Process mining is an important tool to construct innovative RE procedures to follow for product optimization. The proposed theoretical approach is useful to understand how dynamically it is possible to formulate a dynamic RE process integrating data mining.

8.2.1 Process Simulations

RE process simulation is typically executed by adopting business process modeling (BPM) languages integrating data mining algorithms [6]. A Process Mining Integrating Decisions (P-MInD) framework is an alternative tool to implement and design dynamical processes. This approach is adopted to solve RE problems. The logic ports of a process workflow select and switch the best procedure to follow according to acquisition system tolerance, system error, calibration sensitivity, processing precision, and failure prediction. The logic conditions to improve in a process workflow take into account:

- The required precision with respect to the defined tolerances.
- The best patterns of the sensors taking into account collision risks (collision between probes and objects to detect).
- The best patterns to decrease detection time (best points to detect).
- The refinement of points to detect corresponding to a complex region of the object to reconstruct.

Data mining supports the decision about these conditions. In addition, deep learning (DL) techniques are suitable for object recognition and classification in heterogeneous formats processing 3D models, photographs and videos [2, 7]. Process simulation also provides information about the effects of manufacturing processes, thus matching defects with fabrication procedures and parameter setting. FEM numerical tools simulate working conditions and, consequently, are suitable for the redesign of components: concerning for example thermal insulation composites system for non-residential buildings, thermal and fluid dynamics numerical simulations are compared with experimental investigations, thus defining the best layout of the polyvinyl chloride (PVC) composite panels for thermal efficiency [8]. Sensors are important for testing components in the real environment, and are essential for RE. Moreover, the design of the sensor allocation layout and numerical modeling are fundamental for the design optimization of the components. Table 8.3 lists some important advantages of process simulations in RE.

Table 8.3 Advantages of process simulations in RE.

Typology of process simulation	Advantages
Classification of procedures to perform for object detection techniques	Complex objects can be reconstructed by three-dimensional computer-aided design parts, by focusing on the detection of complex areas. The procedures to follow must be classified in order to optimize the reconstruction. The process simulation is representative for the choice of the detection technique to adopt
Piece classification by deep learning technique	Classification automatism enables the recognition when the object identity is unknown. The deep learning technique facilitates the classification when different types of objects must be recognized before their reconstruction
Process mining for reverse engineering decision making	Process mining can identify the degree of complexity of the object to detect, by executing decision tree algorithms or artificial neural networks, suggesting the program to adopt for the probe movement and control. The degree of complexity is first defined by a pre-classification of the objects to detect
Process mining selecting manufacturing procedures	Artificial intelligence algorithms combine detected defect information with the executed manufacturing process thus modeling a process mining workflow. This combination allows the association between precision degree and parameter setting program
Process simulation as physical effect simulations	The simulations of physical effects are important for the redesign of components. Thermal, fluid dynamics, mechanical, electronic and electric simulations help to analyze the physical effects on the piece to redesign, thus optimizing the design process
Process simulation in sensor detection systems	The simulations of working conditions facilitate the reproduction of real environmental situations. This aspect is important for the allocation optimization of sensors to be placed in hard environmental conditions (high temperatures, high stress or strain conditions, strong vibrations, etc.). This simulation is executed together with the process simulation of the reproduction of physical effects
Signal process simulation	The sensor signals are primarily reproduced in bench tests in order to simulate in scale the sensing conditions. Signal processing simulations are useful for the redesign of the sensing areas to increase sensing efficiency. In order to reconstruct the whole production scenario, the signal processing simulations are executed together with the simulations of physical effects

8.2.2 Process Mining Actuation and Digital Aspects Concerning Decision Support Systems Implemented by Data Mining Algorithms

Process mining is defined as the process suggested by a data mining algorithm. The DT models match the decision-making process well, allowing high dimensional data to be processed. DTs are widely used in various fields of application of data mining as they have numerous advantages. They are easily implementable and are characteristically transformed into a set of production roles, they do not make any preventive assumptions about the nature of the data, they are successfully applicable to real problems, and they are able to build models from datasets that contain numeric and categorical data. In RE applications, DTs are suitable for the definition of the optimized process to follow. DTs are well-known structures in supervised learning. The building of models on the basis of DTs has the purpose of dividing a dataset into a series of steps. The results of the model can be represented by a set of roles by a tree structure. A DT is a simple structure in which non-terminal nodes represent tests of one or more attributes and terminal nodes reflect the resulting decisions. DTs are easy to understand and to transform into roles to optimize RE processes. In the classification algorithms, and in particular those for the construction of trees, the input data, also called training sets, consist of records which each have multiple attributes or characteristics. In addition, each record is labeled with a special class label. DTs are the easiest way to classify "objects" into a finite number of classes. They are constructed by repeatedly dividing records into homogeneous subsets. The subdivision produces a tree hierarchy, where the subsets of records are called nodes and the final ones are called leaves. In particular, the nodes are labeled with the name of the attributes, the branches of the tree are labeled with the possible values of the attributes, while the leaves are labeled with the different methods describing classes. An object is classified following a path along the tree that leads from the root to a leaf. The paths are represented by the branches of the tree that provide the tools for building a set of roles. The construction of trees is not done randomly but is carried out in compliance with certain criteria, in order to try to optimize parameters, and to decide the data splitting conditions. The various existing algorithms differ mainly in the type of heuristic function as entropy that is used to choose the splitter fields of the tree. By considering *n* classes, the entropy is defined as:

$$H(S) = -\sum_{i=1}^{n} P_i \log_2 P_i \tag{8.1}$$

where P_i indicates the percentage of classified examples. The entropy associated with a training set is defined as a measure of the information content of the input data of the algorithm for the generation of the DT. If entropy has a high value, then there is a high "disorder" in the entry record space so it is rather difficult to classify a record as belonging to a certain class. Therefore, starting from a situation of maximum disorder in which $H(S) = 1$ or from a high entropy value, the records are partitioned on the basis of a particular attribute value, thus leading to a new entropy value less than the starting one. The role of the partitioning is defined by the minimum entropy condition. A function allowing the decrease in entropy to be measured, and therefore of disorder, is the *information gain G* defined as:

$$G(S,A) = H(S) - \sum_{\text{Attribute values}} \frac{|S_A|}{S} H(S_A) \tag{8.2}$$

This function is the additional information about the class attribute, provided after the partitioning process of the records. The choice of the attribute *A* is oriented toward the maximization of the information gain, and therefore on the decrease in the maximum entropy. However, there may be problems related to the use of information gain; in fact, *G* has very high values corresponding only with attributes that are highly informative and that therefore help to identify with good probability the class to which the records belong. To avoid this problem, it is possible to use a measure of the dispersion obtained by using the *SplitInfo* function defined as:

$$\text{SplitInfo}(S,A) = -\sum_{i=1}^{n} \frac{|S_A|}{S} \log_2 \frac{S_i}{S} \tag{8.3}$$

Another function useful to define criteria for the tree splitting is the *GainRatio* (normalized gain), defined as:

$$\text{GainRatio}(S,A)=\frac{G(S,A)}{\text{SplitInfo}(S,A)} \tag{8.4}$$

This function provides trees that are not too "specific". The C4.5 algorithm, evolution of the Iterative Dichotomiser 3 (ID3) algorithm, and currently most used, makes use of this criterion for the generation of the branches of the tree. There are several algorithms for building DTs. The algorithms most implemented in commercial packages are ID3, Classification and Regression Tree (CART), C4.5, and chi-square automatic interaction detection (CHAID). The ID3 algorithm is the precursor of C4.5. CART is of particular interest because there are several commercial variants, and it is the first system introducing regression trees. Essentially, regression trees take the form of DTs whose leaf nodes have numeric rather than categorical values. CART is very similar to the C4.5 algorithm but there are some differences to highlight. The first important difference is that CART always performs binary divisions of data. A second difference is that CART uses the data test to achieve algorithm tuning, while C4.5 uses only the training data to create the final tree structure. CHAID is another DT creation algorithm found in commercial statistical packages. CHAID uses the statistical chi-square test to determine the variables suitable for the construction of the DT. Each algorithm for the construction of DTs generally consists of two phases: in the first phase the tree is generated and in the second one the pruning is accomplished. Specifically, in the first phase full and very complex trees with a large number of roles are obtained. To avoid overfitting, it is necessary to define a series of criteria to simplify the DTs eliminating redundancy identified by the less significant branches. The CART algorithm proceeds in this way: it starts from the totality of the records belonging to the training set and begins a binary division into classes. The division criterion is based on information gain and therefore aims to obtain the maximum information gain at each subsequent division. Since the CART algorithm performs a binary division, the concept of information gain is particularized in the concept of diversity. This quantity is a function of the fraction of records classified in each class. By indicating with P_c the fraction of records classified in class c in terms of binary classification, it is possible to define a diversity function $D(c_1, c_2)$, assuming values in [0,1], and providing a measure of how much the records are equally distributed in the classes. In the two limit cases the records are, respectively, concentrated only in a class if $D = 1$ or divided equally between the two classes if $D = 0$. Since the algorithm goal is to accomplish a classification by positioning the records in pre-established classes, it allocates the records between the different predefined classes and does not consider those fields or those values allowing an unbalanced division of the records. The algorithm selects the splitter fields so that in each node the diversity function is minimized and therefore maximizes the following function:

$$F = D_{\text{before split}} - \left(D_{\text{left child}} - D_{\text{right child}}\right) \tag{8.5}$$

To effectively apply the tree model to other datasets, it is necessary to eliminate some branches, decreasing the overall error of the tree. To determine which subtree to use, the adjusted error rate is analyzed:

$$\text{Error}_t = E_t + \alpha \cdot \lambda_t \tag{8.6}$$

where the index t indicates the particular subtree and λ_t is the number of leaves in the subtree t. The adjusted error rate takes into account the number of leaves λ_t and the weight α of this number. For $t = 0$ the tree is complete. The pruning algorithm is structured as follows:

1) Set α and calculate Error_t for each subtree containing the root.
2) Increase α from 0 to 1 until for some subtree t_1 results $\text{Error}_{t1} \leq \text{Error}_{t0}$; all branches that are not part of t_1 are deleted.
3) Proceed in the same way until the root is reached, thus obtaining a set of subtrees $\{t_1, t_2, \ldots\}$.
4) The error of all subtree trees is calculated using a dataset, and the subtree with minor error is chosen.

Each node produces two branches that grow independently and can be calculated in parallel. The parameter that allows to understand if the DT built correctly classifies, or even if it is wrong, is

$$\varepsilon_i = \frac{n_i}{n_i^*} \tag{8.7}$$

where n_i^* is the total number of records of the training set ending in leaf i and n_i is the number of records correctly classified in i. Equation (8.7) associates with each leaf i the error ε_i. The error associated with the whole tree is then defined as:

$$E = \sum_i \frac{n_i}{N} \varepsilon_i \tag{8.8}$$

The problem of overfitting can be controlled by pre-pruning. There are three possible methods for determining the optimal tree size:

- Use a separate set of examples to evaluate the tree to construct (adopting training and validation set).
- Use all the data available for training by applying a statistical test to determine whether or not to keep a node.
- Use an explicit measure of the tree complexity.

It is possible to translate each DT into a set of production roles, having the following form:

- IF antecedent conditions.
- THEN consequent conditions.

These conditions define the values or ranges of values of the output attributes. The technique for transforming a DT into a set of production roles is described as follows: a role is created starting at the root node, and following one of the possible branches up to the leaf node; the antecedent condition of a role is given by the combinations of the values of the attributes that are encountered along the pattern; the consequent condition of the corresponding role is the value of the leaf node.

However, there are several ways to switch from trees to roles; many of these variants tend to simplify or eliminate existing roles, this is done to eliminate uninteresting or redundant roles.

DTs can be integrated into more complex data mining data flow. Process mining techniques are processes integrated by data mining decision making improving switching decisions. Figure 8.4a,b shows two examples of DT and ANN models integrated into a basic process, respectively. These examples show schematically how it is possible to integrate into a standard BPM workflow an AI algorithm enabling logic conditions oriented toward production optimization. The DT and ANN models are suitable to integrate into a process workflow the defect classification, by automatically suggesting the working conditions. The workflow can represent the machine processing of a component, where the DT or the ANN switches the parameter setting according to the defect classification, defect prediction, or a specific role.

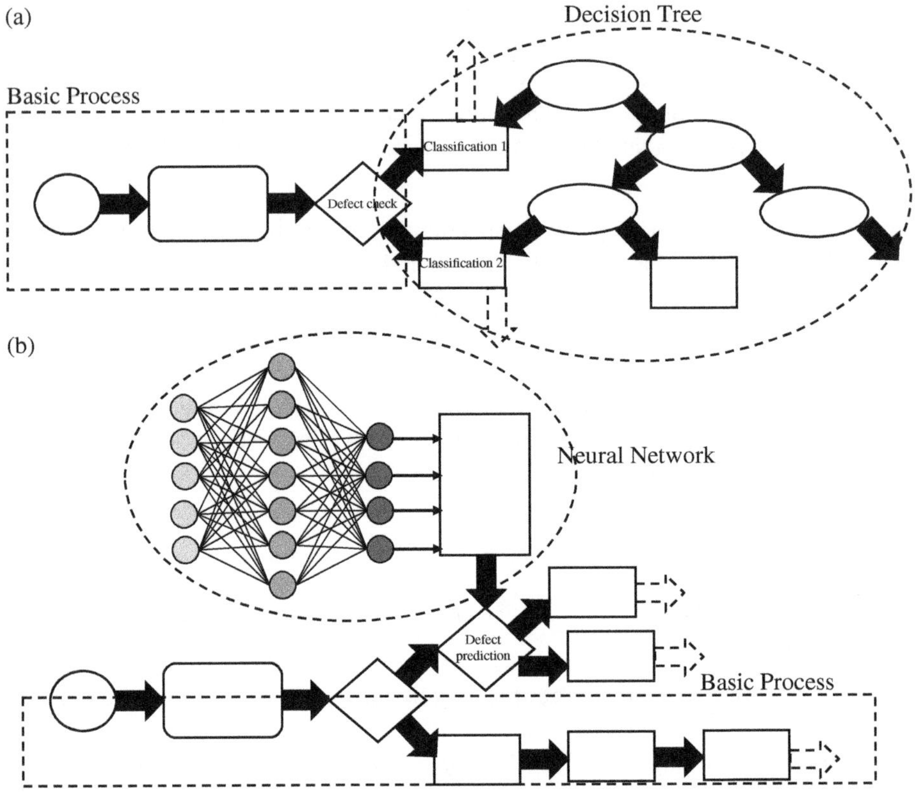

Figure 8.4 (a) Process mining and a fully integrated DT data flow selecting a production process. (b) Decision making switching process by ANN output.

8.3 Reverse Engineering and Self-Learning Automatic Working Piece Classification

In this section, an example of the RE approach to the processing procedure is discussed. The proposed approach is based on the prediction of working defects, in order to optimize the piece according to tolerances. By considering a milling process, the geometric measurement of each single hole of a specific piece can be visualized in graphs allowing to estimate the measured value exceeding the tolerances (under a minimum and over a maximum). Figure 8.5 shows an example of a theoretical graph, where the solid line indicates the measured value while the dotted line indicates the theoretical value (standard working condition of the milling machine). The upper center line (UCL) and lower center line (LCL) limits shows the region of acceptability of the measurement to graphically evaluate which geometric measurement of the hole is out of the tolerance.

The graph in Figure 8.5 represents an advanced production process dashboard, since it permits to control how the process parameters evolve over time and defines when the process performance is unacceptable with a consequent increase in risk. The redesigning process of a piece must consider both tolerances and machine failures. The UCL and LCL control limits indicate the upper and lower control limits, respectively, and it is supposed that a certain characteristic of the production process output has mean μ, and standard deviation σ, both known. The UCL and LCL values are simplified as:

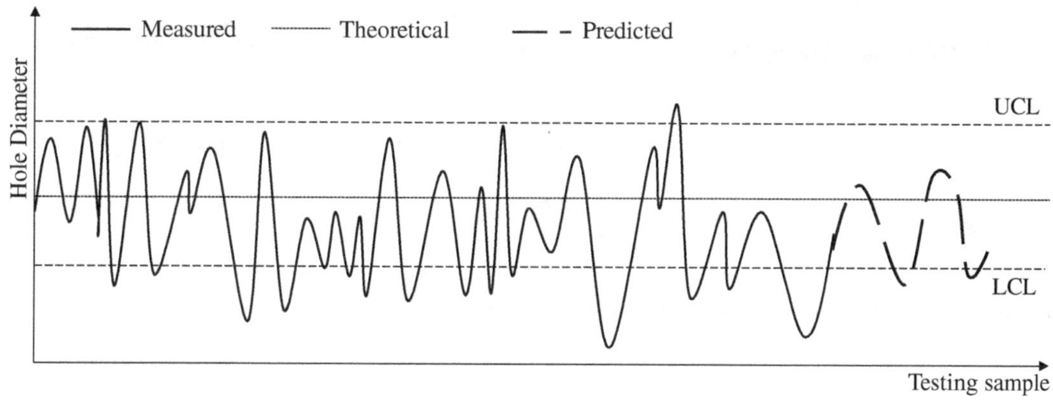

Figure 8.5 Measured, theoretical and predicted values of hole diameters.

$$\text{UCL} = \mu + 3\sigma / \sqrt{n} \tag{8.9}$$

$$\text{LCL} = \mu - 3\sigma / \sqrt{n} \tag{8.10}$$

where n is the sample number. From graphs similar to Figure 8.5, it is also possible to evaluate how the measurement evolves over time. By a comparison between the measured and theoretical values, it is possible to evaluate whether the measurement can be classified as a defect or not, based on the design tolerances. Furthermore, it is possible to count the number of times that the measured value exceeds the design tolerances. The control process takes into account different specific factors, such as:

- machinery not working properly
- operator errors
- defective materials

By assuming the probability that the materials are defective is negligible, and by considering that most of the production processes use CNC machines for the machining of the pieces, it is concluded that this approach defines when the machinery needs maintenance or when a piece must be redesigned according to the new required precision and updated tolerances. The prediction of the failures can be estimated by forecasting the fraction of defective holes using a recurrent neural network and comparing the results with the previously discussed limits of acceptability. The long short- term memory (LSTM) neural network is suitable for defect prediction. The use of a simple three LSTM level network allows the temporal features of the data series taken as input to be analyzed in order to obtain information on recurring data patterns. The activation function typically used is the Rectified Linear Unit. The LSTM algorithm is appropriate to update the data in the database, and the related prediction results are stored into a database table. Figure 8.5 also shows the predicted results completing the failure risk analysis. According to the theoretical example, the predicted results estimate a dangerous oscillation between the LCL threshold and the theoretical values thus suggesting to set new tolerances according to the desired precision. If the tolerance to set is too high, it is necessary to carry out machine maintenance. Figure 8.6 shows the block diagram of a generic controlled system, including the automatisms and the PLC interface. The system can tune machine parameters when a defect is measured or predicted. This control and actuation concept can be applied for other production machines (milling, cutting, treatment, smoothing, etc.), by setting different parameters (power, temperature, pressure, etc.).

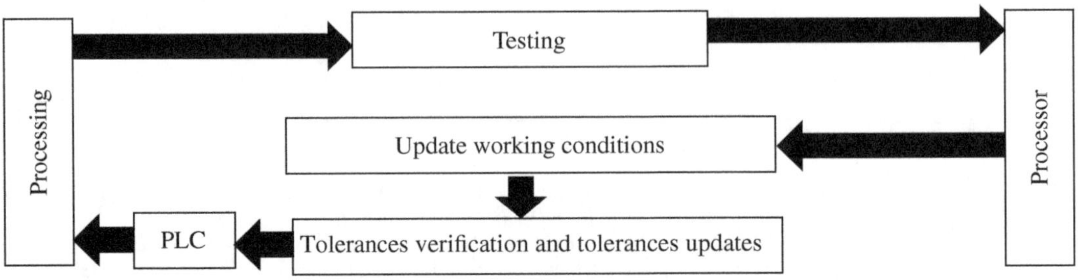

Figure 8.6 Feedback system updating tolerances and machine parameters during piece processing.

The RE approach in this case is oriented toward real-time checking and prediction of the defect, involving defect classification and prediction of machine failures. In this way, RE is directly applied to the parameter setting avoiding production defects in self-modality.

8.4 Tools Supporting RE: AR and Image Processing for Size Measurement

The growing interest in smart technologies follows the growth of the mobile device market (smart watches, smartphones, etc.). All smart devices are emerging very quickly in the consumer market, however their applications in the industrial world are still limited. The use of a smart measuring device based on an integrated wearable platform composed of AR visors, camera, embedded processors, and wireless communication devices, having submillimeter precision, represent innovations in RE, and can consistently innovate the production. The critical technological factors are related to the precision of the measuring devices, the repeatability of the measurement and its robustness to the perturbations due to the industrial environment. In the market there are some applications based on AR dedicated to the measurement of object measurements, however these applications are mostly gaming applications. The real issue is to increase the image resolution in order to apply correctly image processing tools suitable for measurement. Mobile technology facilitates the common uses of RE tools for example using an operation system supporting advanced AR platforms and viewer systems. AR can be addressed to control deformations in mechanical parts, for measurement in different prospective views (at different angles of the object to be measured with respect to the camera of the viewer), and in general for the mechanical defect check of 3D objects. The block diagram of the architecture of an advanced AR platform addressed for RE is shown in Figure 8.7. The system architecture includes:

- An AR viewer able to visualize information content superimposed on the real scene; it is interfaced with the data processing unit via High-Definition Multimedia Interface (HDMI) input.
- A camera sensor used for frame detection of the object to be measured (it is interfaced with the data processing unit via HDMI output).
- A data processing unit for the pre-processing of the detected frames and for local data storage (local memory buffering) before transmission on the server.
- A data transmission unit used to transmit data frames to the server via the Wireless Highway Addressable Remote Transducer (HART) protocol, which is a wireless sensor networking technology based on the HART protocol.

The AR wearable device is interfaced with:

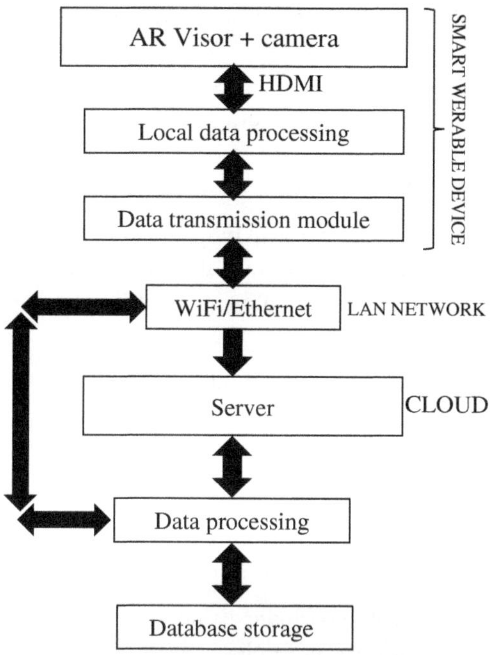

Figure 8.7 Architecture of a smart visor used for piece measurement with the image vision technique.

- A server managing data processing.
- A numerical data processor engine enabling image processing; the image processing can be performed in the cloud or locally by adopting graphics processing units (GPUs) for architectures based on total local data processing.
- A database archiving and organizing the detected images and the measurements (server interfacing).
- A company local area network (LAN) for data flow and data exchange.

Using AR, the operator acquires a frame of the object to measure by a camera placed on the visor. The frame is transmitted to the server and then to the numerical processing unit. Through image processing, and pattern analysis operations, geometrical features of the object are extracted and its size is measured. The mismatch existing between the features of the unit under test and the corresponding features of the reference unit, and the comparison of this mismatch with threshold values, permit to establish if the piece is compliant or not. A voice command can facilitate the interaction with the visor unit, thus accelerating the AR inspection process. A further improvement of the compliant piece check is the adoption of AI classification automatically identifying possible defects.

8.5 RE in Micrometric Scale: RE Approach for Photonic Crystals

Optical circuits are characterized by very small dimensions typically in the micrometric range. RE methodologies require very high accuracy, because a very small defect can generate a very high error in the optical device output. The RE procedures applied to micrometric devices such as PCs are mainly related to:

- Image measurements by atomic force microscopy (AFM), scanning electron microscopy (SEM) or transmission electron microscopy (TEM).
- Numerical simulation of 3D CAD models reproducing technology tolerances.
- Numerical simulations predicting device outputs as a function of possible geometrical defects (output sensitivity response).
- Numerical simulations of device behavior reproducing input and output signals.

The analyzed circular PC [9] is characterized by holes in the central region indicated by the letter A and other holes arranged in circular layouts indicated by the letter B. Specifically, the circular PC layout consists of air holes etched in a gallium arsenide (GaAs) material following circular concentric patterns. The RE is improved by changing:

- The diameters of the inner and outer holes according to the technological tolerances.
- The relative position of the inner holes in order to generate cavity modes.
- The distances between the circular layouts of the outer holes.
- The hole diameters according to technological limits.
- The excitation port and by analyzing the output at every other port of the 3D model.

Other characteristics adjusted in the RE redesign process and improving resonance efficiency are:

- All the outer holes arranged in the circular layout have the same periodic distance.
- The inner holes have a diameter different to the diameter of the outer holes.
- The inner holes are allocated around the central region and have a decreased radius.
- The inner holes are shifted to improve cavity mode resonance.
- The 3D configuration consists of a membrane made by an active air/GaAs/air waveguide containing four layers of InAs/InGaAs embedding quantum dots (QDs) emitting around a working wavelength of 1.28 μm.

In order to compare numerical results with experimental ones, it is necessary to design and implement an experimental setup suitable to measure cavity modes supported by the vertical emission of QDs: a backscattering micro-photoluminescence (μPL) setup is suitable to collect energy transmitted and radiated by the central microcavity [9]. The circular patterns are fabricated through electron beam lithography (EBL) and inductively coupled plasma (ICoP) etching. The membrane-type slab is obtained by selective wet etching of the Al/Ga/As sacrificial layer. The PC device can be modeled by enclosing it in a box interfaced with six input/output ports where signals pass (3D numerical modeling). A generalized scattering matrix (S-matrix) describes the power fraction transferred by a laser source excitation: this excitation is able to transmit or reflect at each port a power fraction. The S-matrix for the analyzed six-port optical structure is theoretically defined by:

$$\begin{pmatrix} o_1 \\ o_2 \\ o_3 \\ o_4 \\ o_5 \\ o_6 \end{pmatrix} = \begin{pmatrix} S_{11} & S_{12} & S_{13} & S_{14} & S_{15} & S_{16} \\ S_{21} & S_{22} & S_{23} & S_{24} & S_{25} & S_{26} \\ S_{31} & S_{32} & S_{33} & S_{34} & S_{35} & S_{36} \\ S_{41} & S_{42} & S_{43} & S_{44} & S_{45} & S_{46} \\ S_{51} & S_{52} & S_{53} & S_{54} & S_{55} & S_{56} \\ S_{61} & S_{62} & S_{63} & S_{64} & S_{65} & S_{66} \end{pmatrix} \cdot \begin{pmatrix} i_1 \\ i_2 \\ i_3 \\ i_4 \\ i_5 \\ i_6 \end{pmatrix} \tag{8.11}$$

where the magnitudes of o and i signals are normalized to 1 W of power; $|i_i|^2$ is the excitation power at port i, and $|o_i|^2$ is the power of the transmitted or reflected field at the same port. The single element of the S-matrix is defined as:

$$S_{i,j} = \left.\frac{o_i}{i_j}\right|_{i_k=0 \text{ for } k \neq j} \tag{8.12}$$

The scattering parameter $S_{i,j}$ is defined by exciting port j with a power signal and evaluating the reflected power amplitude coming out from port i. The excitation power on all ports, except the jth port, are set to zero: this physical condition is equivalent to consider all ports connected with matched loads thus avoiding signal reflections. The $S_{i,i}$ reflection coefficient (reflection coefficient at the port i) is defined when all other ports of the 3D model are terminated in matched loads. The simulations are used mainly for the redesign of the device: a sensitivity response of the scattering parameters representing reflection at each port is defined by changing for example the radius of the inner holes by a few nanometers according to technological precision. Sensitivity analysis is then important to define the gap of possible variations of geometrical parameters by obtaining an efficient output signal predicted by the 3D simulations. Figure 8.8a shows the 3D reconstruction of a SEM image of a circular PC with a central microcavity region, where the inner holes (A) have a smaller diameter than those of the external holes (B). An example of the cavity resonant mode is represented in the simulation in Figure 8.8b where the concentric lines model the hole patterns. RE approaches are fundamental because fabrication processes are usually imperfect. Figure 8.8c shows an example of an imperfect etching process: the hole profiles may assume a conical configuration thus changing the resonance and the frequency responses of the whole PC device. Figure 8.8d explains the importance of RE for design, fabrication, and experimentation results: RE combines all the results to optimize the fabricated device. Table 8.4 summarizes the main RE procedures useful to improve device efficiency and quality.

An important aspect for RE optimization of optical devices is the laser source model. The laser beam is modeled as a focused Gaussian beam characterized by a minimum waist in a defined point

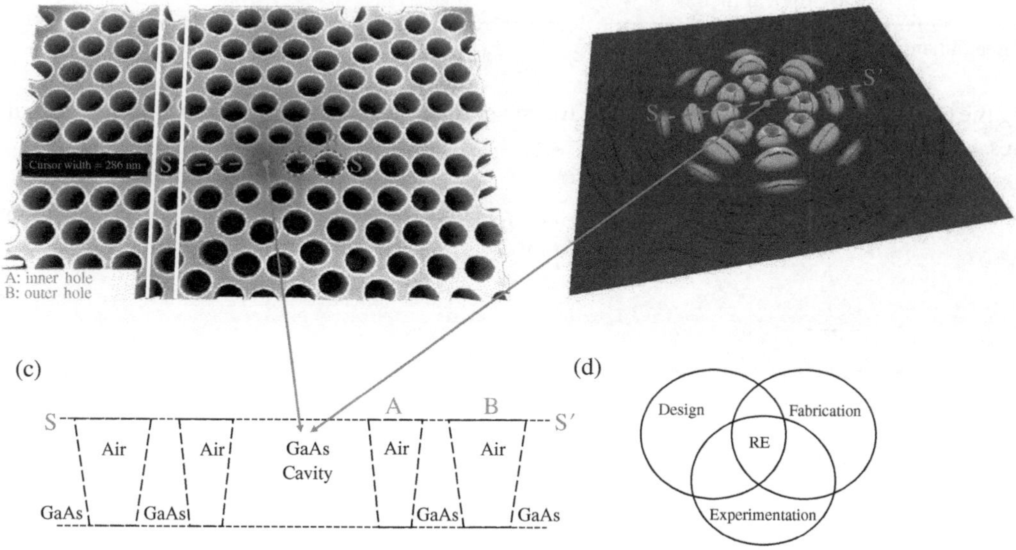

Figure 8.8 RE applied to PC structures: (a) 3D reconstruction of a SEM image of a circular PC; (b) example of FEM simulation enhancing cavity modes; (c) defects in etching process (cross section of the device); and (d) showing the importance of RE in the design, experimentation and fabrication process.

Table 8.4 RE procedure for design optimization of a micrometric optical device.

RE procedure	Description
3D modeling	3D modeling is fundamental in order to realize the first prototype according to fabrication tolerances. The modeling takes into account the perfect layout without possible defects. The 3D model is the ideal device to fabricate
FEM simulations	The simulations are very important to check the ideal frequency responses of the devices, and to analyze the effect of the imperfections on the ideal conditions. FEM simulations define the sensitivity response versus the geometrical and physical parameters: the etching process can be imperfect, by generating conical layouts or elliptical holes. The parametric analyses are useful to speed up the simulations and to observe the resonances, the frequency shift of the device response, and the variation of the device efficiency. The procedure to follow for a parametric analysis is described by the following steps: • Variation of the geometrical parameters by fixing the materials (materials commonly used in the market). • Determination of the main layout individuating the main specifications (working frequencies, frequency of resonances, etc.). • Slow variation of the geometrical parameters by optimizing the device efficiency (sensitivity analysis). • Increase of the numerical resolution and meshing in order to estimate with higher precision the device responses and the efficiency parameters (simulation refinement).
Technology characterization	The images in microscale support the check of the fabrication imperfections and, consequently, allows to understand the variations of the device responses
Laser beam source modeling	The laser source modeling must replicate the spectroscopy experimental setup matching simulations and measurement results. The study of the laser beam coupling is important to optimize the full system source/device. The experimental tests are important for the redesign of the whole system taking into account integration of optical components (lens, fiber optic, tapered regions guiding optical modes, etc.)

3D, three-dimensional; FEM, finite element method.

along the direction of propagation [10]. For a transverse electric (TE) polarized beam the analytical expression of the laser beam is:

$$\mathbf{E}_y(x,y,z;t) = E_0 \frac{w_0}{w(z)} e^{-(r/w(z))^2} \cos\left(\omega t - kz + \eta(z) - r^2 \frac{k}{2R(z)}\right) \mathbf{e}_y \tag{8.13}$$

with

$$w(z) = w_0 \sqrt{1 + \left(\frac{z}{z_0}\right)^2} \tag{8.14}$$

$$\eta(z) = \operatorname{atan}\left(\frac{z}{z_0}\right) \tag{8.15}$$

$$R(z) = z\left(1 + \left(\frac{z}{z_0}\right)^2\right) \tag{8.16}$$

where w_0 is the minimum waist, ω is the angular frequency, r is the radial cylindrical coordinate, k is the wavenumber, z_0 represents the electric field peak position, and $R(z)$ is the spherical wave radius. A useful parameter for efficiency estimation is the quality factor defined as:

$$Q = 2\pi f \frac{E_n}{P} \tag{8.17}$$

where f is the frequency, E_n is the energy, and P is the power stored in the device by exciting the optical slab with the laser beam modeled in Eq. (8.15). The quality factor Q is decomposed in the vertical direction and into the parallel one as:

$$\frac{1}{Q} = \frac{1}{Q_{\perp}} + \frac{1}{Q_{//}} \tag{8.18}$$

$Q_{//}$ is mainly indicated for the estimation of in-plane waveguiding devices, whilst $Q_{\perp}$ is representative for devices emitting a radiated field in the vertical direction. RE is important to characterize and optimize the physical behavior of the device concerning mode guiding or radiated field or both. RE takes into account the comparison between measurement results and numerical ones, in order to find the best device layout.

8.6 RE for the Production of Pipeline Components

Different techniques can be applied for the strategic production of pipeline components. Automatic classification and choice about specific applications favor the production of new components according to standards and circuit compatibility. Specifically, industry research takes into account the design and development of an intelligent RE engine suitable for the selection of features of different types of hydraulic components, by considering wear analysis and AI wear prediction. The entry of different input data facilitates the choice of appropriate circuit components, providing the most suitable design configurations for engineering applications. The classification modeling must consider the technical regulations (ISA-IEC, ASME, ISO, UNI, etc.), fluid or gas types, the types of connections of fluids, the operating and environmental conditions, regulation type, physical and chemical compatibilities, dynamic fluid properties, etc. The basic algorithm is structured in several sequential phases describing the following properties (Figure 8.9):

- Calculation of the fluid dynamic coefficients (main classification).
- Materials and compatibility classification.
- Calculation of the physical limits by means of numerical simulations (fluid speeds and other factors influencing the pipeline operation conditions).
- Identification of material limits according to the specific application (wear analysis and prediction).
- Identification of other characteristics based on the application context and on the models used for the design (update of the modeling).

The algorithm outputs for an automated process are:

- Calculation of new specifications of the optimized component.
- Validation of the choice by means of measurements and numerical results.
- Optimized CAD drawings (*dxf*/*dwg*).
- Production cost analysis.

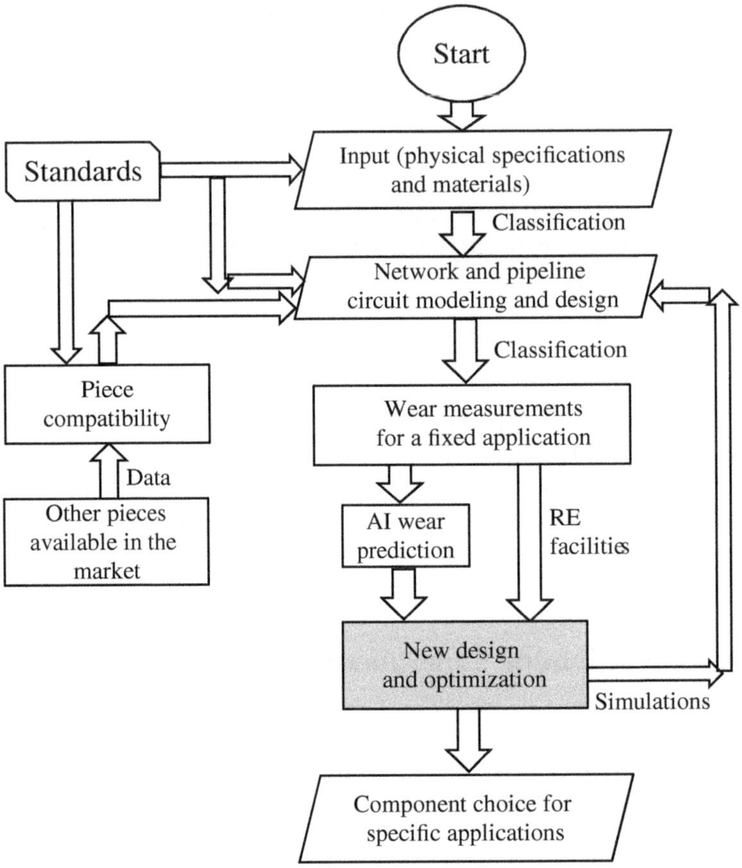

Figure 8.9 Classification and automatization of the best choice of a pipeline component.

The flowchart in Figure 8.9 is scalable: other specifications must be added during the RE phase, including compatibility and wear resistance. The new component to produce is automatically classified for the specific application. The support system for the choice of components introduces a RE process able to combine different elements for a specific pipeline circuit, thus suggesting the guidelines for fabrication. Furthermore, the possibility to combine automatically different components for a specific pipeline circuit introduces new engineering services for companies working in pipeline design. RE facilities are also useful for:

- Parameterization, modeling and study of the behavior of the fluid (turbulent, laminar, transient flow, subsonic, transonic, supersonic, hypersonic flow field); the modeling of a fluid starts from an already existing and physically similar component, knowing physical characteristics such as temperature, pressure, and specific volume.
- Parameterization and modeling of the components following the approaches analyzed in the preliminary analyses.
- Design of the geometric layouts and study of the dynamic fluid behavior (the design is important to define the new procedures and new approaches to follow for the production processes).
- Comparison of numerical models with analytical models.
- Analysis of the fluid behavior in certain network layouts (pipeline layouts).

The classified components are stored in a MySQL database which is structured in order to facilitate their search through appropriate filtering operations using queries. The database is structured based on the component attributes useful for the characterizations to follow for the selection. The use of the database potentially accelerates the support operations for the customer, and the automatic choice of design solutions in engineering services. The design of a new component or the optimization of an old one is executed when:

- More compatibilities are required.
- New technologies and materials are available in the market.
- A new pipeline network must be integrated into an old one.
- A strong component wear is checked.
- The components are inappropriate for specific operating conditions.

8.7 RE in the Precision Manufacturing Process for Thin Film Devices

RE methodologies applied in manufacturing thin film production processes are very important for the standardization of fabrication procedures and for the improvement of the devices. Many variables influence thin film generation processes in modern multilayered devices. In order to find the best approach to follow, it is necessary to analyze experimental results of physical behaviors and technological limits. Two case studies are analyzed in order to introduce the RE approach to follow for the production of innovative micrometric antennas and energy harvesting devices.

8.7.1 Ring MEMS Manufacturing

The fabrication of MEMS requires different technological steps. Concerning molybdenum/aluminum nitride/molybdenum (Mo/AlN/Mo) MEMS behaving as antenna [11, 12] and as energy harvesting devices [13], the fabrication process involves sputtering deposition, optical lithography, and a controlled wet etching process [12]. A multilayered freestanding MEMS structure, if excited by a voltage signal, moves itself by means of the piezoelectric effect of the AlN layer. The relationships between the stress **T**, the strain **S**, the electric field **E**, and electric displacement field **D** are:

$$\begin{aligned} \mathbf{T} &= C_E\mathbf{S} - {c_O}^T\mathbf{D} \\ \mathbf{D} &= c_O\mathbf{S} - \varepsilon_r\mathbf{E} \end{aligned} \tag{8.19}$$

with the elasticity matrix C_E (in Pa), the coupling matrix c_O (in C m^{-2}), and the permittivity matrix ε_r of the AlN material defined, respectively, as:

$$C_E = \begin{pmatrix} 4.1e^{11} & 1.49e^{11} & 9.9e^{10} & 0 & 0 & 0 \\ 1.49e^{11} & 4.1e^{11} & 1.49e^{11} & 0 & 0 & 0 \\ 9.9e^{10} & 1.49e^{11} & 3.89e^{11} & 0 & 0 & 0 \\ 0 & 0 & 0 & 1.25e^{11} & 0 & 0 \\ 0 & 0 & 0 & 0 & 1.25e^{11} & 0 \\ 0 & 0 & 0 & 0 & 0 & 1.25e^{11} \end{pmatrix} \tag{8.20}$$

$$c_O = \begin{pmatrix} 0 & 0 & 0 & 0 & -0.48 & 0 \\ 0 & 0 & 0 & -0.48 & 0 & 0 \\ -0.58 & -0.58 & 1.55 & 0 & 0 & 0 \end{pmatrix} \tag{8.21}$$

$$\varepsilon_r = \begin{pmatrix} 9 & 0 & 0 \\ 0 & 9 & 0 \\ 0 & 0 & 9 \end{pmatrix} \tag{8.22}$$

When the voltage source overcomes a threshold limit the MEMS begins to deteriorate. Figure 8.10a–b shows the effects when a voltage signal is applied over the threshold limit which is a function of the Mo/AlN/Mo layer thickness. The voltage signal is applied to move the piezoelectric antenna improving radiation properties, but a high input voltage can destroy the device. The correct definition of voltage thresholds and tolerances is a RE goal.

The technological steps realizing ring MEMS are graphically shown in Figure 8.11. The AlN thin film is sandwiched between two Mo layers (Figure 8.11a) behaving as electrodes. The AlN and Mo layers are grown on a substrate of a thin SiO_2 sacrificial layer on a thick (100) silicon substrate. Molybdenum is chosen as the electrode material because of its capability of guiding a preferential (002) orientation of the AlN film, and because it is characterized by a low electrical resistivity behaving as a conductive material. The AlN and Mo films are grown by direct current magnetron sputtering using a LLS EVO Unaxis sputtering system. The technological steps are:

- A positive photoresist layer is deposited on the whole multilayered structure (Figure 8.11b).
- After a mask alignment the structure is irradiated by UV light from the top (Figure 8.11c,d).
- $SiCl_4$-based plasma by ICoP dry etching using a photoresist mask is used to remove the upper Mo layer (Figure 8.11e).
- $SiCl_4$-based plasma by ICoP dry etching using a photoresist mask is used to remove the AlN layer (Figure 8.11f).
- The lower Mo layer is etched by a H_2O_2-based solution (Figure 8.11g).
- A hydrofluoric acid (HF)-based solution is used to remove the SiO_2 sacrificial layer (Figure 8.11h).
- A rolled up freestanding structure is automatically realized by the releasing of the rings from the substrate surface (Figure 8.11i).

(a) (b) (c)

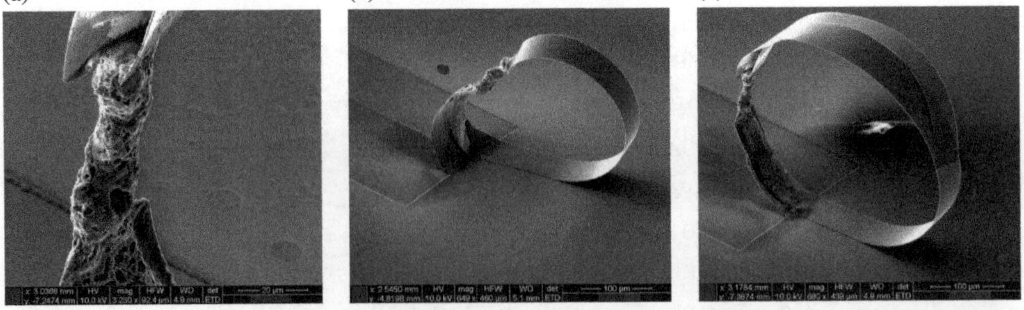

Figure 8.10 SEM images of damaged ring MEMS: (a–c) effects of a high voltage signal applied on the Mo/AlN/Mo MEMS.

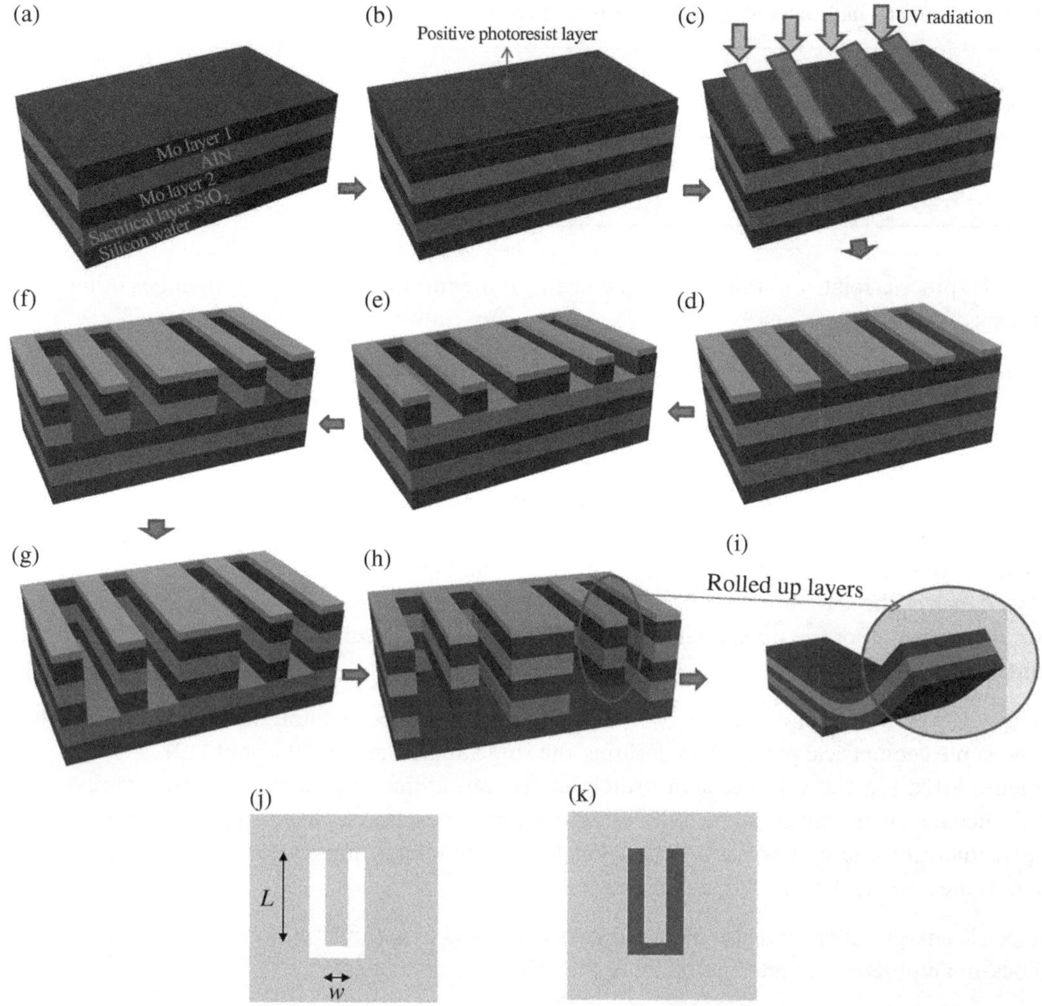

Figure 8.11 Technological phase for freestanding ring MEMS: (a–h) phases of the etching process; (i) free standing layout obtained after the etching processes; (j) geometrical mask of the freestanding part; and (k) effect of the etching process.

The first Mo layer and the AlN layer provide a high residual stress improving a specific radius of curvature of the layered structure. By changing the first Mo layer thickness it is possible to control this curvature. Other geometrical parameters influencing the curvature are the total length L and the width w (Figure 8.11j related to the step in Figure 8.11d and Figure 8.11k related to the step in Figure 8.11e). The RE approach involves the optimization of the growth conditions in order to achieve the best compromise between adhesion stability, crystallographic quality, and high residual stress of all the films. The RE procedure is related to the:

- Choice of the geometrical parameters.
- UV exposure radiation times.
- Setting the parameters of the ICoP process.
- Possibility to improve helicoidal layouts for specific radiation patterns simulated by a FEM tool.
- Times of the dry etching processes based on the use of solutions.

Table 8.5 Ring and helicoidal MEMS layout versus Mo top layer thickness.

Mo top layer thickness (Mo layer 1) (μm)	Length (μm)	Layout
100	250–1000	Ring
70	250–1000	Helicoidal

The RE process related to the antenna behavior of the multilayered MEMS, includes other specific aspects such as:

- Optimization of each step of the fabrications process by analyzing step by step the images in microscale of the production process in order to check the freestanding layout.
- Optimization of the micro-antenna design by considering the technological limits and the measurements.
- Definition of new tolerances about working conditions and stress testing results.
- Redesign of the electrical contacts of the device by ensuring the best electrical linking.
- Definition of the error margins between the design and fabrication process.
- Avoidance of the short circuit effect due to touching both Mo layers.
- Characterization of the tip spatial displacement due to the piezoelectric effect according to the antenna layout.

The study of the antenna layout is based principally on experimental observations. Table 8.5 shows some geometrical parameters defining the ring and the helicoidal layout [12].

Figure 8.12a illustrates the torsion principle, generating the helicoidal layout in Figure 8.12b. The helicoidal layout can generate a short circuit condition thus avoiding the effect of piezoelectric displacement or energy harvesting behavior. RE of the energy harvesting efficiency takes into account the optimization of the following aspects:

- Electric energy generation at low frequencies according to natural vibration forces [14].
- Lock-in amplifier parameters enhancing the voltage output signal.

(a)

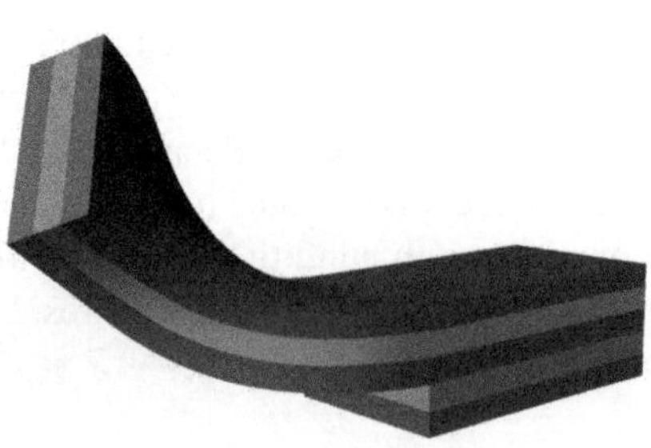

(b)

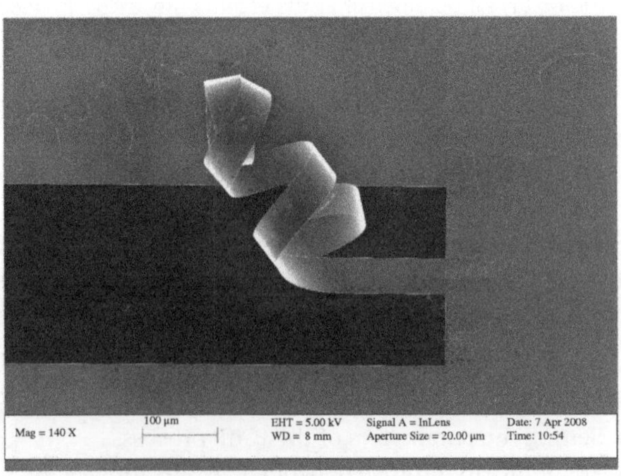

Figure 8.12 (a) Torsion allowing helicoidal layout. (b) SEM image of an helicoidal layout.

- Stabilization of the output signal.
- The possibility to vibrate an array of MEMS Mo/AlN/Mo freestanding devices in order to increase the energy conversion efficiency.
- Control of the shaker apparatus used for energy harvesting MEMS characterization.
- Avoidance of the short circuit effect which prevents the generation of electrical current.
- Engineering of the electrical connections for the measurement setup.

In all the experimental observations it is particularly important to study the physical effects induced by changing the production manufacturing parameters. Many parameters influence the final device behavior. AI classification algorithms are able to combine different parameters into classes useful for RE oriented toward optimizing the parameter combinations, allowing production repeatability of devices having the same specifications as the designed one. In particular, the classes define:

- The combination of parameters of the growth machine for the optimization and repeatability of thickness with a very high precision.
- The combination of the clean room procedures and processing conditions (exposure time, time using solutions or solvent, etc.).
- Possible layout parameters obtained by combining possible L and w geometrical parameters.
- The combination of all the parameters described above within the working range defined by the technological limits.

Concerning electrical connectivity, a standard way to connect conductive layers is by via hole connection: the via hole in a printed circuit board (PCB) consists of an electrical connection between two electrical conductive parts of an electronic circuit. By considering a multilayered structure with the top layer as a conductive layer, the via hole assumes the configuration in Figure 8.13. Typical antenna patch manufacturing takes into account source linking by via hole connection. The coaxial feeding approach via hole connection is implemented to excite the top layer with two different signal sources: both a THz and a kHz signal are transferred to the top conductive layer along two different patch lines [12], according to the layout in Figure 8.14a,b.

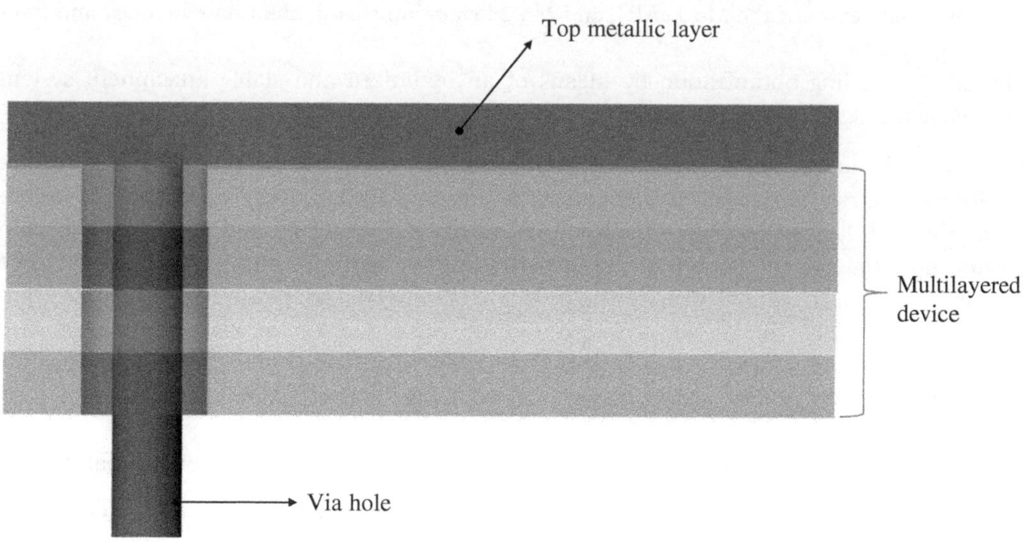

Figure 8.13 Via hole connection of the top metallic layer.

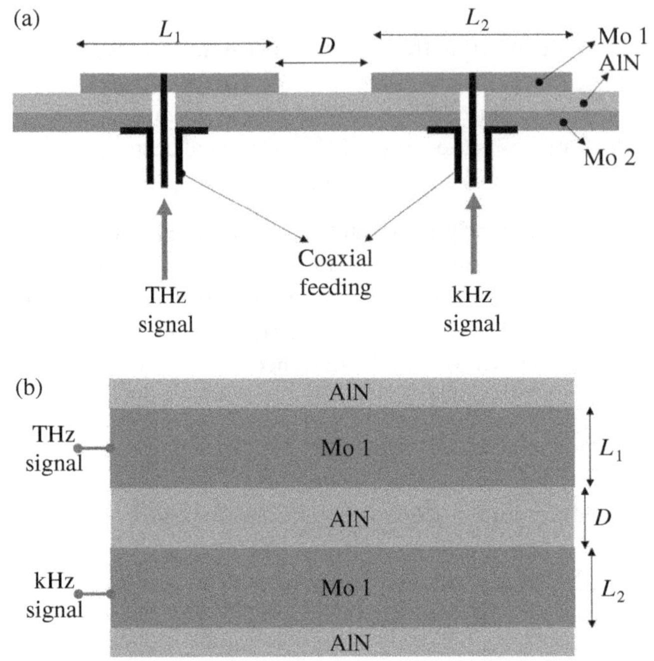

Figure 8.14 (a) Cross section of the coaxial feeding layout exciting by THz and kHz signals the Mo top layer 1. (b) Top view showing the two excited paths.

In the case of the Mo 1 layer freestanding configuration behaving as antenna [12], the RE approach is oriented to define:

- The best dimension of the distance D, avoiding THz and kHz signal undesired coupling which can disturb the propagation (FEM simulation estimates the distance tolerance).
- The best dimensions L_1 and L_2 transferring the maximum power to the Mo top layers.
- The best thicknesses of the Mo 1, AlN, and Mo 2 layers improving antenna efficiency and free-standing layouts.
- The coaxial feeding optimization by means of an optimized and stable attachment system increasing the signal versus the noise.

If two metallic layers are to be electrically connected by very thin metallic tips, a solution to apply for the electrical surface measurements is the three-tip configuration in Figure 8.15a,b: considering the multilayered Mo/AlN/Mo structure in Figure 8.11a, tips 1 and 3 touch the top Mo layer and tip 2 touches the bottom Mo layer permitting the current–voltage characteristic to be measured. Precision is required to put the tip probe into a micrometric aperture: typically 50×50 μm apertures are considered for antenna array layouts [12], thus requiring a measurement system with a microscope and micrometric tip movement systems (Figure 8.15c). For repetitive measurements it is preferable to realize micrometric wire welds transferring the electrical linking on external pads as shown in Figure 8.15d.

The RE approach for electrical measurements of a multilayered structure is related mainly to:

- The realization of apertures on the top conductive layer with dimensions according to the available experimental setup and the application field.

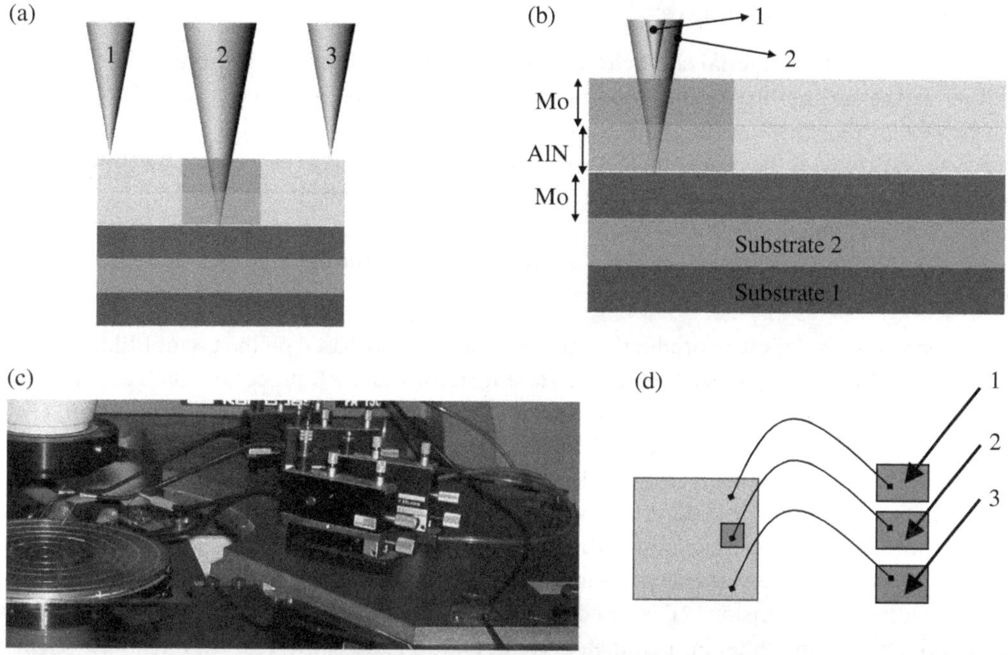

Figure 8.15 (a) Front view and (b) side view of a three-point connection layout with three tips touching a multilayered device having two conductive layers. (c) Measurement probe station composed of a central microscope and three tips with micrometric up and down movement. (d) Example of a layout with micro welds supporting the three-point measurement approach (the numbers 1, 2, and 3 refer to the three tip probes used for the *I–V* characterization).

- The choice of a two- or three-tip measurement configuration according to the symmetry of the device layout.
- The possibility to realize more stable measurements, by touching external pads linking the bottom Mo layer.

8.7.2 Thin Film Diamond Antenna

The RE approach in thin film devices is focused mainly on the redesign of the device starting with the analysis of technological limits. The analyzed case is a diamond antenna based on a planar monopole layout obtained by growing diamond film by means of the optimized microwave plasma enhanced chemical vapor deposition (MWPECVD) technique [15]: the antenna prototype is shown in the photograph in Figure 8.16. The polycrystalline diamond (PCD) growth is shown in the inset in Figure 8.16, where various grain sizes are observed. Due to the different working frequencies, the diamond device can operate simultaneously as a Si-V optical antenna and as a GHz antenna transmitting part of the acquired data in a C-band communication system. The substrate pretreatment is done by spraying 100 pulses of a thin nanodiamond (ND) dispersion on a cleaned *p*-doped silicon (100) substrate (500 μm thick). The dispersion is made by quantity (30 mg) of ND having a diameter of 250 nm in 30 ml of apolar solvent (1,2-dichloroethane). Before the spray, the dispersion is sonicated for 30 minutes. A laser cut metallic mask is used in order to spray only the regions where the diamond (layout) will be grown by the MWPECVD technique.

In this case the RE process involves:

- FEM simulation of the material characterization (mechanical, thermal, electromagnetic, etc.).
- Optimization of the growth working conditions (temperature, pressure, plasma conditions, etc.).
- Optimization of the mask layout by means of the 3D FEM simulations able to estimate the best reflection coefficients at the input port (S_{11} coefficient indicating a low reflection at the input port).
- Optimization of the linking between the coaxial cable and the layout characterized by the surface electrical current.
- Optimization of the thickness of the growth layer influencing the electrical current distribution in the bulk material.

The RE approaches to follow in production processes are mainly based on the use of FEM simulation tools. The FEM technique is adopted in different steps of the RE processes, such as material behavior characterization, rapid prototyping, sensor characterization, and sensor optimization. The material characterization by FEM is useful in order to understand the material behavior under strong stress conditions (high temperatures, high pressures, etc.). Figure 8.17 shows an example of the diamond material behavior as deformation displacements along the spatial domain with different temperatures. This numerical characterization is important to understand the working conditions of the sensor to be realized, and because it is difficult to characterize experimentally the material in microscale dimension. RE is based in this case on establishing the deformation behavior of the material by varying different parameters (temperature, stress, mechanical forces, etc.). The deformation trend behavior changes with the chemical composition, the applied fabrication process procedure, and with the geometrical layout (thickness, volume subject to the application of forces, joints of materials, etc.). The first step of the RE is to optimize the material and the basic layout for the prototyping. Rapid prototyping is a successive production phase: after the choice of the materials to use (produced materials, substrates, etc.), the modeling and the design of the sensor indicate the successive steps to perform. The device to fabricate must be also experimentally characterized in order to compare the numerical results of the designed prototype. If good matching between numerical and experimental results is checked, the design is stable, and the model is a reference for the optimization process and for the parametric analysis. Figure 8.18 shows the comparison between

Figure 8.16 Photograph of the diamond planar antenna sensor. Inset: optical image of a zoomed part of a diamond edge boundary.

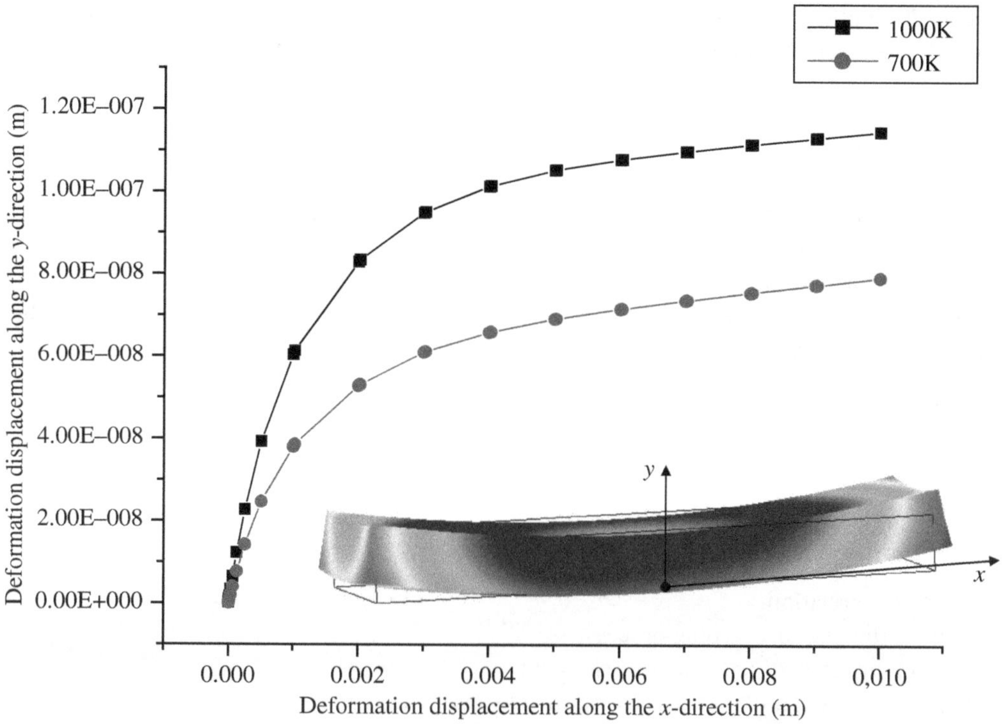

Figure 8.17 Deformation displacement along the x- and y- directions with temperatures of 700 and 1000 K. Inset: example of a 3D deformation.

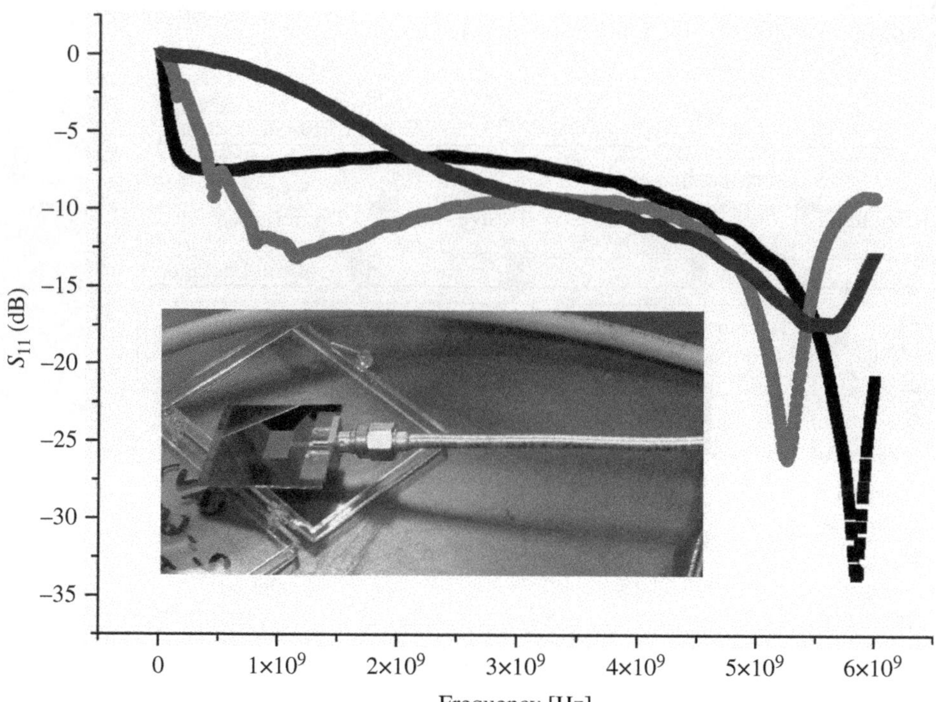

Figure 8.18 Measured S_{11} responses of different antenna prototypes. Inset: photograph of a prototype antenna.

measured scattering S_{11} results of different integrated diamond antennas. The measurement is a fundamental step of the RE approach in the frequency behavior analysis of the samples.

8.8 Advanced RE Processes in Industry 5.0

In Industry 5.0, some facilities are RE methodologies able to optimize the production of new devices and new components. Rapid prototyping techniques are especially important for the first realization of prototypes, and RE is fundamental for prototype optimization addressing the production processes for standard fabrication procedures. Figure 8.19 provides a full framework of the RE, structured in the following different fields:

- FEM techniques
- AI
- Experimentation

The FEM technique is an important tool for the formulation of advanced RE processes in industry and supports the industrialization processes. FEM techniques are suitable for:

- Material characterization.
- 3D design for the first prototyping process.
- Device design by simulating the sensor behavior.
- Sensor optimization (simulations of the optimized sensor).

AI represents an important tool for advanced RE. In each step of the FEM technique and measurement, it provides information and data to process for:

- Material classification.
- Tolerance definition addressing the fabrication optimization.

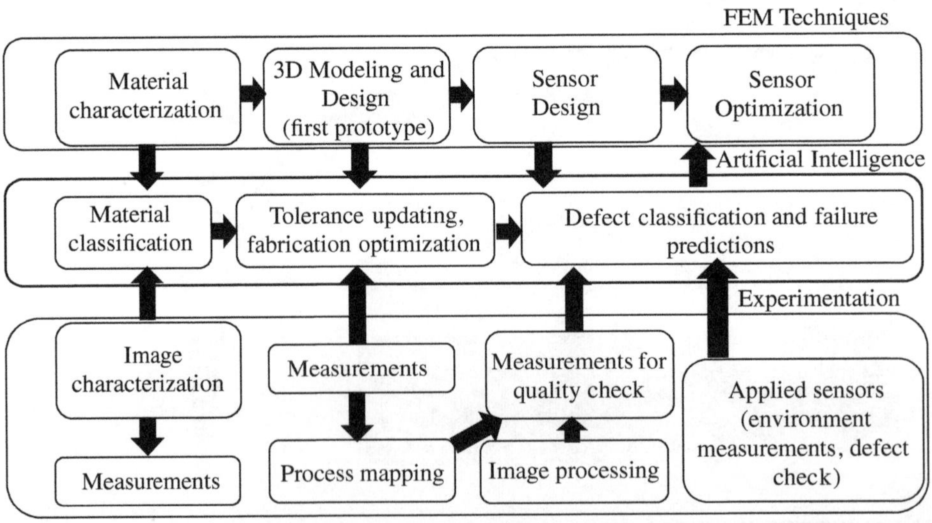

Figure 8.19 Advanced RE processes in Industry 5.0: production of a new sensor.

- Defect classification and defectivity prediction of the new fabricated product.
- Machine failure predictions due to defectivity analysis.

The experimentation phase is particularly important for the optimization process. Experimentation includes:

- Production characterization by images tracing the production processes, checking material implementation, and measuring geometrical parameters.
- The test of the new production processes to map related to fabrication of new devices.
- Measurements for checking product quality provided by the new processes, by means of image vision or other measurement methods.
- Sensor measurements of the production environment and defect check.

The flowchart in Figure 8.19 represents the evolution of the RE processes in the new era of Industry 5.0 by highlighting all the facilities concerning FEM techniques, AI, and experimentation. The new concept of RE is mainly based on the adoption of AI, which facilitates the choice of the materials and the fabrication optimization according to the defect and the quality analysis. The defect classification and the sensor failure prediction are suitable for the quality assessment, quality assurance, and quality control.

The automatization of the RE processes addresses production for a self-adaptive production process enabling self-tuning of the product quality. The automatisms in advanced RE are related to different production processes. Table 8.6 lists the most important automatisms.

Table 8.6 Possible automatisms in RE.

Automatism process	Description
3D object reconstruction	The use of laser scanners enables the 3D computer-aided design reconstruction of a used device. The possibility to reconstruct accurately new objects is also appropriate to recognize possible defects. The self-learning classification algorithms automatically recognize the defects by providing important information for the redesign process
FEM simulation	FEM simulations are executed for different prototypes in parametric modality, automatically providing a wide range of numerical results. The outputs of the FEM tools are used for the prototype optimization by taking into account tolerances and sensitivity analysis
Measurement process	The measurements to perform must be planned and automatized by following precise steps. The *in-line* measurement approaches are suitable for a continuous real time check of a device or component production
Test process	The test processes can be automatized by analyzing during the test the device response. The tests performed to analyze risks and limits of the component or device are static processes. The self-adaptive tests defined by the real time test adjustment are dynamic and change continually during the time, based on the observed results
Manufacturing processes	The manufacturing processes are automatized by setting the updated production machine parameters. The updating of machine parameters is the result of the optimization process. Additive manufacturing processes are advanced solutions for the programming of the manufacturing processing

3D, three-dimensional; FEM, finite element method.

8.9 RE in Nanocomposite Production Processes

Implementation of NMs in sensors requires an accurate design of the materials and the preliminary study of the physical behavior as a result of the technology used. RE processes for nanocomposite sensor fabrication are mainly focused on the technological aspects. The improvement of the technology can optimize the sensor capabilities, by passing from a basic prototype to a tailored one. The first pressure sensor prototypes based on NMs take into account the superposition of a polydimethylsiloxane (PDMS) polymer with gold (Au) nanofillers constituting a nanocomposite cladding [16, 17]. The basic principle of pressure detection is based on the optical intensity decrease at the output of the optical fiber when a pressure force is applied on the top of the sensor. In the first prototype sensor in Figure 8.20 [16], the nanocomposite cladding is placed directly on a tapered optical fiber which is semi-embedded into the PDMS polymer. In this case the RE approach takes into account many aspects of the sensor efficiency, such as:

- The adhesion of the PDMS-Au cladding when a pressure force is applied.
- The concentration of Au nanoparticles.
- The thickness of the PDMS-Au cladding.
- The optimization of the chemical processes of the PDMS polymer and of the PDMS-Au nanocomposite cladding.
- The light coupling efficiency due to the unfixed PDMS-Au cladding.
- The sensitivity optical response versus the applied pressure forces.

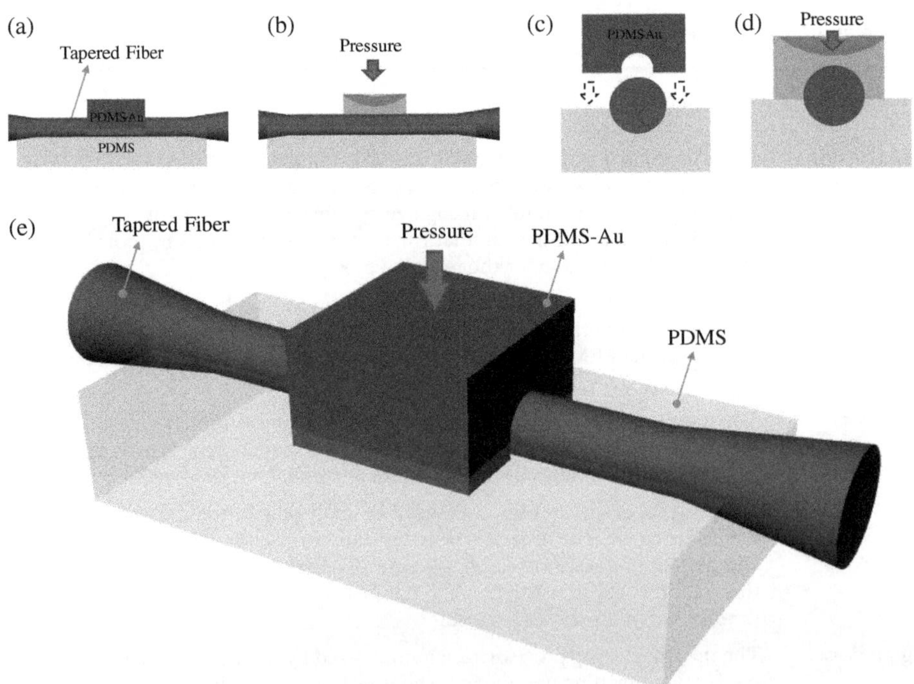

Figure 8.20 First prototype of PDMS-Au pressure sensor. (a) side view; (b) side view and PDMA-Au configuration after the application of pressure forces on the NM; (c) front view of PDMS-Au placement on the optical fiber tapered region; (d) front view configuration after the application of pressure forces; and (e) 3D CAD modeling of the first pressure sensor prototype.

- The optimization of the tapering of the middle part of the optical fiber.
- The possibility to reproduce pressure forces with weights to place on the PDMS-Au cladding for the measurements.
- The experimental setup able to couple the source light into the input of the optical fiber with the best efficiency.
- Characterization by numerical simulations of the light coupling effect in the nanocomposite PDMS-Au materials.
- The definition of physical limits matching with the application fields.
- The design of optoelectronic circuits enhancing the output light intensity, thus increasing the signal-to-noise ratio.
- Measurement repeatability.
- Electromagnetic and mechanical response time.

The RE approach provides more information about possible improvement of the first sensor prototype by enhancing the critical aspects.

Most of inefficiencies are improved by the second prototype shown in Figure 8.21, where the PDMS-Au is directly deposited on the PDMS substrate, optimizing mechanical stability of the whole sensor. The RE processes of the second prototype are different from the first prototype, and are related mainly to the optimization of technological aspects. The RE involves in this case the analysis of a reduced part of the specifications, such as:

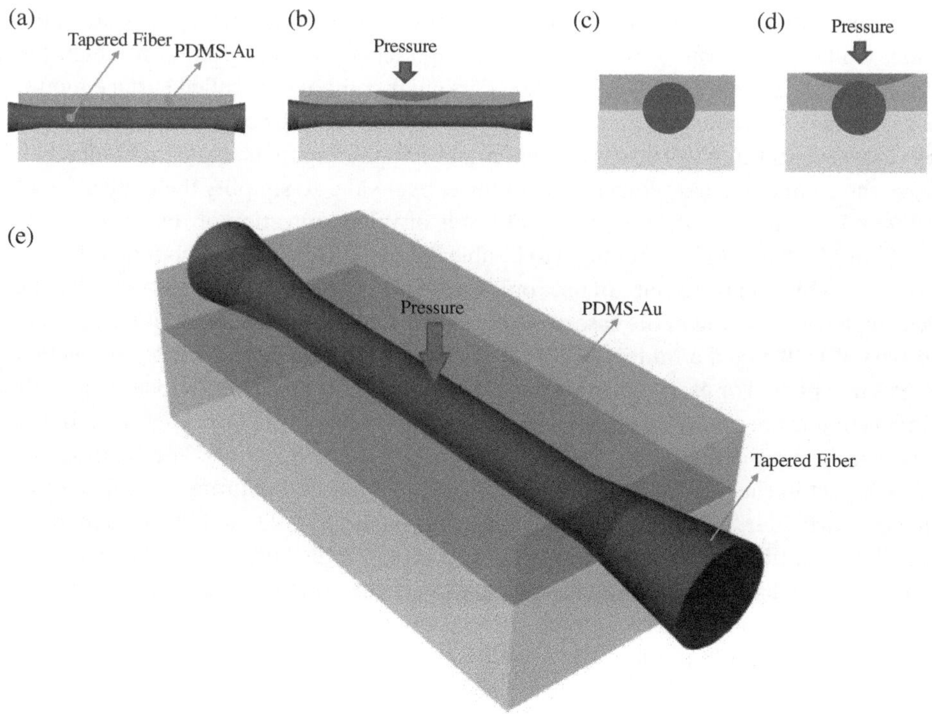

Figure 8.21 Second prototype of PDMS-Au pressure sensor. (a) side view; (b) side view and PDMA-Au configuration after the application of pressure forces on the NM; (c) front view of PDMS-Au material deposition on the whole device; (d) front view configuration after the application of pressure forces; and (e) 3D CAD modeling of the second pressure sensor prototype.

- The thickness of the deposited PDMS-Au cladding.
- The optimization of the chemical process of the nanocomposite PDMS-Au cladding.
- The extension of the application fields by tuning sensor sensitivity and mechanical properties, with gold nanoparticle concentration variation.
- Electromagnetic and mechanical response times changed with the improvement in technology.

The RE approaches change with the adopted technology and are updated continually after each prototype optimization process, as proved by the two examples discussed. The proposed examples are useful to understand the approach to follow step by step when preliminary limits are overcome with new optimized prototypes.

8.10 RE in Electronic Board Production

The printed circuit is the most used and most stable mechanical support for connecting electronic components. The technologies of production of printed circuits have undergone radical changes over the last few decades. A PCB consists of an insulating support on which thin copper strips (tracks) are made to connect the electronic components soldered to the same support. The components are assembled according to two techniques: through hole technology (THT) and surface mount technology (SMT). The first technique is the oldest and consists of assembling the components by welding their terminals on a copper area called the pad, by passing the terminals through holes made on the insulating support. The face on which the components are arranged is called the component side, while the opposite face, where the components are welded and on which the tracks are present, is called the copper side. In SMT, particular components called surface mount devices (SMDs) of reduced dimensions and with terminals adhering to the container (package) of the component are used. In this case the component is placed adhering to the plate and welded on the same side, so the component side coincides with the copper side. To simplify the realization of a printed circuit and to minimize its area, different levels of interconnection of the components can be used (both in THT and SMT technology) to be able to cross different tracks, but on different planes. In the case in which a printed circuit uses only one face for the realization of the tracks, the circuit is called single-sided, while in the case of two faces the circuit is called double-sided. There is also a third type of PCB called a multilayer circuit in which different levels of interconnection are made on the same plate. For example, in the case of personal computer motherboards usually four-layer printed circuits are used: one for the voltage common collector input voltage, one for the ground, and two for the considerable interconnections between the components. The creation of a PCB is a delicate aspect in the construction of electronic equipment and requires careful analysis by the designer. In particular, for the realization of a PCB it is necessary to apply a considerable inventive effort linked to the experience acquired by the various PCBs built. For this reason, it is always better to try to realize a first prototype PCB starting from simple single-sided circuits in THT technology. The creation of a printed circuit is preceded by a delicate design phase, in which the choices made play an important role in order to obtain a good final result. The construction of a PCB starts from the realization of an electrical diagram in which all the components and the interconnections between them are represented. Already in this phase the choices made can affect the final result; in fact, some interconnections present on the wiring diagram can be modified in a preventive way, already thinking about the consequences for the PCB. The RE processes to perform can be matched with interconnection reconfiguration improving the first prototype. For example, appropriate configurations of the input/output pins of a microcontroller are chosen in order to minimize the crossings between the wiring diagram connections. After the realization of the wiring diagram, it is possible to move on to the PCB project. In this phase, the position of the components in the printed circuit is chosen, and the connections between them are made by means of the tracks. This phase is simplified by the use of appropriate CAD tools. The realization of the layout can be difficult, especially for circuits in which several components with many terminals are used; in fact, the best arrangement of the components and the paths of the tracks must be chosen, minimizing the number of intersections. This phase is called routing and represents an important RE step to follow. The combinations of component position, paths, and minimized areas have different solutions, most often dictated by the skill of the designer. RE takes into account these solutions by introducing jumpers, in order to "jump over" the tracks corresponding to the intersections. The jumpers are made by inserting a normal piece of cable (usually 0.5 mm) between two pads previously inserted in the layout. In the most critical cases, in the presence of many intersections, a double-sided plate is used, in which the tracks are created on the two surfaces. In this case the electrical connections between two sections of the circuit take place through a pass-through connection, which is made by inserting a piece of cable (always 0.5 mm thick) through the plate and welding the two ends to two overlapping pads (Figure 8.22).

After the realization of all the interconnections between the components, the obtained path is called a master, and in the case of a double-sided layout, two masters are obtained. It is useful to observe that the dimensions of the tracks must be adequate: both for the flowing electrical current passing through them (wider tracks for higher currents), and for the type of process used for the realization of the PCB (the tracks cannot be too small, at most, 0.4 mm tracks can be made). Therefore, it is possible to create tracks with a minimum width of 0.4 mm but it is always a good idea to use thicker thicknesses (about 1 mm). A graph similar to that shown in Figure 8.23 can be used to select the most appropriate thicknesses.

The RE aspects are then correlated to the geometrical parameter and simultaneously to electrical current flowing in the circuit. After the layout design, the next step to follow is the PCB construction phase which requires a lot of attention. In this phase it is necessary to transfer the created master (or the masters in the case of a double-sided plate) on the plate, creating the copper traces for the passage of the current. Initially, as for the rapid prototyping process, it is recommended to print the master on a sheet of paper in order to check all the connections by comparing them with the wiring diagram, and to check if the dimensions of the components are the real ones, especially for SMD components where usually the tolerance is poor.

The realization of a PCB, that is, the transformation of the master made of copper placed on an insulating support, requires several steps, such as:

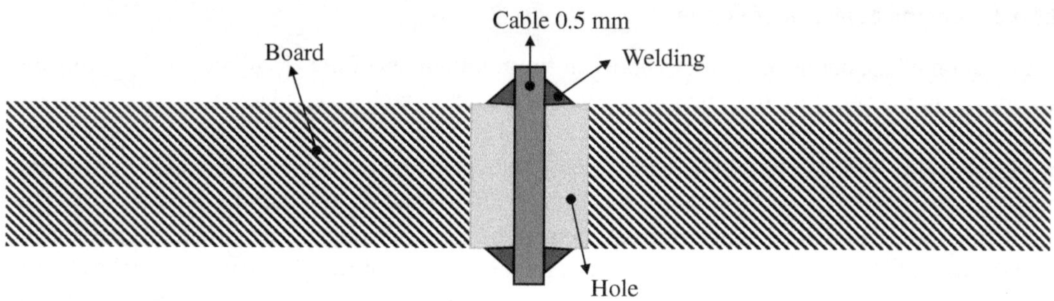

Figure 8.22 RE: jumper solution.

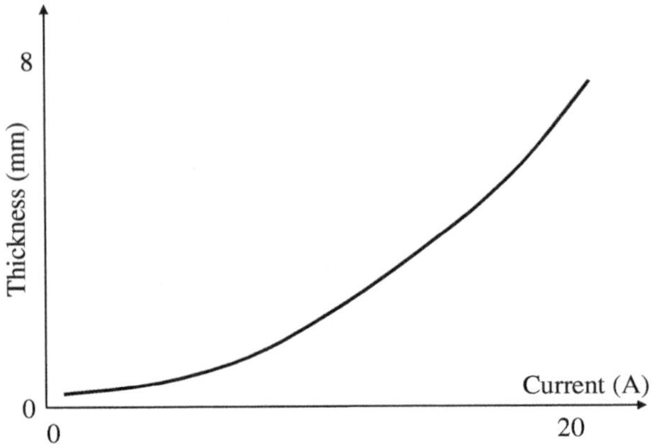

Figure 8.23 Thickness trend versus electrical current.

- Transfer of the master to the copper plate.
- Chemical attack of copper.
- Drilling and finishing.

The steps are optimized by following RE procedures defined by the adopted technological steps.

8.10.1 Transfer of the Master to the Copper Plate

Starting from the master, it is possible to transfer the printed layout onto a copper plate using various "artisanal" techniques: use of press-n-peel (PnP) sheets, photoengraving, or transfers. The first PnP technique is simple and immediate. The PnP sheet, or more commonly called a blue sheet, is a particular plastic sheet of A4 size with a smooth plasticized face and a rough face, both blue in color. PnP is a sheet that can be printed by any laser printer (only on the rough side) and has the property of releasing the toner when heated to a temperature of about 100 °C. Using this sheet it is possible to bring the master realized on the copper plate. Initially the master to be transferred on the PnP sheet must be printed, using the maximum definition of the printer. After printing the master, the piece of blue sheet is detached from the sheet of paper and placed on a copper plate (with the printed side in contact with the copper), which has previously been cleaned. The next step is the adoption of a double-sided bromograph with vacuum pump for the exposure of photosensitive materials.

8.10.2 Chemical Attack of Copper

The creation of the connection tracks between the components of the circuit occurs by removing the copper not covered by the previously created master, which takes place through a chemical oxidation–reduction process. This reaction is caused by liquid salts which attack the copper. The previously created master has, in fact, the purpose of protecting the copper that will later form the layout tracks. A solution of ferric chloride ($FeCl_3$) can be used for the etching process, requiring about 10 minutes for a fixed temperature. When all the exposed copper is exported, it is possible to remove the plate from the solution and clean it under a stream of running water. After the chemical attack, it is possible to note the absence of copper in the areas not covered by the blue card. The development method is given in detail in the following steps:

1) Darken the room where the board will be created, since light could attack the unexposed copper and therefore alter the design phases.
2) Place the plate to be engraved inside a plastic container, combine 7 g of caustic soda with 1 l of water (this solution must cover the entire plate).
3) Leave the card in the solution for 5–10 seconds.
4) Shake the solution to increase the reaction rate (small bubbles are observed resulting from the reaction; for this reason, it is necessary to carry out the development in a well-ventilated environment).
5) Immerse the board in $FeCl_3$ solution.
6) When all the copper not covered by the master has been eliminated, it is possible to remove the plate from the solution using tweezers.
7) Finally, the plate must be washed under running water.

8.10.3 Drilling and Finishing Processes

After the chemical attack, it is necessary to remove the blue layer and any transfers from the plate using abrasive paper. To facilitate the operation, it is advisable to use ethyl alcohol. Then it is possible to cut the parts outside the layout, and drill the pitches using a small drill. After making the PCB, the components are positioned on the plate and soldered. Generally, it is recommended to start with the assembly of smaller components, for example SMD ones, in order to easily manage the welding phase.

References

1 Kuo, T.C. and Zhang, H.-C. (1995). Design for manufacturability and design for "X": concepts, applications, and perspectives. In: *Proceedings of IEEE/CPMT Electronics Manufacturing Technology Symposium*, Austin, USA (2–4 October 1995). Piscataway, NJ: IEEE.

2 Dehtiar, J., Durupt, A., Bricogne, M. et al. (2018). Deep learning for big data applications in CAD and PLM – research review, opportunities and case study. *Computers in Industry* 100 (1): 227–243.

3 Manzella, J. and Mutafelija, B. (1992). Concept of re-engineering life-cycle. In: *Proceedings of the Second International Conference on Systems Integration*, Morristown, USA (15–18 June 1992). Piscataway, NJ: IEEE.

4 Chen, H., Cui, Y., Qiu, R. et al. (2018). Image-alignment based matching for irregular contour defects detection. *IEEE Access* 6 (1): 68749–68759.

5 Rucco, M., Giannini, F., Lupinetti, K., and Monti, M. (2028). A methodology for part classification with supervised machine learning. *Artificial Intelligence for Engineering Design, Analysis and Manufacturing* 33 (1): 100–113.

6 De Smedt, J., Hasić, F., Vanden Broucke, S.K.L.M., and Vanthienen, J. (2017). Towards a holistic discovery of decisions in process-aware information systems. In: *Business Process Management*, Lecture Notes in Computer Science, vol. 10445 (eds. J. Carmona, G. Engels and A. Kumar). Cham: Springer https://doi.org/10.1007/978-3-319-65000-5_11.

7 Wu, Z., Yao, T., Fu, Y. et al. (2017). Deep learning for video classification and captioning. In: Frontiers of Multimedia Research (ed. S.-F. Chang), 3–29. Association for Computing Machinery and Morgan & Claypool, ISBN: 978-1-970001-07-5, https://doi.org/10.1145/3122865.3122867.

8 Dipierro, G., Massaro, A., and Galiano, A. (2020). Characterization of an efficient thermal insulation composite system for non-residential buildings: a comprehensive theoretical, numerical and experimental approach. *International Journal of Current Advanced Research* 9 (5C): 22177–22185.

9 Massaro, A., Errico, V., Stomeo, T. et al. (2008). 3-D FEM modeling and fabrication of circular photonic crystal microcavity. *IEEE Journal of Lightwave Technology* 26 (16): 2960–2968.

10 Massaro, A., Tasco, V., Todaro, M.T. et al. (2009). FEM design and modeling of chi(2) second harmonic enhancement in circular photonic crystal. *IEEE Journal of Lightwave Technology* 27 (19): 4262–4268.

11 Caratelli, D., Massaro, A., Cingolani, R., and Yarovoy, A. (2012). Accurate time-domain modeling of reconfigurable antenna sensors for melanoma skin cancer detection. *IEEE Journal of Sensors* 12 (3): 635–643.

12 Massaro, A., Cingolani, R., and Passaseo, A. (2011). 3D FEM modelling and technology of piezoelectric ring MEMS antenna. *Progress In Electromagnetics Research C* 23 (1): 123–135.

13 Massaro, A., De Guido, S., Ingrosso, I. et al. (2011). Free standing piezoelectric rings for high efficiency energy harvesting at low frequency. *Applied Physics Letters* 98 (5): 1–3.

14 Massaro, A. (2015). Energy harvesting aspects of irregular vibrating forces acting on piezoelectric devices. In: *Innovative Materials and Systems for Energy Harvesting Applications* (eds. L. Mescia, O. Losito and F. Prudenzano), 143–158. Hershey, PA: IGI Global.

15 Cicala, G., Massaro, A., Velardi, L. et al. (2015). DLC integrated GHz antenna for aerospace. In: *Proceedings of 2015 IEEE Metrology for Aerospace (MetroAeroSpace)*, Benevento, Italy (4–5 June 2015). Piscataway, NJ: IEEE.

16 Massaro, A., Spano, F., Cazzato, P. et al. (2011). Real time optical pressure sensing for tactile detection using gold nanocomposite material. *Journal of Microelectronic Engineering* 88 (8): 2767–2770.

17 Massaro, A., Spano, F., Ekuakille, A.L. et al. (2011). Design and characterization of nanocomposite pressure sensor implemented in tactile robotic system. *IEEE Transactions on Instrumentation and Measurement* 60 (8): 2967–2975.

9

Rapid Prototyping

The chapter is focused on the approaches enabling rapid prototyping. In particular, aspects are discussed concerning modeling, design and prototype fabrication, by enhancing the procedures to follow for a pre-industrialization process starting with material implementation. The tools of the three-dimensional (3D) design modeling, the simulation, and the basic experimentations are introduced to understand the procedures to follow for the prototyping process. Rapid prototyping is also addressed for the possibility to merge different technologies to improve processes for water leakage underground inspections based on the use of different technologies such as Ground Penetrating Radar (GPR), infrared (IR) thermography, and unmanned aerial vehicles (UAVs). Particular attention is focused on antenna prototyping by highlighting all aspects involving design, advanced technologies, electromagnetic characterization, and possible applications. Moreover, as examples indicating the approaches to follow, three patent schemes enhancing research aspects for rapid prototyping are described. Artificial intelligence (AI) is included in the innovative methodologies to apply in the Industry 5.0 environment, combining design, modeling and simulation in auto-adaptive processes, and quickly generating a device or a mechanical component for the pre-industrialization. An important issue in the industry research is to design innovative materials as nanocomposite materials (NMs), which can improve physical or sensing properties. The goal of this chapter is to provide and hypothesize methodologies and approaches for rapid prototyping of advanced devices and systems implemented by standard, millimetric and nano technologies. This chapter discusses specific case studies of advanced technologies, tools, and applications, by enhancing possible innovations in prototypes as functions of the device properties or of the material to design and improve. Particular attention is paid to the prototyping of compact antennas, optical probe systems, micro- and nano-structured devices, and NMs improving physical properties, or implementing functionalities such as pressure sensing. All the proposed examples are addressed to provide methods and criteria to start production of innovative technologies in industries mainly oriented toward smart sensor and nanodevice fabrication, by adopting some basic tools suitable for rapid prototyping processes. The optimization process of the first prototypes is the first step of the pre-industrialization process of a designed innovative device. A self-adaptive production system controlling the working process addresses the pre-series production of the prototype to the advanced Industry 5.0 industrialization process. The rapid prototype is a fundamental step for a new concept of flexible production.

Electronics in Advanced Research Industries: Industry 4.0 to Industry 5.0 Advances, First Edition. Alessandro Massaro.

9.1 Rapid Prototyping Tools and Microscale Electronic Systems: Methodological Approaches

Rapid prototyping tools allow to accelerate the production of the first prototype to fabricate. A basic tool is the computer-aided design (CAD) tool. The CAD tool is usually integrated into a simulation tool providing software suitable for rapid prototyping matching with physical effects and improvements. The materials used for the prototyping of devices and components are processed by the available technology of the industry or by production services. Optoelectronic components and devices typically involve advanced technologies and, consequently, are applicable for a new production. Different tools for optoelectronic systems are listed in Table 9.1.

CAD, computer-aided design; CNC, computer numerical control; 2D, two-dimensional; 3D, three-dimensional; PCB, printed circuit board.

CAD modeling represents the first main step for the rapid prototyping procedure, which involves the preliminary simulation of the device or of the piece to fabricate. In this section, different examples are proposed of prototypes fabricated by following the rapid prototype procedure: design, simulation, and preliminary measurement matching of all the results. The proposed examples facilitate the comprehension of the tools to adopt for specific high technological devices having nanometric and micrometric dimensions, and highlight how to analyze each prototype by different methodologies and approaches.

9.1.1 Photonic Crystal Pillars for Filtering and Optical Resonance

The characterization of photonic crystal (PC) devices requires specific numerical and spectroscopy tools. Figure 9.1a–d shows an example of square lattice resonating nanopillars suitable for optoelectronic sensors: a tapered end optical fiber allows to couple a light source with the nanopillar structure enhancing or filtering the signal at the output of the periodic structure; the signal passing the nanostructure is coupled with an output optical fiber able to experimentally characterize the optical frequency behavior. The 2D and 3D CAD models are useful in this case for the design of the micro/nano layouts, and of the experimental setup indicating the input and output signal to analyze.

The high resolution of PCs represents a particularly important research challenge, especially for the implementation of plasmonic resonators suitable for many applications such as plasmonic resonator, gas sensor, biosensors, and photovoltaic cells. A very important aspect in rapid prototyping is the error margin related to the resolution of the fabrication process: in fact, the imperfect etched profiles due to the fabrication process can generate consistent variations if compared with the expected results. High resolution gold pillars have a resonant behavior in a frequency band ranging from 350 to 800 nm. In order to compare experimental frequency band results with numerical ones, a 3D simulator finite element method (FEM) tool can be executed. The FEM tool is based on the concept of a unit cell box enclosing the PC where each side of the unit cell is a waveport which implements the input and output signals. The 3D FEM simulation is able to confirm the resonance conditions in agreement with experimental results, thus proving the fabrication accuracy and resolution. Typical parameters used for the structure characterization are the absorbance and the transmittivity response (transmittivity from the input tapered fiber to the output as illustrated in Figure 9.1d). The absorbance is estimated as $A = -\log10(I_T/I_0)$, where I_0 is the input intensity and I_T is the transmitted intensity. The absorbance represents the resonant energy of the pillar-type PCs as the equivalent electromagnetic energy enhanced in the plane containing the

Table 9.1 Overview of the main rapid prototyping tools.

Tool	Description
2D CAD	Design for photolithography and for printed circuit. The 2D CAD layouts are designed by 2D simulators or are extracted by the 3D models
3D CAD	Design for CNC machine processing and 3D print fabrication. CNC production is typical of the Industry 4.0 framework. In automated production, the optimized 3D CAD design of a prototype is automatically processed by the production machines through the CNC
Signal simulator	Signal simulators are adopted for example to execute electric and electronic circuits, checking the use of the whole system made by different elements and components. The signal simulator is important for the experimental setup setting and for the pre-design of the first device or component
Numerical tools	Numerical tools implement different equation models in the following fields: • electrical • chemical • electronic • mechanic • thermal • fluid dynamic • electromagnetic • other specific physical behaviors Concerning for example electromagnetic sensors, the time domain modeling is well suited for problems involving transient analysis, wave scattering and waveguide coupling: finite-difference time-domain and transmission line matrix, and hertzian potential formulation are used also to solve devices at high frequencies. In the frequency domain, the finite element method and wave expansion method are suitable for the characterization of the frequency behavior of the devices [1]. The choice of the correct numerical simulators is related to the equation model to solve
3D printer	The 3D printer is suitable for the fabrication of micrometric and nanometric structures, and as an alternative technique for the rapid prototyping of small devices
Lithography tools	Lithography tools are typically used for thin film devices. Electron beam lithography is adopted for photonic crystal realization. Other lithography techniques can be applied for the rapid prototyping of small devices
Electronic PCB tools	PCBs are designed by different tools generating output for the manufacturing process
Basic experimental setups	A basic experimental setup is used to primarily check the matching between experimental and numerical results. Homemade setups facilitate the rapid measurements for a quick check of the new device characteristics
Materials for optoelectronic devices	Innovative optoelectronic devices and micro-/nanosensors can be fabricated by nanocomposite materials which are integrated into compact devices. The technologies adopted for the realization of these sensors, and for their characterization, require a very high precision. Spectroscopy tools and "ad hoc" designed experimental setups are applied for the preliminary check of the devices. For optical characterization, discrete optics experimental setups are typically designed
Thin film devices	Thin micrometric films are grown or deposited on industrial substrate for the realization of sensors working typically in the GHz range. Tools for the devices appertaining to this class are selected according to the specific application

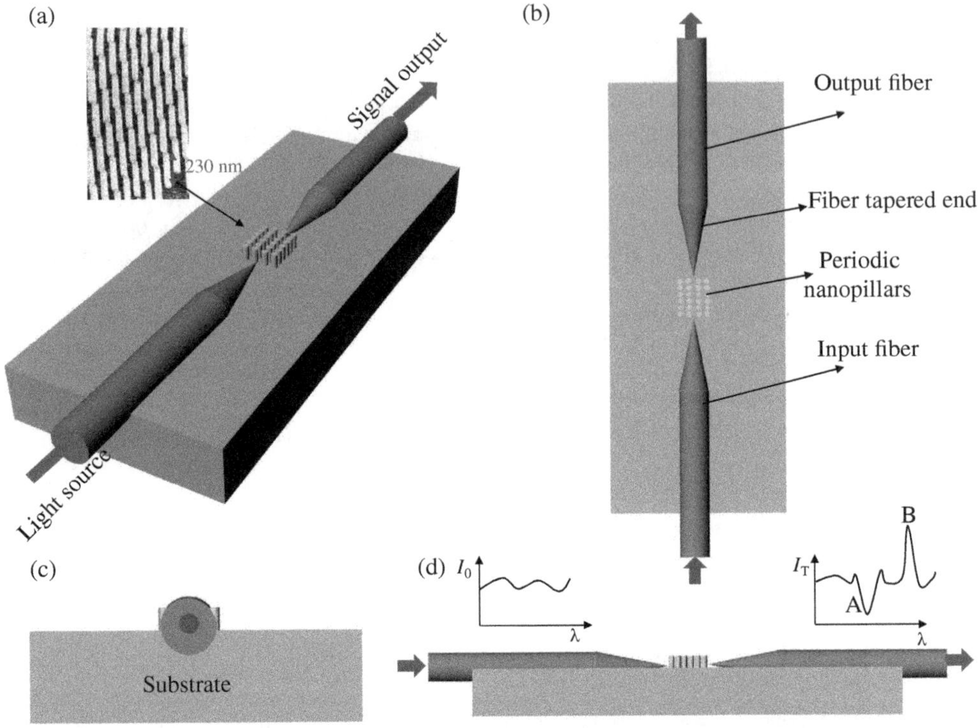

Figure 9.1 Example of a 3D CAD modeling of an optoelectronic device based on 60 nm diameter of nanopillar PC resonator: (a) perspective view (inset: SEM image of a pillar structure); (b) top view; (c) side view; and (d) front view. A represents a filtering condition and B is a mode enhancement case.

periodic layout. Figure 9.2 shows different repeatable measurements of the absorbance physical effect in a specific band, showing a filtering behavior as a stopband condition (the flat plot is the reference response of the substrate without the pillar structure). The absorbance is due to the energy trapping effect as resonance of the pillar structure: the light coming from an external light source is coupled with the periodic structure, and, at a specific wavelength, it is trapped on the whole periodic structure. The setup of the inset in Figure 9.2 is typical to observe the transverse resonance of the device: at the input of the reading photodetector the light intensity decreases due to the transverse resonance, describing the filtering behavior and consequently the energy absorbance. Otherwise, the resonance generates the opposite effect to enhance, as for active materials, the input light which is gained at the output of the structure. Considering this last case, by measuring the device by a nitrogen-cooled indium gallium arsenide (InGaAs) optical multichannel analyzer (OMA), it also possible to find resonant mode gaining at the output of the light signal (see the transmittivity response and the electric field resonance calculation in Figure 9.3 obtained by the setup configuration in Figure 9.1). A technique to fabricate metallic nanopillars is the electron beam lithography technique. The 2D simulations using the wave expansion method (WEM) approach facilitate the preliminary simulations defining the layout design finding resonant and cavity modes for structures with a central defect. The 3D simulation orients the simulation modeling on the real structure. The 3D simulation is characterized by a higher computational cost compared with the 2D case. For this reason, it is important to simplify the structure into a 2D one, to simulate, by properly setting the boundary conditions, the main frequency behavior. The steps to follow for the preliminary design are:

- 2D CAD modeling and simulation of the PC layout finding the best combination of physical material properties, such as refractive indices and spatial distribution of the pillar improving the frequency behavior.
- 3D CAD modeling and simulation, defining slab thicknesses and pillar heights, and optimizing the filtering and the resonance behavior.

The 3D modeling provides more information about the nature of the resonance supporting the ideation of the experimental setup to adopt for the characterization. Two resonance typologies occur in a PC structure: a transverse resonance in the plane of the 2D plane distribution of the periodic structure and a vertical resonance typical for optical antennas, where the electromagnetic field is enhanced in the direction vertical to the 2D plane distribution. The 2D simulation is also useful for a preliminary check of the cavity resonant modes. Figure 9.4a shows the 2D transverse plane layout of the periodic square lattice pillar structure, with a missing central pillar indicating the defect central region. Figure 9.4b illustrates the WEM simulation indicating the presence of a cavity mode confined in two stop bands. The cavity mode is confined inside the central region as proved by the electric field distribution in Figure 9.4c. The 3D model checks if the cavity modes produce energy trapped inside the cavity as transverse resonance modes or emit radiation in the vertical direction indicating antenna behavior.

The approach to follow for a rapid prototyping of a PC pillar-type device is based on the following steps:

- Definition of the application field (biomedical, industrial, etc.).
- Definition of the variables to detect according to the application field.
- Choosing of materials and technology.
- 2D simulation for the preliminary layout.
- 3D simulation for the full device design according to the application field.
- Preliminary measurements to check filtering or field gain behavior.
- Check of the sensing by adopting an appropriate experimental setup.

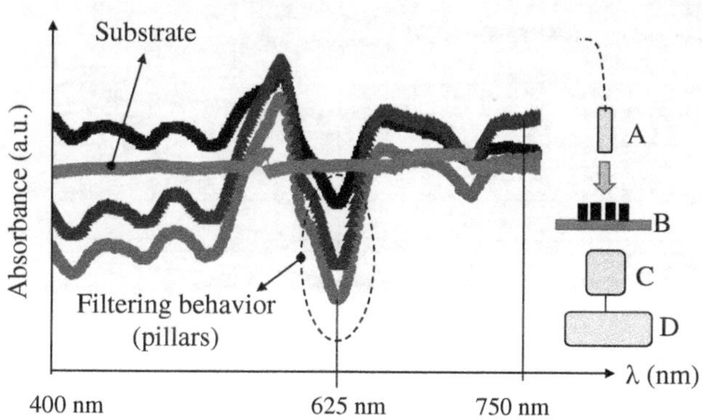

Figure 9.2 Absorbance enhancing the filtering behavior of the pillar type structure. Inset: experimental setup configuration (A, optical source; B, pillar structure; C, photodetector; and D, electronic signal analyzer).

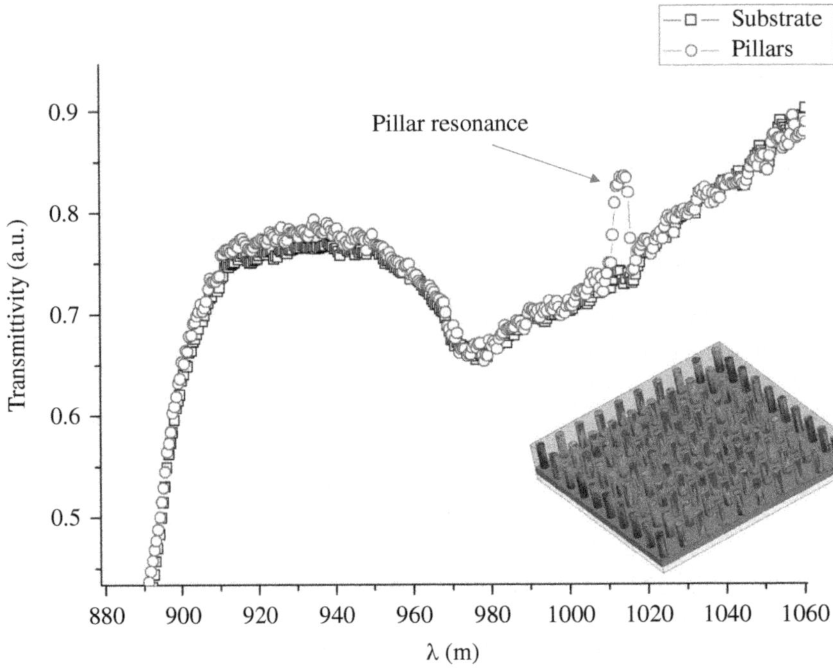

Figure 9.3 Measured transmittivity of the periodic pillar structure (inset: 3D FEM simulation reproduction proving the resonance of the mode peak enhanced by the periodic structure by the electric field distribution in the pillar periodic structure).

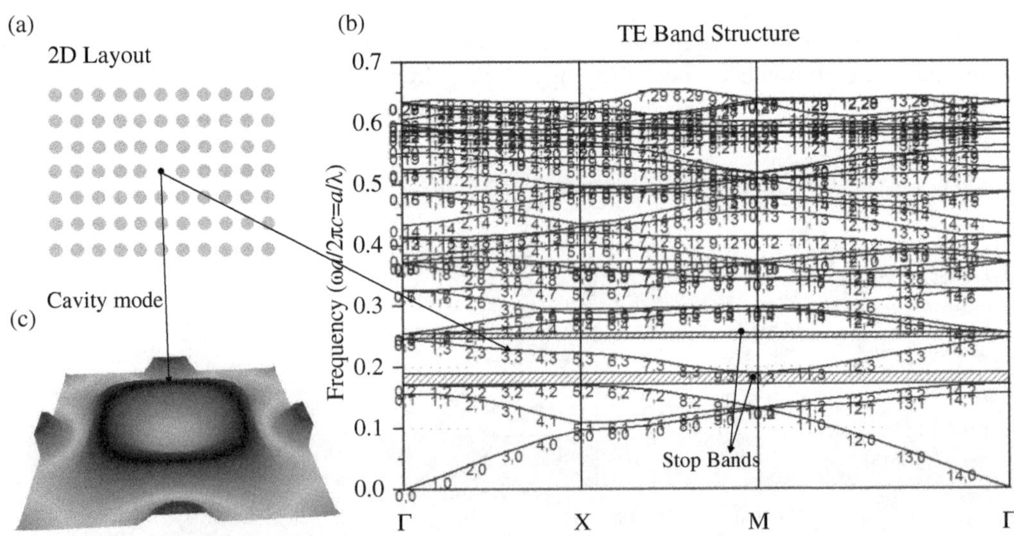

Figure 9.4 (a) Top view of the pillar layout (2D plane distribution of pillars); (b) transverse electric (TE) square lattice modes, and cavity mode localized in the central defect structure simulated by the WEM method; and (c) cavity mode resonance in the central defect region of the pillar periodic structures (electric field distribution).

9.1.2 Thin Film Microelectromechanical System Prototyping and Photolithography Approach

As a second example, Figure 9.5a–d shows a microelectromechanical system (MEMS) 3D design behaving as a radio frequency pressure sensor working around $f_0 = 11.19\,\text{GHz}$ [2]. The 3D modeling for the simulation is usually simplified to quickly check the resonant frequencies. Figure 9.6 shows a 3D FEM simulation proving the resonance at the working frequency: the structure is simplified by defining the boundary conditions at the lateral sides of the structure, and at the input and output ports where the radio frequency wave travels, and by neglecting in the simulation the substrate contribution which increases the computational costs. In the 3D CAD modeling, the complete application of the sensor is shown with the pressure force modeled as a weight placed on an electrical isolation material. In this case it is difficult to simulate the pressure force effect which theoretically changes due to the local deformation of the layered structure. Moreover, the application of the pressure force also theoretically changes the cladding permittivity, changing from 1 (relative permittivity of the air) to the value of the material applied on the top and in the middle part of the periodic structure. For these reasons the preliminary simulations are mainly focused on checking of the resonant frequencies. The comparison of the simulated results with experimental ones, together with the application of applied forces inducing a perturbation of the physical characteristics of the central part, provide important information about the frequency behavior of the prototype: at the working resonant frequency, and with the application force, the device stops transmission at the output port for a specific frequency band [2]. The first prototype does not provide any information about the lifecycle of the sensor, but is sufficient for the sensitivity analysis, thus defining the best working conditions such as typology of pressure force to apply and location of the application force, fixing conditions of the input and output ports. For a rapid realization, after layer growth by means of a suitable deposition technique, the layout is printed on plastic paper as shown in Figure 9.7a which is then placed on the multilayered growth structure to apply the bromograph photolithography process, realizing the device in Figure 9.7b. A bromograph is a standard machine to use to accelerate rapid prototyping processes.

The rapid prototyping procedure includes the definition of the best operating parameter conditions, such as power and irradiation time of the bromograph.

The approach to follow for a rapid prototyping of the MEMS-type device in Figure 9.5 is based on the following steps:

- Definition of the application field (scenario of the pressure forces).
- Definition of the thin film multilayered structure to realize by the lithography process.
- Simulation of a 2D or simplified 3D layout, checking device resonance at GHz frequencies.
- Imaging check of the preliminary prototype (resolution, defects, etc.).
- Engineering of the experimental setup measuring the frequency response (coupling element for the input and output port linking to the frequency analyzer).
- Measurement checking and comparison with numerical results to optimize the whole device (resonant modes, frequency band gap, pressure sensing effect, etc.).

9.1.3 Thin Film GHz Microstructures by the Photolithography Approach

The lithography approach is suitable for the fabrication of other GHz devices, such as integrated in-plane antennas. Different layouts of integrated antennas are shown in Figure 9.8a–c, which can be realized by using the quartz mask in Figure 9.8d in the photolithography process. For a better resolution in the micrometric scale, it is necessary to design a more stable mask to use for the

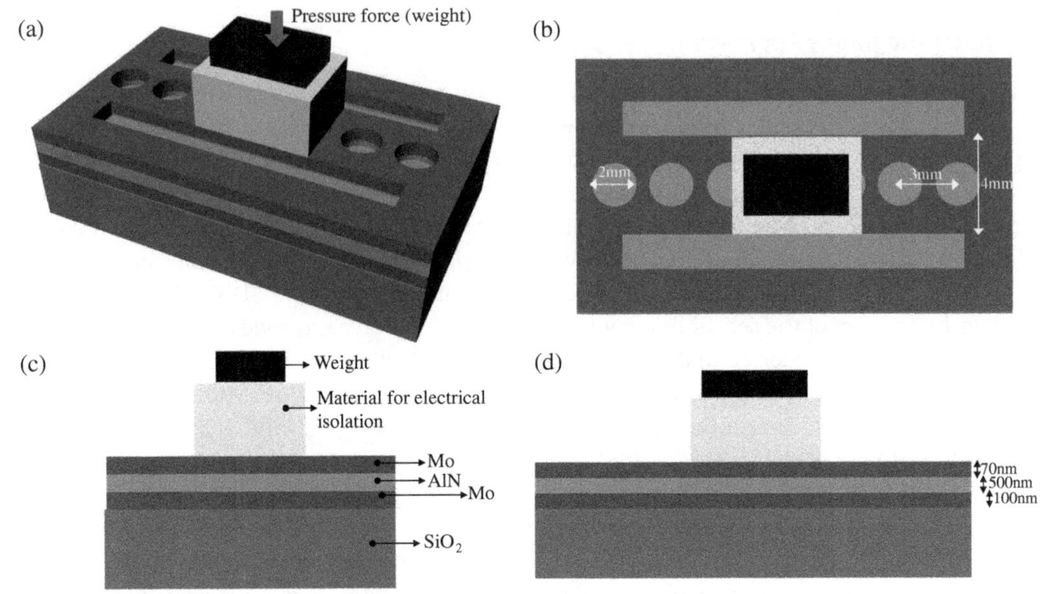

Figure 9.5 Example of 3D CAD modeling of a MEMS pressure sensor: (a) perspective view; (b) top view; (c) side view; and (d) front view. The upper layer of the integrated structure is molybdenum (Mo), the middle layer is a thin film of aluminum nitride (AlN) and the bottom layer is a Mo film on the silicon substrate. All the structure is growth on the silicon substrate.

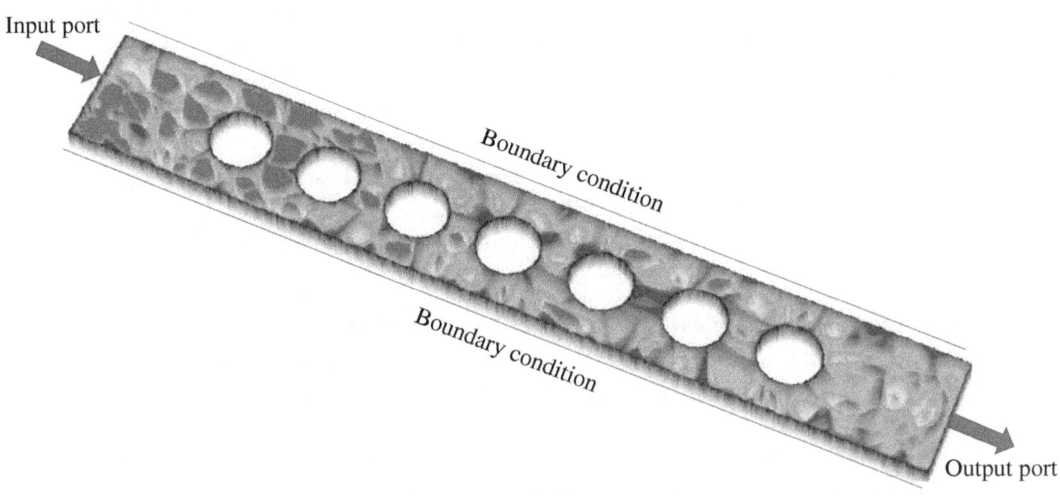

Figure 9.6 Simplified modeling and FEM simulation of the periodic structure in Figure 9.5.

lithography approach. The quartz mask is also applied for photolithography processes realizing gold (Au) micrometric structures in-plane by a laser irradiation technique applied on a chitosan including a gold precursor (chloroauric acid salt [$HAuCl_4$]) [3]. Particular parameter settings for microantennas are laser fluence (adopting an ultraviolet [UV] pulsed laser irradiation enabling photoreduction and photofragmentation) and the laser irradiation time (using a Nd:YAG laser) [4]. Different techniques such as UV–visible spectra, Fourier transform infrared (FTIR) spectra, atomic force microscopy (AFM) images, scanning electron microscopy (SEM), and X-ray photoelectron spectroscopy (XPS) are able to characterize the integrated antenna sample from the technological

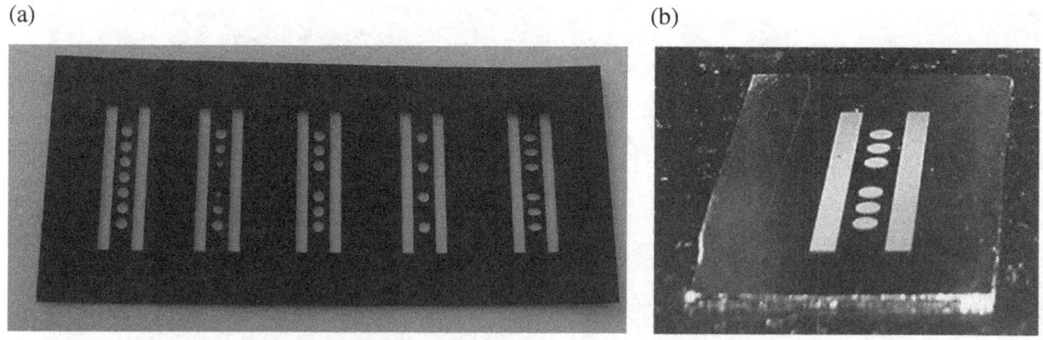

Figure 9.7 (a) Printed mask on a plastic paper to be applied for bromograph photolithography. (b) Photograph of the first prototype MEMS device. The hole diameter is 2 mm, the period is 3 mm, and the width is 4 mm.

point of view. The compact antennas are typically integrated as patch or freestanding devices [5]. After the realization, it is important to define the antenna application field: gold patch-type antennas with thin metallic film behave as plasmonic probes detecting near field, and freestanding structures are applied for local detection as in biomedical applications [6]. The experimental spectroscopy setup completes the rapid prototyping procedure by providing the first check of the working conditions. Figure 9.9a–d shows the preliminary micrometric antenna arrays: the preliminary check by microscopic images is important to check the laser texturing effect by analyzing the diffraction effect for the device resolution. This preliminary verification is fundamental to optimize the laser technique, and the methodology for the thin chitosan-Au film deposition.

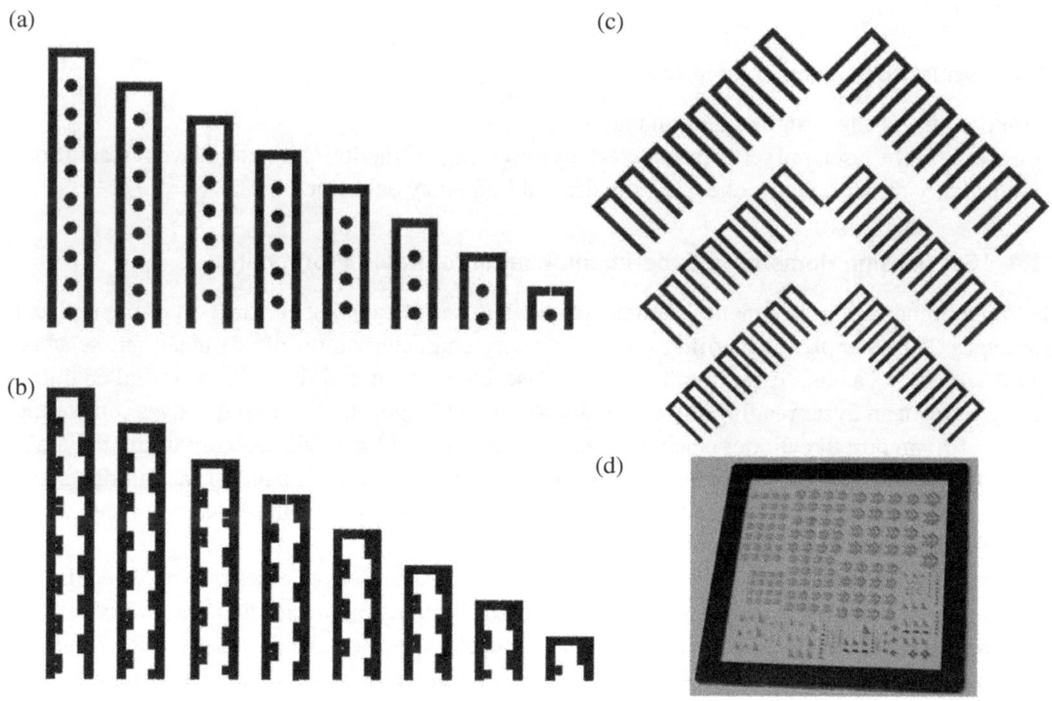

Figure 9.8 Microantenna patch-type layouts: (a) layout with holes; (b) layout with periodic edges; (c) array of microantennas; and (d) quartz mask with antenna array layout.

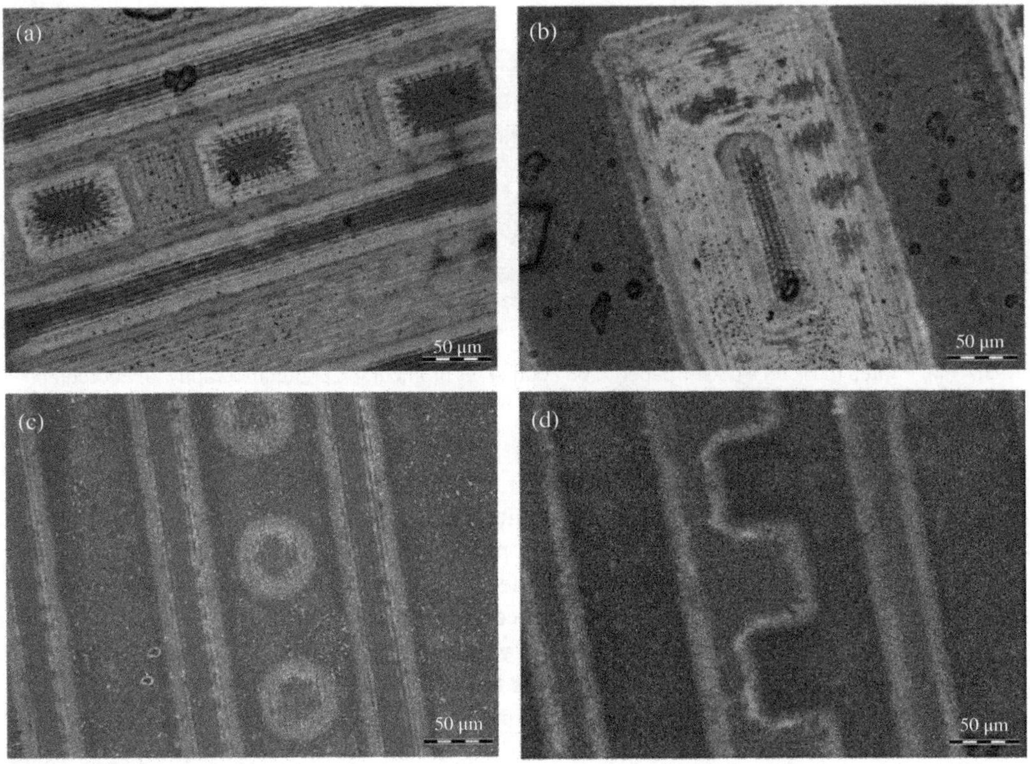

Figure 9.9 (a–d) Diffraction effect: preliminary profiles of integrated gold antenna arrays by the laser texturing approach.

The laser textured antenna samples are optimized by:

- Simulations for the design of antenna layout.;
- Checking of the laser and setup parameters, by observing diffraction effects and layout resolution.
- Preliminary measurements checking the desired frequency behavior.

9.1.4 Gas Sensing Homemade Experimental Setup for Rapid Prototyping

The rapid homemade experimental setups are usually used for a quick verification of the realized prototypes. For example, concerning the preliminary characterization of advanced gas sensors, Figure 9.10 shows a basic experimental setup to use, constituted mainly by: (i) a central chamber connected to an analyzer reading current (I) and voltage (V) signals, (ii) different valves controlling gas flow, (iii) an output cylinder closing the circuit, (iv) an input gas cylinder containing the main gas, and (v) a T-junction allowing gas mixing. The gas sensor sample is fixed in the middle of the central chamber and connected to the $I(V)$ analyzer by thin electric cables passing a mechanical support which prevents gas escaping.

The setup scheme in Figure 9.11a is suitable to measure the effect of gases mixing. Specifically, this setup considers five input gases going into a closed chamber and measuring the surface electrical current of a gas sensor. The gas surface absorption changes the current density **J** as a function of the gas or mixing of gases injected into the closed chamber. Electrical connections are attached to the sample inside the chamber, and outside on the amperometer ports. Figure 9.11b is a photograph of a homemade realized prototype setup for gas mixing detection.

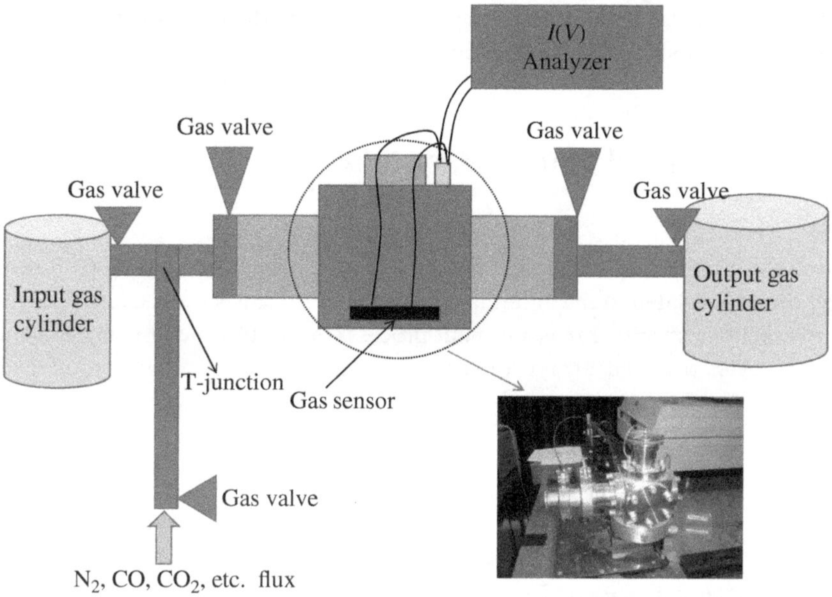

Figure 9.10 Basic experimental setup for gas sensing detection (inset: photograph of the realized experimental setup).

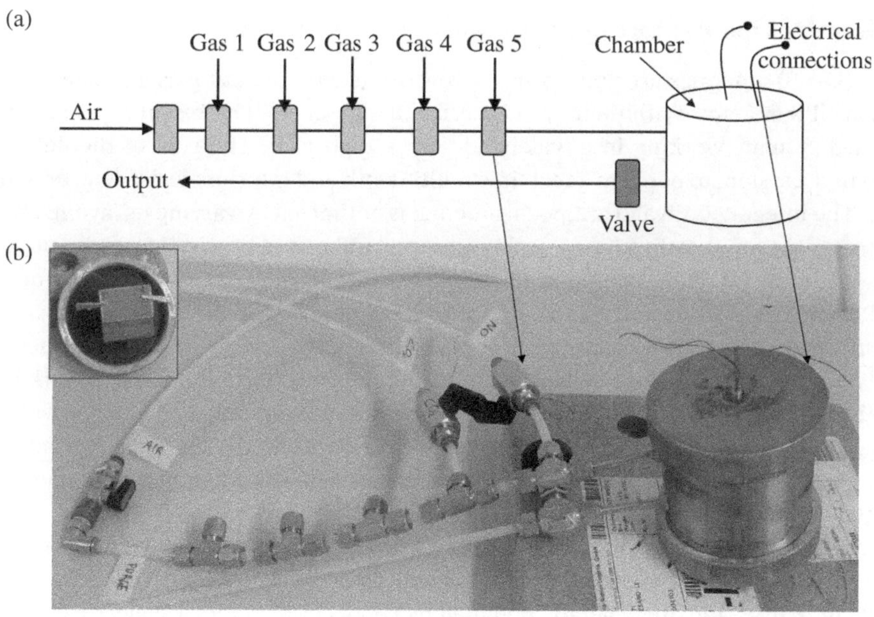

Figure 9.11 (a) Scheme experimental setup for gas mixing sensing detection (the air input is used to discharge the gas residues after a preliminary measurement). (b) Photograph of the realized experimental setup (inset: electrical connection of the nanocomposite sensors to check the gas sensing).

The homemade experimental setup is primarily designed according to the measurement type to per and the sample technology (morphology, dimensions, etc.)

9.2 Examples of Antenna and Detection System Rapid Prototyping

Rapid prototyping involves different advanced and standard technologies. Concerning GPR systems, the prototyping process is related to the antenna device, the electronic systems, and the systems of technologies to execute a specified measurement process. Innovative metamaterials, thin film antennas and technologies oriented toward underground leakage detection are prototyped based on the specific application field.

9.2.1 GPR Antenna Design for UAV Integration System

The GPR antenna systems present different advantages such as:

- Nondestructive and noninvasive investigation.
- Speed of execution.
- Numerous fields of application (engineering, geology, archeology, agronomy, environment, etc.).
- Versatility of the technology (the antennas are designed and built to have specific operating characteristics such as the frequency behavior of the electromagnetic field).
- Extreme variability in the shape and size of the device to be designed.
- Wide range of investigation (from millimeters to meters).
- Cost efficiency.
- Possibility to reconstruct the 3D image frequency behavior using software.

Ultra-wideband (UWB) antennas are applied in microwave image processing as a nondestructive technique as in GPR systems. With their low cost, Vivaldi antennas [7] represent a good alternative for integrated systems working in a wideband with a high gain. The goal of the design methodology is to find a system to perform a real time multi-depth and multi-material detection in GPR applications. The integrated Vivaldi antipodal antenna is optimized by varying its layout, and by adding metamaterials for metalized vias as in substrate integrated waveguide technology. Consider the design of an integrated antenna having dimensions $73.71 \times 48 \times 1.07\,\text{mm}^3$ with a Rogers Duroid RO4003C substrate ($\varepsilon_r = 3.55$). Figure 9.12a shows the designed layout and Figure 9.12b illustrates the theoretical behavior of the S_{11} coefficient, where the trough indicates a possible working frequency near 10 GHz. According to the dimensions, the antennas can work in a wide frequency band between 3 GHz and 10 GHz. As an example, the simulated antenna provides directivity $D = 9.25\,\text{dB}$ and gain $G = 8.95\,\text{dB}$ at frequency $f = 6.76\,\text{GHz}$. The theoretical trend in Figure 9.12b indicates the condition of maximum power transferred to the antenna corresponding to the main trough position.

An innovative aspect of the proposed antenna layout is due to the presence of metamaterials. These materials are characterized by electrical and magnetic properties not found in any of the materials available in nature, and are generally obtained by means of metallic or dielectric inclusions in a homogeneous substrate. Consider a volume V of dielectric material including a

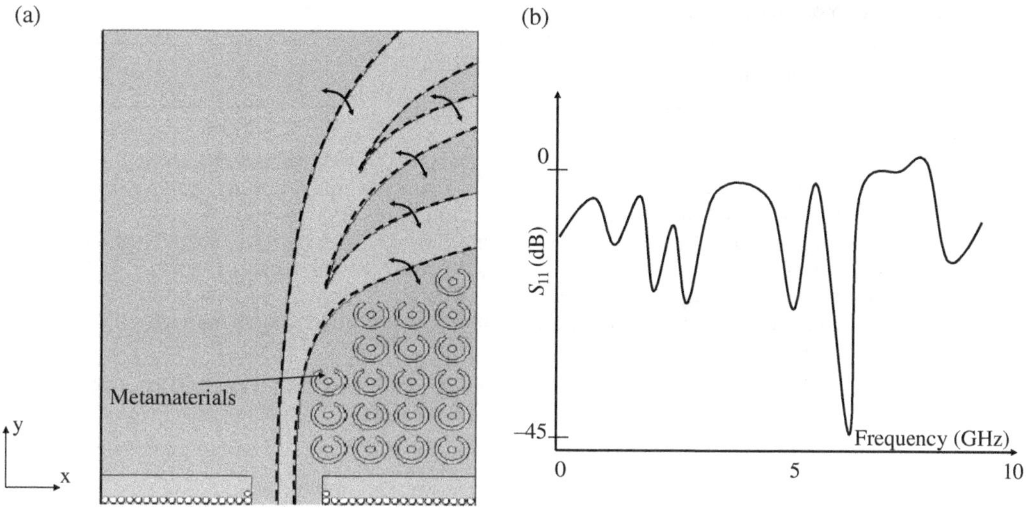

Figure 9.12 (a) Example of a layout of Vivaldi antennas and (b) reflection coefficient S_{11} theoretical trend, defining the working frequency as the main S_{11} trough (main frequency dip).

conductor or a dielectric with different physical properties, defining the effective electric and magnetic permittivities, respectively:

$$\varepsilon_{\text{eff}} = \frac{D_{\text{Average}}}{E_{\text{Average}}} \tag{9.1}$$

$$\mu_{\text{eff}} = \frac{B_{\text{Average}}}{H_{\text{Average}}} \tag{9.2}$$

where D_{Average} is the electric polarization vector, B_{Average} is the magnetic polarization vector, E_{Average} is the electric field, and H_{Average} is the magnetic field. The new obtained material is considered to be homogeneous and is characterized by the effective electric and magnetic permittivities, ε_{eff} and μ_{eff}, respectively. Metamaterials are periodic structure (or quasi-periodic structures) able to increase the radiated power in antennas especially when smart devices are considered. The rapid prototyping is typically performed by fixing a commercial substrate, and by varying the layout profiles as indicated in the example in Figure 9.12a, where the metamaterial layout is defined as working at a desired frequency. The compact UWB antennas are suitable for integration in UAVs. According to the weight limit, the UAV shown in Figure 9.13a is equipped with different elements and sensors. Figure 9.13b and Table 9.2 describe some possible elements to integrate in UAVs.

UAVs are appropriate for agriculture and in agri-photonics applications [8], enabling a decision support system (DSS) for example supporting hydric stress analysis by adopting the following image processing procedure:

- Radiometric image detection.
- Data mining processing of the radiometric images (K-means processing determining clusters characterized by different water contents of the plants).

Table 9.2 Description of UAV elements.

Element	Description
Microprocessor/ processor	A smart and compact processor manages all sensor data, by synchronizing the transmission to the external unit. The processor is connected to a MC implementing a firmware for multi data management. A specific operative system (for example Windows 10 64-bit, OS X, and Linux OS) facilitates the sensor programming. The processor is connected by a USB port to the Global Positioning System, Bluetooth, and 4G units enabling data transmission, to a designed PCB managing multiple sensors. The MC is directly connected to a smart multispectral camera, to a power bank (power supply), and to a motherboard managing Tx and Rx GPR antennas
Power bank	Different power banks, supply low voltage (5 V) electrical energy to different elements such as the PCB, GPR motherboard, and processor
PCB	The PCB is linked to a specific MC managing different sensors and photodetectors
Sensors and photodetectors	Different sensors and photodetectors acquire a lot of information in different application fields where precision detection is required. Data fusion and artificial intelligence techniques are applied
Multispectral camera	A multispectral camera detects signals in the RGB, NIR and SWIR bands. Examples of multispectral bands (with wavelength ranges in μm) are: • Band 1 (0.43–0.45) • Blue (0.45–0.51) • Green (0.53–0.59) • Red (0.64–0.67) • NIR (0.85–0.88) • SWIR 1 (1.57–1.65) • SWIR 2 (2.11–2.29) • Band 8 (0.50–0.68) • Band 9 (1.36–1.38) • Thermal infrared 1 (10.60–11.19) • Thermal infrared 2 (11.50–12.51) The camera also behaves as a hyperspectral camera: hyperspectral imagery consists of much narrower bands (10–20 nm), where the image is characterized by hundreds or thousands of bands. The multispectral bands are constituted by 3–10 wider bands, and hyperspectral bands are typically constituted by hundreds of narrow bands. Specific bands are selected by single photodetectors
GPR motherboard	The GPR motherboard manages the RX and the TX modules for the transmission and detection of the radar wave
TX module	The TX module is constituted by a digital-to-analog converter unit, a filtering stage, an amplifier stage, and the transmitter UWB antenna
RX module	The RX module is constituted by an analog-to-digital converter unit, a filtering stage, an amplifier stage, and the transmitter UWB antenna
External processor	An external processor processes all data coming from the unmanned aerial vehicle unit, executing the fly pattern programs and the precision navigation control. The software enabling the decision support systems is implemented in the external processor The external processor is required for data processing requiring a high computational cost, especially when a lot of data are collected by the multisensor system
USB cables	Cables with USB input and output ports connect different elements

GPR, Ground Penetrating Radar; MC, microprocessor; NIR, near infrared; PCB, printed circuit board; RGB, red, green, and blue; Rx, receiver; SWIR, short wave infrared; Tx, transmitter.

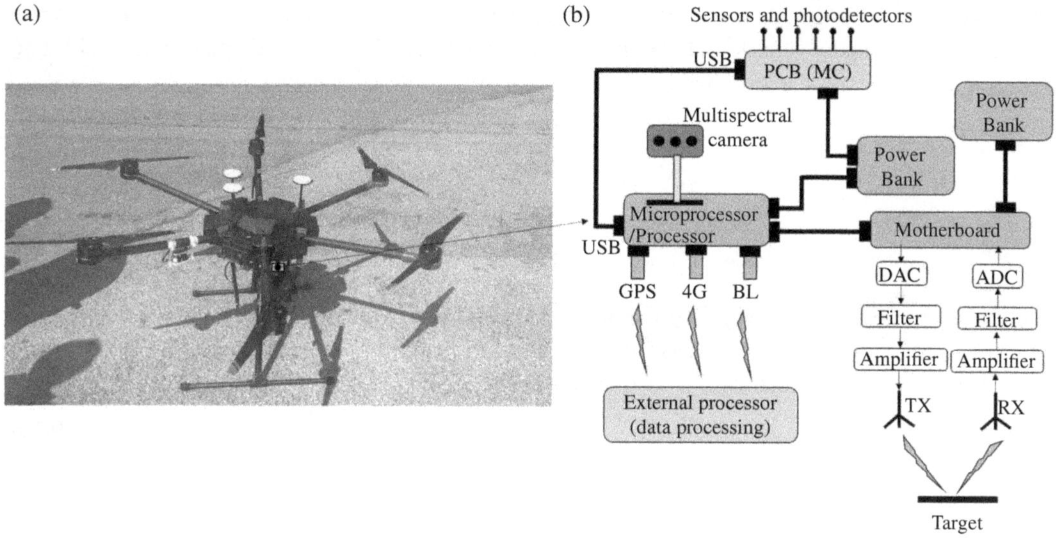

Figure 9.13 (a) UAV equipped with sensors. (b) Electronic scheme of the components of the UAV equipment.

- Image post-processing (by setting the pixel intensity threshold it is possible to extract and to enhance a part with lower or higher water content).
- Procedures suggested by the outputs of the post-processing results (decision making for irrigation, fertigation, or plant treatment).

The multispectral images are also used for estimating the ripeness of fruits, and for the detection of fluorescent-coated insects by exciting the target with a UV light source fixed on the UAV. The basic principle of GPR detection is simply explained by the geometrical ray model. Figure 9.14 shows the geometrical ray model representing the GPR signal trajectory. The model is suitable for the setting of the transmitter/receiver (Tx/Rx) inclination angle and, consequently, of the distance between the two antennas. For both antennas, the target analysis is a function of the reflection and refraction properties of the terrain where the target is placed. The geometrical parameters, like the inclination angles, are estimated by considering the following Snell equation:

$$\sqrt{\varepsilon_1}\sin(\theta_1)=\sqrt{\varepsilon_2}\sin(\theta_2) \tag{9.3}$$

where the air permittivity is $\varepsilon_1 = 1$. The distances are calculated by trigonometric formulas. The distance D is the distance of the UAV from the terrain: for fixed inclination angles of the Tx and Rx antennas the drone should fly a distance sufficient to detect the target placed at a certain depth. The permittivity ε (F m^{-1}), the permeability μ (H m^{-1}), and the conductivity σ (S m^{-1}) define the dielectric characterization of the medium where the radar wave travels. For a linear, homogeneous and isotropic medium, these parameters are coupled into the following constitutive relationships:

$$\bar{B}=\mu\bar{H} \tag{9.4}$$

$$\bar{J}=\sigma\bar{E} \tag{9.5}$$

$$\bar{D}=\varepsilon\bar{E} \tag{9.6}$$

where B (Wb m^{-2}) is the magnetic flux density, H (A m^{-1}) is the magnetic field, J (A m^{-2}) is the current density, E is the electric field (V m^{-1}), and D (C m^{-2}) is the electric displacement field. The dielectric parameters are generalized as:

$$\varepsilon = \varepsilon' - j\varepsilon'' \tag{9.7}$$

$$\mu = \mu' - j\mu'' \tag{9.8}$$

$$\varepsilon'' = \frac{\sigma}{\omega} \tag{9.9}$$

where ε' is the absolute dielectric constant, ε'' is the dielectric loss factor, μ' is the absolute magnetic constant, μ'' is the magnetic loss factor, and ω is the radian frequency ($\omega = 2\pi f$). In free space the permittivity and the permeability are provided by:

$$\varepsilon = \varepsilon_r \cdot \varepsilon_0 \tag{9.10}$$

$$\mu = \mu_r \cdot \mu_0 \tag{9.11}$$

where ε_0 is the free space permittivity ($\varepsilon_0 = 8.854187817 \cdot 10^{-1}$) and μ_0 is the permeability in free space ($\mu_0 = 4\pi \cdot 10^{-7}$). Finally, the loss tangent is defined as:

$$\tan\delta = \frac{\varepsilon''}{\varepsilon'} \tag{9.12}$$

The defined parameters are important for the setting of the GPR apparatus and for target discrimination. Soil parameters are experimentally defined and modeled at microwave frequencies [9–12].

The rapid prototyping procedure of the GPR UAV system is summarized by the following steps:

- 3D CAD modeling defining the main spaces and mechanical assemblies.
- 3D FEM simulation defining antenna layout and working frequencies for a best Tx/Rx antenna system efficiency.

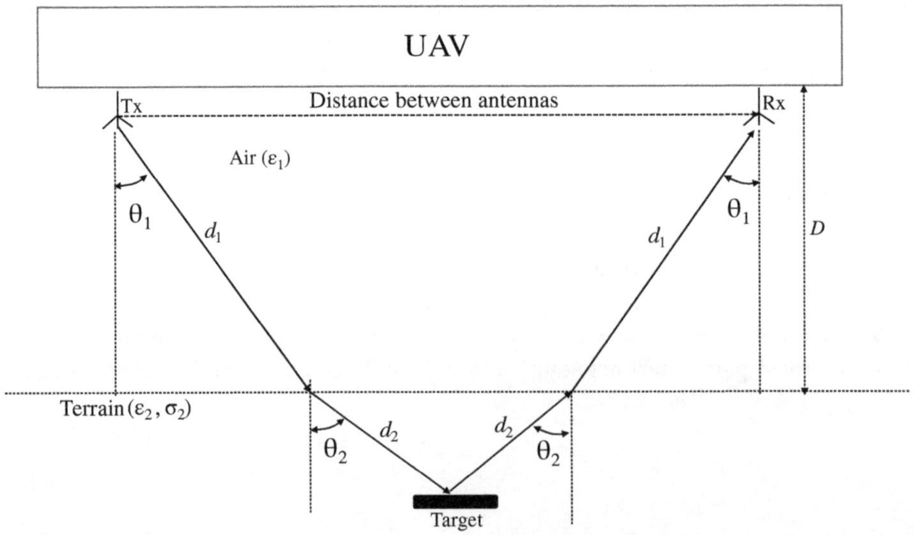

Figure 9.14 Geometrical model for GPR signal transmission and reception, determining the best distance between the Rx and Tx antennas.

- Design of the best geometrical model system and of the best target discrimination procedure.
- Design of electronic elements of the whole prototype electronic sensor system, including cabling and connectors as subminiature version A (SMvA), USB, or others.
- Signal simulation concerning the whole architecture.
- Basic firmware development managing the signal transmission and signal synchronization.
- Bench test of all the assembled components.
- Mechanical part design and development of the part fixed on the UAV unit able to support the electronic equipment (all distances and the antenna orientation are defined in this stage).
- Choice of the UAV unit supporting the whole estimated weight and with stability whilst flying in high wind forces.

The system optimization is performed after the preliminary tests are carried out on the whole UAV unit equipped with all of the components.

9.2.2 Example of an Underground Water Leakage Detection System Integrating GPR, UAV, and Infrared Thermal Imaging: System Prototyping

A ready to use system o detect underground water leakage can be made from commercial sensors installed and configured on a UAV unit or other systems supporting detection. A fully noninvasive technological system for the intelligent detection and localization of water leaks involves automated thermogram and radargram image processing of the monitored pipelines, and alerting in cases of anomalous detections of possible water leakage. As technology support of GPR for leakage inspections, a high sensitivity IR thermal camera is used fixed on a pole and remotely managed in movement and detection, or fixed on a mobile unit, or assembled on a remote-controlled UAV. The whole system is capable of georeferencing water losses in an urban or extra-urban aqueduct hydric network. The technological system is supported by a georadar device capable of satisfying other required design specifications. With regard to the use of the latest technologies, it is possible to realize the fully noninvasive system detecting water leakage shown in Figure 9.15.

Figure 9.16 shows the functional diagram related to the entire monitoring system. Real-time monitoring is carried out on areas in which leaks have been pre-localized (anomalous signals, anomalous flow data, etc.), or on areas deemed "critical". With regard to automated fixed pole workstations, the operator of the inspection unit remotely controls the leakage in real time providing an "alert" signal as an output if the first processing has exceeded a certain threshold probability (threshold defined by fixing a minimum and a maximum temperature value). Alternatively, this first image acquisition process is performed by the UAV, or by the mobile unit equipped with GPR and an IR camera. Only after the alerting signal, an operator is sent equipped with another high-resolution thermal imaging camera and tablet interface in order to send new images to be processed to the data acquisition unit (external processor). During this second phase (waiting for a second feedback), other data are collected into a database through the georadar and other devices already supplied such as geophones or data logger. If the second and more detailed data processing provides a positive response, then the classic intervention procedure to control and repair the pipeline leakage is enabled.

Georadar is also used as an independent inspection tool. The basic principle used to read a water leakage is that the radar wave resonates corresponding to the leakage region by storing energy, and the water locally varies the permittivity and conductivity of the terrain. If the resonance is strong, the water layer shields the part below it. Figure 9.17a shows an example of water leakage detection from observing a radargram: as shown in Figure 9.17b, the wave resonance is not observed after adjustment of the pipeline, thus confirming the previous leakage detection. In order to enhance image processing results, the K-means algorithm is applied directly to the radiometric images.

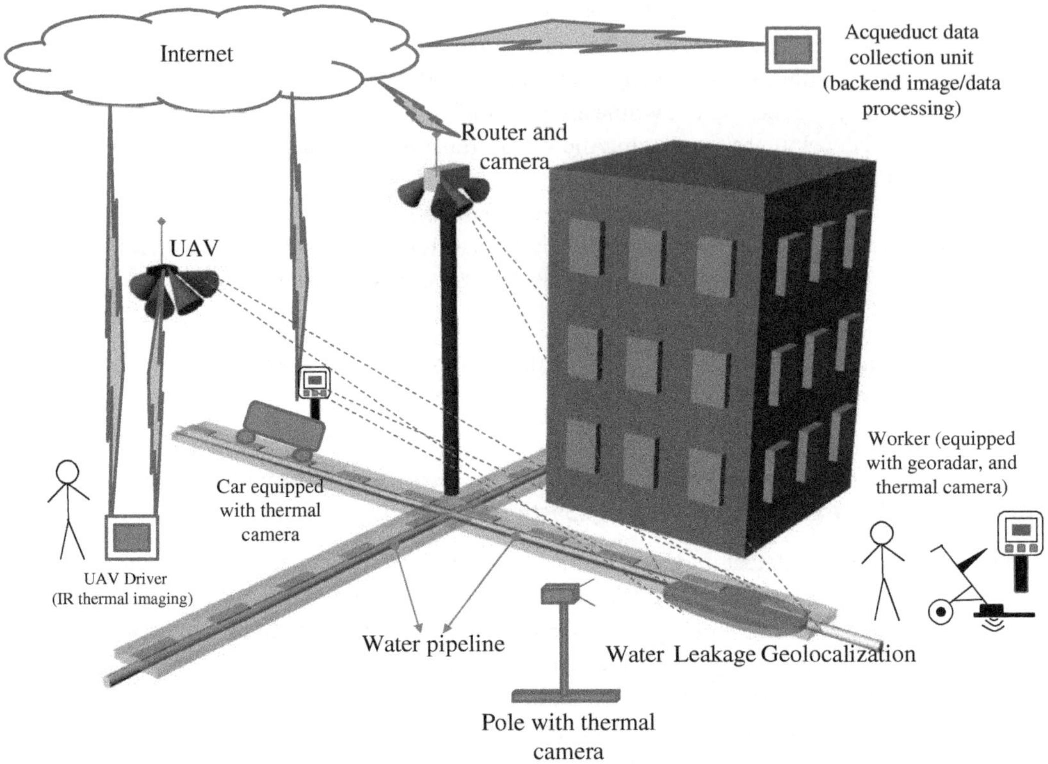

Figure 9.15 Scheme of ready to use prototype technology for underground water leakages using UAV, IR thermography, and GPR.

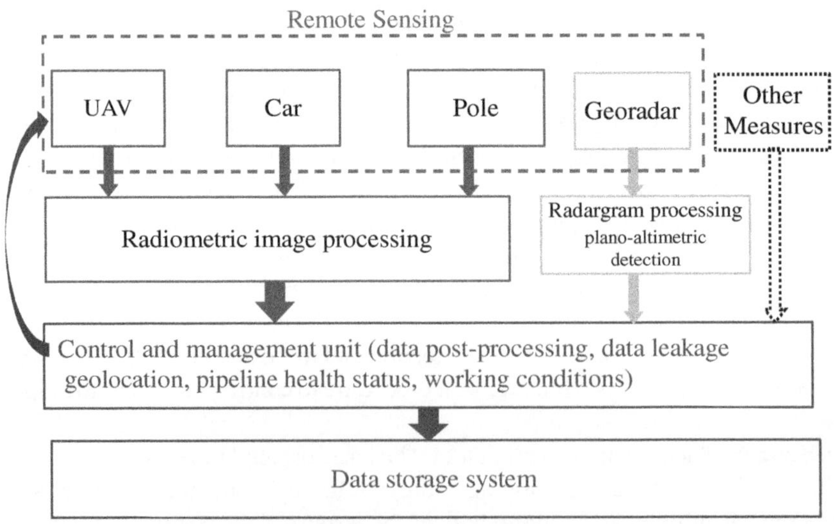

Figure 9.16 Engineered processes for aqueduct inspection activities.

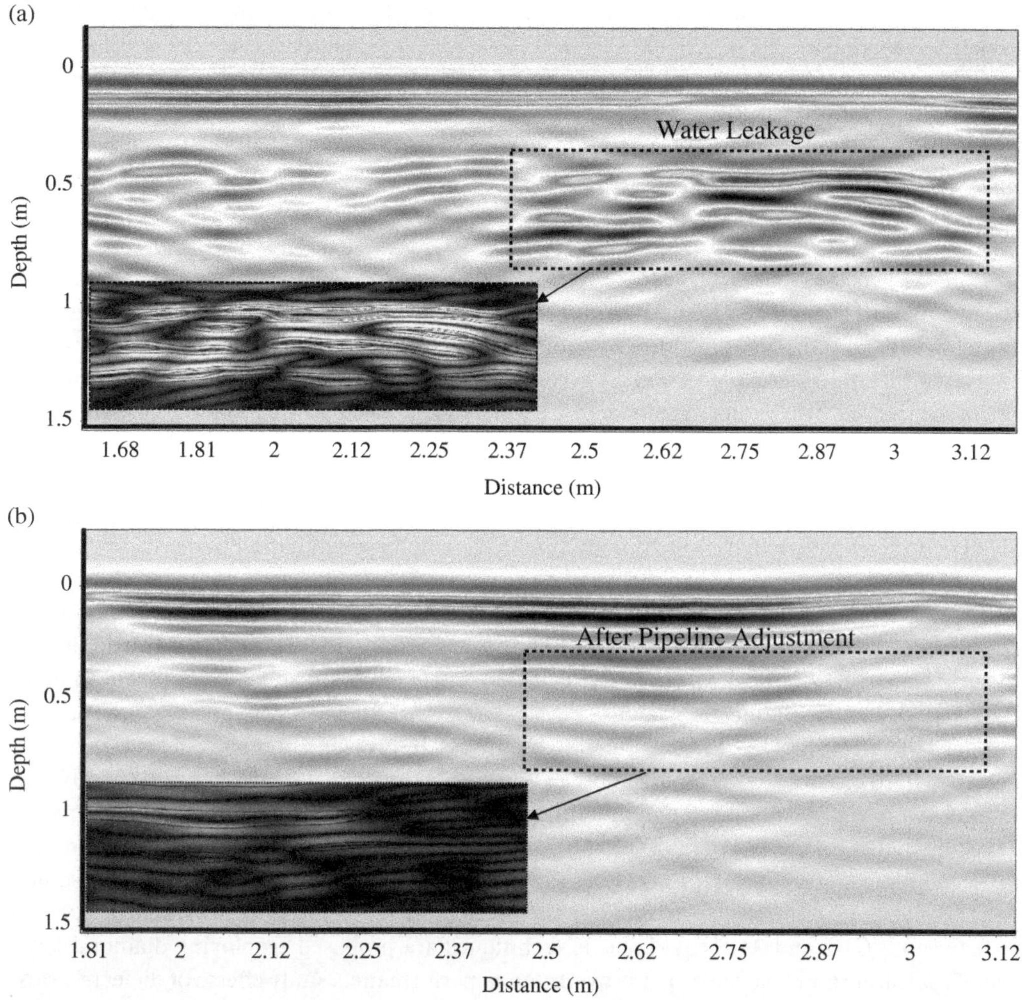

Figure 9.17 (a) Radargram detecting water losses in a pipeline, indicating leakage by water layers formed in the terrain as an anomalous wave resonance (inset: edge extraction by image post-processing enhancing leakage). (b) The same radargram after pipeline adjustment (inset: edge extraction by image post-processing defining a non-leakage condition). The find edges function adopts a Sobel edge detector to highlight sharp changes in intensity in the subimage selection.

The rapid prototyping in this case consists of the whole system integrating all the facilities of the noninvasive water leakage inspection. The system prototyping is composed of the following standard elements and technologies:

- Thermography applied on a UAV, car, and pole fixed station to check urban and suburban roads.
- Georadar applied on a car, UAV, or a manually controlled machine, and able to facilitate the reading of IR thermography results.
- Application of AI algorithms facilitating the automatic detection of the leakage by means of the application of a long short-term memory (LSTM) network or convolutional neural network (CNN).

- Software platform indicating geolocalized inspections and 3D plano-altimetric pipeline layout reconstructed using GPR.
- Automation for the parameter setting (camera and GPR setting, and software setting by means of threshold definition suggested by the DSS).
- Optimization of the new inspection water leakage process by merging all the functions of the systems and by defining the operator roles.

9.2.3 Integrated Diamond Patch-Type Antennas and Applications

Antennas at optical frequencies is an interesting topic in plasmonic research especially concerning the coupling of third-harmonic-upconverted radiation [13], the implementation in layouts for beam steering as spiral gold antenna [14], and for gold thin layer compact microstructures integrated on silicon substrates and behaving as nanoantennas [15]. In a super high frequency communication system (3–30 GHz), innovative antennas can be used for wirelessly monitoring neuron activity at a working frequency of 3 GHz [16], as an omnidirectional metamaterial source integrating 3D gradient-refractive-index lenses which have been simulated by 3D far-field radiation patterns at 5.5 GHz [17], and as metamaterial-lined horn antennas exhibiting a radiation pattern showing significantly reduced sidelobes and back lobes working at 5.8 GHz [18]. Diamond material has been proposed as a good candidate for optical telecommunication, showing a marked χ^3 nonlinear effect [19]. Moreover, diamond material [20, 21] and nanodiamonds (NDs) [22, 23] exhibit N-V (nitrogen-vacancy)/Si-V (silicon-vacancy) active color center emission suitable for biochip implementation [23]. In GHz applications, diamond has been used for MEMS/nanoelectromechanical system (NEMS) antenna [24, 25] and as chip interconnects, improving antenna gain [26]. Diamond materials are used in slot antenna for mixer optics working in the submillimeter band (500–750 GHz) obtaining low noise [27]. Moreover, antenna transmission gain of an on-chip dipole antenna pair is improved by thin diamond layers [28], suitable for intra-chip wireless interconnection, providing a high resistivity substrate. Thin diamond layers as a substrate alternative to Duroid, gallium arsenide (GaAs) and silicon (Si) can be used to increase the performance of a P-shaped wearable device in a range between 2 GHz and 3 GHz [29]. Novel techniques are proposed employing diamond layer between Si substrate and heat sink to improve antenna performance, study effects of dielectric variety, diamond thickness, substrate resistivity and antenna pair separation on transmission gain. Chemical vapor deposited diamond films are implemented to generate high-power terahertz (THz) radiation for the photoconductive switch-arrayed antenna [30]. A stable procedure based on the microwave plasma enhanced chemical vapor deposition (MWPECVD) technique and on ND layer deposition improving growth is adopted to realize compact GHz diamond antenna layouts behaving as patch antenna. A goal is to match the enhanced color center emission property of the diamond material together with its C-band (4–8 GHz) efficient radiation property in a unique antenna. The examined diamond antenna design is based on a planar monopole layout [31] growing a diamond layout by means of an optimized MWPECVD technique [32]. The antenna prototype is illustrated by CAD with different perspectives in Figure 9.18a–c, and the related 3D FEM simulations are illustrated in Figure 9.18d and e. Figure 9.18f shows the unit cell modeling used for the FEM simulations defined by six input and output ports. The polycrystalline diamond (PCD) growth is characterized by various grain sizes. Following the abovementioned technological scenario, all the phases are described concerning rapid prototyping design, realization and characterization of a Si-V emitter diamond antenna for GHz transmission. The diamond-based antenna, able to enhance the Si-V active emission, is tailored to transmit efficiently at GHz frequencies with a narrow pass band. Due to the diversity of the different working frequencies, the diamond-based device can operate simultaneously as a Si-V optical antenna and as a GHz antenna, transmitting part of the acquired data in a C-band communication system. The diamond material is grown by the MWPECVD technique on a Si material accomplishing a monopole antenna layout. Measurements and simulations of the S_{11} scattering parameter prove a low reflectivity at the input port around 5.6 GHz, thus defining a working frequency.

The antenna is designed by considering the dimensions given in the monopole antenna layout [21] shown in Figure 9.19a, extracted to the CAD model in Figure 9.19b. A key parameter used in the design, is the S_{11} scattering parameter indicating the reflectivity at the input port of the antenna: a low reflectivity, corresponding to low S_{11} value, indicates maximum power transferred to the antenna, and this means that the whole device is able to transmit most of the input signal power. The antenna is modeled and simulated by 3D FEM, based on the concept of the input/output wave ports in Figure 9.18f where the input and output signals are entered and read, respectively. The FEM simulation in Figure 9.20a and b is provided at the working frequency around $f_{0\text{FEM}} = 5.65$ GHz, obtaining a low value of S_{11} (−35.0 dB) thus representing a good working condition. A similarly low value ($S_{11} = -33.7$ dB) is measured around $f_{0\text{measured}} = 5.69$ GHz by means of a vector signal analyzer. A good agreement with a weak deep shift of about $\Delta f_0 = f_{0\text{measured}} - f_{0\text{FEM}} = 0.04$ GHz between numerical and experimental has been observed, confirming the good approach followed for the rapid prototyping. As for band-notched monopole antennas, a narrow pass-band is found. The S_{11} measured response of the same sized *p*-Si substrate without the diamond layout shows a flat response ranging from 10 to 15 dB in the frequency band $1 < f < 6$ GHz. This proves that the best impedance antenna matching for $Z = 50\,\Omega$ (impedance of the coaxial cable connecting the source with the antenna) at $f_{0\text{measured}}$ is achieved only by considering the diamond layout effect. The layout is also important to couple and to direct the radiated electric field: Figure 9.18d shows the calculated

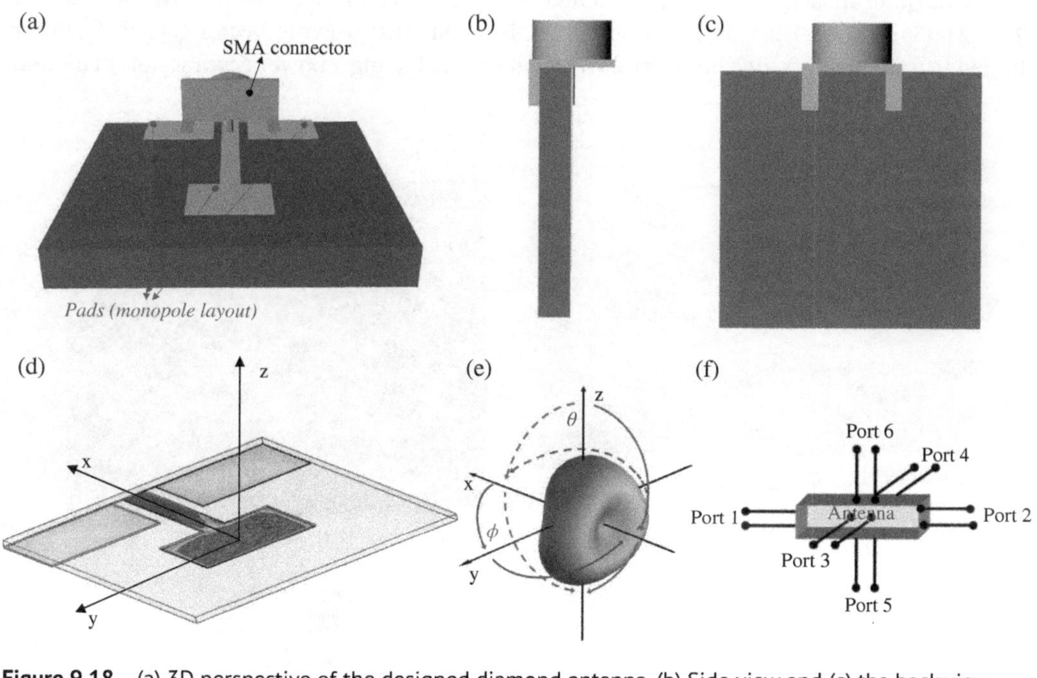

Figure 9.18 (a) 3D perspective of the designed diamond antenna. (b) Side view and (c) the back view perspectives of the antenna. (d) 3D FEM simulation proving the surface electrical current distribution. (e) 3D radiation pattern. (f) 3D unit cell modeling with the six ports defined.

electric field coupled with the layout and the 3D FEM (surface density current $\mathbf{J} = \sigma\mathbf{E}$), and Figure 9.18e shows the radiation pattern of the emitted far field. Cross sections of the 3D radiation pattern (2D radiation patterns) are reported in Figure 9.21a–f and in Figure 9.22a–f, where a symmetry with respect to the z-axis (between positive and negative θ values) of the radiation lobes is checked. The 2D radiation patterns are obtained by the 3D plot fixing the θ value and varying the ϕ parameter. The 3D FEM simulation is performed by an appropriately designed FEM tool able to calculate the frequency responses at the ports enclosing the spatial domain of the antenna sensor. The 3D far field radiation pattern is calculated by evaluating the total electric field radiated by the diamond layout. During the simulations, the following dielectric permittivity values are considered: $\varepsilon_{\text{DiamondCVD}} = 3.5$ and $\varepsilon_{\text{Silicon}} = 11.9$. The input port is defined by a rectangular cross section joining the external diamond boxes with the central channel connected to the internal part of the coaxial cable (central pad). The radiation lobe analyses are important for the design of a complete Tx and Rx communication system where the Rx antenna must be placed corresponding to the maxima of the radiation lobes. For the analyzed antenna, different maximum radiation lobes are present suggesting the use of different Rx antennas for a point to a multipoint wireless system. The analysis of the radiation lobes is useful for the design of a new prototype of a multi-array antenna implemented on the same substrate.

Concerning adopted technology, the Raman and photoluminescence spectra provided the diamond film emitting characterization at room temperature. The substrate pretreatment is performed by spraying 100 pulses of a thin ND dispersion on a cleaned p-doped silicon (100) substrate (500 μm thick). The dispersion is made by quantity (30 mg) of ND having a diameter of 250 nm in 30 ml of apolar solvent (1,2-dichloroethane). Before the spray, the dispersion is sonicated for 30 minutes. A laser cut metallic mask is applied in order to spray only the regions for the diamond layout growth. The MWPECVD completed the diamond growth process on the ND sprayed layer. Experimental data are acquired using a National Instruments PXI-based Vector Network Analyzer model PXIe-5632 (300 kHz–8.5 GHz bandwidth), National Instruments Vector Signal Analyzer calibrated in a frequency range between 1 MHz and 6 GHz, having a power accuracy of ±1 dB and

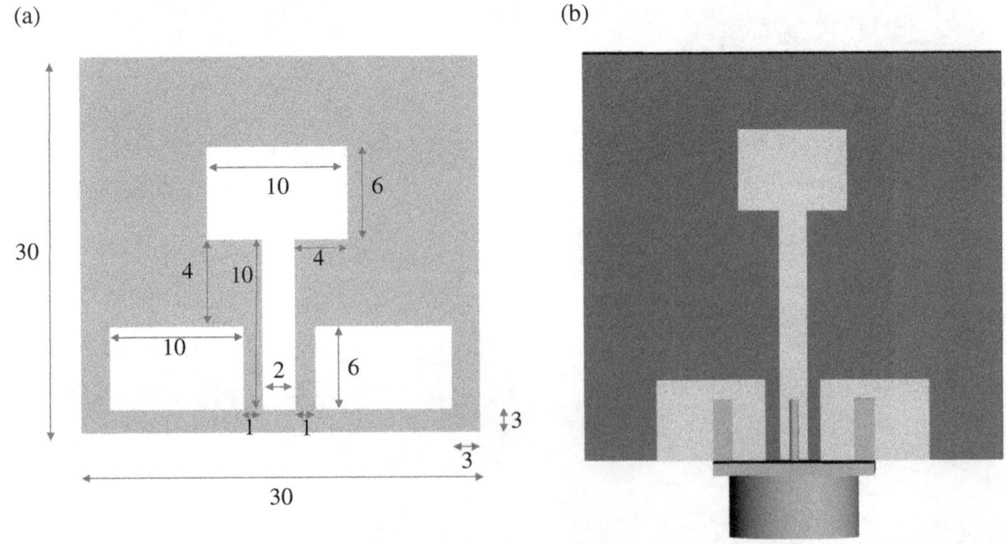

Figure 9.19 (a) Dimensions in millimeter scale of the antenna layout; (b) equivalent layout designed on the antenna surface.

a power step resolution of 0.01 dB. JOHNSON 142-0701-806 RF Coaxial, Straight Jack, 50OHM SMA connectors are used to couple the first port of the Vector Analyzer with the antenna sensor. In order to couple the source with the diamond layout, the connector is welded on the top surface of the diamond film by means of conductive silver glue allowing a rapid way to perform primary measurements.

The rapid prototyping procedure of the diamond antenna is structured by the following steps (first phase):

- First design by means of a 2D and 3D CAD model tool defining layouts and input cable ports.
- 3D simulation to calculate the amplitude and phase signal, defining resonant frequency and working band.
- 3D simulation for the radiation pattern calculation.
- First prototype realization following the designed layout used to realize the mask.
- Insertion of the connector linked to the radiating pad and to the ground parts using a conductive silver glue (the optimized prototype must be realized by a controlled technique ensuring low losses at the interconnections).

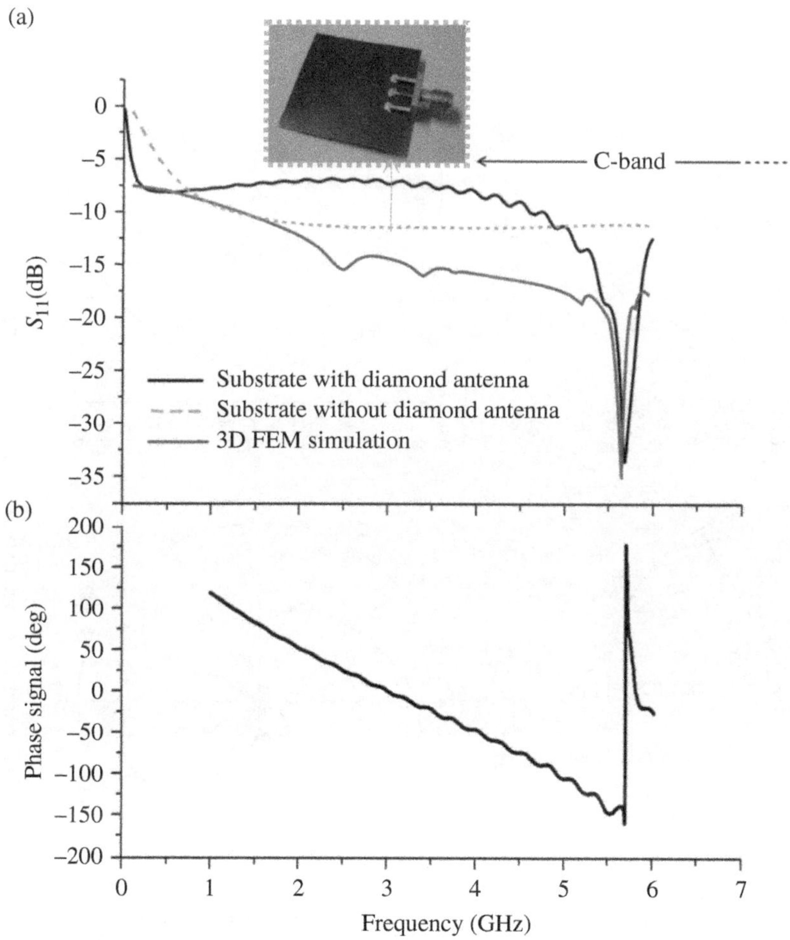

Figure 9.20 (a) Comparison of the S_{11} signals between numerical and experimental results. (b) Phase signal corresponding to the resonant frequency.

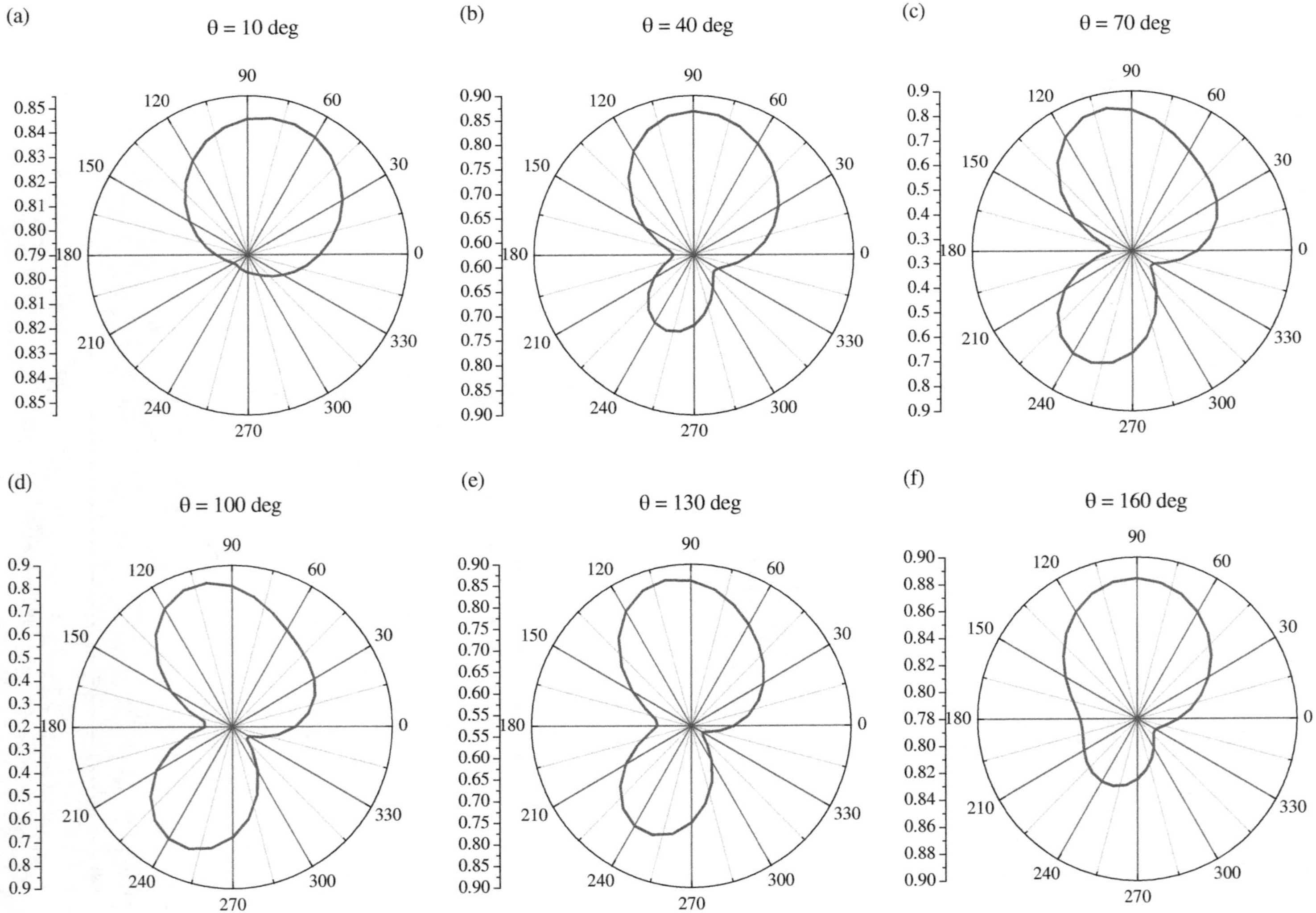

Figure 9.21 2D FEM normalized radiation patterns calculated by fixing θ values at (a) 10°, (b) 40°, (c) 70°, (d) 100°, (e) 130°, and (f) 160° by varying ϕ angles (from 0 to 360°). The plots represent the cross sections of the 3D radiation pattern shown in Figure 9.18e.

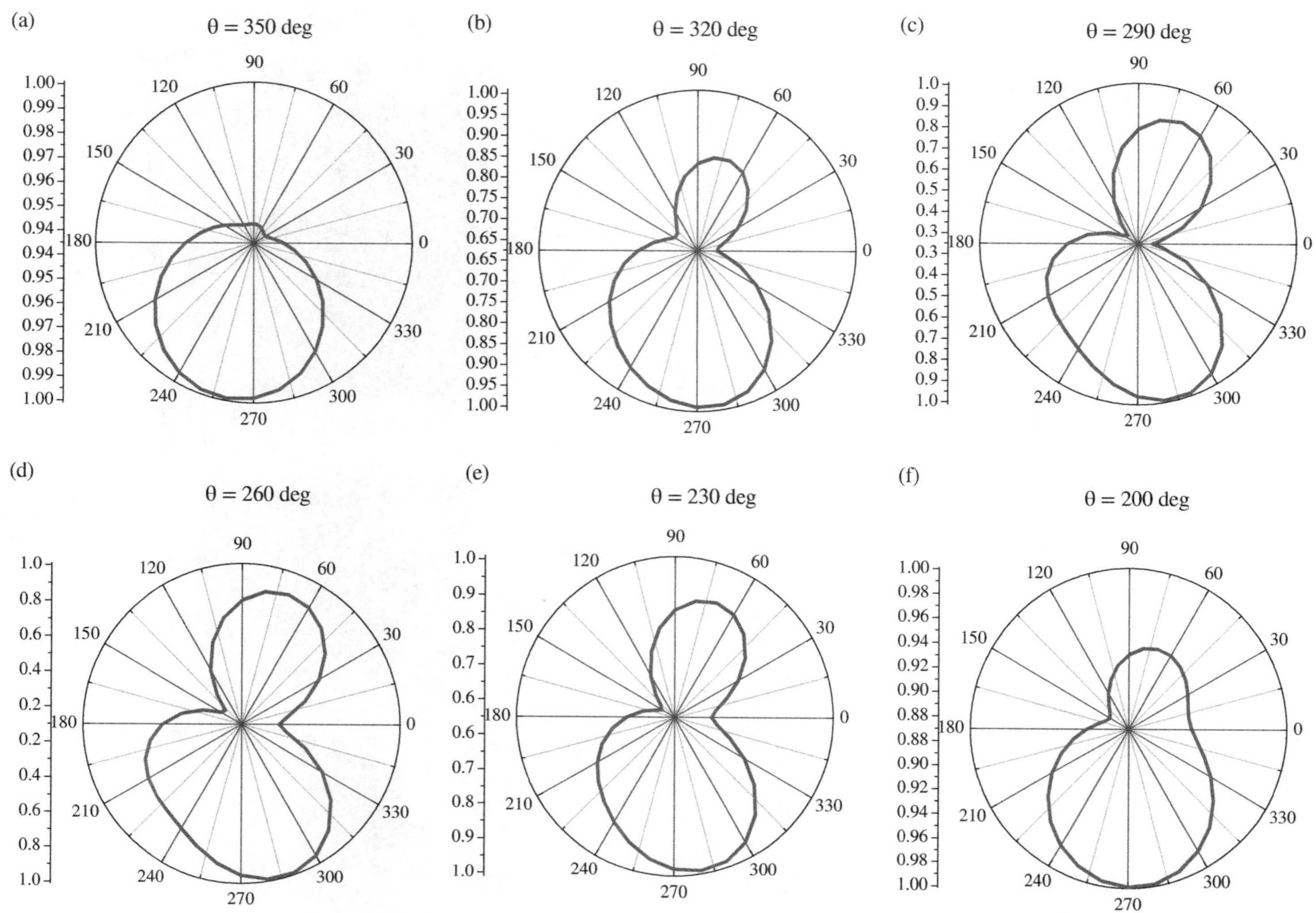

Figure 9.22 2D FEM normalized radiation patterns calculated by fixing θ values at (a) 350° (equivalently −10°), (b) 320° (equivalently −40°), (c) 290° (equivalently −70°), (d) 260° (equivalently −100°), (e) 230° (equivalently −130°), and (f) 200° (equivalently − 160°) by varying ϕ angles (from 0 to 360°). The plots represent the cross sections of the 3D radiation pattern shown in Figure 9.18e.

The second phase of the first prototype procedure involves:

- Preliminary measurements.
- Technology optimization of all the techniques.

Rapid prototyping is executed by comparing different fabrication processes. Figure 9.23a and b shows two alternative technologies for the design of diamond antennas. Figure 9.23a shows the technology used for a diamond-like carbon (DLC) antenna [33]. This technology involves the use of a metallic mask for a direct deposition of the 50 nm thick DLC layer on a *p*-doped silicon (100) substrate (500 μm) by adopting the ion beam sputtering (IBS) technique. IBS is a sputtering technique performed with an ion cannon [34]. For the proposed DLC antenna, a Kaufman ion source with a diameter of 2.5 cm generating an argon (Ar) sputtering beam impinging on a graphite target is adopted. The Ar source is managed through a digitally controlled gas flow system. The other working conditions are: substrate holder at room temperature, beam current $I_{Beam} = 40\,mA$, beam voltage $V_{Beam} = 1500\,V$, $I_{acc} = 1.7\,mA$, $\Phi(Ar) = 2.5\,cm$, t (deposition time) = 67 minutes. The DLC antenna works around 5.47 GHz. The diamond antenna in Figure 9.23b is fabricated by means of a hybrid technique involving:

- A first step of an initial spray deposition of NDs dispersed in a solution (by a pulsed spray technique).
- A second step by MWPECVD diamond growth enabled by the sprayed layer and improving the characteristics of the antenna in Figure 9.20a and b.

In order to check the sprayed deposition resolution of the first prototype sample in Figure 9.24a, a microscope providing the boundary resolution information shown in Figure 9.24b is used.

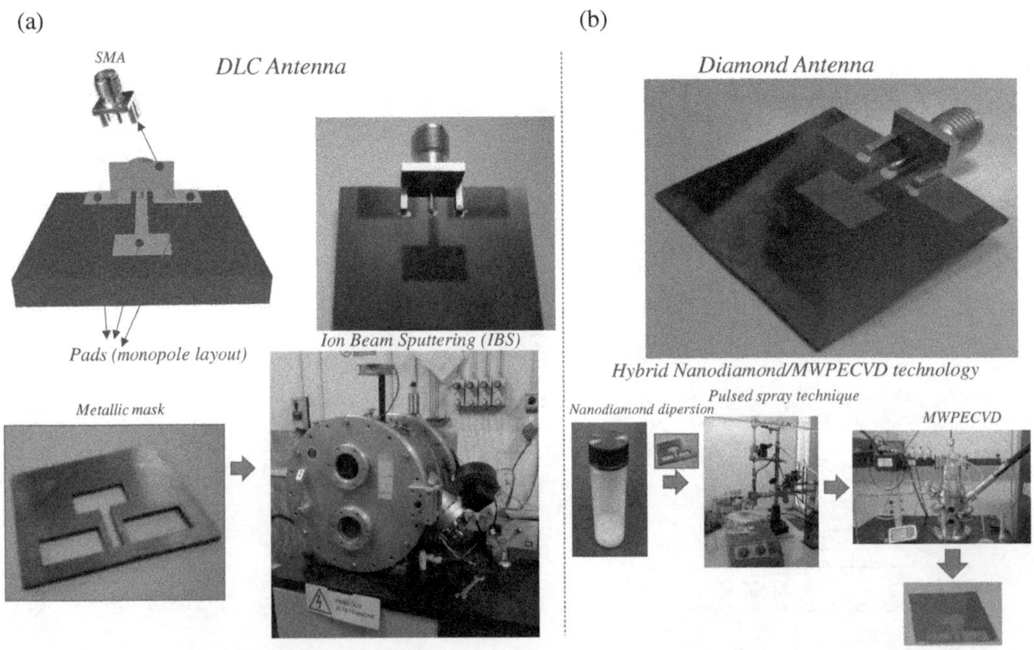

Figure 9.23 Examples of technologies for the realization of diamond antenna. (a) DLC antenna and (b) diamond antenna.

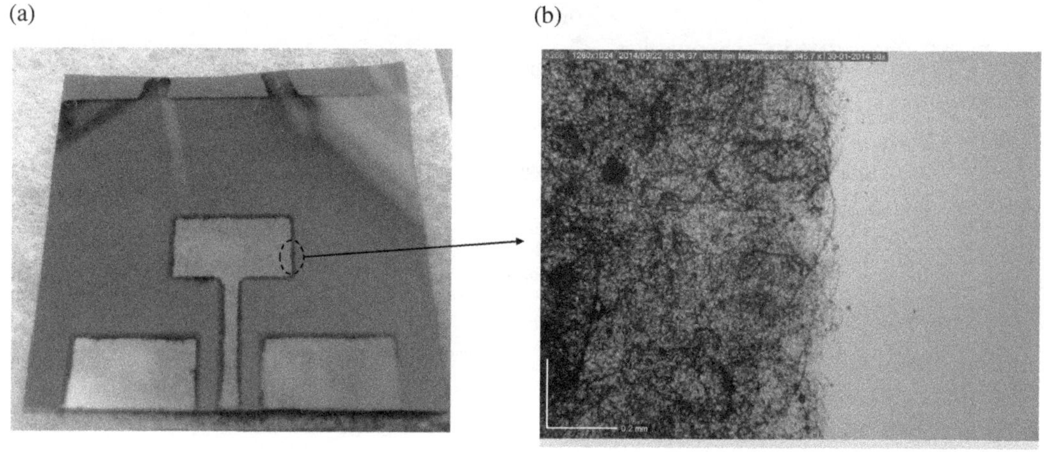

Figure 9.24 (a) NDs sprayed on the antenna layout and (b) microscope image of the antenna boundary.

Before the fabrication of the first prototype, different preliminary samples are measured to characterize the superficial electrical currents. Consider for example a layout made by spraying NDs, the measurements are performed at different contacting points of the sample, such as:

- Substrate surface.
- Regions with high density of NDs.
- Regions with low density of NDs.
- Boundary regions between ND layer and substrate surface.

This choice is due to the inhomogeneous distribution of the diamond material on the substrate, as usually happens for the realization of the first prototype. The measurements of the electrical current at different sample points is useful for the optimization of the fabrication technique. Figure 9.25a shows an image of a first antenna sample with different inhomogeneous areas. The measurements of the current–voltage characteristics are acquired by two tips placed in the three most significant regions (region A with low density of NDs, and regions B and C with higher density of NDs). As expected, Figure 9.25b shows a lower electrical current intensity due to the inhomogeneous ND deposition, compared with the measurements in Figure 9.25c and d related to more homogenous areas. The measurements performed are important to validate the basic principle of surface electrical enhancement by ND deposition checked by simulation results. In order to improve the surface density current **J**, DLC technology is implemented.

The diamond antenna has suitable properties to be applied in high temperature environments such as on engines or pipelines. Figure 9.26 shows an example of an array of patch antennas allocated on a pipeline working at high temperature. The design in this case is expanded to a layout of an antenna array by defining a specific pattern for the source excitation of all patch-type antennas arranged on the same substrate, where the Tx and Rx units of the whole system are on the same substrate. The design involves in this case the signal simulation of the whole array system defining the following specifications:

- Radiation diagram of the combined effect of all patch antennas.
- The sensing condition considering the specific applications.
- The sensing spatial discrimination (definition of the transmission signals).
- The signal synchronization.
- The electromagnetic compatibility aspects (interference analysis).

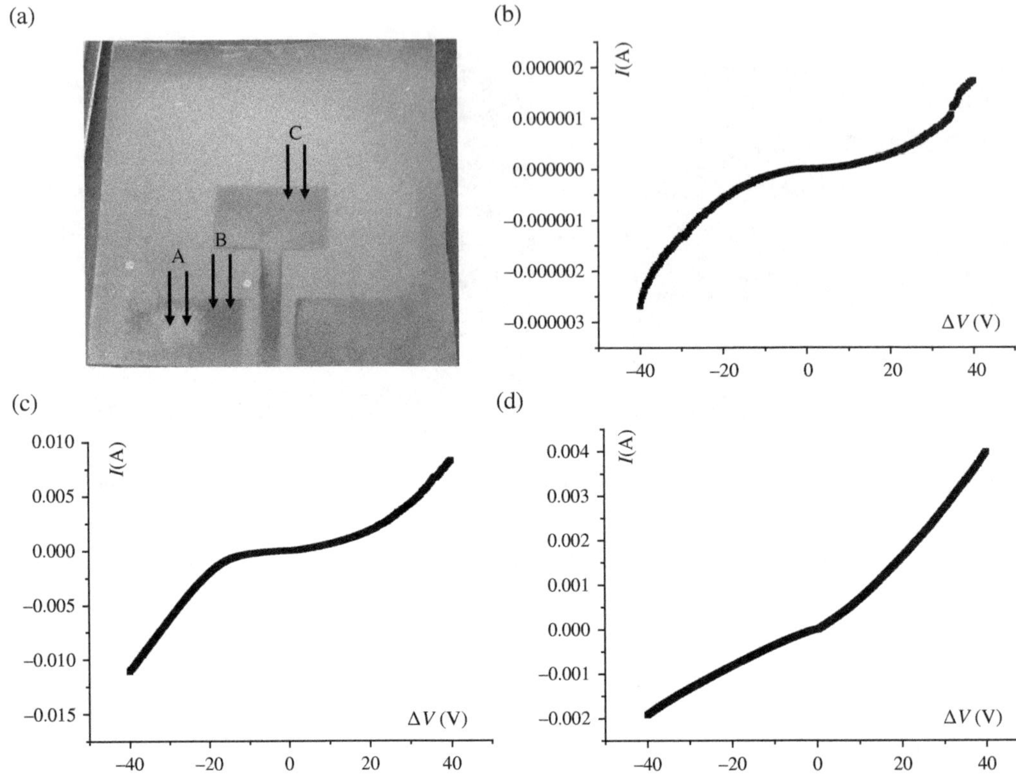

Figure 9.25 (a) Antenna sample; (b) *I*(*V*) characteristic measured in position A; (c) *I*(*V*) characteristic measured in position B; and (d) *I*(*V*) characteristic measured in position C.

Analytical and numerical tools are fundamental to design the full system. The simultaneous simulation of different antennas requires a high computational cost and, consequently, a suitable processor. For the simulation of a complete array, the parallel computing technique can be adopted thus partially avoiding the problem of the computational cost.

As a further application of an antenna sensor, Figure 9.27a and b shows an example of the diamond antenna being used in signal security applications. The shown example takes into account different components constituting the whole Rx and Tx apparatus including decoder, microcontroller, buffer, memory, antenna source, antenna connection, and optical fiber source. The proposed example represents a case of a system working simultaneously at optical and GHz frequencies.

The scheme in Figure 9.27a represents an innovative application of the dual frequency behavior of a diamond Rx antenna. A GHz signal is coupled to the Rx antenna. A Tx antenna having the same layout as the Rx antenna (same working frequency at $f_0 = 5.69\,\text{GHz}$ characterized by a low reflection coefficient S_{11}) transmits a signal having the function to activate Si-V transmission: the GHz signal arrives at the Rx antenna, and is coupled with the optical signal consisting of a coded signal. The Si-V emission at $\lambda = 738\,\text{nm}$ enables signal decryption (decoded word). The antenna encryption and decryption behaviors are also described in Figure 9.27b with the following working steps:

- **Step I:** an optical fiber transmits continuously an encrypted signal.
- **Step II:** the optical signal is coupled to the Rx antenna (reading of the coded signal); the signal is reflected on the central pad when the Si-V emission is activated by the GHz excitation.

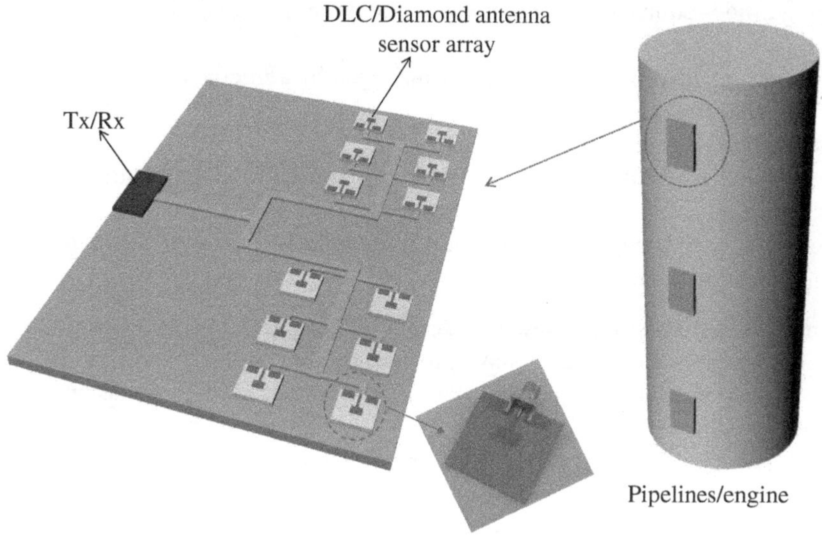

Figure 9.26 Array of patch antennas allocated on a pipeline or engine part working at high temperatures.

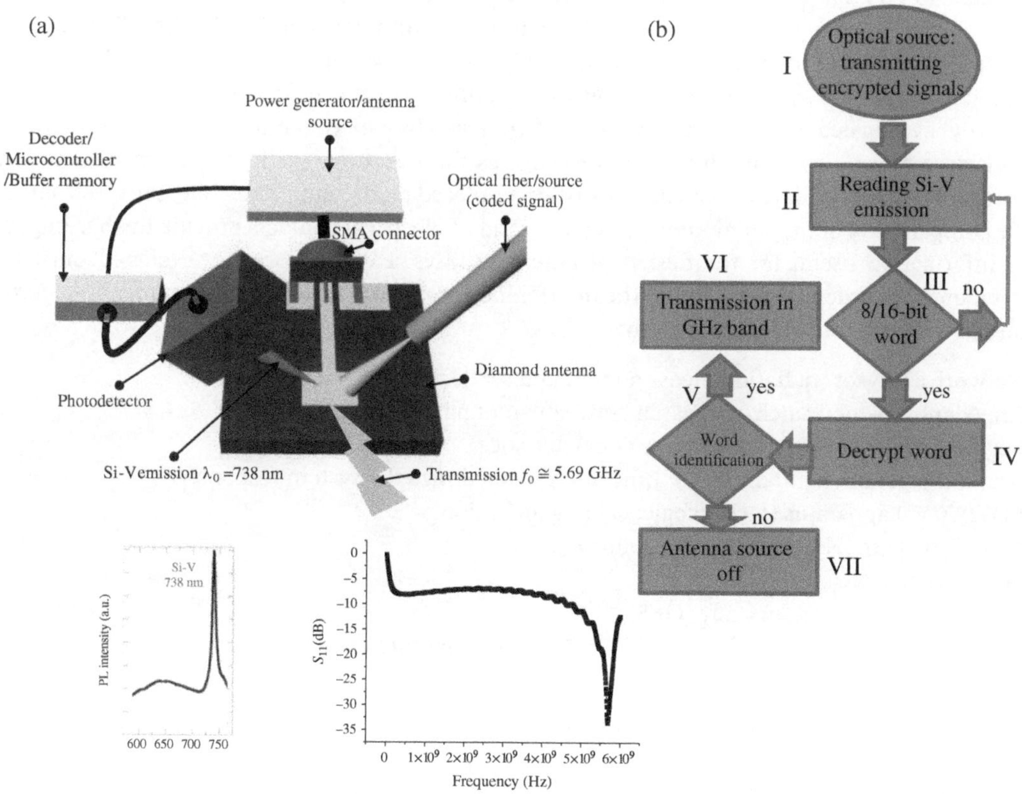

Figure 9.27 (a) Optoelectronic circuits for encryption and decryption of an 8/16-bit word. (b) Flowchart describing the encryption and decryption procedure.

- **Step III:** if the reading time is equivalent to a reading of a 8- or 16-bit signal, then the signal is decrypted.
- **Step IV:** the decryption is enabled by a photodetector managed by a microcontroller and by a memory buffering.
- **Step V:** the signal is checked.
- **Step VI:** if the signal is correctly decrypted, the word is identified and transmitted in the GHZ band.
- **Step VII:** if the word is not correctly identified, the Rx antenna is on standby awaiting the retransmission of another signal to decrypt.

The rapid prototyping procedure is oriented also to redesign the sensor. Concerning the antenna layout optimization, the redesign includes the possibility to develop a novel class of patch antennas based on different layouts. Optimized diamond films can be considered taking into account the capacitive effect between the substrate and diamond antenna layer. The goal is to fabricate a compact coplanar waveguiding (Figure 9.28a–d) working around 2.4 GHz. At this working frequency, and by using initially silicon as substrate, the dimensions of the substrate are of the order of a centimeter. By using low-loss substrates for the construction of integrated antennas, then basic radiating elements exhibit high quality factor (Q), and consequently narrow bandwidth. The use of a thicker and lower permittivity substrate reduces Q and increases bandwidth. Moreover, with a higher relative permittivity dielectric, the size of the antenna must be smaller. For this purpose, it is possible to vary diamond layouts and silicon thickness, by using other silicon substrates or etching the silicon substrate in order to improve an air cavity, tuning Q and bandwidth. The antenna efficiency concerning the source connection, reflectivity, impedance matching, impedance load estimation, and other radiation aspects (return loss, directivity, far-field radiation pattern, etc.) is primarily investigated from a numerical point of view, and by experimental verification. Moreover, the electromagnetic compatibility study investigates the electromagnetic interferences between substrate and diamond layouts arranged in arrays or placed on the same substrate: the electromagnetic compatibility of diamond interconnections and of electrical contacts provide further important information useful for the design of antenna arrays. Layouts correctly designed must be implemented in order to overcome all the interference problem. The following facilities are appropriate for a new layout characterization:

- Network analyzer (radiation measurement and scattering parameters).
- Impedance meter (match/mismatch between antenna and source).
- Digital multimeter (current–voltage characteristics).
- Scattering parameter analyzer (optimization to maximize the system responsivity).
- MWPECVD apparatus (technological characterization).
- Ultrasonic bath (ND dispersion in solution).
- Oscilloscope.
- Optical emission spectroscopy (OES).
- A dual wavelength ($\lambda_1 = 2.1\,\mu m$ and $\lambda_2 = 2.4\,\mu m$) IR pyrometer.
- Power energy meter.

The analyzed examples prove that rapid prototyping approaches are flexible to find more fabrication solutions and efficient layouts, also considering hybrid techniques and according to the available facilities. For this reason, rapid prototyping is representative of the industry research and suggests the need of investment in new production machines or materials.

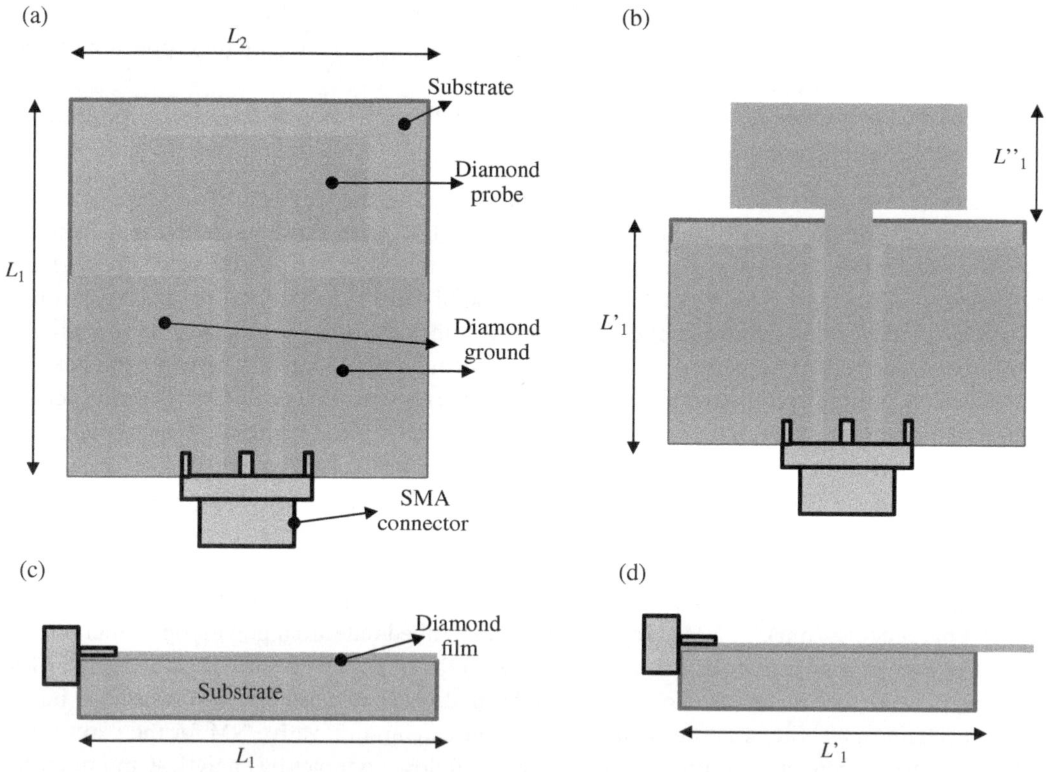

Figure 9.28 Different layouts and perspectives of diamond patch antennas: (a) layout with bigger areas of diamond ground region; (b) layout with freestanding diamond probe; (c) side view of the patch layout of (a); and (d) side view of the patch layout of (b).

9.3 Principles of Mechanical Piece Rapid Prototyping and Innovative Materials

Rapid prototyping of mechanical components is characterized by the following procedures:

- The first step is the choice of correct materials according to the application.
- The components to realize are simulated by numerical simulators characterizing the mechanical properties (stress, strain, displacements, modal analysis and harmonics, spectrum response, vibration responses, acoustics simulations for vibroacoustic characterization, solving linear static behavior, transient analysis, elastic/viscoelastic behavior, etc.).
- A modality is planned to analyze in detail the mechanical behavior of the simulated component by means of parametric simulations (the multiparametric variation highlights mechanical weaknesses thus suggesting the best design).
- The assembled part integrating the single designed component is designed by modeling all the forces acting on the component and the related boundary conditions (constraints, moments, vector forces, etc.).
- The component is engineered by analyzing the output results of all the simulations.

- The 3D CAD design is used for the realization of the mold for the production.
- The mold is re-engineered by means of reverse engineering (RE) techniques, optimizing the design by the wear and defect analyses.
- Quality tests performed during the time provide data and information about mechanical robustness.

The rapid prototyping of mechanical components starts with the study of different aspects concerning the specific application field. In particular, simulations in fluid dynamics or thermodynamics improve the design process according to the implementation in the specific working environment. Concerning NM implementation, it is useful to execute hardness tests using nanoindentation approaches which are applied for small volumes. Nanoindenters work by pressing a small probe into the surface part of the sample, measuring the relationship between the applied pressure force and the resulting probe displacement. This measurement allows the force versus displacement curve to be obtained mechanically characterizing the material in nanoscale. The mechanical characterization of an innovative material, as a NM, is accomplished by the use of numerical simulators: the simulations provide important information for example about interactions between nanoparticles and matter, thus defining the macroscopic behavior in terms of mechanical responses. An innovative mechanical piece can be coated or totally realized by the NM which is a multiphase solid material where one of the phases has one, two or three dimensions less than 100 nanometers (nm). In NMs the nanofillers are embedded into a polymeric or metallic or ceramic matrix, improving mechanical properties such as impact resistance. Further properties can be associated with mechanical ones such as thermal resistance or corrosion resistance influencing mechanical stability. In order to design and fabricate an innovative NM for the realization of mechanical components, it is important to study the following topics by analytical and numerical models:

- Surface forces interacting between embedded nanoparticles and the matrix of the raw material to improve.
- Analyses of the whole mechanical behavior of the NM.
- Explanation of the whole mechanical behavior, by analyzing the single effect of a single nanoparticle embedded into a micrometric volume cell modeling the basic element (unit cell modeling).
- Definition of the limits as external forces to apply.
- Definition of the environment limits as for the definition of the maximum working temperature.
- Definition of the correct nanofiller concentrations tuning mechanical properties (material engineering).
- Definition of the chemical procedures to fabricate the innovative NM according to simulation outputs.
- Preliminary measurements of the fabricated innovative piece.
- Comparison of the preliminary measurements with analytical and numerical ones.
- Re-engineering of the procedure by optimizing all the executed procedures.

Nanocomposite coatings are in the NM class. In many applications where mechanical forces generate accentuated wear of components, the coating usually improves the durability of the piece. In order to produce a coating film, it is necessary to:

- Analyze and model the coating according to the application field and to the available technological facilities defining limits of film thickness, film morphology, and film composition.
- Analyze the technological procedures to fabricate a fixed deposition layer according to the force to apply on the surface of the component to improve.
- Analyze the operating limits (applied forces and other environment conditions).
- Perform a nanoindentation of the coated surface measuring hardness, elastic modulus, flow properties and other surface properties from the nanometer scale to the micrometer scale.

The simulations are fundamental to obtain a preliminary idea of the material to improve providing an understanding of the range limit conditions and material design feasibility.

9.3.1 Example of Diamond Material Implementations

When a material is analyzed, different aspects and properties must be analyzed. Considering diamond and ND materials, Figure 9.29 shows many implementations classified as passive and active components. The passive devices include:

- Mechanical packaging.
- Electromagnetic shields for electromagnetic compatibility.
- Protective coatings or films.
- Sensors and antenna-type devices detecting temperatures, charge, and electrical current.

The active devices include:

- optical active components
- energy harvesting devices
- electronic components

As diamond is a material with strong mechanical resistance, it is easy to consider adopting it to improve other materials. NDs are added as fillers in a polymeric matrix, thus improving mechanical resistance. Implementation of a poly(methyl methacrylate) (PMMA)/ND composite coating is potentially suitable for high hardness, flexural endurance and durability tests. The main goal to design this NM is to characterize the new mechanical and material properties, addressing the research on the chemical process of the PMMA/ND coating. The optimization of the fabrication process consists of tuning the mechanical properties of the PMMA polymer by using different sizes and concentration of NDs. The NDs are introduced uniformly as fillers inside the coating, thus improving hardness and durability properties of the whole material and of its surface. Moreover, acting on the PMMA/ND thickness, the curling effect of the coating can be studied. By treating the NM with UV irradiation, the yellowness effect and the new mechanical properties due to the UV stimulated photo process can be analyzed. The results can be addressed to specific applications such as the realization of touch screens or sun UV resistance coatings. PMMA polymer is prepared by using particular concentrations of a PMMA solution in chloroform obtaining polymer films with thicknesses of the order of tens of micrometers, by using PMMA granules dissolved in chloroform at room temperature, or by dissolving PMMA in dichloroethane solvent for fewer surface fluctuations. Rigid plastics tend to be strong and resist deformation but they tend not to be very tough, which means that they are brittle. The introduction of an innovative PMMA/ND material can modify the mechanical properties by means of the introduction of ND fillers: by varying the

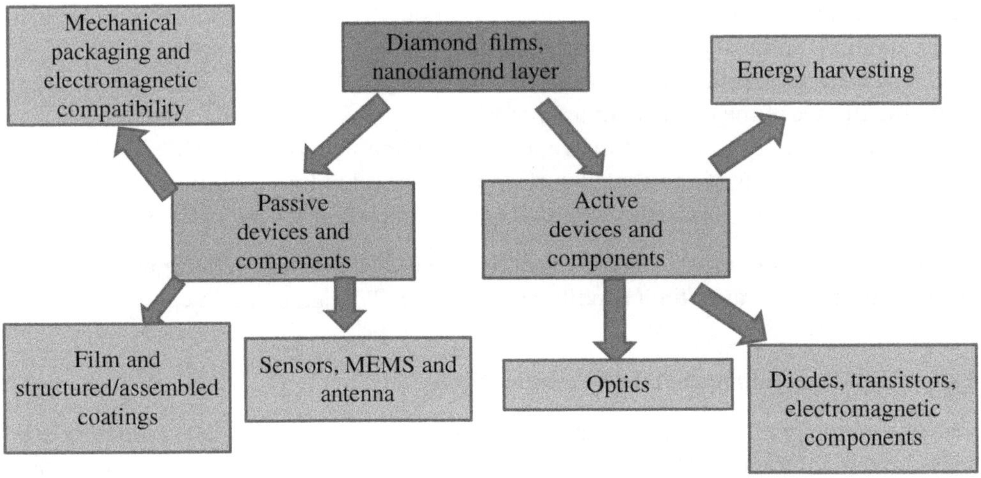

Figure 9.29 Diamond material: properties and possible implementations.

ND concentration of the bulk PMMA material, it is possible to change the ND density in proximity of the coating surface improving mechanical resistance. The surface properties of PMMA polymer may change by UV treatment as an increase of its corrugations and polymer/clay nanocomposites as photo-curable coatings can improve the surface mechanical properties. Moreover, the UV effect also degrades PMMA, and can change the durability of the material. For this reason, an accurate study of the effects of UV on PMMA/ND samples is necessary. The facilities to use for the characterization of the new material are:

- Pencil hardness tester.
- Nanoindenter (mechanical characterization).
- ND treatment facilities (plasma treatment).
- UV light source.
- Chemical facilities.
- Tools for bending tests and measurement of ductility.
- Tools for hardness and flexural endurance tests.
- AFM for surface characterizations.
- Raman spectroscopy (material characterization).
- Transmission electron microscopy (TEM) for surface characterization.
- 3D confocal technique (surface characterization and measurement of PMMA/ND film thickness).
- Optical microscope.
- Sonification for cleaning process.
- Tribometer (friction characterization).
- Properly designed FEM codes for PMMA nanocomposite simulations.

Different solutions can be adopted for chemical preparation. One solution is to dissolve PMMA powder in solvents such as chloroform considering different percentages. NDs of different sizes ranging from 0 to 500 μm can be introduced as fillers into the PMMA polymeric matrix. Ultrasound sonification is suitable in order to primarily clean the NDs. The thickness of the PMMA/ND coatings is controlled by changing the concentration, spinning rate, solvent type (chloroform and dichloroethane) and polymer type. During this stage, different nanocomposite coating samples must be prepared on glass, quartz and silicon substrates. Raman spectroscopy characterization is performed in order to check the diamond component versus the different ND concentrations. Moreover, Young's modulus of the different coatings must be estimated by a nanoindenter in order to have reference values before the UV and/or bending treatments. The TEM and AFM approaches are able to check the nanocomposite surfaces. Finally, tribometer analysis provides information about the surface friction characterization. Figure 9.30a shows the forces acting when the PMMA/ND is forced for a cracking condition and Figure 9.30b shows the basic scheme for a nanoindentation measurement. The research activities involve the following tasks:

- **Curl free UV curable of hard coating material on thin substrate.** Curl is the tendency of a coated film to curve toward the coated side of the film instead of lying flat. This aspect is to be studied for different PMMA/ND coatings deposited on different substrate materials and irradiated by UV. Simultaneously, it is important to also analyze the effect of the increase in temperature on the coating curling. A complete characterization of the UV effect must be performed by analyzing the surface sample by microscope and 3D confocal and AFM techniques.
- **High pencil hardness over 8H with and without UV treatment.** The Pencil Tester is appropriate to test the nanocomposite PEMMA/ND coatings for their hardness and resistance to scratches and wear. The test consists of pencils which have a known hardness grade to be moved over the surface to be tested. The test is typically performed for a fixed angle and pressure load. Passing the pencils for a specified number of times on the PEMMA/ND coating, it is also possible to estimate the wear factor which is a function of the hardness of the pencil used. A hardness scale is used for the pencil measurements. Pencils of increasing hardness values are moved over the surface until one lead damages the surface. Surface hardness is defined by the hardest pencil grade which just fails to damage the surface. Observing pencil track, enough pressure must be applied to damage the coating or to crush the lead pencil point. The tests are repeated until a pencil crash or when a surface crack is observed. The tests must be performed for sample with and without UV treatment.
- **Surface crack phenomena during flexural bending test.** Two types of bending tests of PMMA/ND are considered: three-point and four-point. Bending tests provide information about ductility of PMMA/ND samples and check the nanocomposite coating limits determining the load measurement and its relationship to a load specification. The check involves bending until the material breaks and determines the loads for the deflection limits. After the bending test, the coating surfaces must be analyzed by AFM and optical microscopes. FEM tools help to understand the crack mechanisms.

9.4 Rapid Prototyping and Artificial Intelligence Upgrade

AI plays an important role in advanced prototyping. Starting with the raw material to process, in each step of the first fabrication of the prototype the "AI opinion" can be involved. The scheme of AI adoption in Figure 9.31 integrates all the rapid prototyping processes involved for the pre-industrialization of a component or device. The flowchart is structured as follows by six main phases:

1. The pre-industrialization process starts with the choice of the raw materials to implement, by defining the main physical and chemical properties according to the application field (thermal conductivity, electrical conductivity, dielectric permittivity, Young's modulus, corrosion resistance, etc.).

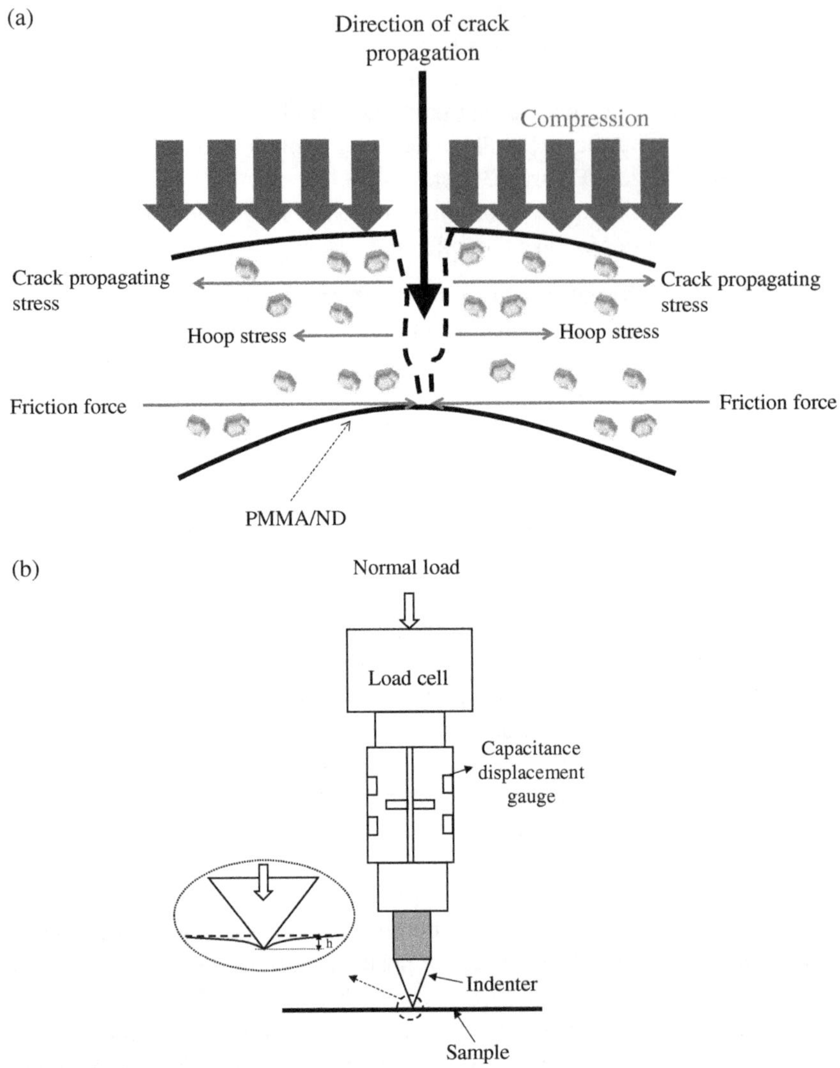

Figure 9.30 (a) PMMA with ND fillers and forces acting during a material cracking condition. (b) Nanoindenter scheme.

2. The material is engineered by improving its properties for example by designing a new NM (material prototyping step).
3. The new material is implemented for a preliminary device for device prototyping, in order to check the main desired properties (sensing sensitivity, parameter gain, biocompatibility, electromagnetic shielding, etc.).
4. The first prototype is realized by using the proper facilities for the manufacturing process; the check of the first realization is adopted for device optimization improving geometrical resolution and mitigating the checked defects.
5. The device optimization process is executed step by step orienting the fabrication process toward the pre-industrialization.

6. The pre-industrialization process consists of pre-series production of the optimized devices controlling during the time quality tests and the device lifecycle.

The design, modeling and simulations are "transversal" to the first five phases of the scheme in Figure 9.31. Specifically:

- The design is related to the material and to the whole device design and it is initially executed by applying analytical modeling, and then by numerically simulating the whole device to analyze all physical properties.
- The modeling is the 2D and 3D model of the device, and the 3D modeling may be the output of the RE process.
- The simulation is applied to the single material and to the whole device, by fixing physical properties and boundary conditions (the choice of the numerical simulator is a function of the physical model to adopt); parametric simulations are useful to quickly find the best physical and geometrical parameters satisfying the device specifications.

In all six phases, AI algorithms are useful for:

- Setting of the parameters of the simulations to run.
- Setting of the parameters of the technological facilities to adopt for the manufacturing process.
- Defect classification in RE processes recognizing each component and wear defects.
- Prediction of the defectivity by analyzing historical quality tests.

The AI engine behaves as a DSS, processing mainly measurements and quality tests performed during the time of the device characterization. The outputs of the AI engine automatize the procedures to follow for each phase of the prototyping processes. The AI effort represents the upgrade of the production processes in a modern manufacturing industry working with innovative devices and components. In this scenario, more data are processed for the defectivity analysis in specific working and environment conditions, thus explaining possible causes of defectivity. RE includes the possibility to check each defect by images or by laser scanners, executing AI for the classification.

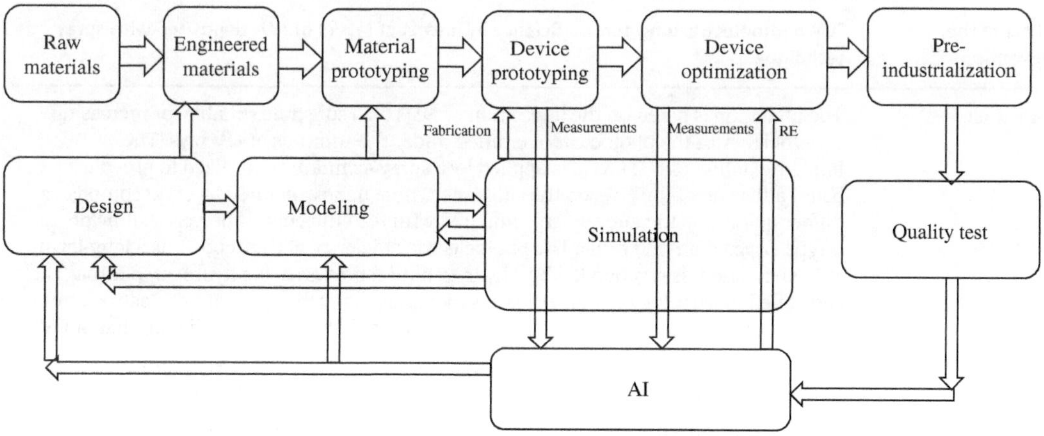

Figure 9.31 AI self-adaptive rapid prototype model oriented toward technology transfer and pre-industrialization.

9.5 Rapid Prototyping Oriented Toward Patent Development

The proposed technical sheets provide information about the preliminary studies of patent prototypes. The patent schemes are discussed in order to highlight how it is possible to set a patent by discussing prototype properties and claims, describing the application fields, the state of the art, physical basic principles, preliminary measurements, and preliminary layouts. The examples highlight all aspects occurring in the validation of a prototyped invention and are discussed taking into account typical templates of the patent documents.

9.5.1 Prototyping of Devices Implementing Nanoparticles

The spray technique [23] is an important and versatile approach for rapid prototyping. Preliminary prototypes and experimentations are fundamental for the development of a patent. As an example, Table 9.3 gives the preliminary description of a patent [42], which is useful to understand the stages, by discussing prototype properties, finding claims, and describing the application fields, the state of the art, physical basic principles, preliminary measurements, and preliminary layouts.

9.5.2 Prototyping of an Optoelectronic Device Based on a Nanocomposite Tip

In this subsection, the preliminary study of the patent introducing the invention of an endoscope nanocomposite tip is described [43]. Table 9.4 shows the main features of the invention.

9.5.3 DNA Lab-on-Chip

PCs are typically used for optical resonators. An application of PCs is in lab-on-chip for DNA detection systems [69]. Table 9.5 shows the main features of this invention.

Table 9.3 Scheme of the patent.

Title of the invention	**"UV photoelectric long-term efficiency by means of layers of NDs deposited with spray technique"**
Abstract	The invention is based on the deposition of NDs on a substrate suitable for increasing the efficiency of the photocurrent emitted under the stimulus of UV rays. The implementation of ND layers is applied by a spray technique optimized to provide a better adhesion of the NDs on the substrate. These nanostructured layers accomplish a higher quantum efficiency when compared with the efficiency of polycrystalline or *n*-type doped diamond films. The photoelectric efficiency of the deposits is a long-term efficiency, and it is regenerated by a hydrogenation process using hydrogen plasma. The controlled morphology of the ND grains increases the capacity to extract electrons from the surface. The cost of making these materials is reduced if compared with that of CVD
Keywords	Sprayed ND layers, photoemission aging efficiency, UV energy harvesting, thermoionic energy harvesting, photo-emitted regenerated material
Main technical problem solved by the invention	The problem solved by the invention is to have a low-cost diamond-based material that is more efficient than polycrystalline diamond from the point of view of the photocurrent stimulated by UV rays, and that at the same time maintains its efficiency during the time. This property is useful if the material is used in environments that are hardly accessible or inaccessible to humans such as in aerospace applications

Table 9.3 (Continued)

Title of the invention	**"UV photoelectric long-term efficiency by means of layers of NDs deposited with spray technique"**
Main discussion (complete and concise description of the invention, highlighting how it allows to solve the limits of the known solutions; also included are all the features and elements that are considered new and inventive, specifying the possible alternative personifications and related applications; to explain better the invention drawings of the first prototypes, photographs, tables, and laboratory tests are attached)	The physical principle of the invention is shown in Figure 9.32a,b. UV rays impact the exposed surface of the deposit of NDs and generate an emission of electrons in the form of a stimulated current. The NDs are initially dispersed in a solvent (water, dichloroethane, alcohol, etc.) and are then deposited on the substrate by means of a spray atomizer capable of controlling the homogeneity of the deposit. By comparing the surface exposed to UV rays of the ND deposit with that of a film deposited using the CVD technique, it is observed that the equivalent surface of the first case is larger, thus increasing the effectiveness of electron emission. The spray technique is used in order to adhere the deposit by van der Waals forces to the substrate. In fact, by depositing the ND powder without the use of solvent, the powder is easily removed at every small mechanical disturbance. In addition, the atomizer can control the quantity of NDs to be deposited through controlled pulses that allow to open and close the atomizer in well-defined time windows. The preliminary measurements showed that the photocurrent efficiency of a sprayed diamond deposit was found to be greater than that of diamond film (CVD growth) of different thickness and different concentration of nitrogen (N_2) (*n*-type doping). In all measurements the *p*-type silicon substrate (Si-*p*) was used. This efficiency is also found to be greater than polycrystalline diamond (CVD growth). Figure 9.33 shows the theoretical trend of the measurements performed. The NDs used in the experiment have a size ranging between 0 nm and 250 nm with a significant graphite component (verified by Raman analysis). As effective deposit alternatives from the point of view of the photocurrent, the equivalent exposed surface is changed by acting on the number of shots of the atomizer (controlled sprays), considering a larger graphitic component and taking into account NDs of different sizes. One way to increase the graphitic component of the diamond is to treat the diamond before exposure to UV rays by means of laser pulses at a well-defined fluence ($J\ cm^{-2}$). To further increase the equivalent exposed surface, a substrate with a sawtooth profile can also be used on which the NDs are sprayed. Instead, to increase the efficiency of the material, other types of nanoparticles in dispersion or deposits of NDs in multilayer films of different types can be integrated. The geometrical layout of the substrate can also be paraboloid type, thus facilitating the collection of UV rays. The choice of the substrate type is made based on the desired type of adhesion of the deposit. The deposited ND samples are pretreated by means of the hydrogenation process and measured after 45 days while maintaining approximately the original photoelectric efficiency: at wavelength of 150 nm the original efficiency is 20%, a value above those found in polycrystalline diamond (which are at most 12% at 140 nm), and after 45 days it becomes 16%. Hydrogenation treatments are carried out to regenerate the initial photoelectric efficiency
Basic state of the prior art: the research context in which the invention was conceived is briefly illustrated (brief description of the known techniques and technologies, and of their limitations); the main bibliographic references on the state of the art are explicitly indicated	Diamond grown by CVD technology with a different microcrystalline structure is used to generate photocurrent [35]. In particular, N_2-doped diamond film is used for the simultaneous generation of photocurrent and thermionic current at different temperatures [36]. Lithium (Li) can be deposited on a diamond film surface suitably treated with oxygen in order to provide a low working function, thus favoring an easier extraction of electrons [37]. The simultaneous insertion of Li and N_2, however, makes the diamond grains electrically inactive [38]. Instead, the polycrystalline diamond showed some relevant properties in terms of photoemission showing maximum efficiencies around 12% at 140 nm [39]. To increase the electrical conductivity of the diamond, a pulsed laser can be applied to the diamond, thus triggering the graphitization process [40]. The spray deposition technique, on the other hand, is inspired by an optimized approach for the deposition of carbon nanotubes [41], and then is applied to the deposition of NDs [23]

(Continued)

Table 9.3 (Continued)

Title of the invention	**"UV photoelectric long-term efficiency by means of layers of NDs deposited with spray technique"**
Advantages and disadvantages of the invention: a list of the main advantages associated with the invention is provided, as well as any disadvantages and limitations (legal provisions, pre-existing third-party rights, marketing authorizations, etc.)	The main advantages of the invention are: • Deposit facility. • Low cost of construction. • Possibility of regenerating the photocurrent efficiency. • Possibility of reusing the deposit (material recycling). • UV photoelectric efficiency. • Thermionic efficiency. • Long time durability of efficiency. • Possibility of integration in more complex storage systems that can integrate more elements of energy harvesting (thermoelectric effect, betavoltaic, photovoltaic, etc.). The main disadvantage is that CVD machinery for hydrogenation (regeneration and pretreatment processes) is required
Development stage of the invention	A first material of NDs deposited on *p*-type silicon is experimentally analyzed in terms of photoelectric efficiency, exhibiting a better efficiency than that found for diamond films. In addition, the aging of the material is analyzed in terms of efficiency with positive results
Possible future developments of the invention	Future developments concern the possibility of using the deposited ND layers on materials already efficient from the point of view of energy harvesting for other effects of generation of alternative current as in solar thermal and photovoltaic energy, piezoelectric vibration, thermionic, wind, and thermoelectric cases. Since the Sun emits photons over a wide range of frequencies, covering those of UV light in all three bands (UV-A [400–315 nm], UV-B [315–280 nm] and UV-C [280–100 nm]), and due to the absorption by the ozonosphere of about 99% of the UV rays that reach the earth's surface being UV-A (in fact almost 100% of UV-C and 95% of UV-B are absorbed by the earth's atmosphere), the material can be applied to efficiently generate energy for aerospace applications
Details about when and in what situation was the invention ideated	The ideation happened during a scientific research activity between different institutes and departments about the ND spray technique
Applications of the inventions, its potential industrial and commercial uses, reference market	The patent is appropriate for companies working in the energy and aerospace sectors supporting transfer technology

CVD, chemical vapor deposition; ND, nanodiamond; UV, ultraviolet.
Source: Based on [42].

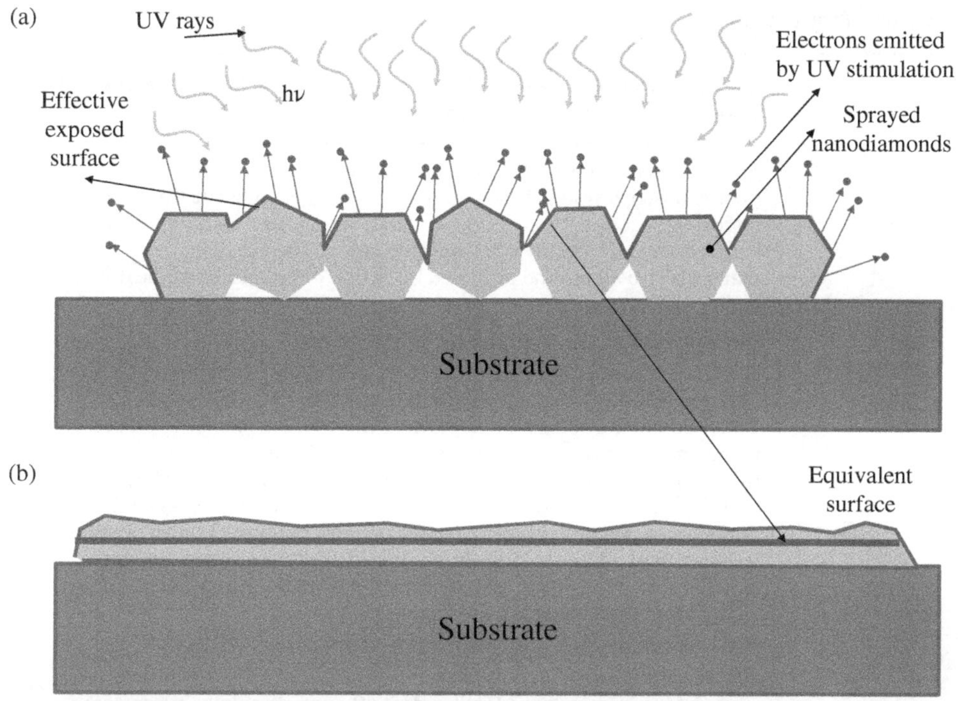

Figure 9.32 Scheme of the first prototype: (a) Principle of photocurrent stimulated by UV rays. (b) Equivalent exposed surface of sprayed NDs compared with the exposed surface of a CVD-grown diamond film.

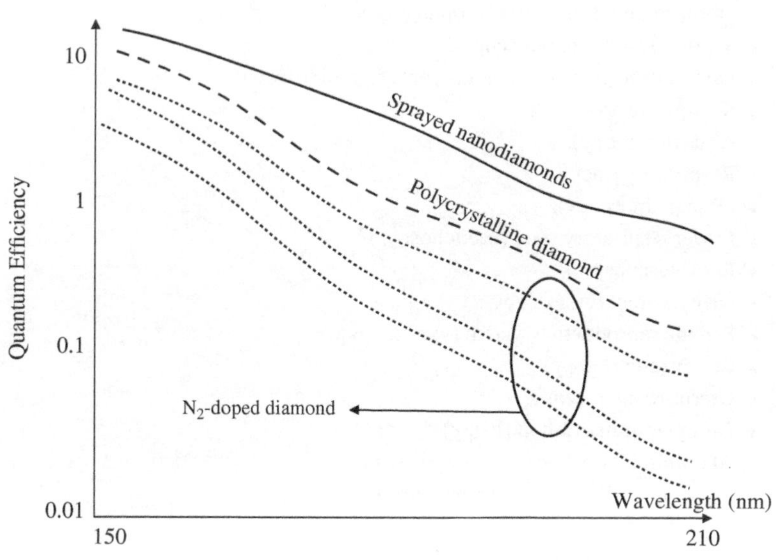

Figure 9.33 Experimental trend: comparison of the current photoelectric efficiency between diamond films grown by CVD technique with *n*-type doping (with different percentages of N_2), polycrystalline diamond film, and sprayed ND deposit.

Table 9.4 Scheme of the patent.

Title of the invention	**"Nanocomposite tip integrated in an endoscope system for 3D object detection systems and for tissue's anomalies detection"**
Abstract	The proposed invention concerns an innovative tip integrating an optical fiber endoscope system based on NMs. The nanocomposite tip enhances the optical light coming from an optical fiber source, and couples the light backscattered by a generic 3D detected target. The use of biocompatible materials also allows the proposed invention to be used in medical endoscopic applications
Background of the invention	**Background of the skin cancer application** A work analyzed optical fiber probes for skin cancer detection by micromachined oblique incidence reflectometry [44]. In particular, the absorption in the optical range provides information about the tissues: diffuse reflectance is used for the extraction of optical properties in noninvasive real time diagnosis and monitoring of superficial lesions [44–46]. Moreover, some authors studied the possibility to obtain optical information about the depth of the lesion by correlating the measured reflectance with the depth of the lesion both in a Monte Carlo simulation environment and in laboratory experiments [47]. Other classic approaches concern the post processing and algorithms of the images of the skin lesion [48–50]. **Background of the endoscopy approach** Endoscopes are instruments used to examine the interior of an organ or cavity of the human body. Unlike most other medical imaging devices, endoscopes are inserted directly into the organ. In particular, endoscopy is appropriate for: • Gastrointestinal tract. • Esophagus. • Stomach and duodenum (esophagogastroduodenoscopy). • Small intestine (enteroscopy). • Large intestine/colon (colonoscopy, sigmoidoscopy). • Rectum (rectoscopy). • Anus (anoscopy). • Respiratory tract. • Nose (rhinoscopy). • Lower respiratory tract (bronchoscopy). • Ear (otoscope). • Urinary tract (cystoscopy). • Female reproductive system (gynoscopy). • Cervix (colposcopy). • Uterus (hysteroscopy). • Fallopian tubes (falloposcopy). • Abdominal or pelvic cavity (laparoscopy). • Interior of a bone joint (arthroscopy). • Organs of the chest (thoracoscopy and mediastinoscopy). • Amnion (amnioscopy). • Fetus (fetoscopy). • Plastic surgery. • Orthopedic surgery. • Hand surgery (endoscopic carpal tunnel). • Epidural space (epiduroscopy).

Table 9.4 (Continued)

Title of the invention	"Nanocomposite tip integrated in an endoscope system for 3D object detection systems and for tissue's anomalies detection"
	Other endoscopic applications concern pre-visualization of infrastructures, engine inspection, internal inspection of complex technical systems, examination of improvised explosive elements, and surveillance via a tight space. A basic scheme of an endoscope for acquisition of images is found in the patent in [51]. The research in advanced endoscope systems is also related to the micro-/nanoscale [52]. **Background of 3D image reconstruction (reconstruction of a 3D object)** The reconstruction of 3D images can be achieved by ultrasonic systems [53] and for medical applications [54]. Optical fiber systems can be used for the recognition of large tissue volumes by means of fluorescent agents [55] and for macroscopic image detection through miniature and flexible probes [56], including robotic [57] and automotive [58] applications. **Background of the nanocomposite materials** Polymer composites [59, 60] are manufactured commercially for different applications such as sporting elements, aerospace components, automotive, and biomedical probes. In recent years, there has been a strong emphasis on the development of polymeric nanocomposites, where the dimension of the filler material is of the order of a nanometer. With microscopy and scanning probe microscopy tools, scientists can see the nature of the surface structure with a very high resolution. Simultaneously, by means of simulations, the rapid growth of computer technology has facilitated the knowledge of the NM properties characterizing physical properties. In addition, many polymer nanocomposites are fabricated and processed in ways similar to that of conventional polymer composites, making them particularly attractive from a manufacturing point of view
Patent state of the art	The basic endoscope system for optical images described in [51] concerns the basic principle of medical image detection. The proposed invention does not concern the system but the tip of the system designed in order to enhance the light, and also to detect information of the analyzed tissue by means of data post-processing. Other systems such as an ultrasonic system [53] or optical probe combined with a fluorescent agent [55] are applied for the detection of a 3D image (detection of a volume by a fluorescent agent). The invention is designed to be integrated in an endoscope system. For this reason, it is ideated to be used in traditional detection approaches based on endoscopy. The NMs [61] are used also as optical amplifiers [62], as chemical sensors [63], for the enhancement of electrical and mechanical properties [64], for the detection of cells [65], for the coupling of optical waveguides [66], for medical devices [67] and, finally, integrated in optical fibers for filtering applications [68]. All the listed patents do not include the possibility to integrate the NMs in a tip for endoscopy as for the invention, where the light coupling and the light enhancement properties of the NMs are combined in an endoscope tip suitable for the detection of 3D objects, for tissue anomalies and in general for industrial applications. The information about the detected object is in the optical reflectivity of the detected object, and does not require fluorescent agents. Moreover, the optical signal is enhanced at the output of the source fiber and again enhanced at the input of the receiver fibers (double enhancement) with a large acceptance angle (large area to detect)

(Continued)

Table 9.4 (Continued)

Title of the invention	**"Nanocomposite tip integrated in an endoscope system for 3D object detection systems and for tissue's anomalies detection"**
Object of the invention	The object of the invention is the nanocomposite tip endoscope type system illustrated in Figure 9.34a and b. An optical fiber (optical fiber source) is coupled with an external light source. The external light source is the input of the endoscope system. The light then arrives in a nanocomposite sample, made by a polymer with gold nanoparticles, which enhances the light at the output of the optical fiber source. Figure 9.34c shows the basic physical principle of the nanocomposite tip: the enhanced light shoots the object or the tissue (targets) and is reflected (backscattered light) into the same nanocomposite sample which enhances the light again. The enhanced reflected light is then coupled to an optical fiber bundle (fiber receivers) which transports the information of the target (optical intensity and optical absorbance) to a system able to post-process the signal. The fiber receivers are arranged around the optical fiber source. The 2D scheme in Figure 9.34d explains (by arrow) the amplification concept associated with the nanocomposite tip, where the input electric field signal is enhanced at the output (Figure 9.34e). The realized first prototype probe is shown in the photographs in Figure 9.35a and b
Claims and advantages	The main claims of the patent are: • Backscattering and reflection optical real time sensing by nanocomposite tip by means of the light resonance and light enhancement of the NM. • Increase of the NA of the optical fiber. • Tip integration with medical endoscope systems. • Skin scope detection system of melanoma. • Reconstruction of a 3D object by a fiber optic scope system with nanocomposite tip in environments with light (with noises of artificial light). • Algorithm for melanoma skin detection systems based on post-processing of the skin scope's data (low error of the diagnosis). • Different tip layouts for efficient detection systems. • Detection of the color of a 3D object by the nanocomposite tip. • Easy implementation in a robotic system without the use of integrated CMOS, CCD sensors. • Disposable endoscopic probe. • Automotive sensor for the detection of an obstacle (useful for the parking of a car).

Table 9.4 (Continued)

Title of the invention	"Nanocomposite tip integrated in an endoscope system for 3D object detection systems and for tissue's anomalies detection"
	A real time sensing is performed by means of the controlled dispersion of gold nano inclusions. The tailoring of the electromagnetic interaction is size dependent and determined by the dimensions and the concentration of the nanofillers inside the polymer matrix. According to these parameters it is possible to enhance the light at a particular working wavelength. The acceptance angle is related to the refractive indices of the core, cladding, and medium surrounding the fiber. The NA is a measurement of the ability of an optical fiber to capture light and is used to define the acceptance cone of an optical fiber. The nanocomposite tip allows to increase the NA of the fiber tip by facilitating the light coupling and the detection of a target. Concerning application in the melanoma detection system, the methods proposed in the literature are based on the images of the skin melanoma and the information about the depth of the lesions is affected by many errors (absorbance spectra). By enhancing the optical field, the nanocomposite tip facilitates the detection of the asymmetry of the melanoma (important parameter) and of the lesion's depth with a low diagnosis error. The probe is applied as indicated in Figure 9.36a as a noninvasive detection system. The preliminary experimental results in Figure 9.36b prove the ability to detect the Clark level indicating the depth of the skin lesion: in the case in Figure 9.36b the sensitivity of the detection of a melanoma lesion's depth is experimentally simulated by using graphite solution in isopropanol solvent. The signal coming from the receiver fibers is transformed to a voltage signal by a photodetector. The endoscope nanocomposite tip can also be applied for the monitoring of the extension of the lesion on the surface of the skin. Preliminary measurements checked the sensitivity of the 3D endoscope system to detect a grayscale bar, thus also providing information about color variation (Figure 9.37). The nanocomposite endoscope tip can detect different kind of objects in three spatial dimensions (3D), by placing the tip at a distance of an order ranging from 1 to 50 cm. By using only a system of fibers without NM, the object must be attached to the tip; this represents an important advantage of the use of the nanocomposite tip. This maximum distance can be increased by acting on the power of the light source, on the concentration of nano inclusions, and on the layouts of the tips. The preliminary experimental results are performed in a laboratory with neon lights which represent a "background noise." For this purpose, by using a powerful light source, the system can be used for a robotic system as an "optical eye." Figure 9.38a shows the probe system detecting a sphere with the tip placed a distance *d* from the sphere (Figure 9.38b). The tip measures the optical field reflected by the sphere by means of a scanning process performed in spatial discretized steps (2D grid in Figure 9.38c). The reflected optical field intensity assumes the form in Figure 9.38d.
Conclusion (main advantage)	The invention is in the tip of an endoscope system behaving as an active material. This tip is obtained by the use of controlled NMs able to enhance the optical field reflected by the object or a tissue to detect. The double enhancement (of the light at the output of the fiber source, and again enhanced at the input of the receiver fibers) and the good light coupling allow to optically detect a 3D object in the presence of light noise and for large distances

CCD, charge-coupled device; CMOS, complementary metal oxide semiconductor; 2D, two-dimensional; 3D, three-dimensional; NA, numerical aperture; NM, nanocomposite material.
Source: [43].

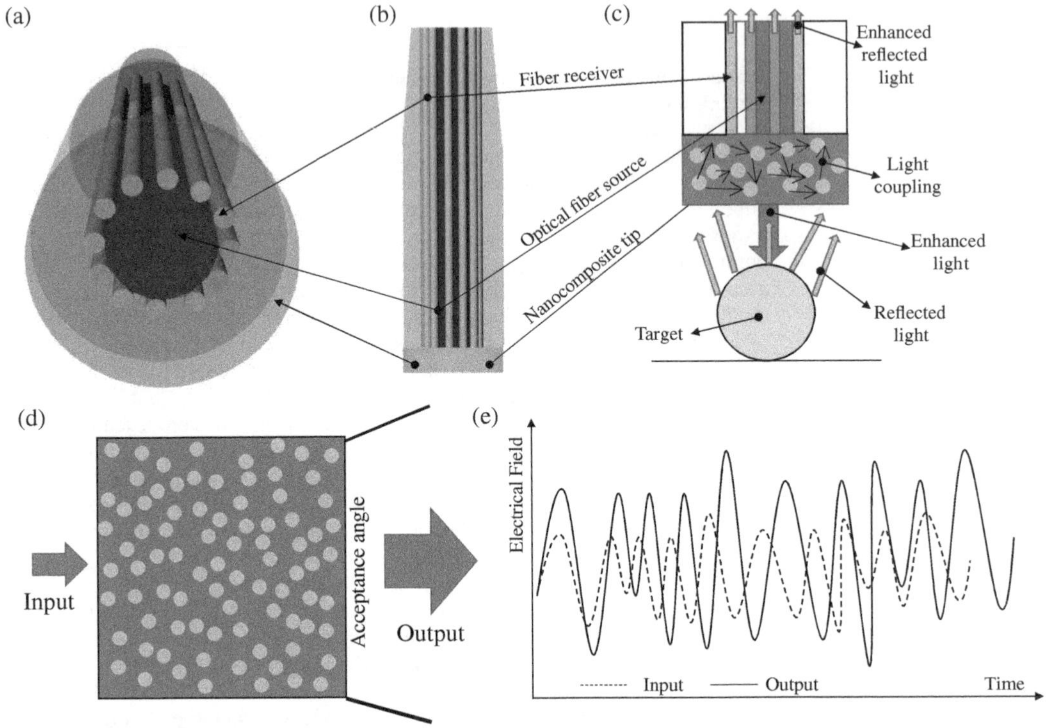

Figure 9.34 (a) 3D endoscope optical fiber system with nanocomposite tip. (b) Another perspective of the 3D model. (c) Basic principle of the endoscope system. (d) Nanocomposite tip section and acceptance angle. (e) Theoretical trend of the input and output electric field.

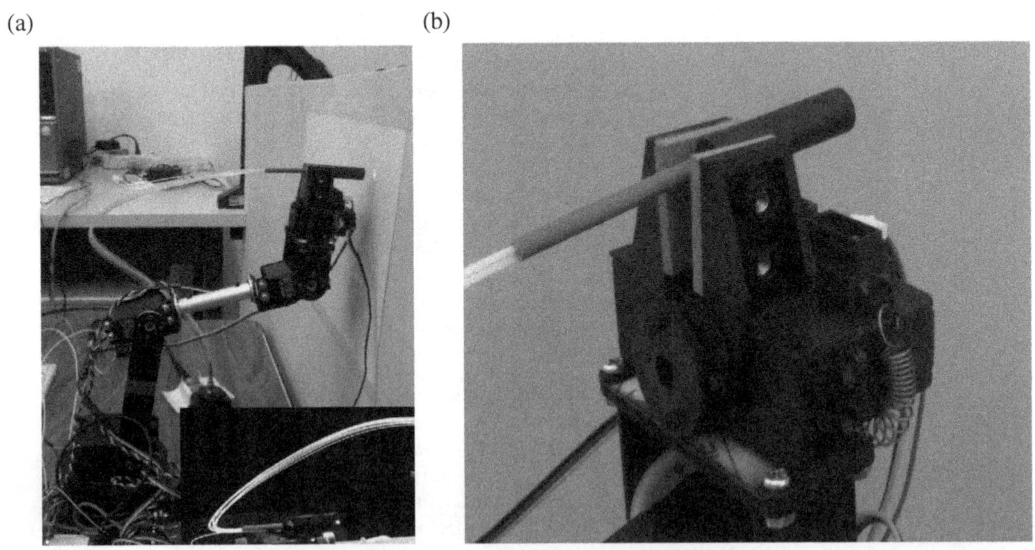

Figure 9.35 (a) 3D endoscope probe controlled by a robotic arm. (b) Zooming of the probe.

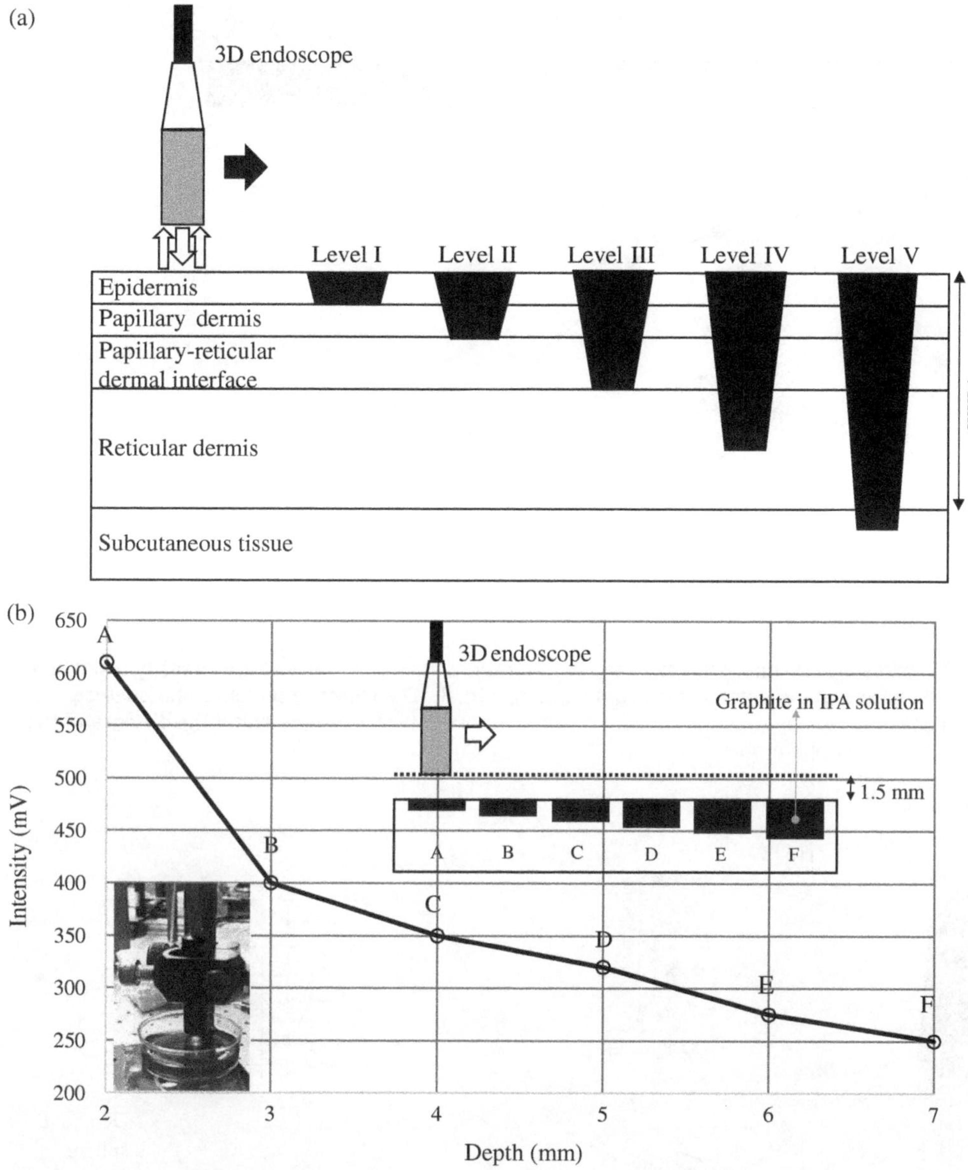

Figure 9.36 (a) Clark level model (skin melanoma detection). (b) Voltage intensity of the reflection response measured by the 3D endoscope system simulating the model of Clark levels. Top inset: experimental setup where the melanoma is simulated by a solution of graphite in isopropyl alcohol (IPA) solvent. Bottom inset: photograph of a measurement performed using a graphite/IPA solution.

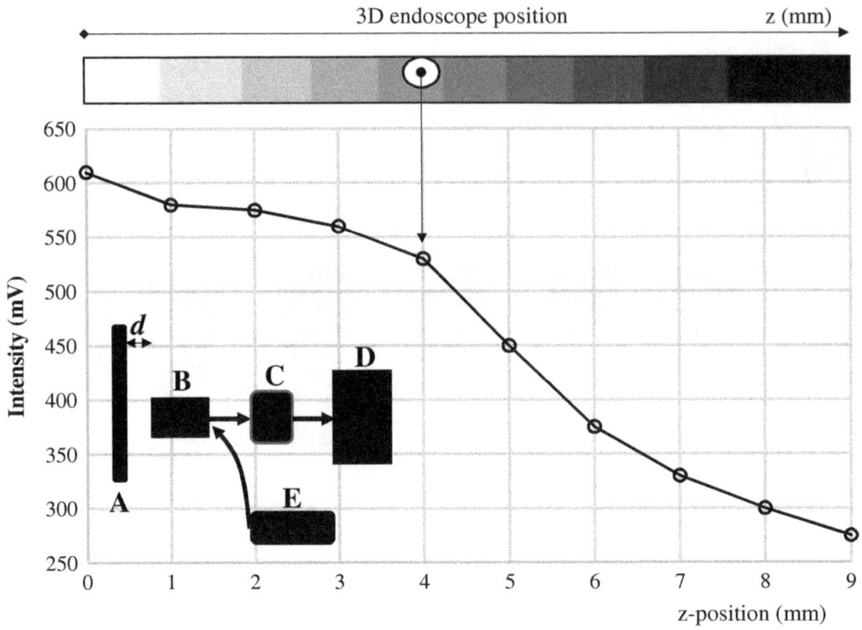

Figure 9.37 Endoscope system: grayscale reflectivity response measured as voltage intensity. Inset: block diagram of the basic experimental setup: A, grayscale screen; B, 3D endoscope probe; C, photodetector; D, electronic signal intensity analyzer; E, light source in input of the probe system; and *d*, the distance of the screen from the endoscope probe..

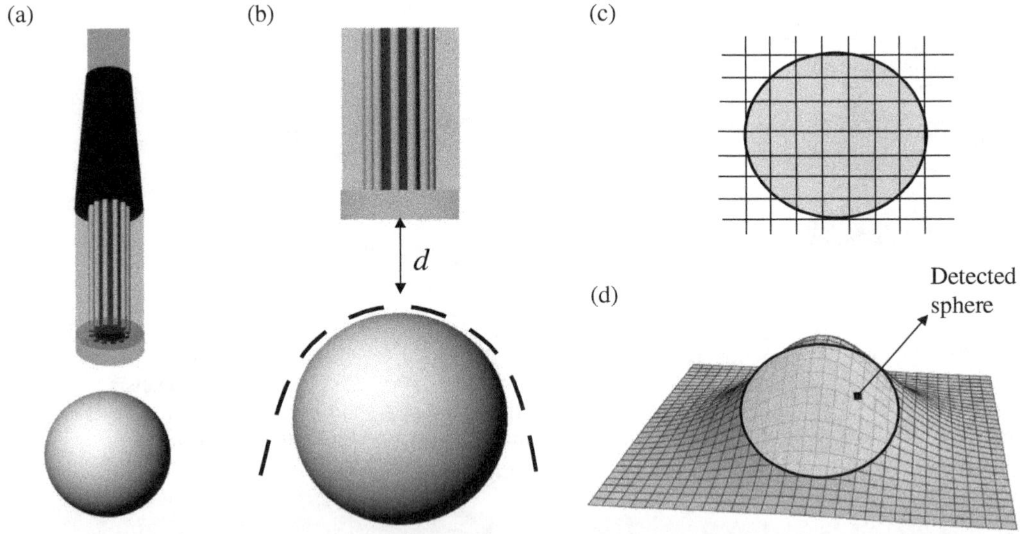

Figure 9.38 (a) 3D endoscope system detecting a sphere. (b) Frontal perspective of the detection probe system. (c) Grid modeling scanning movement discretization of the probe system. (d) Reflection signal reconstruction of a detected sphere.

Table 9.5 Scheme of the patent.

Title	**"Photonic crystal resonators array chip for improved optical sensing of (bio)molecules in genomics and proteomics"**
Abstract	The patent provides a method about the optical read-out optimization for bio-molecular detection systems (DNA chips, protein chips, microarray, lab-on-chip systems) in genetic/protein applications. The method is based on arrays of PC resonators. The new approach provides the efficiency of the detection of bio-samples marked with luminescent substances: the detection efficiency considers the accuracy of the analysis (sensitivity of the analysis), the S/N ratio, and finally the quick optical read-out process
Keywords	PCs, biomolecules, biochips, optical sensing, luminescent probes, resonant cavities, spectral analysis, spectral tailoring
Patent goals	The main goal of the patent is the detection efficiency of the optical signal coming from the biosensor chip based on the luminescence emission. This goal is achieved through arrays of PCs which resonate in the emission band of the marker associated with the biomolecule target. A proper design of these PC arrays and the spectral analysis of the emission band provide the association of each bio-target with an emission band, and permit the separation of the source signal, separation of the reflected and diffracted signals (reflection/diffraction noise) from the useful signal. This concept is used to increase the source efficiency of bio-sample detection, and to increase the emitted intensity of the optical marker localized on the PC surface [70]. The patent proposes to utilize the PC resonators to enhance, with specific selectivity, the luminescent marker-emission only in the resonant spectral bands. By fabricating more PC resonators on the same chip, each one characterized by a wavelength resonance, and by localizing on each resonator a specific bio-recognition sample (for example a different oligonucleotide known sequence), it is possible to localize spatially the biomolecule targets and to associate univocally each target with an emission band. Through the analysis of the total emission band collected on the whole read-out area, it is possible to detect the presence of each single target of the analyzed bio-sample. In comparison with the traditional detection systems, based on the spatial scanning of the optical read-out area, the innovative reading approach permits to: • Decrease the time of the analysis, in fact, the detection is done simultaneously on the whole read-out area in which all the targets are localized; this characteristic allows to accomplish more analyses simultaneously as in a multiplexing process, by reducing the detection time through only the spectral analysis (a single scanning detection signal is considered); it is possible to realize more different resonators and consequently detect more targets, having the resonant bands inside the marker emission band (spectral contributions coming from different PC cavities). • Decrease drastically the reading error generated by the diffused, reflected and diffracted source light. • Increase the detection sensitivity through the enhancement of the luminescent marker signal (the signal enhancement is generated by the coupling with the PC and is useful in both cases of spatial and spectral detection).

(Continued)

Table 9.5 (Continued)

Title	**"Photonic crystal resonators array chip for improved optical sensing of (bio)molecules in genomics and proteomics"**
	In the patent, the unambiguous assignment of an emission band to each target does not require the spatial disposition or the use of different fluorescence markers emitting in different bands: the unambiguous detection is guaranteed by using for each bio-target properly designed PC cavities emitting in specific wavelengths. In this manner, each PC resonator can modify the luminescence of the common marker. This approach is extended by simultaneously utilizing in the same chip fluorescent substances with different emission bands, in order to increase the number of analyses to perform at the same time, and consequently the target number to detect. Moreover, the proposed technique can be utilized in different spectral ranges, (starting from the UV to IR), by utilizing for example an organic fluorescent substance or colloidal nanocrystals. The optical properties of the structure can be modified with good accuracy in each desired frequency range by simply changing the geometry and the PC hole dimensions. Moreover, a further advantage of the patent prototype is the possibility to combine the enhancement of the selective emission band with the annulment of the source noise in the region where the detected signals (properties of the PC cavity resonance effects) are collected: the S/N ratio, defined as the ratio between the light intensity of the useful signal and the source noise regarding diffraction, diffusion, and scattering effect of the excitation; this can be furthermore increased by the PC resonance providing good signal detection sensitivity
State of the art and limits of the solutions already known, and introduction of the main patent specifications	The development of micro- and nanofabrication technologies, biomolecular patterning and microelectromechanical systems are addressed to the creation of real miniature laboratories applied to genomic and proteomic applications. The fields of application of these biochips are broad, and there are currently many terminologies used to indicate these devices such as gene-chip, gene-array, DNA microarray, protein chip, and lab-on-chip. Basically, these chips, developed both in simple formats and integrated in more complex devices or architectures, are planar structures developed on different materials (commonly glass or plastic), on which biomolecules are immobilized (by chemical modification of the surface or by synthesis in situ), such as DNA, proteins, and cells capable of carrying out a selective recognition of target molecules [71]. Biochip technology finds extensive use in gene and protein expression studies (genomics and proteomics) in various fields such as experimental and clinical diagnostics, biomarker discovery, and pharmacogenomics. Indeed, different types of chip are applied for a very wide range of applications, such as enzymatic assays [72], immunochemistry assays [73], polymorphism recognition analysis in gene variants [74], sequencing of nucleic acids [75], and also for the realization, on micro-volumetric scales, of DNA ligation and amplification reactions [76–78]. Given the high specificity in the hybridization reaction between oligonucleotide sequences, chips based on biomolecular interactions between DNA elements are developed faster than protein-based chips. The market of the latter, in fact, although of enormous interest in the scientific world, has suffered a slowdown compared with the new technology, given the complexity of the interactions that are the basis of the biorecognition mechanism between molecular protein species [79].

In general, in a chip, the recognition event between the (bio)molecule and the target species must be translated into a detectable and measurable signal. The physical transducers that allow the signal conversion can be of different types, and are based on different operating mechanisms: optical, electrical, electrochemical transduction, or devices sensitive to changes in mass, current or frequency [80, 81]. In the vast field of biosensing, however, optical sensors play a pivotal role. Optical detection is essentially based on the measurement of the luminescence that develops in the biorecognition site following the binding event between interacting species (for example, between complementary oligonucleotides). Typically, the most common biochip formats require one of the two biomolecular species to be conjugated to a fluorescent marker. Optical transducers commonly make use of simple optical detection schemes: typically, the fluorescence signal derives from a fluorophore conjugated to the target biomolecule, and the detection of the signal occurs at the end of the biorecognition reaction; in the “reading” phase, the read-out region is suitably illuminated by a light source capable of exciting the markers linked to the target analytes, and a detector collects the signal emitted by the markers themselves. These approaches, while allowing a purely qualitative revelation on the presence or absence of the target biomolecule in the solution under examination, are of great interest in various fields of application, since they allow to carry out, in a single measurement, a parallel screening on a very high number of different target biomolecules. The operating principle of these biochips is usually based on the spatially controlled immobilization on suitable substrates (to be treated by various chemical–physical procedures) of matrices of different biorecognition molecules (especially probe), so that it is possible to associate to each region of the substrate a specific probe molecule able to selectively recognize the target analyte. The approach described above represents the most commonly used strategy, especially for commercial applications, but obviously other types of biochip are possible, which provide operating strategies and detection schemes based on different or more complicated principles, depending on the specific analysis to perform. Some biochips allow the monitoring of specific biomolecular interactions in real time. For example, the use of optical fibers for signal transduction has a wide use, as well as the development of new detection strategies based on fluorescence resonance energy transfer, quenching, molecular beacons, total internal reflection fluorescence, surface plasmon resonance, phosphorescent probes, or the design of new fluorescent markers. These strategies led to a notable improvement in the detection limits of the signal [82]; however, the growing interest in the use of such chips in real diagnostic applications requires an ever-increasing sensitivity of analysis to be able to quickly and reliably reveal ever lower quantities (concentrations) of analytes ofpathogenic interest.

(Continued)

Table 9.5 (Continued)

Title	**"Photonic crystal resonators array chip for improved optical sensing of (bio)molecules in genomics and proteomics"**
	The patent proposes an innovative optical device, based on an appropriate coupling of PC technology with nano-biotechnologies (biochemistry/biophysics, surface chemistry, patterning of biomolecules, bioimaging), which allows to improve traditional technology, in terms of detection sensitivity, improve the S/N ratio, with the possibility to carry out highly parallel analyses in short times, and based purely on spectral analyses. The optical transduction device proposed for the determination of biomolecules with high sensitivity characteristics and in multiplexing format, adaptable in general to genomic and proteomic analyzes, is essentially constituted by an appropriate array of PC resonators, on which are immobilized, with precise spatial control, specific probes for the analytes to be determined (for example, DNA, proteins, ligands, etc.). Figure 9.39a shows a 3D layout of the patent chip. The target analytes can be labeled directly by synthesis or indirectly with one or more suitably chosen fluorophores. The basic scheme of the device functionality is shown in Figure 9.39b–d, which shows the general case for the determination of specific biomolecules (proteins, ligands, etc.). In the case of implementation of a biochip for DNA sequence analysis, it is possible to couple to the optical read-out part an appropriate microreactor in which to carry out DNA amplification reactions (PCRs). This technology, in its various modalities (reverse transcriptional PCR, multiplex PCR, real-time PCR, quantitative PCR [83]), also provides an important diagnostic tool, being able to be used for analysis of genotyping and phenotypic expression of pathogenic or viral antigens [84]. The proposed device is based on an original optical detection scheme: the detection is, in fact, carried out by collecting the emission spectrum from the entire bio-recognition area of the chip (Figure 9.40a). With the use of an appropriate matrix as shown in Figure 9.40b, consisting of different PC resonators, a different resonance peak is associated with each bio-recognition element of the device (probe species), and therefore with each specific target analyte. By optical peak detection (generated following optical coupling between the PC and the fluorophore conjugated with the target species), it is then possible to deduce the presence or the absence of specific (bio)target molecules sought in the analyzed sample. In particular, with a single analysis of the emission spectrum (collected from an area comprising several different resonators), it is possible to carry out the parallel recognition of multiple analytes, increasing the information collection speed. The PC also permits to break down the excitation radiation which is diffused or reflected in the direction of the collection of the signal. This effect, associated with the increase in the emission of the useful signal due to resonance, makes it possible to significantly increase the overall S/N ratio, by avoiding possible reading errors of the detector.

Recently, some solutions for the optical read-out of biochips based on PCs have been proposed thus validating the idea of using these kinds of devices. PCs, due to a discontinuity of the refractive index in one, two or three directions, facilitate the engineering of photonic modes, the creation of highly efficient mirrors, waveguides, and optical micro- and nanoresonators of extreme spectral purity and efficiency.

Optical read-out techniques based on PC structures are classified according to the type of optical detection scheme.

In the case of "label-free" detection, in particular in the absence of luminescent markers on the target analytes, PC resonator arrays are characterized by resonance bands which are modified by the presence or absence of target analytes: the analytes, in fact, locally modify the refractive index of the layer overlying the PC, thus shifting its resonance frequency. The detection of these displacements is carried out by means of reflectivity or transmittivity measurements on the single resonator. The analysis of the read-out region takes place, consequently, by carrying out repeated scans, sequentially, on each resonator.

In the case of detection based on optical markers, however, a significant increase in the emission of fluorophores occurs when there are localized cavity modes in the PC. In this case, the electromagnetic field corresponding to the wavelengths of the excitation source is locally increased on the surface cavity of the PC due to the engineering of resonant modes or "leaky" modes. The effect of the PC is, therefore, to increase the intensity of the useful signal in a frequency selective way. The relationship between useful signal and noise coming from diffusion or diffraction of the source is considerably improved, but it is however necessary to resort to spatial scanning of the whole PC matrix.

(Continued)

Table 9.5 (Continued)

Title	**"Photonic crystal resonators array chip for improved optical sensing of (bio)molecules in genomics and proteomics"**
Technical description of the invention	In this invention, PC technology is applied to the optical read-out region of the biochip (bio-recognition region), with the excitation and emission of luminescent markers bound to target analytes (such as DNA, proteins, etc.). These analytes interact specifically and selectively with the bio-recognition elements immobilized on the surface of the resonator. The 1D, 2D or 3D PC consists of optically nonlinear materials and structured geometries in optical micro- and nanoresonators (regions where light is trapped and where the emission in the vertical direction occurs at specific frequencies and by means of optical modes). A peculiar feature of PCs is the ability to control the propagation of light by the 1D, 2D or 3D periodicity in materials that are transparent at the wavelength of interest. The light is trapped in the PC by introducing, for example, a defect in the crystal. Therefore, PC technology, applied to biochip technology, has several advantages such as: (a) The possibility of varying, through an accurate design of the material and geometry, the resonant wavelength of each resonator of the matrix (consequently modifying the emission spectrum of the fluorophore associated with the target analyte in a particular way); this feature provides the opportunity to achieve the optical reading no longer or not only on the basis of spatial discrimination, but also in a spectral way. (b) Increase of the emission efficiency of the fluorophore in specific spectral bands through the Purcell effect, the quality factor, and the S/N ratio. (c) The possibility of designing the resonator and the optical detection system in such a way that the optical imaging system (detector) collects only the signal emitted by the fluorophore and not the whole signal (diffuse, reflected, or diffracted signal); this is achieved by tuning the angular radiation pattern as a function of the wavelength, thus spatially decoupling the excitation radiation from the emission band. Concerning point (a), an external radiation source, affecting the PC matrix in an optimal way, excites the fluorophore associated with the recognized analyte, which emits an amplified signal typical of the marker. The variation from pixel to pixel (i.e. from analyte to analyte) of these frequencies emitted simultaneously from the matrix allows to carry out a highly parallel analysis on an arbitrarily large number of different analytes, also achieving an increase in the recognition speed of the examined samples. With regard to points (b) and (c), the accurate design of the geometry and the materials necessary for the realization of the PC allows to obtain, in certain directions, a band of forbidden energy for photons. As already mentioned, it is possible to enable the presence of some localized modes, i.e. confined in a small region such as, for example, that of a crystal defect. These cavity modes, oscillating within the resonance region, are selectively amplified. The PC cavities are therefore sized in such a way as to easily separate the useful signal of the biological sample from the noise and generate very narrow optical amplified peaks that are easily distinguishable due to the wavelength selectivity of the detection system. A further spatial decoupling between the radiation coming from the excitation source and the radiation due to the emission of the marker can be carried out by sending the excitation to the read-out region through appropriate wave guides. In this case, the PC is designed in such a way as to take part of the excitation radiation from the guiding region and conduct it toward the resonator. The marker located on the cavity is then excited by the guided modes (which, as guided modes, cannot therefore be diffused toward the detector), and its emission is modulated and amplified by the resonance effect.

Advantages of the proposed invention	The main advantages of the patent solution are: 1. The possibility of using a resonant PC micro- or nanocavity in the emission range of a fluorescent marker. The cavity or resonator is made by using 1D or 2D or 3D PCs. In particular, by using a 2D PC, these resonators are fabricated on a guiding layer, on a substrate with a different refractive index, or possibly on a suspended guiding layer, in a membrane configuration. The final configuration of the device can be of high mechanical stiffness (using, for example, PC resonators in Si on air, Si on SiO_2, TiO_2 on SiO_2, Si_xN_y on air, Si_xN_y on SiO_2, or GaAs on Al_xGa_1-xAs) or with high flexibility using plastic materials such as PC in PMMA on air, or PMMA on polydimethylsiloxane. 2. The possibility of implementing a single wavelength or multiple wavelength detection system (spectral multiplexing). This system characterizes a spectral and a spatial discrimination. 3. Increase of the data acquisition speed through spectral analysis of the acquisition over the entire detection area. 4. Unique attribution of a specific emission spectrum to each target (bio) analyte to be searched in the unknown sample (specific emitted wavelengths). 5. Unique attribution of a peculiar emission spectrum to each target (bio) analyte to be searched in the unknown sample, using a single common marker, or multiple luminescent markers to further increase the degree of parallelism of the system. 6. Emission band selectivity by means of high quality factor PC resonator arrays. 7. Amplification of the emission spectrum of the marker conjugated with the target (bio)molecule, by coupling light with photonic cavity modes located in the bandgap. 8. Increase of the S/N ratio: (a) reduction of the noise signal due to undesired effects of the source due to the frequency-discriminated detection and the modulation of the angular pattern of radiation; (b) increase of the useful signal by coupling the emission of the marker with the resonant modes of the PC cavity. 9. The possibility of using more complex design systems of the PC pattern in order to combine the resonance properties between guided and leaky radiating modes.

1D, one-dimensional; 2D, two-dimensional; 3D, three-dimensional; IR, infrared; PC, photonic crystal; PCR, polymerase chain reaction; PMMA, poly(methyl methacrylate); S/N, signal/noise; UV, ultraviolet.

Source: Based on [69].

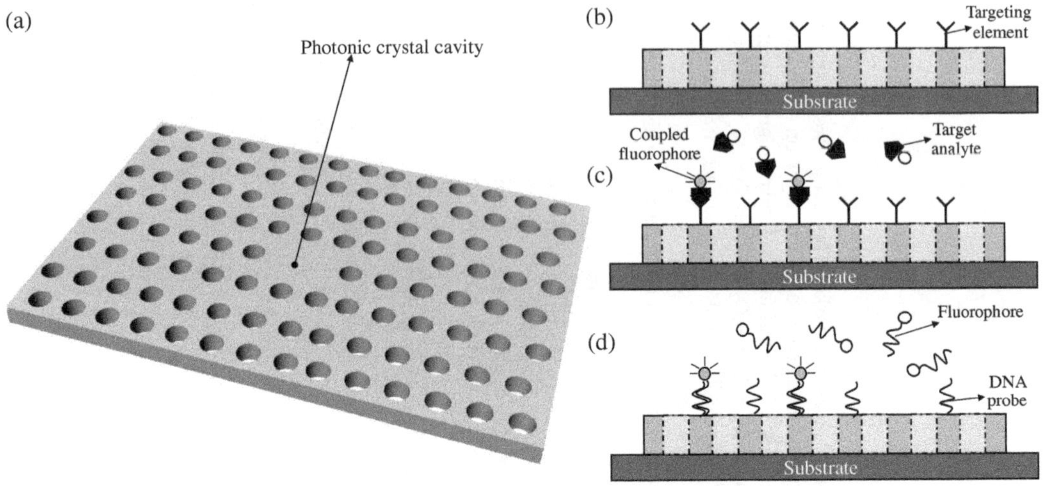

Figure 9.39 (a) 3D PC cavity resonator. (b–d) Basic principle of DNA detection.

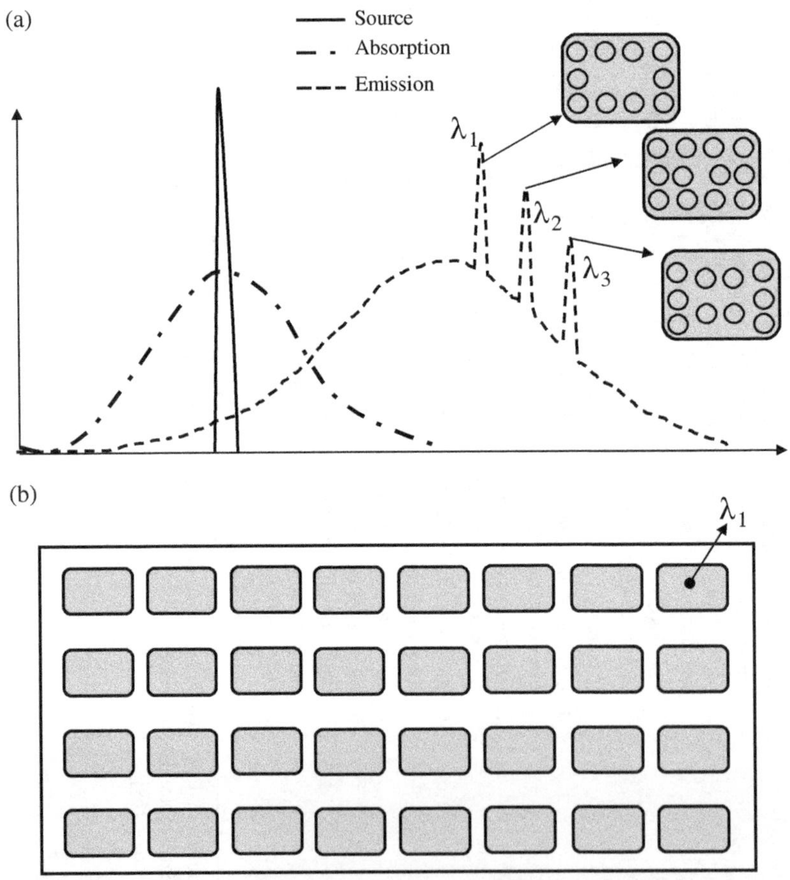

Figure 9.40 (a) Example of the emission signal of the fluorophore, laser source and the absorption spectrum enhancing three wavelength resonances. (b) DNA lab-on-chip matrix integrating different PC devices.

9.6 Nanocomposite Artificial Skin Rapid Prototyping Process

Pressure sensors by NMs are an interesting research topic. A related application field is in artificial skin [85]. A preliminary prototype of an artificial skin pressure sensor is shown in Figure 9.41a, where a polydimethylsiloxane (PDMS)-gold (Au) material is designed and fabricated to improve pressure sensitivity for robots. In order to guide the light coupling, different gold plasmonic resonators are structured on the external surface of the PDMS-Au bulk material. The resonators are generated by laser or bromograph texturing technique generating the square micrometric patterns shown in Figure 9.41b on the PDMS-Au surface. The main function of the plasmonic resonator is to collect the light corresponding to the gold micropatterns when a pressure is applied. The first prototyping process involves the preliminary study of the flexible PDMS-Au bulk, by defining the thickness for the robotic artificial skin sample, the laser texturing setting, and the preliminary realization of a multi-tapered optical fiber ensuring light coupling along the whole prototype. The resonators are textured near each tapered region in the central region of the first prototype for a first check of the light coupling principle due to an applied source. Then, the tapered single optical fiber is changed to the multi-tapered fiber matrix shown in Figure 9.42 allowing the pressure sensing region to be enlarged to check the spatial distribution of the applied pressure force.

Finally, the prototyping converged into the compact solution shown in Figure 9.43 suitable for robotic integration: a light source excites the PDMS-Au bulk by an optical fiber embedded into the same bulk, and an output fiber collects the signal at the output of the plasmonic resonators transmitting the signal into a photodetector.

The scheme of the preliminary experimental setup is shown in Figure 9.44a where a white lamp source excites a multi-tapered optical fiber directing the light on the plasmonic circuits textured on the sample surface. When a pressure is applied on the sample, the light is strongly coupled to the corresponding spatial resonator focusing the output light on the optical fiber probe connected to the photodetector reading a gained voltage signal. A typical trend of the measurement is shown in Figure 9.44b, which plots the photodetector voltage versus the pressure force (expressed in kPa). The device prototyping for this case study is summarized as follows:

- Fabrication of the PDMS-Au bulk material improving light coupling of an embedded tapered fiber connected to a lamp source (choice of the best gold percentage).

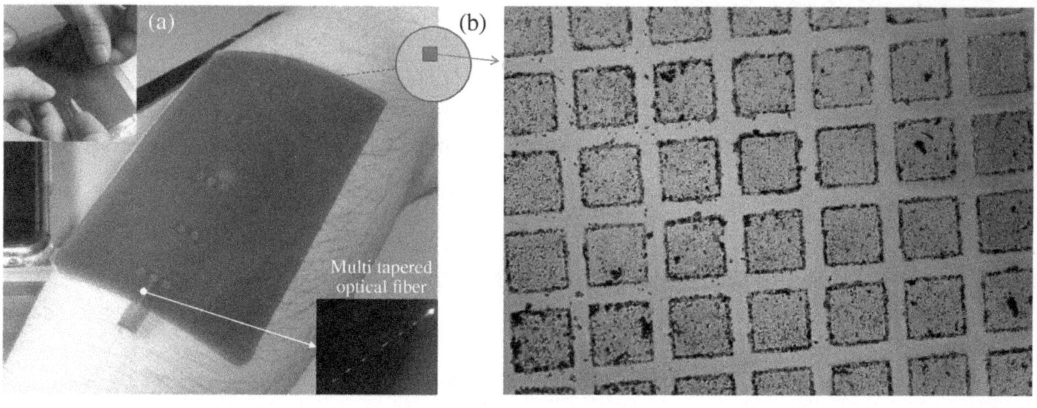

Figure 9.41 (a) First prototype of an artificial skin sensor. (b) Plasmonic gold micropattern generated by the texturing technique.

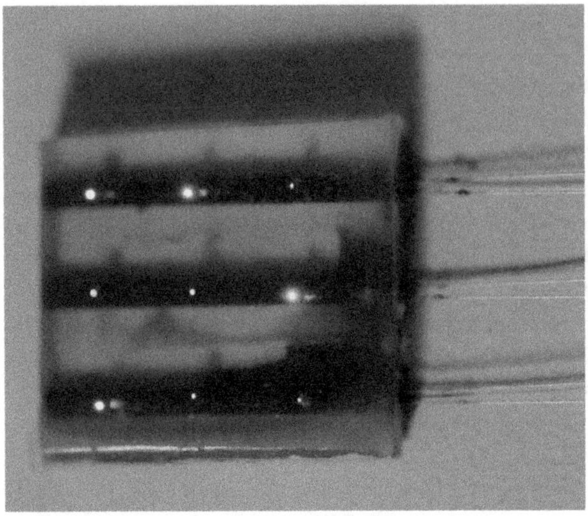

Figure 9.42 First prototype of a fiber bundle matrix emitting light from different tapered regions.

Figure 9.43 Photograph of the lighted embedded prototype connected to a photodetector.

- Basic design, modeling and simulation of the first prototype proving light coupling into the bulk material.
- 2D and 3D FEM simulations of the plasmonic resonances.
- 2D and 3D FEM simulations proving light emission of the plasmonic resonators.
- Simplified 2D modeling of the pressure effect on the bulk material.
- 2D simplified simulation of the light coupling considering the model simulating applied pressure forces.
- Fabrication of a millimetric thick bulk material for flexible applications.
- Texturing of micrometric plasmonic resonator on the PDMA-Au bulk surface.

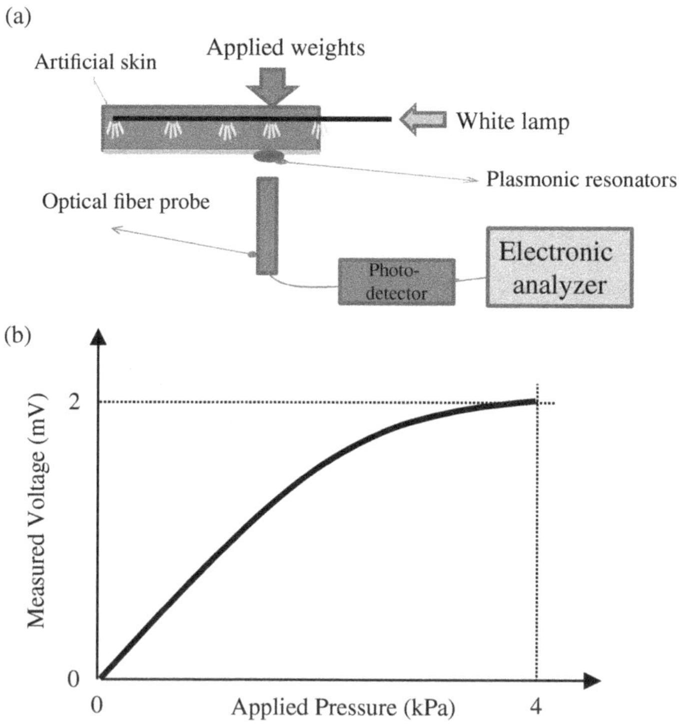

Figure 9.44 (a) Experimental setup scheme. (b) Theoretical trend of pressures as measured voltage versus the applied pressure.

- Check of the plasmonic resonators (check by microscope).
- Improvement of the multi-tapered input optical fiber embedded into the PDMS-Au bulk.
- Finding new solutions for optical fiber array coupling in larger bulk regions.
- Design and realization of compact sensors with input and optical fiber directly integrated in the prototype device.
- Measurement and testing to check pressure sensitivity.
- Optimization of the compact prototype device (thickness of the bulk material, best embedding of the optical fibers).

References

1 Massaro, A. (2010). Theory, Modeling, Technology and Applications of Micro/Nano Quantum Electronic and Photonic Devices, 1e. Transworld Research Network.

2 Massaro, A., Venanzoni, G., Farina, M. et al. (2011). Experimental pressure sensing and technology of piezoelectric microwave/RF MEMS filters. *International Journal of Microwave and Wireless Technologies* 3 (5): 587–593.

3 Massaro, A., Spano, F., Caratelli, D. et al. (2011). New approaches of nanocomposite materials for electromagnetic sensors and robotics. *International Journal on Measurement Technologies and Instrumentation Engineering* 1 (1): 55–72.

4 Spano, F., Massaro, A., Blasi, L. et al. (2012). In situ formation and size control of gold nanoparticles into chitosan for nanocomposite surfaces with tailored wettability. *Langmuir* 28 (8): 3911–3917.
5 Massaro, A., Cingolani, R., and Passaseo, A. (2011). 3D FEM modelling and technology of piezoelectric ring MEMS antenna. *Progress in Electromagnetics Research C* 23 (1): 123–135.
6 Caratelli, D., Massaro, A., Cingolani, R., and Yarovoy, A. (2011). Accurate time-domain modeling of reconfigurable antenna sensors for melanoma skin cancer detection. *IEEE Journal of Sensors* 12 (3): 635–643.
7 Mehdipour, A., Mohammadpour-Aghdam, K., and Faraji-Dana, R. (2007). Complete dispersion analysis of Vivaldi antenna for ultra wideband applications. *Progress in Electromagnetics Research* 77 (1): 85–96.
8 Massaro, A., Savino, N., and Galiano, A. (2020). Agri-photonics in precision agriculture. In: Proceedings of IEEE ICTON 2020, Bari, Italy (19–23 July 2020). Piscataway, NJ: IEEE.
9 Cerquera, M.R.P., Montaño, J.D.C., and Mondragón, I. (2017). UAV for landmine detection using SDR-based GPR technology. In: Robots Operating in Hazardous Environments (ed. H. Canbolat), 25–58. London: IntechOpen.
10 Hallikainen, M., Ulaby, F.T., Dobson, M.C. et al. (1985). Microwave dielectric behavior of wet soil-part 1: empirical models and experimental observations. *IEEE Transactions on Geoscience and Remote Sensing* 23 (1): 25–34.
11 Chaudhari, P.R. and Ahire, D.V. (2014). Variation of dielectric constant of Indian soils with their physical parameters at CBand. *International Journal of Innovative Research in Science, Engineering and Technology Microwave Frequency* 3 (8): 15396–15400.
12 Hu, Q., Shao, Y., and Guo, H. (2003). Microwave dielectric behavior of moist salt soil-experimental observations and improved dielectric models. In: Proceedings of the IEEE International Geoscience and Remote Sensing Symposium, Toulouse, France (21–25 July 2003). Piscataway, NJ: IEEE.
13 Aouani, H., Rahmani, M., Navarro-Cia, M., and Maier, S.A. (2014). Third-harmonic-upconversion enhancement from a single semiconductor nanoparticle coupled to a plasmonic antenna. *Nature Nanotechnology* 9 (1): 290–294.
14 Rui, G. and Zhan, O. (2014). Highly sensitive beam steering with plasmonic antenna. *Nature Scientific Reports* 4 (5962): 1–5.
15 Lin, K.T., Chen, H.L., Lai, Y.S., and Yu, C.-C. (2014). Silicon-based broadband antenna for high responsivity and polarization-insensitive photodetection at telecommunication wavelengths. *Nature Communications* 5 (3288): 1–10.
16 Diaz, R.E. and Sebastian, T. (2013). Electromagnetic limits to radiofrequency (RF) neuronal telemetry. *Nature Scientific Reports* 3 (3535): 1–7.
17 Xu, H.-X., Wang, G.-M., Tao, Z., and Cui, T.J. (2014). High-directivity emissions with flexible beam numbers and beam directions using gradient-refractive-index fractal metamaterial. *Nature Scientific Reports* 4 (5744): 1–10.
18 Lier, E., Werner, D.H., Scarborough, C.P. et al. (2011). An octave-bandwidth negligible-loss radiofrequency metamaterial. *Nature Materials* 10 (1): 216–222.
19 Hausmann, B.J.M., Bulu, I., Venkataraman, V. et al. (2014). Diamond nonlinear photonics. *Nature Photonics* 8 (1): 369–374.
20 Muller, T., Heoo, C., Pingault, B. et al. (2014). Optical signatures of silicon-vacancy spins in diamond. *Nature Communications* 5 (3338): 1–7.
21 Aharonovich, I. (2014). Silicon magic. *Nature Photonics* 8: 818–619.

22 Vlasov, I.I., Shiryaev, A.A., Rendler, T. et al. (2014). Molecular-sized fluorescent nanodiamonds. *Nature Photonics* 9: 54–58.

23 Cicala, G., Massaro, A., Velardi, L. et al. (2014). Self-assembled pillar-like structures in nanodiamond layers by pulsed spray technique. *ACS Applied Materials & Interfaces* 6 (23): 21101–21109.

24 Gaidarzy, A., Imboden, M., Mohanty, P. et al. (2007). High quality factor gigahertz frequencies in nanomechanical diamond resonators. *Applied Physics Letters* 91 (203503): 1–5.

25 Sumant, A.V., Auciello, O., Liao, M., and Williams, O.A. (2014). MEMS/NEMS based on mono-, nano-, and ultrananocrystalline diamond films. *MRS Bulletin* 39 (6): 511–516.

26 HE, X., Zhang, M., Li, J., and Qi, S. (2010). Improvement of integrated dipole antenna performance using diamond for intra-chip wireless interconnection. In: Proceedings of IEEE ICICDT-10 Conference, Grenoble, France (2–4 June 2010). Piscataway, NJ: IEEE.

27 Zmuidzinas, J., Ugras, N.G., Miller, D., and Gaidis, M. (1995). Low-noise slot antenna SIS mixers. *IEEE Transactions on Applied Superconductivity* 5 (2): 3053–3056.

28 He, X., Zhang, M., and Li, J. (2011). Characteristics of on-chip dipole antenna using diamond for intra-chip wireless interconnect. *Science China Technological Sciences* 54 (1): 1035–1043.

29 Hazra, R., Ghosh, C.K., and Parui, S.K. (2013). Effect of different semi conductive substrate materials on a P shaped wearable antenna. *International Journal of Advanced Research in Computer and Communication Engineering* 2 (8): 3071–2074.

30 Yoneda, H., Tokuyama, K., Ueda, K. et al. (2001). High-power terahertz radiation emitter with a diamond photoconductive switch array. *Applies Optics* 40 (36): 6733–6736.

31 Naji, D.K. (2013). Compact broadband CPW-fed taper-shaped monopole antenna with L-slots for C-band applications. *International Journal of Electromagnetics and Applications* 3 (6): 136–143.

32 Cicala, G., Nitti, M.A., Tinti, A. et al. (2011). Effect of properties of MWPECVD polycrystalline diamond films on photoemissive response. *Diamond and Related Materials* 20 (8): 1199–1203.

33 Cicala, G., Massaro, A., Velardi, L. et al. (2015). DLC integrated GHz antenna for aerospace. In: Proceedings of the 2nd IEEE International Workshop on Metrology for Aerospace, Benevento, Italy (3–5 June 2015). Piscataway, NJ: IEEE.

34 Cola, A., Lupo, M.G., Vasanelli, L., and Valentini, A. (1992). Ti/GaAs Schottky barriers prepared by ion beam sputtering. *Journal of Applied Physics* 71 (10): 4966–4971.

35 Salvatori, S., Brugnoli, E., Rossi, M.C., and Pinzari, F. (2001). Field- and photo-emission properties of CVD-diamond with different microcrystalline structure. *Diamond and Related Materials* 10 (3–7): 852–857.

36 Sun, T., Koeck, F.A.M., Stepanov, P.B., and Nemanich, R.J. (2014). Interface and interlayer barrier effects on photo-induced electron emission from low work function diamond films. *Diamond and Related Materials* 44 (1): 123–128.

37 Fox, N., Martin, T. L., and O'Donnel, K. M. (2013). Low work function diamond surface and radiation energy converters using same. US Patent 2012/0244281 A1.

38 Othman, M.Z., May, P.W., Fox, N.A., and Heard, P.J. (2014). Incorporation of lithium and nitrogen into CVD diamond thin films. *Diamond and Related Materials* 44 (1): 1–7.

39 Laikhtman, A., Avigal, Y., Kalish, R. et al. (1999). Surface quality and composition dependence of absolute quantum photoyield of CVD diamond films. *Diamond and Related Materials* 8 (2–5): 725–773.

40 Konovov, V.I. (2012). Laser in micro and nanoprocessing of diamond materials. *Laser & Photonics Reviews* 6 (6): 739–766.

41 Melisi, D., Nitti, M.A., Valentini, M. et al. (2013). Radiation detectors based on multiwall carbon nanotubes deposited by a spray technique. *Thin Solid Films* 543 (1): 19–22.

42 Cicala, G., De Pascali, G., Massaro, A. et al. (2017). High-efficiency nanodiamond-based ultraviolet photocathodes. Patent WO 2017/051318A1.

43 A. Massaro, F. Spano, P. Cazzato et al. (2012). Punta per un dispositivo di rilevamento ottico di un oggetto tridimensionale, particolarmente per uso in campo medico, e dispositivo di rilevamento ottico comprendente una tale punta. Patent TO2011A001178.

44 Garcia-Uribe, A., Balareddy, K.C., Zou, J., and Wang, L.V. (2008). Micromachined fiber optical sensor for in vivo measurement of optical properties of human skin. *IEEE Sensors Journal* 8 (10): 1698–1703.

45 Marquez, G. and Wang, L.V. (1997). White light oblique incidence reflectometer for measuring absorption and reduced scattering spectra of tissue-like turbid media. *Optics Express* 1 (13): 454–460.

46 Pavlova, P., Borisova, E., Avramov, L. et al. (2010). Investigation of relations between skin cancer lesions' images and their fluorescent spectra. *Laser Physics* 20 (13): 596–603.

47 Mazzoli, A., Munaretto, R., and Scalise, L. (2010). Preliminary results on the use of a noninvasive instrument for the evaluation of the depth of pigmented skin lesions: numerical simulations and experimental measurements. *Lasers in Medical Science* 25 (1): 403–410.

48 Rajab, M.I., Woolfson, M.S., and Morgan, S.P. (2004). Application of region-based segmentation and neural network edge detection to skin lesions. *Computerized Medical Imaging and Graphics* 28 (1–2): 61–68.

49 Mete, M. and Sirakov, N.M. (2010). Lesion detection in demoscopy images with novel density-based and active contour approaches. *BMC Bioinformatics* 11 (6): 1–13.

50 Taouil, K., Chtourou, Z., and Romdhane, N.B. (2010). A robust system for melanoma diagnosis using heterogeneous image databases. *Journal of Biomedical Science and Engineering* 3 (6): 576–583.

51 Hachioji, H.H. and Takikawa, K.K. (1989). Optical endoscope having image transmitting cable. US Patent 4866516A.

52 Singhal, R., Orynbayeva, Z., Kalyana Sundaram, R.V. et al. (2011). Multifunctional carbon-nanotube cellular endoscopes. *Nature Nanotechnology* 6 (1): 57–64.

53 Becker, D., Wray, T., and Hart, J. (2008). Ultrasonic intracavity probe for 3D imaging. US Patent 20080228081.

54 Sato, Y., Miyamoto, M., Nakamoto, M. et al. (2001). 3D ultrasound image acquisition using a magneto-optic hybrid sensor for laparoscopic surgery. In: Medical Image Computing and Computer-Assisted Intervention – MICCAI 2001, Lecture Notes in Computer Science, vol. 2208 (eds. W.J. Niessen and M.A. Viergever), 1151–1153. Berlin: Springer.

55 Godavarty, A., Regalado, S.A., Jayachandran, B., and Zhu, B. (2010). Hand-held optical probe based imaging systems. US Patent 2010/0010340 A1.

56 Yelin, D., White, W.M., Motz, J.T. et al. (2007). Spectral-domain spectrally-encoded endoscopy. *Optics Express* 15 (5): 2432–2444.

57 Nakazawa, K. and Suzuki, C. (1991). Development of 3-D robot vision sensor with fiber grating: Fusion of 2-D intensity image and discrete range image. In: Proceedings of the International Conference on Industrial Electronics, Control and Instrumentation IECON '91, Kobe, Japan (28 October–1 November 1991). Piscataway, NJ: IEEE.

58 Klein, A., Schroderi, A.H., Oppermanni, H. et al. (2010). Integration of a multifunctional and multispectral optical sensor for automotive applications using surface mountable planar optical interconnect. In: Proceedings of the Electronic System-Integration Technology Conference (ESTC), Berlin, Germany (13–16 September 2010). Piscataway, NJ: IEEE.

59 Gacitua, W., Ballerini, A., and Zhang, J. (2005). Polymer nanocomposites: synthetic and natural fillers a review. *Maderas. Ciencia y Tecnología* 7 (3): 159–178.

60 Hussain, F., Hojjati, M., Okamoto, M., and Gorga, R.E. (2006). Review article: polymer-matrix nanocomposites, processing, manufacturing, and application: an overview. *Journal of Composite Materials* 40 (17): 1511–1575.

61 Buretea, M., Empedocles, S., Niu, C., and Scher, E.C. (2007). Nanocomposites. US Patent US7228050B1.

62 Empedocles, S. (2009). Nanotechnology-enabled optoelectronics. US Patent 7515333 B1.

63 Vaidya, B. and Drechsler, U.W. (2008). Nanocomposite material for direct spectroscopic detection of chemical vapors. US Patent 2008/0278708 A1.

64 Chu, B. and Hsiao, B.S. (2010). Nanocomposite fibers and film containing polyolefin and surface-modified carbon nanotubes. US Patent 7652084 B2.

65 Chen, F.F. (2010). Multimodal imaging probes for in vivo targeted and non-targeted imaging and therapeutics. US Patent 2010/0183504 A1.

66 Lee, H.W.H., Chin, A.H., and Pfenninger, W.M. (2006). Optical devices with engineered nonlinear nanocomposite materials. US Patent 7020372 B2.

67 Parsonage, E., Horn, D.J., Chen, J.J. et al. (2009). Medical devices comprising nanocomposite. US Patent 7591831 B2.

68 Yadav, T. and Kostelecky, C. (2007). Optical filters from nanocomposites. US Patent 7238734 B2.

69 Aoki, K., De Vittorio, M., Stomeo, T.et al. (2010). A method of identifying a target analyte using photonic crystal resonators and related device. US Patent 2010/0028898 A.

70 Mathias, P.C., Ganesh, N., Chan, L.L., and Cunningham, B.T. (2007). Combined enhanced fluorescence and label-free biomolecular detection with a photonic crystal surface. *Applied Optics* 46 (12): 2351–2360.

71 Fan, J.-B., Chee, M.S., and Gunderson, K.L. (2006). Highly parallel genomic assays. *Nature Reviews Genetics* 7 (8): 632–644.

72 Hadd, A.G., Raymond, D.E., Halliwell, J.W. et al. (1997). Microchip device for performing enzyme assays. *Analytical Chemistry* 69 (17): 3407–3412.

73 Wang, J., Ibanez, A., Chatrathi, M.P., and Escarpa, A. (2001). Electrochemical enzyme immunoassays on microchip platforms. *Analytical Chemistry* 73 (21): 5323–5327.

74 Dunn, W.C., Jacobson, S.C., Waters, L.C. et al. (2000). PCR amplification and analysis of simple sequence length polymorphisms in mouse DNA using a single microchip device. *Analytical Biochemistry* 277 (1): 157–160.

75 Scherer, J.R., Kheterpal, I., Radhakrishnan, A. et al. (1999). Ultra-high throughput rotary capillary array electrophoresis scanner for fluorescent DNA sequencing and analysis. *Electrophoresis* 20 (7): 1508–1517.

76 Cheng, J., Shoffner, M.A., Mitchelson, K.R. et al. (1996). Analysis of ligase chain reaction products amplified in a silicon glass chip using capillary electrophoresis. *Journal of Chromatography A* 732 (1): 151–158.

77 Kopp, M.U., Mello, A.J., and Manz, A. (1998). Chemical amplification: continuous-flow PCR on a chip. *Science* 280 (5366): 1046–1048.

78 Daniel, J.H., Iqbal, S., Millington, R.B. et al. (1998). Silicon microchambers for DNA amplification. *Sensors and Actuators A: Physical* 71 (1–2): 81–88.

79 Bodovitz, S., Joos, T., and Bachmann, J. (2005). Protein biochips: the calm before the storm. *Drug Discovery Today* 10 (4): 283–287.

80 Wang, J. (2000). From DNA biosensors to gene chips. *Nucleic Acids Research* 28 (16): 3011–3016.

81 Murphy, L. (2006). Biosensors and bioelectrochemistry. *Current Opinion in Chemical Biology* 10 (2): 177–184.

82 Piunno, P., Krull, U., Hudson, R. et al. (1995). Fiber-optic DNA sensor for fluorometric nucleic acid determination. *Analytical Chemistry* 67 (15): 2635–2643.

83 Speers, D.J. (2006). Clinical applications of molecular biology for infectious diseases. *Clinical Biochemistry* 27 (1): 39–51.

84 Domiati-Saad, R. and Scheuermann, R.H. (2006). Nucleic acid testing for viral burden and viral genotyping. *Clinica Chimica Acta; International Journal of Clinical Chemistry* 363 (1–2): 197–205.

85 Massaro, A., Spano, F., Missori, M. et al. (2014). Flexible nanocomposites with all-optical tactile sensing capability. *RSC Advances* 4 (6): 2820–2825.

10

Scientific Research in Industry

This chapter proposes different solutions oriented toward the realization of scientific units in industry, the formulation of a research project, the definition of innovative topics as patent claims, and methodologies to follow for project management. Different research topics are analyzed including informatics, sensors, electronics, and mechatronics. Moreover, the procedures to formulate and structure a research project integrating advanced facilities are discussed, thus providing different useful schemes for the setting and planning of research activities, including the aspects related to the business plan associated with a research project addressed for production or for realization of a new advanced product or machine. Specifically, this chapter discusses the main guidelines to construct an internal research unit for a company, the main guidelines to formulate a patent oriented on mechatronics, the setting of a research project oriented toward technologies with specific application fields, the setting of a research project involving the realization of advanced inspection machine by means of an innovative Industry 5.0 production line plant, and the scheme setting of a project management activity. The proposed topics are useful for companies working with innovative technologies and can be useful to activate a specific research and development (R&D) unit working on a new product to industrialize. This chapter is structured to provide important elements for the industry researcher about the writing of a research project involving new products, new services, and new processes. Some examples are provided to understand how to set an industry project, and to follow all the stages and the results during its development. Different approaches are discussed to highlight the main keywords and expected results. The identification of defined goals, work packages, deliverables, and milestones allow to define during the project design the main checkpoints to verify and to validate. Other aspects such as risk analysis and possible working synergies are suitable to improve the correct development of the project activities. This chapter provides different templates useful for the researcher to write an industry research project as a principal investigator. Other proposed methodologies to manage projects are adopted by the project manager (PM) to follow and check all the tasks and activities of the project during its development. The topics discussed in this chapter are also referred to innovative technologies and related self-adaptive production and control systems for flexible advanced production lines typical of Industry 5.0 environments. Different case studies are described to introduce the framework of research projects.

Electronics in Advanced Research Industries: Industry 4.0 to Industry 5.0 Advances, First Edition. Alessandro Massaro.

10.1 Guidelines to Construct an Advanced Research Unit in Industry in the Electronic and Mechatronic Field

A research unit industry must be structured by multidisciplinary skills by defining a strategic research plan according to the industry business models. The R&D unit can be addressed to the research of innovative materials, micro-/nanocomposite sensors, new algorithms for control and actuation, and new decision support systems (DSSs) involving electronic and mechatronic automatisms. Concerning DSSs, artificial intelligence (AI) algorithms are important topics for the upgrade of the whole industry production, innovative production machines or innovative products integrating advanced sensors or new sensors based on new detection concepts. The realization of a new product enables a new production process to be improved by machines installed into a network which must be flexible and dynamic following the smart manufacturing principle that employs computer integrated manufacturing (CIM). The following steps must be followed to plan a R&D advanced research unit.

- **Identification of the teams to involve.** Different research application fields must be considered in order to define a main research topic *in-line* with the market needs, and with the industry strategic plan. Different teams can be involved for the realization of a unique big research project composed of different subprojects. Each subproject must be developed by a multidisciplinary team constituted by engineers, economists, computer scientists, oriented toward electronic and mechatronic topics. Each team must be coordinated by a PM who follows all stages of the project development. The project team must include the following main specialists:
  - Administrative human resources managing the financial aspects (important for big research projects).
  - The main technological specifications including the working conditions delineating the application domain.
  - Engineers (specialized in the topics of the research project).
  - Developers (including web designers).
  - Researchers (important for scientific improvements).
  - Data analysts.
  - Data scientists.
  - Upgraded technicians.
  - System architects.
  - Tester.

- **Choice of the basic technology and of the materials to adopt.** A new product must be constituted by innovative technologies and innovative materials possibly realized in the same industry to decrease the costs. The development of innovative mechatronic technologies and of innovative materials improving physical properties in the same industry of the R&D unit facilitates the upgrade of the new product and the rapid prototyping following advanced reverse engineering (RE) approaches. The advanced technologies and the innovative materials must be characterized in the laboratory first by experimental tests planned before writing a project, according to the company business plan and the industry strategies. During the choice of the basic technology, it is important to define possible claims for a possible patent, thus improving the value of the research. During the preliminary evaluation of the possible claims, it is important to define all possible applications defining subproducts derivable from the main ideated product.
- **Research planning.** The research activities must be planned in agreement to the time indicated in the business plan and to the planned releases. The research unit must also plan technical and

financial risks, possible exit strategies, possible partners and collaborations, the use of software tools, the experimental setup, and the production machines oriented toward a flexible production implementing control and actuation facilities. The innovative research includes elements oriented toward an auto-adaptive production system and AI data processing improving product quality.

- **Availability of a structured laboratory.** The experimental results are very important for the beginning and for the continuation of the project. The R&D unit must adopt the same laboratory facilities in order to use consistent results for the research project. The laboratory includes all the other software tools necessary for the design and development of the new product. With respect to mechatronic innovations, it is important to plan automatized tests, matching with environmental conditions of the specific application. The laboratory tests are important to validate the possible initial hypothesized claims of the patent.
- **Control unit.** The risk analysis and the evaluation of deliverable and milestones facilitate the prevision of a possible industrialization plan. An executive external unit (external to the R&D unit) must "translate" the partial results as possible opportunities, defining the new research line to follow according to the strategic marketing of the company. The commercial point of view is fundamental for the definition of the new perspectives of the research.
- **Quality check unit.** For products as results of internal research, the related industrialization process must be controlled by a quality check unit able to define and to update the lifecycle of the product. In this unit, specialists in RE and redesign of components must work. In this scenario, the researcher also works in order to understand and predict usury and failures of a robotic component or machine integrating mechatronic and sensors.
- **Product specialist.** A human resource of the R&D team is selected as product specialist supporting human resources addressing marketing by planning demonstrative presentations of the ideated product, enhancing the technical characteristics.

Research project activities mainly appertain to one or more of the following classes:

- **Basic or fundamental research**, which is considered an experimental or theoretical activity with the aim of increasing knowledge for a specific application or use.
- **Industrial research**, which is a planned research or critical investigations aimed at acquiring new knowledge and skills to be used to develop new products, processes or services or to make a significant improvements in existing products, processes, or services; this research includes the creation of components of complex systems, the construction of prototypes in a laboratory environment or in an environment equipped with simulation interfaces, and the construction of pilot lines able to validate generic technologies.
- **Experimental development**, defined as acquisition, combination, structuring and use of existing knowledge and skills of a scientific, technological, commercial and other nature in order to develop new or improved products, processes, or services; in this class are the activities for conceptual definition (proof of concept), the planning of new products, processes or services; experimental development includes the construction of prototypes, demonstration, realization of pilot products, testing, and validation of new or improved products, processes or services carried out in an environment that reproduces real operating conditions where the primary objective is the contribution of further technical improvements to products, processes, and services that are not substantially final; experimental development therefore includes the development of a commercially usable prototype or pilot product which is necessarily the final commercial product and whose manufacturing cost is too high to be used for demonstration and validation purposes only; experimental development does not include standard changes or periodic changes to existing products, production lines, manufacturing processes and services and other ongoing operations, even when such changes represent improvements.

- **Applied research,** which includes industrial research, experimental development or any combination of the two types of activity.

10.2 Guidelines to Formulate a Patent

The guidelines related the formulation of a patent about electronic and mechatronic prototypes are fundamental to hit the target quickly. It is important to write in a short time the patent deposition and publication allowing the pre-industrialization of the new ideated product. The guidelines follow the proposed procedure:

- **Definition of the new technology and research topics.** The analysis of the state of the art must be addressed to simultaneous knowledge of the new technologies and of the updated scientific studies associated with the new technologies. This preliminary study is fundamental to find a possible matching between technology and scientific research topics, thus allowing the definition of the patent claims. The state of the art provides the "proof of concept" of the new idea to analyze.
- **First on/off experimentations.** Rapid prototyping is important to check experimentally and quickly new ideas based on the new advanced technology. The on/off tests allow to understand if the patent idea can be furthermore studied and developed. Bad results of preliminary experimentations generally stop the creation of the specific R&D unit.
- **Claims definitions.** Based on the preliminary test results, a list is defined of possible claims matching with the results found in the first stage.
- **Claims update.** After the definition of a set of claims, the study of the state of the art is updated by choosing the most significant element. This facilitates the revision process of external specialists assigned for the patent proposal evaluation.
- **Proposal evaluation.** The first external proposal evaluation can reject the patent idea or request other information redefining the possible patent keywords. The R&D unit must provide all the results and clarifications explaining possible doubts. At this stage it is very important to provide new experimental results proving the novelties introduced. After this stage, a response is provided for a possible patent publication.

  Concerning mechatronic elements, important elements to define claims are:

- Definition of the new technology to implement for the realization of a specific mechatronic element.
- Definition of the logic inherent to the control and the actuation process to implement.
- Definition of technological limits.
- Definition and simulation of the full scenario where the new component will work.
- First design analyzing output results.
- First prototyping.
- First laboratory test.
- Simulations of all possible inputs and outputs of the mechatronic system (signals, physical effects, noises, compatibilities, parameters sensitivity, inducted effects, etc.).
- Possible adaptation with other existing systems (modularity and universality of the solution).
- Estimation of efficiency parameters by comparing the hypothesized results with ones found in the state of the art.
- Forecast of possible variations of the solution.

10.3 Guideline to Propose Technological Advances for Public Entities and in Industry 5.0 Research Project

In this section, different schemes useful to follow for the ideation of a research project involving automated technologies are provided.

10.3.1 Setting of a Research Project of Underground Water Leakage

As an application example schemes are provided for the setting of a research project related the supply of electronic components and devices for a noninvasive system detecting pipeline underground water leakage. The proposed schemes are guidelines to write a technical project. Table 10.1 discusses the main technical aspects to formulate the project by considering a project for underground water leakage system as discussed in Chapter 9 (Section 9.2.2).

Table 10.1 Example of a scheme of the technical proposal.

Project goal	The R&D objective consists of the technical design of innovative solutions for a new system addressed to the localization of water leaks of underground pipelines, the monitoring of the status of the pipelines, and the geo-referenced survey of the urban hydric network. Specifically, the proposed example involves the simultaneous use of UAVs, thermal cameras and GPR
Detailed description of the innovative proposed solution	In this part, the technical and applicable characteristics of the project solution are listed and described, enhancing performance requirements compared with competitor ones. For the specific case study, technologies enabling water leakage detection, calibration procedures of the instrumentation, and the related standards are described. Moreover, the following are added: • The conceptual and functional schemes of the preliminary design. • The list of the actors involved. • The user interface schemes (frontend system). • The backend of the whole information system. The actor description is related to: • Actor name and main functions. • Details of the actor functions. • Motivations about the required actor. Table 10.1.1 reports a functional scheme of the project requirements (classes of requirements) and of the use case classes indicated by identification codes. An example of the use case scheme is reported in Table 10.1.2. The scheme is reproduced for each use case code. By describing system actors, it is also important to add a summary of the technical specifications and the functional requirements of the technical design elements. Table 10.1.3 gives a summary of the attributes highlighting the methodologies and the advantages. Concerning innovation, important topics to introduce in the project are related to mechatronic systems to be applied mainly for the control and actuation of all devices. It is preferable to use software development kits to implement for the specific case uses

(Continued)

Table 10.1 (Continued)

Technical drawings	After a description of project requirements, elaboration of the technical drawings of the innovative solution proposed is presented. The development of the design of the device system is described, and of its components, by highlighting the aspects of particular innovation with respect to the state of the art, and providing an indication of any components protected by patents. In order to understand possible claims derived from the executive design, a comparison table is made (Table 10.1.4). Concerning the specific case, the drawings will be related to the UAV system, to the fixed pole with sensors, and to other electronic and mechatronic adaptions (energy supply, cable adaptation, functional schemes by Unified Modeling Language, etc.)
Detailed definition of the methods implementing the innovative solution	The development of the design of the device and its technical components are described, highlighting the aspects of particular innovation with respect to the state of the art, by providing indications of any components protected by patents and by industrial property. The technical datasheet of the main components to assemble, the costs, and the different quotes are also provided
Description of the proposed modalities for the experimental stage	The testing phases are listed by describing the testing modalities and the test planning
Description of the organization of research activities	The organization and technical–scientific skills of the research team engaged in the project are described. Table 10.1.5 is a summary of human resource engagement. The personal training procedures are described in Table 10.1.6. In particular, the number of hours of programming that will be used for technology transfer training is provided, as for the example shown in Table 10.1.7.
Project implementation plan	The Gantt diagram presented in the first project phase, related to the prototype development and testing activities, is updated. A Gantt diagram similar to the diagram in Table 10.1.8 (where M indicates the month and WP is the work package) is provided. The work packages are defined as: • WP1: Mechanical support assembly and installation. • WP2: Assembly and installation of devices. • WP3: Prototype realization. • WP4: Backend implementation. • WP5: Frontend and software implementation. • WP6: Bench testing. • WP7: Testing leakage detection zone A. • WP8: Testing leakage detection zone B. • WP9: Testing leakage detection zone C. The project milestones follow the scheme in Table 10.1.9.

GPR, Ground Penetrating Radar; UAV, unmanned aerial vehicle.

Table 10.1.1 Code of classes or requirements.

R01-Class of requirements 1	**Code**	**Description**
R01	R01–01	//
R01	R01–02	//
R01	R01–03	//
R02-Class of requirements 2	**Code**	**Description**
R02	R02–01	//
R02	R02–02	//
R02	R02–03	//
R02-Class of requirements 3	**Code**	**Description**
R03	R03–01	//
R03	R03–02	//

Table 10.1.2 Use case template scheme.

USE case R01–01		
Description	Description of the function of the use case	
Pre-condition	Pre-condition enabling the specific use case	
Post-condition	Success	Condition verified successfully
	Failure	Condition not verified
Triggering event	Description of the event triggering the use case	
Primary actor	Actor involved in the use case	
Basic scenario	Basic scenario of the event happening	
Alternative scenario	Alternative scenario when the event is not triggered	

Table 10.1.3 Example of a framework describing technological advantages.

Technical elements	Specifications and requirements	Application methodologies and advantages
Device 1	Radiometric measurement	Infrared thermal camera
Device 2	GPR	Radargram supporting thermograms for underground water leakage detection
Device 3	UAV	Large area to inspect in less time
Device *n*	//	//

Table 10.1.4 Comparison of patent and project facilities.

Function description	Project prototype	Patent
Element 1	✓	✓
Element 2	✓	✓
Element 3	x	✓
Element *n*	x	✓

Table 10.1.5 Summary of the tasks assigned to professional units.

Professional unit	Assigned task	Minimum number of units
Design engineer	Design of construction details, construction management and prototype testing. Monitoring of joint works for better achievement of results	1
Systems engineer developer	Development of the backend and of the database system	2
Software developer	Code developer	2
Web designer	Design of the control panel graphics. Frontend design	1
Operator 1	Use of device 1	1
Operator 2	Use of device 2	2
Operator 3	Use of device 3	1
Project manager	Project fund management and reporting activities	1
Administrative secretary	Administrative activities related to the project	1
Qualified technician	Prototype construction	1
Trainer	Staff training	1

Table 10.1.6 Example of a training planning.

Training typology	Description and modality	Period
Use of device 1 (TT1)	Testing of device 1	After month 8 and until closure of the project
Use of device 2 (TT2)	Testing of device 2	After month 8 and until closure of the project
Use of device 3 (TT3)	Testing of device 3	After month 12 and until closure of the project (after a training course)
Use of software (TT4)	Tests and instructions on data entry and access to the archiving database. Training on managing the access of the system administrator. Analysis of predisposition to use the data system through the servers	From the early stages of implementation to use in the experimental phase

Table 10.1.7 Technology transfer hours for the specific modules.

Training module	Number of hours
TF1	15
TF2	10
TF3	20
TF4	10

Table 10.1.8 Example of a Gantt diagram.

Work package	M 1	M 2	M 3	M 4	M 5	M 6	M 7	M 8	M 9	M 10	M 11	M 12
WP1												
WP2												
WP3												
WP4												
WP5												
WP6												
WP7												
WP8												
WP9												

Table 10.1.9 Project milestones.

Milestones	Description	WP involved	Reporting period
10) Devices	Design and development of the devices. Testing	WP3, WP6	After 8 months
11) Device software	Software controlling devices. Testing	WP4, WP6	After 8 months
12) Control panel and data storage software	Panel control software development	WP5, WP6	After 8 months
13) Image detection systems	Image processing	WP1, WP2, WP6	After 8 months (first release), after 12 months (final testing)
14) Testing		WP6, WP7, WP8, WP9	After 12 months (final testing)

Together with the technical proposal the business plan is written. A template of the parts of a business plan is shown in Table 10.2.

Table 10.2 Example of a business plan.

Part I: the proponents	
Activity sector	Description of the business sector in which the individual company intends to operate
The entrepreneurial group	Description of the companies involved in the project: • Description of the technical profile. • Indication of the number of employees divided by technicians, administrators, and researchers. • Revenue of the last three years. • Main differences if compared with the competitors

(Continued)

Table 10.2 (Continued)

Part II: the innovative solution	
Abstract	Brief description of the technical project including: • How the prototype idea was born. • What it consists of. • What is new in the prototype compared with the state of the art. • What has been done so far
Time to market	Information about: • The industrial development times downstream of the prototype. • The time to market for the introduction on the market. Specifically, the following are described: • Current conditions. • Market goal for the nest 5 years (including the project period). • Budget and plans. • Metric and adjustments
Part III: the market	
The competitive environment	Description of the market area to which the innovative solution is addressed, with indication of the economic size of the market and geographic distribution. Specifically, how the devices proposed for water leakage detection are located in the market is described
The customers	Who and how many potential customers are described, together with where they are and for what reasons they should buy from the proponents
The competitors	A list of direct competitors is provided, indicating their location, strengths (elements of success), and weaknesses (limits and criticalities) as regards the technological aspects
Part IV: strengths and weaknesses, opportunities and threats	
SWOT analysis	An analysis is proposed, discursively or using a scheme, of strengths, weaknesses, opportunities, and threats of the project. Normally, the project is supposed to have some strengths and inevitable weaknesses that logically respond to the opportunities and threats that exist in the market. A scheme to follow is the SWOT analysis shown in Table 10.2.1.
Part V: economic and financial analysis	
The investment plan	Table 10.2.2 reports the investment plan scheme. Table 10.2.3 shows the revenue plan.

SWOT, strengths, weaknesses, opportunities, and threats.

Table 10.2.1 Basic scheme of SWOT analysis.

Strengths	**Points of weakness**
List of strengths	List of points of weakness
Opportunities	**Threats**
List of opportunities	List of threats

Table 10.2.2 Template of the investment plan.

Investment	Budget				
	Year 1	Year 2	Year 3	Year 4	Year 5
Costs of management and protection of intellectual property rights					
Costs deriving from the filing of the patent application					
Maintenance costs					
Taxes					
Total					
Production cycle design costs					
Costs deriving from the definition of the production cycle					
Costs arising from the acquisition of the equipment necessary to produce the device/product/process					
Total					
Industrialization and production costs					
Costs deriving from the large-scale production of the device/product					
Costs deriving from product and process certification					
Total					
Operating and maintenance costs					
Costs deriving from the operating activity of the device/product					
Costs of intervention and repair of the device/product/process					
Total					
Promotion and marketing costs					
Costs deriving from the promotion campaign for the launch of the device/product/process					
Costs deriving from the promotion campaign for the marketing of the device/product/process					
Total					
Distribution costs					
Costs arising from distribution					
Costs deriving from the supply of the device/product/process to the final customer					
Total					
Disposal costs					
Costs deriving from the disassembly and disposal of some components at the end of their lifecycle					
Costs deriving from the disassembly and disposal of the entire device/product/process at the end of its lifecycle					
Total					

Table 10.2.3 Template of the revenue plan.

Revenues	Values				
	Year 1	Year 2	Year 3	Year 4	Year 5
Revenues from the national market					
Revenues from the sale of the device/product/process to the main customers in the national territory					
Total					
Revenues from the foreign market					
Revenues from the sale of the device/product/process to the main customers in the reference foreign markets					
Total					
Revenues from alternative markets					
Revenues from the sale of the device/product/process in markets related to different applications					
Total					

Table 10.3 shows an example of a simplified use case table for the specific topic of the project.

The same scheme is used in other research topics such as in telemedicine [1], thus confirming the generic use of the template. Table 10.4 reports an example of the scheme of specifications, requirements about the water leakage system.

Table 10.5 gives the SWOT analysis for the specific project.

10.3.2 Setting a Research Project Involving Technologies for Hydrogeological Risk Monitoring

In this subsection, the setting of another executive project mainly focused on hydrogeological risk detection is proposed. Table 10.6 summarizes the related project setting approach.

Table 10.3 Example of a simplified use case table.

USE Case R02-01		
Description	Loading radargram	
Pre-condition	Confirmation of detection of water loss from thermal images or plano-altimetric surveys is required	
Post-condition	Success	The radargram is loaded by a specific software tool
	Failure	The radargram is not correctly loaded
Triggering event	Image detection by georadar	
Primary actor	Remote operator and inspection operator	
Basic scenario	Raw georadar image loading for image processing	
Alternative scenario	The image is not correctly acquired and acquisition of another radargram is performed	

Table 10.4 Example of a scheme of the business plan.

Technical elements	Specifications and requirements	Application methodologies and advantages
Device 1: fixed station with thermal camera	• Technological innovation (patent idea which is exploited throughout the lifecycle analysis of the prototype). • Integration into industrialization processes. • Data geolocation. • Low environmental impact (photovoltaic power supply). • Technical feasibility (use of already certified and tested components). • Detection of water leaks in meshed water distribution (pipeline diameter ≥80 mm). • Reproducibility of the system by replacing an additional non-very high-resolution thermal camera with the use of image processing software. • Ability to innovate in the area of operation	• Remote thermal image detection. Geolocation identified during the positioning phase of the station. • Noninvasive diagnostic method. • Environmental sustainability (possibility of moving the fixed inspection station to other places to be inspected). • Limitation of inspection worker number in the case of remote monitoring. • Assembly of components by using the experience of all the partners (pipes to be monitored, evolution of the loss, type of soil, device configuration, etc.). • Zero cost of maintenance and of disposal. • Video surveillance function. Possible analysis also of the sewer status (losses). • Training. • Possibility of use during the day and night
Device 2: thermal camera fixed on a car	• Technological innovation (thermal imaging camera of the latest technology). • Geolocation. • Minimal environmental impact. • Industrialization. • Reduction of the overall cost (unit of personnel employed). • Technical feasibility (insertion of already certified and tested components). • Detection of water leaks in hydric network distribution (pipeline diameter ≥80 mm)	• Very high thermal gradient detection performance (<0.02 °C). Geolocation by mobile device. • Zero cost of disposal. • Easy to use. • Possibility of inspecting several kilometers of water network in a very short time (vehicle transit times). • Zero cost of the solution. • Zero cost of disposal. • Possible analysis also of the sewer status. • Possibility of use during the day and night. • Training
Device 3: Unmanned aerial vehicle unit	• Technological innovation (inspection procedures). • Geolocation. • Low environmental impact. • Reduction of costs on inspection workers. • Technical feasibility (certified drone use). • Consistency with the factors expressed in the feasibility study (actual use). • Integration with thermal imaging procedures and with GPR systems. • Integration with the actual network inspection procedures. • Detection of water leaks in network water distribution (pipeline diameter ≥80 mm)	• Launch of use in the experimental phase in order to identify the guidelines of use for the entire lifecycle of the device. • Zero cost of disposal. • Possibility to inspect several kilometers of network in a very short time. Possible analysis also of the sewer status (losses). • Basic training

(Continued)

Table 10.4 (Continued)

Technical elements	Specifications and requirements	Application methodologies and advantages
Supporting devices: georadar	• Technological innovation (latest generation of GPR). • Geolocation (GPS unit). • Industrialization. • Completeness of functional and performance requirements (pipeline and subsoil health status). • Plano-altimetric inspection. • Possibility to attach on a car (radargram acquisition with the car in motion). • Minimal environmental impact (noninvasive detection method). • Technical feasibility (use of already certified and tested components). • Detection of water leaks in network water distribution (pipeline diameter ≥80 mm)	• Acquisition of three-dimensional images for plano-altimetry. • Comparison with thermal images for confirmation of leak detection. • Zero cost of disposal. • Industry specialists to support the analysis of water absorption peaks resulting from a leak. • Very low environmental impact. • Additional validation method by thermal image-based experimentation. • Device ready for use. Reproducibility of the noninvasive solution using devices of any brand (industrial reproducibility). • Possibility of monitoring even in rain and snow conditions. • Basic training
Combined analysis procedure of use of the three devices	• Completeness of functional requirements. • Quality and completeness of analysis. • Ability to innovate in the area of operation. • Repeatability analysis of leak detection by using different devices	• Well-defined intervention procedures managed by the control panel. • Joint analysis of data by the industries participating in the project (total expertise in software, hardware implementations, and knowledge of pipelines and water networks)
Comparative analysis procedure of the data acquired using traditional techniques	• Completeness of functional requirements, feasibility of integration with the business processes. • Ability to improve the operational environment by limiting costs during the lifecycle of the project. • Use of traditional data to identify an analysis procedure that is the closest to practical cases. • Optimization of the experimental results by comparing the results of the tradition tools. • Cost limitation on any additional time required for system setup. • Possibility of establishing the state of health of a pipeline by means of a specific analysis algorithm (e.g. checking limestone obstructions)	• Analysis of algorithms for determining the state of the pipelines (internal limestone formation) developed on the basis of data analyses. • Possibility of not abandoning traditional methods (cost reduction) through the use of traditionally acquired data. • Calibration of the thermographic, radargram system, and of the detection parameters using traditional knowledge. • Easier insertion of the thermographic/georadar system in the current operational context

Table 10.4 (Continued)

Technical elements	Specifications and requirements	Application methodologies and advantages
Image processing analysis procedures	• Technological innovation (decision-making algorithms applied on a scientific basis). • Industrialization (integration with open-source tools). • Completeness of functional and performance requirements. • Added value to the latest technology instrumentation, currently available on the international market. • Technical feasibility (open-source tools). • Reproducibility of the analysis results by repeated application of the algorithms. • Ability to innovate in the operation area	• Image processing through segmentation processes, innovative filtering and use of open-source tools. • Analysis of "hidden" losses not immediately perceptible from the thermal image. • Estimation of the extent of the lost volumes of water and the plume extent over the territory. • Activation of any operating procedures from the control panel. • Supply of innovative software for the diagnosis of the state of the pipeline from images. • Applicability of the software also to radargrams. • Cost reduction of inspection personnel. • Training
Graphic control panel and dashboards	• Consistency with the factors expressed in the feasibility study. • Completeness of functions relating to georeferencing and monitoring the status of pipelines in real time. • Immediate decision-making for circuit activation and deactivation. • Ability to innovate in the operation area. • Efficient operator management (reduction of management costs through a software platform capable of geolocalizing the operators)	• Control and management of operational interventions in accordance with the operational procedures discussed in the preliminary project. • Integration of the analysis procedures on the state of health of the pipeline in a single working environment. • Communication between the telematic unit and the operational unit (more efficiency in interventions decreasing costs). • Zero cost of use, maintenance and disposal. • Training
Archiving database	• Technical feasibility (integration with the actual information systems). • Current need to digitize information (planimetries, field plans, data associated with the water network). • Development of a new platform for managing the health status of the pipeline	• Cloud storage of databases related the hydric networks to be inspected and already inspected. • Data collecting of further data records. • Platform collecting historical data. • Zero cost of the solution in terms of installation, use and disposal. • Training

GPR, Ground Penetrating Radar.

Table 10.5 Example of SWOT analysis for a water leakage detections system.

Strengths	Points of weakness
Immediate compensation of critical issues Critical situations may relate to hydric losses that are no longer controllable and which therefore may require the purchase of water and supply electricity immediately. The introduction of the proposed innovative system was conceived in the design phase as a support to the current monitoring system. This will result in immediate additional analysis tools which, in addition to traditional systems, will reduce the water leakage risk. The decrease of the direct costs is not immediately quantifiable, and the criticalities of the water system cannot be predicted. The use of more accurate monitoring procedures decreases the occurrence of criticalities	***Technical weakness in the context of a commercial system for universal use, acquisition of thermal images in rain and snow conditions*** This weakness is studied during the feasibility phase of the project in order to define limits about thermograms and radargrams of the experimentation. Of particular interest is the analysis of the thermal gradient induced by the rain (environmental rainy conditions)
Joint operations of several noninvasive monitoring systems The thermal monitoring system is combined with the GPR and traditional acoustic ones for a greater certainty of the detection of leaks in underground water pipelines, with the aid of image processing software	***Learning and insertion in an operational context*** The full use of the system requires, like all innovative services, a learning phase in order to increase the aptitude to use the proposed system
System development through know-how of hydric networks The industries involved in the project provide their background in the water pipeline network for the optimal installation of devices, and for data interpretation	***Qualified operators for the use of the UAV and georadar*** It is required to identify specialists for the use of infrared thermography with UAV and GPR
Single centralized data processing, data management, control and storage platform Telemetry and operating units transmit and manage data by means of a single information portal based on a cloud database access	
Easy opening of the application at national and international level due to the ease of operation of noninvasive methods Noninvasive methods are not subject to particular restrictions or constraints for public administration	
Opening in different markets • Possibility of offering "flexible" software packages based on the type of application. • Basic software packages created with application in thermography, and in GPR, to offer modular products. • Possibility of exploiting the evolution of the fixed stations (prototype station). • During the final and executive design phases, the other functionalities of the prototypes are taken into consideration for a possible patent with specific claims	

Opportunities	Threats
• Hardware and software to be present in an international market as "ready to use" tools. • Proposal of new commercial and development channels. • Possibility to improve the design. • Proposal of international projects, presenting scientific, technological and innovative topics as a "business card." • Upgrade of the hydric network management system useful for all aqueducts. • Introduction of a big data system implementing "disaster recovery" function (data backup)	***Threats of the international market*** Big enterprises of the international market are able to "absorb" the market. Technological and scientific innovations certainly represent a guarantee to ensure the introduction of the proposed technologies in the market. Furthermore, the innovative system can open up new commercial channels thus increasing company revenues ***Use of other "ready to use" having no scientific innovation*** Traditional ready to use technologies are of temporary help in the search for leaks and become obsolete in the medium and long term ***Threat linked to the "copy" of the proposed system*** A patent with several claims further ensures the protection of the project prototypes ***Technology currently available but with the risk that in the future it will no longer be available*** This aspect is already considered from the risk analysis of the technical project. This threat is analyzed in the implementation phase of the image processing software, by taking into account the case of images detected from different low-, medium- and high-resolution devices. This analysis allows the implementation of tools that are as flexible as possible for the technology currently available, and therefore remaining "in line" with market needs. It is noted that the feasibility study already shows the current national and international market context associated with noninvasive investigations (thermal and radar system)

GPR, Ground Penetrating Radar; UAV, unmanned aerial vehicle.

Table 10.6 Project setting approach.

Project abstract	In this project a technological system is proposed for noninvasive monitoring and a DSS software. The DSS is able to detect losses of pipelines providing information on the hydrogeological risk. The technologies concern the use of thermography installed on UAVs and cars, georadars, and other sensors useful for the prediction of hydrogeological phenomena. The DSS is also adopted to forecast weather data, indicating risk levels of geolocalized flood risk for cultivated land and water quality by chemical analyses. To facilitate the collection of information, images and data are acquired also by a mobile application
Partnership and scientific panel	The project partnership is summarized in Table 10.6.1. The contact person for each partner of the grouping is provided as shown in Table 10.6.2. Moreover, the project locations, the end user representatives and the research laboratories involved are listed
Financial plan	A financial plan is provided by the summary in Table 10.6.3 of approved costs for each company.
Project goals	Project goals are generally associated with technologies allowing the achievement of the desired results. Table 10.6.4 lists the objectives to be pursued by carrying out the activities.
Activities and phases	The planned and selected activities for a specific application are described. The design phases are also indicated, such as: • **Planning phase:** planning jointly with all the subjects participating in the project of (i) the experimental activities, (ii) the dissemination of results, (iii) documentation of the involvement of users/citizens, (iv) the procedures for using the platform, and (v) the operational monitoring/intervention procedures. • **Concept design phase:** in this phase, starting from a pre-evaluation of the opportunities offered by the proposed technology, the impact of using the technology for monitoring procedures is described. • **Prototype design phase:** the prototype platform is designed once the opportunities have been determined, orienting this design toward a better usability of the proposed facilities. • **Innovation design phase:** the operating procedures of the new services are designed after a further reassessment of opportunities, and the experiences of using these services are evaluated in order to optimize the experimental development. • **Commercialization:** the business plan for the commercialization of prototype components is planned
Interaction model between actors	The interaction model between all actors involved in the project are summarized in Table 10.6.5.
Prototyping	More details are provided about prototyping and customization of the project solutions
Testing plan	A formulation of a testing plan proves the real feasibility of the project. Table 10.6.6 provides an example of a testing plan.
Demonstration laboratories	The project is validated during its development by living laboratories and by experimentation with end users
Economical values of results	Analysis is performed for the economic exploitation of the results obtained from the experimentation

Table 10.6 (Continued)

Interaction with final users	The ways of involving end users for the experimentation activity are detailed involving: • The design of products and services. • The development of products and services. • The validation and testing of products and services. • The evaluation of future market perspectives for products and services. • The monitoring and possible revision of the project activities. • The provision of organizational assistance and logistical support services by public bodies or subjects of the socio-economic systems
Methodologies and tools	Methods and tools are described to keep the motivation and interest of end users during and after the experimentations, by focusing the attention on: • Carrying out focus groups in the various stages of the experiment. • Creation of permanent panels, forums of citizens, and consumers. • Organization of periodic workshops open to the general public
Experimental aspects	The ways in which this objective is to be achieved are described, highlighting the type of end user subjects (public body, socio-economic association, technological district, or production district), and the achievable target (number of citizens, consumers, and families)
Evaluation of achievable results	The achievable results are evaluated by: • Describing the added value of the experimental approach. • Calculating the costs of the living laboratory approach (laboratories performing with public participation). • Demonstrating the potential for reuse of the results at the end of the experimentation activities
Availability of infrastructure and technologies and experience of candidate companies	The following are indicated: • Availability of laboratories, expertise, and know-how. • The training activities. • Elements of coherence between the technological field of the project and the experience of the proposing companies
Management of project activities	The management of the project activities includes methods of involvement in the experimentation of all the subjects
Additional services	Additional services include the verification of industrial protection of project results, and the support for the licensing of patents and other contractual forms. It is described if: • The project plans produce results for only the public domain. • The project provides an activity to verify the conditions and terms of protection of the results. • The project provides the reuse of results of previous research activities, in the form of patents and other forms of property rights. Moreover, additional services also concern the preparation of communication actions for the developed technologies and applications
Qualifying activities	This section includes the expression of interest in being part of national and international networks, and the enhancement of digital innovation processes for: • The reuse of available solutions already developed and used by public administrations (open source) in the smart city and community line. • The development and use of solutions or services that enhance the use of public documents and data (open data) in the knowledge community line. • The development of services and applications that exploit broadband and ultra-broadband digital infrastructures in the business community line

(Continued)

Table 10.6 (Continued)

Relevance of the methods and solution prospects for experimentation	In this section it is indicated if: • The project provides the provision of the prototype experimentation to implement services for the final user. • The project involves the use of quantitative and qualitative methods to estimate end user satisfaction. • The project involves the collection of broad-spectrum information about usability, functionality, possible specializations and the level of social acceptance of new products and services. • The project involves the use of social facilitation by means of social community tools commonly used by mobile apps. • The project takes into account the end user feedback for the optimization of the platform
Expected results	The expected results are indicated as: • Results, deliverables, and milestones concerning the planning phase, the design phase, the prototyping phase, the innovation design phase, and the marketing phase. • The Gantt diagram
Showing results (planning)	The modality is planned to show the results
Business model	The business plan is proposed by analyzing costs and revenues
Results indicators	Some result indicators are described by defining the initial and final values to be achieved at the end of the project. The indicators are useful to prove the correct implementation of the proposed solutions. Some indicators are: • Software technology development index. • Hardware technology development index. • Efficiency index of the results achieved.

Table 10.6.1 Summary of the main partner information.

	Small/medium/large enterprise	Company name	Registered office (city)	Headquarters (city)
Lead company	SME	Name 1	City 1	City 1
Partner 1	SME	Name 2	City 2	City 4
Partner 2	SME	Name 3	City 3	City 5
Partner *n*	Research unit	Name 4	City 5	City 6

Table 10.6.2 Contact information of the partners involved in the project.

Company	Contact person	Phone	Email
//	//	//	//

Table 10.6.3 Project financial prospect.

A	A1	A2	A3	B	C1	C2	C3	C4	C5	D	E	T
1												
...	...	...	...	...	...	...	...	...	...	...	...	...
Total												

A, company name; A1, project management personnel (≤7% of the total eligible costs per single partner); A2, employee technician; A3, non-employed staff; B, instrumentation and equipment; C1, purchase of licenses and/or software development (≤25% of the total project); C2, development and registration of patents; C3, specialist consultancy (includes training costs ≤5% of the total project); C4, contracted services acquired by end users (≥5 and ≤25% of the total project); C5, contracted services acquired by universities and/or public and private research centers (≥5 and ≤25% of the total project); D, other costs; E, overheads; and T, total costs.

Table 10.6.4 Relationships between project goals and associated technology.

Goal	Associated technology
Prediction of hydrogeological risks for agriculture and citizens	• DSS prediction-based algorithms. • Weather application programming interface (historical and forecast data)
Leak detection in real time (small and large volumes)	• Thermal imaging cameras from cars and UAV. • Radargrams from cars and UAV. • Alerting system by mobile APP and portal
Water quality (possible risk of contamination)	Laboratory facilities related to the biochemical analysis of water
Protection and improvement of agricultural production	Use of information exchange forums for risk prevention (contamination, disease, drought, innovative irrigation solutions, etc.)
Optimization of intervention planning	DSS for drafting the risk map (daily update)

DSS, decision support system; UAV, unmanned aerial vehicle.

Table 10.6.5 Interactions between actors involved in the project.

User group	Design functionality
Partners/suppliers/ consultants	Technical support (platform development) and technology transfer in: • Use of the platform. • Data entry procedures. • Procedure for reading water analysis data. • Unmanned aerial vehicle use. • Thermography use (standards, specific training courses for leak detection). • Control of the portal functions (creation of open data databases, creation of forum interfaces, registration of reports, registration of maps, etc.)
Farmers/citizens/ inspection staff	• Farmers: reporting water leaks or risks of contamination with polluted water or lands. • Citizens: reporting of canal occlusion, repeated and localized flooding, subsurface subsidence, etc. • Inspection personnel (end users): photographic and video reporting of leaks to send to the control room connected in real time with the monitoring platform

(Continued)

Table 10.6.5 (Continued)

User group	Design functionality
End users	Real-time monitoring of the territory and planning of interventions/ warnings through appropriate communication channels (established in the co-design phase)
Researchers	Researchers are important for scientific validation of the used technology, and for the interpretation of the thermographic images, and water quality monitoring data. They contribute to the dissemination of results in scientific journals and for the formulation of economic indicators updating the business plan

Table 10.6.6 Example of a testing plan.

Test type	Users and methods
Thermographic survey test	Thermographic operators using a car set up with external fixing brackets for the thermal imaging camera and a certified drone. An appropriate mobile APP is implemented for these operators thus facilitating the thermographic test
Citizen reporting test	Citizens through mobile APP report flooding and other dangerous situations associated with hydrogeological risk. The reports are available to the Civil Protection and to other subjects (for example the Fire Brigade and other intervention units)
Water quality tests	Investigations are carried out on particular pollutants (such as heavy metals, pesticides, other contaminations, etc.), indicating the quality of the water. The results are shown to the end users
Test alerting	Critical situations are simulated with alerting tests through web and mobile APP in synergy with the Consortium of Reclamation and Civil Protection
Farmer reporting	Farmers send their reports so that end users and other entities can verify the quality of the water (for example in case of pollution of irrigation water)
Portal test	Citizens are able to access the portal to view the uploaded open data contents, to be informed about events. They also participate in forums and in dedicated areas by login and password

APP, application.

Table 10.7 reports some schemes concerning the expected classified results:

Table 10.7 Classified results.

A	Analysis and understanding of the end user also through specific co-design phases
B	Definition of the interaction model between the different actors involved
C	Prototyping and customization of solutions
D	Testing and experimentation of new technologies in real applications meeting the actual needs of the end user
E	Demonstration and presentation in public of the developed prototype solutions
F	Analysis for the economic exploitation of the results obtained from the experimentation.

The work package layout is summarized in Table 10.8.

Table 10.8 Outputs and deliverables related to each work package.

Work package	Deliverables and outputs	Percentage of deliverables to be made public	Membership activity
WP1: Planning of water networks to be inspected of the specific irrigation network	Technical documentation of the water networks that will be analyzed in the first experimentations	100	A, B, C, D, E, F
WP2: Specific design and implementation of the platform	Documentation of the platform's framework and of data management (database management, creation of workspaces, forums, portal sections, etc.). Release of the first version of the platform prototype (after the first 6 months). Release of the optimized version of the platform (after 18 project months)	80	A, B, C, D
WP3: Mobile APP design	Technical documentation of the design of the mobile APP, and related technical data sheets of thermographic devices. Release of the first version of the mobile APP (after the first 6 months). Release of the optimized versions of the mobile APP (after 18 planning months)	80	A, B, C, D, E
WP4: Implementation of alarm devices for end users	Release of acoustic and visual devices connected to the alarm system.	100	A, B, C, D, E
WP5: Water quality	Release of technical documentation, procedures, and analysis of the waters analyzed in the first experimental phase	100	A, B, C, D, E, F
WP6: Economic indicators	Release of documentation of the analysis of economic indicators related to the project outputs. Economic analysis of the impact of the used technologies	100	F
WP7: Scientific enhancement and validation	Scientific reports and publications related to the best project results and innovations. The papers will be inserted in the open data system of the platform	100	A, B, C, D
WP8: Dissemination of results	Release of documentation to be distributed on all communication and dissemination channels	100	A, B, C, D, E, F

APP, application.

A scheme of the milestone plan is shown in Table 10.9.

Table 10.9 Milestone plan.

Milestone	Estimated time	Work package involved
1	At the end of the 6th project month	WP2, WP3, WP4, WP5, WP8
2	At the end of the 12th project month	WP1, WP2, WP3, WP5, WP6, WP7, WP8
3	At the end of the 18th planning month	WP1, WP2, WP5, WP6, WP7, WP8

10.3.3 Setting a Research Project in Mechatronics: Production of a Diagnostic Machine by Means of Industry 5.0 Facilities

The production of a diagnostic machine involves different technologies and sensors to be implemented on board. The production line of this machine must be designed and realized so that its parts can be changed quickly following the new technologies. In this subsection, a template to follow to formulate a research project involving sensors and mechatronics for the realization of an inspection machine prototype for infrastructure diagnostics (named *prototype*), and for the related production line (named *production line*), is discussed. Table 10.10 lists the main specifications to structure a research project.

Table 10.10 Research and development project form.

Research and development form	
Highlights about the project	Title and type of intervention (the envisaged types of interventions and the relative percentages of commitment in industrial research and in experimental development are indicated). Specifically, the project includes two main objectives: • Research and development of an innovative diagnostic machine for infrastructures or buildings inspections, and oriented toward predictive maintenance (*prototype*). • Setting up of a new production site (*production line*) of the prototype, constructed by Industry 5.0 facilities. The model to follow is to add different sensor technologies of the diagnostic machine. All data are transmitted to a control room monitoring the inspection acquisition. All data are transmitted in the cloud to the control room and are stored into a big data system optimizing the training dataset for AI modeling and calculation. The control room must be physically located in the industry, and interconnected to the AI engine of the control room having the following features: • Reading of sensors and images acquired by the prototype. • Simultaneous management of different prototypes (capillary network of several inspection sites located in the territory). • Storing data into the big data system (storage of historical information useful to structure the training dataset of the AI engine).

Table 10.10 (Continued)

Research and development form	
	• Predictive maintenance. • Automatic alerting about worker safety. • Other different analyses. In order to enhance the innovative aspects about the prototype, a list is presented of all the integrated sensors describing their functionalities. Concerning the production line of the inspection machine, some advanced facilities are: • Digital sensor monitoring production (data digitization). • Reverse engineering tools. • Workstation with computer-aided design and computer-aided engineering tools to use for rapid prototyping. • Collaborative robots controlled and managed by AI facilities. • Augmented reality technology supporting laboratory tests and the inspection procedures. • AI for predictive maintenance and for high quality standards (product defect and production machine failure prediction). • Big data system and big data analytics tools. • Simulation process tools. • Additive manufacturing tools for rapid prototyping (3D printing, laser, etc.) with the possibility to process polymeric and metallic materials. • Automated testing machine according to test trend and AI data processing. • Finite element method simulations for rapid prototyping. Particular attention is given to the description of assisted and automated tests of each part of the prototype. The tests can be generic and related to: • mechanical (stress and strain test) • hydraulic • pneumatic • electric • thermal • electronics • mechatronics • other Each production machine is controlled by a programmable logic controller system. The production plant is therefore oriented toward the possibility to (objective of the production line): • Automate test processes. • Automate component re-engineering processes. • Automate all production processes. • Automate the control processes by sensors, and implementation by mechatronics and feedback controls). • Automate predictive maintenance processes.

(Continued)

Table 10.10 (Continued)

Research and development form	
	• Automate the data analysis, and data processing processes oriented toward the pre-industrialization of the prototype fabrication. • Automate data transfer using appropriate protocols (sensor protocols, blockchain, etc.). • Implement an innovative information technology platform and new and efficient information. • Automate strategic operational lines from predictive risk analysis. The use of technologies useful for the processing of both polymers and metals is also envisaged. The polymers have the important purpose of generating aesthetic prototypes, are quick to make, able to provide preliminary information on overall dimensions, with assembly tolerances, constraints, and degrees of freedom. By reproducing a pre-assembled model, it is possible to define the overall functionality of the connected components (functional reproduction of the assembly). The metal components are inherent in the production of the semi-finished product to be assembled in the appropriate area of the industry
Areas dedicated for the production	A layout is provided indicating all the areas of the industries dedicated to all phases of the production (warehouse, assembly, test, design, etc.), and also a scheme about the control and the actuation of each machine in the production line
Aspects related to environmental impact	The description of aspects correlated to the environmental impact are important to acquire a high evaluation score of the proposed project. The topics to discuss are: • Noise vibration harshness development processes. • Laboratory for noise control (useful for the analysis of acoustic impacts). • Energy aspects of the production line.
Aspects related to environmental impact	Furthermore, it is better to use, again for the assessment of the environmental impact, a setup measuring eco-sustainability of the prototype. Particular attention is focused on innovative energy, the pretreatment of waste and its disposal, and on pollution. Specifically, new features are referred to: • Self-powering of motors and sensors by renewable energy sources with integration of innovative accumulators, and innovative materials capable of producing energy (piezoelectric, thermoelectric, etc.). • Pretreatment of waste materials. • Classification of the waste materials, and, on the basis of their classification, planning of the disposal activities suitable for the degree of danger (for example for special waste). • Reduce waste and replacement components by applying predictive maintenance. • The energy production optimization by using a decision support system reading energy consumption and suggesting the best production programs. • The pollution decrease
Description of the final project goal	Description of the final goal refers to: • The product, the process, and the service to be developed.

Table 10.10 (Continued)

Research and development form	
	• The features and performances to be realized. • The description of the operating modes. • The qualitative and quantitative specifications to obtain. • Values of the main operating parameters. Examples of quantitative specifications are: • Reduction in the quantity of the adopted material used for the production of the new prototype. • Reduction in personnel employed in inspection operations. • Reuse rate of the used elements of the prototype. • Reduction of CO_2 emissions or of pollution in general. • Reduction of energy consumption. • Reduction of waste material of the production process. Other important parameters describing the project goals are the values of the main production line processing parameters. In order to quantify the progress toward the industrialization process, the following parameters can be estimated (the main operating parameters to estimate for the production of a single prototype provided in a logical process sequence): • Time spent and human resource to produce each prototype. • Machine time and human resources for testing of a single prototype. • Time spent for the pre-assembly of each prototype. • Machine time and human resource time spent on the pre-assembled test. • Time spent on the assembly and interconnection of the pre-assembled prototype. • Time spent on the assembly of the pre-assembled prototype. • Time spent on testing the entire prototype. • Time of use of the production machine. • Re-engineering times of the prototype due to wear analysis. • Reformulation of the robustness, durability and reliability parameters through the application of predictive maintenance
Risk analysis	The possible main technical–scientific problems to solve are indicated, by structuring a table of criticalities indicating both the technical and the scientific risks. An approach to follow is the risk model provided in [2]
List and costs of research units	The research units are listed indicating possible synergies and their know-how referring to some publications
Project responsible	One project responsible is indicated for technical communications and contacts
Main work packages, activities, and milestones	Implementation of objectives and activities about:
	• The industrial research and experimental development necessary for the achievement of each goal. • Any knowledge, modules, component elements, and results already available in the company or commercially available. The activities are described in work packages structured in tasks. Each work package is structured as follows:

(Continued)

Table 10.10 (Continued)

Research and development form	
	• Title. • Short description (abstract). • Implementing partner. • Typology of activities structured in tasks. • Brief description of each task. • Place of performance
Project timing	This section represents the overall program with a linear time diagram, highlighting the expected dates of completion of the individual implementation objectives, and the investment objective. The Gantt diagram completes the framework description
Reference scenario, expected benefits	The information necessary to evaluate the reference framework (sectorial, territorial, technological, and of the company) is provided. This section contains: • International market analysis. • Potential buyers of the product. • National and local market analysis. • Information about competitors. • Indications about production activities improvements. • Main technological innovations • Relationships between production activities and research activities. • Information about competitiveness recovery of the proposed subject. • The perspectives for using the results in terms of industrial repercussions supporting conditions for competitive development. • Perspectives about the increase in employment. • Indication of the profitability and the degree of profitability. • Costs. The costs are classified into industry research costs and development costs. The costs include: • Human resources (highly qualified personnel for research organization, training, research, patenting, and testing). • Instrumentation and equipment used for the research project (machines in the production line, equipment, and computer sofware). • Costs of contractual research, technical skills, and patents acquired or licensed from external sources, as well as the costs of consulting services and equivalent services used exclusively for the purposes of the research activity (patents and other industrial property rights in research and development). • Overheads directly attributable to the research project (innovation consulting and support services, services for process and organization innovation, technical feasibility studies, patent updating)
News and technical–economic–scientific considerations	Discussion about the degree of interconnection with the research activities envisaged in the project, by providing elements to evaluate the functional connection of the investments with the research activities of the project. Moreover, reasons are provided relating to the choice of the initiative and its location, technical and organizational links with industrial plants, and elements of coherence between the technological field of specialization of the proposal and the scientific production of the working group

Table 10.10 (Continued)

Research and development form	
Eligible costs	In this section the commitment costs are indicated for months/man. The costs are divided among the individual work package activities. Moreover, elements are described that distinguish the adequacy and complementarity of the work group for the implementation of the activities (organizational model, quantity and quality of the resources used, research infrastructures used, etc.). An example is: • **Adequacy:** the staff members employed have the qualifications, skills, and professional experience to carry out the project; in this case, expertise applied to research and development in the field of Industry 4.0. • **Complementarity:** the research and development team is made up of highly specialized profiles, whose multidisciplinary skills are useful for the technological and design issues related to the construction of the work. • **Organizational model:** this includes the collaboration between senior researchers, computer scientists, and developers. • **Quality and quantity of used resources:** the resources are administrative, scientific, and technical (skilled workers); other external resources can be adopted for project development. • **Used research infrastructures:** the infrastructures are internal and external. The internal ones concern the laboratories for experimental and prototype tests within the production line. The external infrastructures are research organizations addressing the proponent about the implementation of the whole project
Verification of the research project results	The research objectives are indicated, the number and type of prototypes and pilot plants to achieve. Moreover, it is explained how the outcome of the entire project is verifiable (to be based on purely technical criteria by indicating the tests to be performed and expected quantitative results)
Technical–scientific interest	Novelties and originalities are described (description of current technologies and comparable solutions used by competitors both nationally and internationally). Other important aspects to discuss are: • The usefulness of acquirable knowledge for product, process, and service innovations that increase competitiveness and favor the development of the applicant. • Transferability of the proposal (description of the possibility of effective realization and industrial exploitation of the results and their dissemination)
Industrial validity of the project	In this section, the following are described: • Strategic coherence and the project management procedures (consistency with the strategic objectives of the company, interaction of the structures involved in the project, selection criteria and project monitoring). • Technological competitiveness (current and technological characteristics of the offer, predictable changes in demand induced by the technology trend, elements useful to analyze the project validity). • Economic effects of the expected results (market size, positions of the applicant in the market, expected revenues and costs about industrialization investments)
Requirements for granting additional facilities	It is discussed whether the results of the project are widely disseminated through technical or scientific conferences or published in technical and scientific journals, or inserted in freely accessible databases (in which the research data are consulted by everyone), or disseminated through free software or open source. The public methods are extensively described for disseminating the expected results both in the regional and extra-regional contexts, attaching the plan for disseminating the expected results

(Continued)

Table 10.10 (Continued)

Research and development form	
Specific objectives of the intervention	The objectives to achieve through intervention implementation are indicated
Coherence of the program with the company structure	The motivations and consistency of the planned interventions matching with the company structure and strategy are described

Concerning the estimation of technical risk, the model proposed in [2] takes into account the risk factor (R) of the project, estimated by:

$$R = P \cdot E \cdot V \tag{10.1}$$

where P is the probability of occurrence of a specific event, E is the event exposure, and V is the vulnerability of the analyzed element indicating the security level for damage.

Table 10.11 shows a summary of risks indicating qualitative and quantitative aspects, the impact as level of risk, and the solutions envisaged for that type of risk. The risk level is defined by the following scale: 1 (no risk), 2 (medium--low risk), 3 (medium risk), 4 (medium--high risk), and 5 (high risk).

Table 10.11 Model risk example.

Type of risk (risk parameters)	Qualitative	Quantitative	Risk level	Solutions (exit strategies)
1. Correct design of production facilities	Yes	f(P, E: P = probability that there is an interruption in the interconnection with the company information system; E = exposure correlated to the geolocation of the inspection site)	2	Dedicated line for data transfer in case of malfunctions (provisional auxiliary line)
2. Correct mechanical assembly of prototype components	Yes	f(P, E, V: P = probability of human error and automated pre-assembly of the prototype; E = correct assembly to avoid breakage and lack of functionality of the piece subjected to mechanical or thermal stress; V = correct assembly)	2	Absolute simplification of the design of the individual component. Optimization of the production line for the abatement of P. As a consequence of the abatement of P, the values of E and V will be lower
3. Creation of a functional and robust prototype demonstrator	Yes	f(P, E, V: P = probability of breakage of the various components during the inspection activity; E = repeatability of inspection events according to the lifecycle of the machine; V = function of the surface irregularity of the target to detect)	2	Further breakage indicators are considered on the basis of the defects found in the experimental phase. The flexibility and dynamism of production will ensure that the risk factor is eliminated over time

Table 10.11 (Continued)

Type of risk (risk parameters)	Qualitative	Quantitative	Risk level	Solutions (exit strategies)
4. Prototype machinery management	Yes	f(P, E: P = probability of incorrect use of the inspection machine due to inexperience of unqualified personnel; E = assignment of unqualified personnel to a specific site activity)	3	Use of smartphones substituted for visor technology in order to speed up and facilitate the use of augmented reality technology by inserting interactive contents into smartphones. At the beginning of the use of the new inspection prototype, it will be necessary to train the staff to further reduce the level of risk
5. Special processing in precision mechanics	Yes	f(P, E, V: P = probability of machining tolerance not achieved; E = malfunction of the single precision machine in additive manufacturing; V = function of the materials used and to be defined during the testing)	1	The statistical analysis technique (analysis of variance [3]) will be used to study how much the inputs influence the outputs and therefore optimize them
6. Precision in inspection operations	Yes	f(P, E, V: P = probability that the first prototype does not work correctly in the inspection processes; E = inspection machine not perfectly working because it has not been suitably tested in the laboratory; V = function of the irregularity of the target to be inspected)	3	Use of other inspection sensors until the innovative technology is implemented
7. Effectiveness of each single technological upgrade module	Yes	f(P, E, V: P = probability that there are operating and environmental conditions that do not facilitate the correct operation of the prototype; E = incorrect functioning of a part leads to incorrect functioning of the whole prototype; V = vulnerability is a function of the technological limits)	3	Simulation of physical behavior in order to understand the detection principle of the adopted sensors
8. Reliability of the decision support system of the control room	Yes	f(P, E, V: P = there are not sufficient and well-structured data for data processing; E = inappropriate defect classification, breakage risks and wrong predictive maintenance; V = data security)	2	Initially statistical calculations are considered (for example linear regression) in order to predict possible trends
9. Errors in inspection automated procedures	Yes	f(P, E: P = probability that there is a particularly irregular target to be inspected; E = inspection site difficult to check)	3	Parameter setting before the inspection procedure

Figure 10.1 gives all the parameters grouped together under "risk impact" (grouping of the values of *E* and *V*), where the probability of risk occurrence is defined on the abscissa axis. All values are grouped for the analyzed case, in a range between 1 and 4.

It is noted that an acceptable risk level is maintained at medium–low and low levels, below the matrix diagonal. In any case, the risk analysis is updated during the project development phase to facilitate the industrialization process. In the example of Table 10.11 there are no risk values 4 and 5, i.e. medium–high and very high, thus confirming the palatability of the technological solutions.

Table 10.12 describes some work packages related to the research of an innovative production line fabricating an advanced inspection device.

Table 10.13 shows an example of the description of deliverables.

Table 10.14 shows an example of the project milestone definitions.

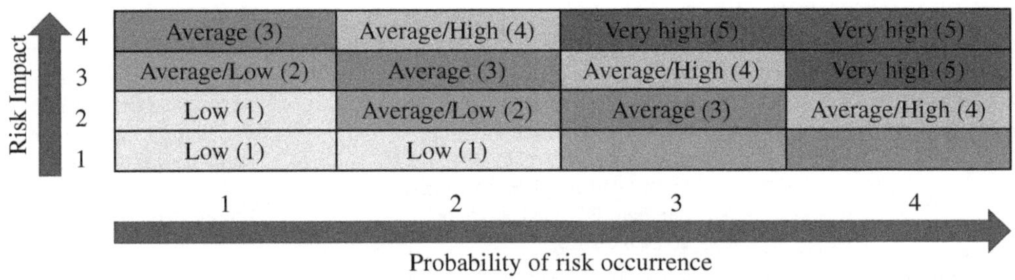

Figure 10.1 Matrix correlating risk impact versus probability of risk occurrence.

Table 10.12 Work package typologies.

Work package	Proposing partner	Goal type	Work package title
WP1	Lead partner	Industry research	State of the art analysis of the technologies to be used for the demonstrator prototype and for production line innovation
WP2	Research units	Industry research	State of the art analysis of technological and scientific upgrade solutions for the prototype demonstrator and reverse engineering techniques
WP3	Research units	Industry research	State of the art analysis for solutions addressed to the pre-industrialization of the patent prototype
WP4	Lead partner	Industry research	Definition of requirements and planning related to the prototype and the production line
WP5	Research units	Industry research	Definition of system requirements and design of the pilot production line
WP6	Research units	Industry research, development	Definition of the variables to be measured
WP7	Lead partner	Development	Development of prototype systems and inspection operating procedures
WP8	Research units	Development	Development of the production line pilot and of the inspection machine technology

Table 10.12 (Continued)

Work package	Proposing partner	Goal type	Work package title
WP9	Research units	Development	Implementation of the experimental production process
WP10	Lead partner	Development	Testing and validation of analysis systems
WP11	Research units	Development	Test and validation of artificial intelligence algorithms, augmented reality technology and all automatic control and actuation systems
WP12	Research units, lead partner	Development	Test and validation of the experimental production process

Table 10.13 Description of deliverables.

Deliverable number	Deliverable title	Work package	Task	Delivery month
D1	Report about the state of the art of materials, hardware and software technologies useful for the realization of the prototype	1	T1.1	M2
D2	Report about the state of the art of automatic technologies suitable for production processes	1	T1.2	M2
D3	Report about the state of the art of technological solutions of the production line	1	T1.3	M4
D4	Report about the state of the art of technological solutions concerning prototype and sensors	1	T1.3	M4
D5	Report about the state of the art of solutions useful for the technological and scientific upgrade of the demonstrator prototype	2	T2.1	M1
D6	Report about the scientific state of the art of optimization methodologies for precision prototypes	2	T2.2	M4
D7	Report about the state of the art of mechatronic solutions aimed at optimizing automatisms and risk management of the production	2	T2.3	M4
D8	Report about the operating procedures for the correct execution of the industrial feasibility study, and the technological transfer of the patent solution	3	T3.1	M1
D9	Report about the constructive aspects of the prototype device which are decisive for the achievement of optimal properties in the inspections	3	T3.2	M3
D10	Criticality analysis document of the prototype utilization	3	T3.3	M5

(Continued)

Table 10.13 (Continued)

Deliverable number	Deliverable title	Work package	Task	Delivery month
D11	Technical document defining the machinery control system	4	T4.1	M8
D12	Analysis document of the functional specifications of the production line and of related processes	4	T4.2	M9
D13	Technical document of the interconnection specifications between the software platform and the prototype device	4	T4.3	M11
D14	Technical document defining the mechanical and the mechatronic systems implemented in the production line, and software solutions for predictive maintenance	5	T5.1	M14
D15	Technical document defining the general management system of the machine	5	T5.2	M17
D16	Requirements analysis document of mechanical components and of finite element method modeling and simulations	5	T5.3	M20
D17	Document indicating the classification of targets to inspect	6	T6.1	M7
D18	Definition document of machine data functional to the implementation of artificial intelligence algorithms	6	T6.2	M23
D19	Document defining reverse engineering practices to be applied to the production line	6	T6.3	M12
D20	Collection of documents and of technical tables related to the facilities in the production line	7	T7.1	M15
D21	Executive tables and layouts of the prototype system components	7	T7.2	M21
D22	Executive design document of the production line solutions implemented in the pilot production line	8	T8.1	M25
D23	Development document of production parts for pre-industrialization	8	T8.2	M26
D24	Document of production line integrated technologies	8	T8.3	M29
D25	Updated process mapping document	9	T9.1	M31
D26	Software code document	9	T9.2	M34
D27	Executive design document of prototype systems with patent claims	9	T9.3	M33
D28	Test document of the parts integrated in the production line	10	T10.1	M29
D29	Verification document of the interconnections of the prototype machine	10	T10.2	M34

Table 10.13 (Continued)

Deliverable number	Deliverable title	Work package	Task	Delivery month
D30	Functional validation document of the implemented software	10	T10.3	M34
D31	Overall system test document	11	T11.1	M34
D32	Production line machine data validation document	11	T11.2	M34
D33	Production system test document	12	T12.1	M34
D34	Data analysis of the prototype system	12	T12.2	M34
D35	Document of the requirements of the new production	12	T12.3	M34

Table 10.14 Milestone definitions.

Milestone number	Milestone title	Work package involved	Delivery month
MI1	State of the art	1, 2, 3	M5
MI2	Requirements and variables to be measured, and first development of the prototype and of the related production line	4, 6, 7	M11
MI3	Measurable system requirements and development (first release)	5, 6, 7	M18
MI4	Reformulation of requirements and development (second release)	5, 6, 7, 8, 10	M27
MI5	Development (third release), testing and testing	8, 9, 10, 11, 12	M34

Table 10.15 lists the modules achievable related to a generic prototype.

Table 10.16 lists the modules achievable related to the innovative production line integrating Industry 4.0 and Industry 5.0 facilities.

Table 10.17 shows an example plan of results dissemination.

Other important information about the project scenario are the indications about the technology readiness levels (TRLs). The TRLs are expressed by the objectives summarized in Table 10.18.

The TRL indicator is fundamental to understand the degree of industrialization of the solutions at the beginning and at the end of the project. There are the following nine main TRL levels:

- **Level 1:** basic principles of fundamental research topics (basic principles observed and discussed).
- **Level 2:** potential applications to be validated by studies (technology concept and formulated application).
- **Level 3:** definition of a proof of concept to be validated and demonstrated analytically and experimentally (description of analytical and experimental critical function, and of characteristic proof of concept).
- **Level 4:** component and/or electronic board validated in an environment laboratory.

Table 10.15 Modules constituting a generic prototype.

Module of prototype	Technological results achievable	Month of achievement
(a1) Smart module	Realization of an integrated system with sensors and control unit for the activation of measurement commands	M8
(a2) Sensor 1	Sensor 1 equipment, calibration, and optimization	M8
(a3) Sensor 2	Sensor 2 equipment, calibration, and optimization	M8, M34
(a4) Sensor 3	Sensor 3 equipment, calibration, and optimization	M8, M34
(a5) Transmission module	Machine data transmission unit (data transmission to the control room)	M8, M34
(a6) Innovative materials	Innovative materials oriented toward innovative prototype (nanocomposite materials, coatings, etc.)	M8, M29
(a7) Innovative components	Claims for patent	M29

Table 10.16 Modules of a generic production line.

Module of production line	Technological results achievable	Month of achievement
(b1) Data transmission network	Digitized data transmission (Industry 4.0 facilities)	M30
(b2) Assembly of production machine line	Preparation of pre-assembly and assembly areas of the production line	M30
(b3) Reverse engineering tools	Reverse engineering tools to be integrated into the whole production line supporting the product quality	M30
(b4) Testing laboratory	Testing laboratory which replicates the real application environments	M30
(b5) Test area for checking components	Execution of specific tests of the realized prototype	M30
(b6) Production area with technological and mechatronic upgrade	Control and implementation of boards for production facilities. The boards will be programmed by means of the artificial intelligence engine selecting dynamically the best production parameters	M30

- **Level 5:** component and/or electronic board validated in a simulated real environment matching with the application scenario.
- **Level 6:** system adequacy validated in a simulated environment (full scenario with all the variables influencing the real application; system, subsystem model and prototype demonstrated in a simulated environment).
- **Level 7:** system adequacy validated in the real space by experimental tests (system prototype demonstration in an operational environment).
- **Level 8:** actual technology completed and qualified through test and demonstrations.
- **Level 9:** actual technology proven through successful use in an operational environment.

Considering the complexity of the research objectives, project partners show a broad range of skills and research experiences. Together they contribute to achieve all objectives. By collaborating

Table 10.17 Activities of results dissemination.

Activity	Action description	Time (month, M)
Diss. 1.1	Definition of the communication plan: definition of objectives, recipients, key messages, and media to be used	M18
Diss. 1.2	Brochures and graphics: after the definition of the project icon, multilingual brochures and documents showing the new product are prepared and processed	M19–M20
Diss. 1.3	Creation and implementation of a web page that describes the scientific results of the project activities	M20–M24
Diss. 1.4	Project results dissemination event: the organization of a project closing conference is an opportunity to present and disseminate the project results to the various subjects directly and indirectly involved in the project, and in general to a wide audience	M24
Diss. 1.5	Results dissemination in scientific journals	M24–M34

Diss., dissemination.

Table 10.18 TRL objectives.

Objectives	TRL at start	TRL at the end of the project
1. Identify, implement and adopt practices	TRL 3: critical function, proof of concept established	TRL 5: testing integrated system (replicated results)
2. Evaluate efficacy of management strategies for mechatronic tool development	TRL 2: applied research and technology concept and formulated application (practical applications formulated)	TRL 5: testing of the integrated system (replicated results)
3. Identify technology interactions	TRL 3: critical function, established proof of concept	TRL 5: testing of the integrated system (replicated results)
4. Identify means of applied innovation for control and actuation	TRL 2: applied research and technology concept and formulated application (practical formulated applications)	TRL 5: testing of the integrated system (replicated results)
5.1. Track potential solutions	TRL 3: critical function, established proof of concept (validation)	TRL 6: detection verified (demonstration in field environment)
5.2. Identify most appropriate approaches and initiatives to counteract inspection risks	TRL 3: critical function, established proof of concept (validation)	TRL 6: detection verified (demonstration in field environment)
6. Produce updated knowledge about electronics and mechatronics	TRL 2: applied research and technology concept of formulated application (practical formulated applications)	TRL 6: verification (demonstration in field environment)
7. Define methods and protocols for field level early detection	TRL 2: applied research and technology concept of formulated application (practical formulated applications)	TRL 6: verification (demonstration in field environment; SWOT analysis to identify and remove risks)
8. Evaluate regulatory principles and best policies	TRL 5: testing of the integrated system (replicated results)	TRL 6: pilot system demonstrator

(Continued)

Table 10.18 (Continued)

Objectives	TRL at start	TRL at the end of the project
9. Platform or network for identification, discussion and implementation of initiatives for policies and regulations, and decision-makers, stakeholders, in a multi-actor approach	TRL 4: testing of preventive measures (development and testing)	TRL 6: verification
10. Technology transfer	TRL 4: testing of preventive measures (development and testing)	TRL 6: verification

SWOT, strengths, weaknesses, opportunities, and threats; TRL, technology readiness level.

Table 10.19 Complementary partners.

	Objectives									
Participants	**1**	**2**	**3**	**4**	**5**	**6**	**7**	**8**	**9**	**10**
1. Name 1	√	√		√		√			√	√
2. Name 2			√		√	√	√		√	√
3. Name 3	√	√	√	√	√	√		√	√	√
4. Name 4	√	√	√		√	√			√	√
5. Name 5	√	√	√	√	√	√			√	√
6. Name 6	√	√	√	√	√		√	√	√	√
7. Name 7	√		√	√	√		√		√	√
8. Name 8		√				√			√	√
9. Name 9		√	√		√	√			√	√
10. Name 10	√	√		√			√	√	√	√
11. Name 11	√			√				√	√	√
12. Name 12	√	√	√	√	√		√	√	√	√
13. Name 13	√	√	√	√	√	√	√	√	√	√
14. Name 14	√		√	√	√			√	√	√
15. Name 15		√					√		√	√
Reference WP	**2**	**3, 4, 6**	**5**	**2**	**5**	**2, 3**	**7**	**8**	**9**	**10, 11, 12**

in a concerted and well-coordinated way, they work to achieve the specific and general objectives in the field of studies, as well as when evaluating mechatronic innovations. To avoid overlaps, the cooperative work is organized by objectives and macro-areas, as shown in Table 10.19.

The identification of costs and benefits are very important to optimize the business plan. The indicators of net present value (NPV) and internal rate of return (IRR) provide important directions about the investment plan [2]. Table 10.20 shows a scheme concerning the evaluation of these parameters.

Table 10.20 NPV and IRR indicators estimated in five years: business plan model concerning Industry 4.0/5.0 tools production.

Costs and benefits analysis and NPV and IRR indicators							
	Year activity	**1**	**2**	**3**	**4**	**5**	**Total values**
Costs	n: Number of Industry 4.0/5.0 tools						
	Tool values						
	Project cost (hypothetical research and development cost)						
	Maintenance costs of tools						
	Total costs						
Benefits	Tools proceeds						
	Tool maintenance proceeds						
	Benefits following traceability services						
	Benefits following predictive maintenance services						
	Total benefits						
	Benefits-Costs						
Discount factor	$(1+i)^n$ where $i = 1.06$						
	NPV = (Benefits-Costs)/Discount factor						
	IRR						

10.4 Innovation Process Projects: Example of a Smart Wine Factory

The proposal of an innovation industry process is structured in different sections such as:

- Summary sheet of the investment project (project costs, main goals, expected results, expected added value due to the introduction of the new production process).
- Partnership description (core business, reference market, motivations, and possible advantages concerning the new process, partner roles).
- Services supporting production processes based on the new technologies and devices (services aimed at the adoption of technological solutions supporting the development of advanced smart-factory manufacturing programs to be performed through intelligent and eco-sustainable production systems).
- Services supporting the innovation of business management processes (services for the adoption of innovative solutions in the field of business management that make it possible to integrate methodologies capable of promoting effectiveness and functionality in the different levels of industry planning and management).
- Qualitative and quantitative indicators concerning the introduction of the new process by the formulation of Key Performance Indicators (KPIs);
- Business plan and SWOT analysis due to the introduction of the new production process.
- Support for the implementation of management systems for information security (services for the implementation of business solutions that encourage the adoption of protection tools in cybersecurity processes).

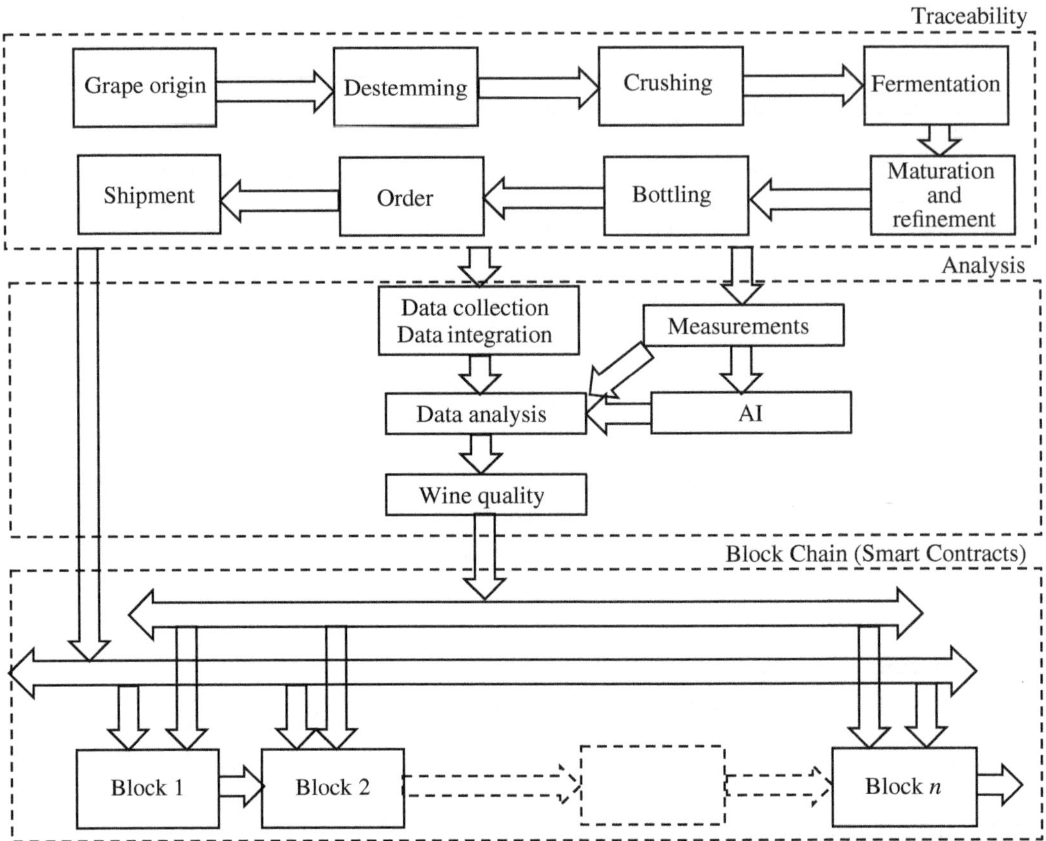

Figure 10.2 Wine production in Industry 4.0 with the upgrade of AI and block chain technology.

- Support services for company certification processes (company certification services aimed at improving the quality, effectiveness and efficiency of a company by also improving its final products and services).
- Project duration (Gantt diagram) and costs.

Figure 10.2 shows an example of an innovation process of a company for wine production where the "As Is" process is based only in the execution of all of the following phases of the wine production:

- grape origin (harvesting of grapes)
- destemming
- crushing
- fermentation
- maturation and refinement
- bottling
- ordering
- shipment

For the specific case study, the traditional process is improved by new technologies enabling digital data collection of production traceability, by AI for quality prediction and supporting data analysis, and by block chain certifying wine quality and enabling smart contracting processes. The project aims to achieve innovative Industry 4.0 objectives for a smart wine factory in which all machines interact with each other to create a network of data collection and data processing. The improvement is related to:

- The redefinition of internal logistic processes of production by using traceability technologies for storing data of all production phases.
- The implementation of a DSS based on AI algorithms predicting product quality parameters and ensuring during the time a high-quality standard.
- The warehouse mapping.
- The implementation of a security network system for production quality certification, based on block chain technology and smart contracting.
- Continuous monitoring of semi-finished wine product.

Concerning the wine quality traceability, the block chain contains basic information such as:

- The geographical areas of production.
- The production type (wine typology and certification).
- The type of soil where the grapes are grown.
- The yield of grapes into wine.
- Wine aging.
- The characteristics of the finished product (acidity, dry extract, minimum alcohol content, organoleptic characteristics).
- Any other qualifications of the wine at the time of marketing.

The upgrade in Industry 5.0 production is mainly achieved by implementing mechatronic boards able to control and automatically refine the production (automatic waste of bad grapes, automatic refinement of the production process, etc.).

10.5 Guideline for Project Management

A project development is followed and managed by a PM. The main work of the PM is to update the detailed Gantt diagram and to check the progress of the task. The management can be performed by filling the following templates tracing the project status:

- Subprojects about the determination of new projects to be executed and followed as "standalone" projects (Table 10.21).
- Information report about the technical and financial risks (Table 10.22).
- Budget check about general information about the budget, expenditures to date and cost variations for each work package (Table 10.23).
- Cost/benefits template referring to the specific work packages (Table 10.24).
- Risk management table describing the full risk description, risk classification and management strategy (Table 10.25).
- Issue management about identification and monitoring of the project issues concerning the unplanned events for each work package involved (Table 10.26).
- Action item inherent tracking and monitor assigned action items as tasks to develop (Table 10.27).
- Deliverable table indicating deliverable status (Table 10.28).
- Milestone table indicating the milestone status (Table 10.29).
- Work breakdown structure (WBS) tracing main task status (Table 10.30).
- WBS structured in weekly activities to monitor, and weekly updated (Table 10.31).
- Contact information about contacts of all the human resources involved in the specific work package (Table 10.32).
- Human roles and responsibilities for each team involved in a project work package (Table 10.33).
- Human resource assignment for different rules and for different project phases such as beginning, planning, execution, and closing (Table 10.34).

Table 10.21 Subproject template.

Subproject			
Main project		**Number**	#
Project manager		**Acronym**	
Scenario			
Advantages and benefits			
Objectives			
Objectives matching with the project	**Matching**	**No matching**	
New deliverables	**Subproject deliverable**	**Due date**	
Main subproject information	**Goal**		
	Deadline		
	Cost		
Further considerations	**Assumptions and constraints**	**Risks**	
		Category	**Risk description**
Success criteria	**1-**		
	2-		
	***n*-**		

Table 10.22 Information report template.

Information report											
Project										**Report number**	**#**
Project manager										**Acronym**	
Template number										**Updated**	
	Deliverables										
	1	**2**	**3**	**4**	**5**	**6**	**7**	**8**	***n***	**Observations**	
Reached up											
Checked											
Cost											
Risk level (1, 2, 3)											
Difficulty (1, 2, 3, 4, 5)											
Management alerts											
Alert type	**Alerting condition**				**Exit strategy**					**Responsible**	

Table 10.22 (Continued)

Information report					
Project				Report number	#
Information about partial results					
Partial result type	**Description**	**Check**	**Date completed**	**Owner**	**Note**
Deliverables	**Title/Description**				
1					
2					
Milestones	**Title/Description**				
1					
2					

Table 10.23 Budget template for work package.

Budget check				
Project name		**Template number**	**#**	
Project manager		**Acronym**		
Work packaged		**Updated**		
Budget status				
Approved budget	**Date of planned expenditures**	**Actual expenditures**	**Possible variations**	
	Planned remaining budget	**Actual remaining budget**	**Variance**	
Additional budget needs				
Budget details				
Internal expenses				
Salaries	**Hourly rate**	**Number of hours**	**Total cost**	**Observations**
Human resource 1				
Human resource *n*				
Other internal expenses			**Total cost**	
Expense 1				
Expense *n*				
Total internal expenses				

(Continued)

Table 10.23 (Continued)

Budget check		
External expenses		
Consulting costs	Total cost	Observations
Consulting 1		
Consulting *n*		
Other expenditures	Total cost	
Expense type 1		
Expense type 2		
Total external expenses		
Total budget		

Table 10.24 Cost/benefits template.

Cost/benefit						
Project				Template number	#	
Project manager				Acronym		
Consultant name				Updated		
Project development and maintenance costs						
Work package	WP1	WP2	WP3	WP4	WP5	Total
Cost of project development						
Human resource salaries						
Other internal expenses						
Consulting costs						
Capital expenditures						
Total project costs						
Cost of support						
Operative costs						
Occasional costs						
Periodic costs						
Total constant support costs						
Total						
Benefits/savings due to the enabled processes						
Type 1						
Type *n*						
Annual savings						
Cumulative savings						
Cumulative costs						

Table 10.24 (Continued)

Cost/benefit			
Project		**Template number**	**#**
Cumulative total net savings			
Future value			
Present value	Description	# Periods	Present value
1-			
n-			
Total			
Future value	Description	# Periods	Future value
1-			
n-			
Total			
Net value (Future-Present)			
Project development cost			
Net value			

Table 10.25 Risk management template.

Risk management														
Project						**Risk number**		**#**						
Project manager						Acronym								
WP involved						Updated								
A	B	C	D	E	F	G	H	I	J	K	L	M	N	O
1														
n														

A: Risk identification number.
B: Risk description.
C: Occurrence probability (scale from 0 to 1).
D: Risk impact (scale from 1 to 5).
E: Risk detectability.
F: Importance (scale from 1 to 3).
G: Risk category or risk class.
H: Triggering event.
I: Risk solution.
J: Exit strategy.
K: Risk management responsible.
L: Risk status.
M: Date of the risk beginning.
N: Date of the possible risk termination.
O: Indications about matrix correlating risk impact versus probability of risk occurrence.

Table 10.26 Issues template.

Issues										
Project				**Issue number**	**#**					
Project manager				**Acronym**						
WP involved				**Updated**						
A	**B**	**C**	**D**	**E**	**F**	**G**	**H**	**I**	**J**	
1										
n										

A: Identification number.
B: Issue description.
C: Impact of the unplanned event.
D: Management strategy.
E: Responsible name.
F: Adopted team.
G: Importance level (scale from 1 to 5).
H: Risk description.
I: Date of the risk beginning.
J: Date of the possible risk termination.

Table 10.27 Action template.

Action								
Project						**Action number**	**#**	
Project manager						**Acronym**		
Work package						**Updated**		
A	**B**	**C**	**D**	**E**	**F**	**G**	**H**	**I**
1								
n								

A: Identification number.
B: Action item.
C: Responsible.
D: Action status.
E: Actors involved.
F: Date of the action beginning.
G: Date of the possible action termination.
H: Action motivations.
I: Observations.

Table 10.28 Deliverable template.

Deliverable							
Project					Deliverable number	#	
Project manager					Acronym		
Involved work packages					Updated		
A	B	C	D	E	F	G	H
1							
n							

A: Identification number (deliverable code).
B: Status.
C: Description.
D: Partners involved.
E: Date of the deliverable beginning.
F: Date of the deliverable termination (as indicated in preliminary Gantt diagram).
G: Possible delay (updating termination date).
H: Constraints and difficulties (risk analysis update).

Table 10.29 Milestone template.

Milestone							
Project					Milestone number	#	
Project manager					Acronym		
Involved work packages					Updated		
A	B	C	D	E	F	G	H
1							
n							

A: Identification number (milestone code).
B: Status.
C: Description.
D: Partners involved.
E: Date of the milestone beginning.
F: Date of the milestone termination.
G: Possible delay (updating termination date).
H: Constraints and difficulties (risk analysis update).

Table 10.30 Work breakdown structure template.

Work breakdown structure										
Project				**WBS number**			**#**			
Project manager				**Acronym**						
Task				**Updated**						
A	**B**	**C**	**D**	**E**	**F**	**G**	**H**	**I**	**J**	**K**
1										
n										

A: Identification number.
B: Task description.
C: Priority level (scale from 1 to 3).
D: Dependencies with other tasks.
E: Task status.
F: Estimated human resource hours.
G: Cost of the whole task.
H: Beginning date.
I: Termination date (as indicated in preliminary Gantt diagram).
J: Possible delay.
K: Observations.

Table 10.31 Work breakdown structure weekly structured template.

A	B	C	D	E	W1	W2	W3	W4	W5	W6	W7	W8	W9	W10	W11	W12
WP1																
Task1.1																
Task1.2																
Task1.*n*																
WP2																
Task2.1																
Task2.2																
Task2.*n*																
WP*n*																
Task*n*.1																
Task*n*.2																
Task*n*.*n*																

A: Indication of task and of related work package.
B: Duration in weeks.
C: Number of members of team 1.
D: Number of members of team 2.
E: Number of members of team *n*.

W, week number.

Table 10.32 Contact information template.

Contact information							
Project					**Project number**	**#**	
Project manager					Project acronym		
Work package					Updated		
A	B	C	D	E	F	G	H
1							
n							

A: Name and surname.
B: Role.
C: Affiliation.
D: Phone.
E: Cell phone.
F: Email.
G: Skype address.
H: Other information.

Table 10.33 Roles and responsibilities template.

Roles and responsibilities								
Project				**Project number**		**#**		
Project manager				Acronym				
Work package				Updated				
A	B	C	D	E	F	G	H	I
1								
n								

A: Name and surname.
B: Membership team.
C: Responsibility level (scale from 1 to 3).
D: Associated deliverable.
E: Other tasks involved.
F: Skills.
G: Key Performance Indicator.
H: Key Performance Indicator prediction (using AI).
I: Comments.

Table 10.34 Resource assignment template.

Resource assignment																	
Project				**Project number**			**#**										
Project manager				Acronym													
Work packages				Updated													
Project roles	End Users	Partners	Sponsor	Consultant	Project Lead	Project Manager	System Analyst	Data Analyst	System Architects	Designers	Trainers	Developers	Technicians	Specialists	Scientific Responsible	Data Manager	Quality Analyst
Project management and requirements																	
Beginning																	
1 Develop preliminary project goals																	
2 Identify business need																	
3 Preliminary budget																	
4 Preliminary return on investment																	
5 Contracts																	
6 Agreements																	
Planning																	
1 Project congruity																	
2 Human resources																	

3	Stakeholder analysis
4	Gantt diagram
5	Project meeting plan
6	Project requirements
7	Business processes definition
8	Deliverables and milestones update
9	Meeting plan
10	Organization structure
11	Training plan
12	Quality check plan
13	Deliverable check plan
14	Risk plan
15	Exit strategies

(Continued)

Table 10.34 (Continued)

Resource assignment			
Project		**Project number**	**#**
16	Cost management plan		
17	Project management plan		
18	Development of technical specifications		
19	Development of system architecture		
20	Upgrade		
21	Finalization of project budget		
22	Finalization of investment prediction		
23	Finalization of the business plan		
Execution			
1	Task development		
2	Meeting plan		

3	SW development
4	HW development
5	New production processes
6	New business processes
7	New SW
8	New HW
9	New technologies
10	Technology transfer
11	WBS matching
12	Agreements updates
13	Risk updating
14	Exit strategy
15	Budget check
16	Deliverable check
17	Milestone check
18	Team change

(Continued)

Table 10.34 (Continued)

Resource assignment			
Project		**Project number**	**#**
19	Results prediction		
Closing			
1	Deliverable acceptance		
2	Evaluation score		
3	Closing contracts		
4	Finalizing budget		
5	Updated business plan		
6	Results check		
7	Tests check		
8	Archive project		
9	Prototype Demonstrator		
10	Publications and patents		
11	Satisfaction check		

HW, hardware; SW, software, WBS, work breakdown structure.

Table 10.35 Human resource time sheet template.

Human resource time sheet													
Project						**Time sheet number**	#						
Project manager						**Acronym**							
Work package						**Updated**							
Hours spent (in months M)													
Names	**M1**	**M2**	**M3**	**M4**	**M5**	**M6**	**M7**	**M8**	**M9**	**M10**	**M11**	**M12**	**Tot.**
Name 1													
Name *n*													
Total													

- Human resource timesheet indicating the time spent by month for each team member to submit effort hours by activity (Table 10.35).
- Work package constraints (Table 10.36).
- Work package decisions (Table 10.37).
- Work package communication plan (Table 10.38).
- Stakeholder analysis identifying stakeholder typologies, their role, and their requirements (Table 10.39).
- Work package opportunities (Table 10.40).
- Work package rule change (Table 10.41).
- deliverable acceptances (Table 10.42);

Table 10.36 Constraints template.

Constraints					
Project			**Constraint number**	#	
Project manager			**Acronym**		
Work package			**Updated**		
A	**B**	**C**	**D**	**E**	**F**
1					
n					

A: Identification number.
B: Main description.
C: Risk level (scale from 1 to 5).
D: Overcoming state.
E: Date of the constraint origin.
F: Comments.

Table 10.37 Decision template.

Decision					
Project			**Decision number**	**#**	
Project manager			**Acronym**		
Work package			**Updated**		
A	**B**	**C**	**D**	**E**	**F**
1					
n					

A: Identification number.
B: Decision description.
C: Human resource involved.
D: Decision date.
E: Responsible name.
F: Comments.

Table 10.38 Communication plan template.

Communication plan							
Project			**Project number** **#**				
Project manager			**Acronym**				
Work package			**Updated**				
A	**B**	**C**	**D**	**E**	**F**	**G**	**H**
1							
n							

A: Identification number.
B: Communication type.
C: Description.
D: Frequency.
E: Responsible name.
F: Participants.
G: Dates.
H: Comments.

Table 10.39 Stakeholder template.

Stakeholder							
Project			Stakeholder number#				
Project Manager			Acronym				
Partner			Updated				
A	B	C	D	E	F	G	H
1							
n							

A: Identification number.
B: Stakeholder description.
C: Business impact (scale from 1 to 5).
D: Associated risk.
E: Constraints.
F: Needs.
G: Responsible name.
H: Comments.

Table 10.40 Opportunities template.

Opportunities						
Project		Opportunity number		#		
Project manager		Acronym				
Work package		Updated				
A	B	C	D	E	F	G
1						
n						

A: Identification number.
B: Origin.
C: Opportunity description.
D: Desired value.
E: Possibility of achievement.
F: Involved partners.
G: Comments.

Table 10.41 Rule change template.

Rule change									
Project					Rule change number	#			
Project manager					Acronym				
Work package					Updated				
A	B	C	D	E	F	G	H	I	J
1									
n									

A: Identification number.
B: Rule change description.
C: Involved partner.
D: Date of decision.
E: Motivations.
F: Agreement.
G: Possible task updating.
H: Possible associated risks.
I: Possible advantages.
J: Comments.

Table 10.42 Deliverable acceptance template.

Deliverable acceptance						
Project				Acceptance number #		
Project manager				Acronym		
Deliverable number				Updated		
A	B	C	D	E	F	G
1						
n						

A: Identification number.
B: Deliverable description.
C: Evaluator.
D: Evaluation score.
E: Evaluation date.
F: Possible integrations (motivations).
G: Comments.

- Pert analysis on WBS tasks [4] (Table 10.43).
- Delphi for opinion convergence evaluated for a specific task [5] (Table 10.44).
- Initial project evaluation (Table 10.45).
- Human resource loading identifying the work effort required, the duration, and other aspects about persons assigned to the project (Table 10.46).
- Project quality metrics identifying critical success criteria, related quality metrics, and their priorities (Table 10.47).

Table 10.43 Pert analysis template.

Pert analysis													
Project							**Analysis number**	**#**					
Project Manager							**Acronym**						
Work package							**Updated**						
A	**B**	**C**	**D**	**E**	**F**	**G**	**H**	**I**	**J**	**K**	**L**	**M**	**N**
1													
n													

A: Identification number.
B: Task.
C: Optimistic duration.
D: Probable duration.
E: Pessimistic duration.
F: Average duration.
G: Optimistic cost.
H: Probable cost.
I: Pessimistic cost.
J: Average cost.
K: Standard deviation (duration).
L: Standard deviation (cost).
M: Motivations of pessimistic estimations.
N: Comments.

Table 10.44 Delphi template.

Delphi										
Project						**Project number**		**#**		
Project manager						**Acronym**				
Task						**Updated**				
Team members evaluation	**Name 1**	**Name 2**	**Name 3**	**Name 4**	**Name 5**	**Name 6**	**Name 7**	**Name *n***	**Tot.**	**Ave.**
Most likely										
Minimum										
Maximum										

Table 10.45 Initial project evaluation template.

Initial project evaluation				
Project			**Document number**	**#**
Project manager			**Acronym**	
Evaluation period			**Updated**	
Goal estimate				
Technology systems impacted				
		1–2	**3–5**	**6 or more**
Number of business requirements	**Less than 50**			
	51–100			
	100 or more			
Cost estimate				
Number of resources involved				
		Less than 10	**11–50**	**51 or more**
Number of business requirements	**Less than 50**			
	51–100			
	100 or more			
Cost estimate				
Number of resources involved				
		Less than 10	**11–50**	**51 or more**
Benefit	**Low**			
	Medium			
	High			

Table 10.46 Human resource loading template.

Human resource loading					
Project			**Document number**	**#**	
Project manager			**Acronym**		
Work package			**Updated**		
Effort hours committed					
ID	**Names**	**Estimate of total effort**	**Activity duration**	**Skill**	**Total**
1					
n					
	Total				

Table 10.47 Project quality metrics template.

Project quality metrics					
Project		**Document number**	**#**		
Project manager		**Acronym**			
Work package		**Updated**			
ID	**Critical success criteria**	**Potential quality metric**	**Priority**	**Metric target**	**Action plan**

References

1 Galiano, A., Massaro, A., Boussahel, B. et al. (2016). Improvements in haematology for home health assistance and monitoring by a web based communication system. In: *Proceedings of IEEE International Symposium on Medical Measurements and Applications (MeMeA),* Benevento, Italy (15–18 May 2016). Piscataway, NJ: IEEE.

2 Massaro, A. and Galiano, A. (2020). Re-engineering process in a food factory: an overview of technologies and approaches for the design of pasta production processes. *Production & Manufacturing Research* 8 (1): 80–100.

3 Montgomery, D.C. (2019). Introduction to Statistical Quality Control, 8e. Wiley.

4 Agyei, W. (2015). Project planning and scheduling using PERT and CPM techniques with linear programming: case study. *International Journal of Scientific and Technology Research* 4 (8): 222–227.

5 Green, R.A. (2014). The Delphi technique in educational research. *SAGE Open* 4 (2): 1–8.

Abbreviations and Acronyms

1D	one-dimensional
2D	two-dimensional
3D	three-dimensional
4M	Manpower, Material, Methodology, Machine
ABC	agent-based computing
AC	alternating current
ACM	Access Control Module
ADC	analog-to-digital converter
ADU	activity diagram UML
AES	Advanced Encryption Standard
AFM	atomic force microscopy
AGV	Automated Guided Vehicle
AHC	agglomerative hierarchical clustering
AI	artificial intelligence
AM	additive manufacturing
AMQP	Advanced Message Queuing Protocol
ANN	artificial neural network
AOV	angle of view
API	application programming interface
APP	application
AR	augmented reality
ARAM	augmented reality aided manufacturing
ARM	Advanced RISC Machine
AUC	area under the curve
BB	Brendel–Bormann
BI	business intelligence
BL	Bluetooth
BLE	Bluetooth Low Energy
BLP	Bell–LaPadula
BPM	business process modeling
BPMN	Business Process Model and Notation
BPR	Business Process Re-engineering
CAD	computer-aided design
CAM	computer-aided manufacturing

Electronics in Advanced Research Industries: Industry 4.0 to Industry 5.0 Advances, First Edition. Alessandro Massaro.

CART	Classification and Regression Tree
CBC	cipher block chaining
CBIR	content-based image retrieval
CCD	charge-coupled device
CD	clearance distance
CFB	cipher feedback
CHAID	chi-square automatic interaction detection
CIM	computer integrated manufacturing
CMM	coordinate measuring machine
CMOS	complementary metal oxide semiconductor
CNC	computer numerical control
CN-LCS	convolutional neural-based learning classifier system
CNN	convolutional neural network
CoAP	Constrained Application Protocol
CoBot	collaborative robot
CPU	central processing unit
CRC	cyclic redundancy check
CRM	customer relationship management
CSI	current-source inverter
CT	computed tomography
CUDA	Compute Unified Device Architecture
CVD	chemical vapor deposition
DAC	digital-to-analog converter
D-AC	discretionary access control
DB	database
DBMS	database management system
DBSCAN	density-based spatial clustering of applications with noise
DC	direct current
DDA	dynamic decay adjustment
DDoS	Distributed Denial of Service
DEA-RTA	Dynamic Encryption Algorithm for Real-Time Applications
DF	data fusion
DFA	Design for Assembly
DFD	Design for Disassembly
DFE	Design for Environment
DFM	Design for Manufacturing
DFT	discrete Fourier transform
DFX	Design for X
Diss.	Dissemination
DL	deep learning
DLC	diamond-like carbon
DLL	Dynamic-Link Library
DMC	data management controller
DoS	Denial of Service
DSS	decision support system
DTP	data transfer process
DT	decision tree

EAP	electroactive polymer
EBL	electron beam lithography
ECB	Electronic Code Book
EHSV	electrohydraulic servo valve
ELAN	Ethernet local area network
EM	expectation maximization
EMF	electromagnetic field
EMFM	electric-magnetic force microscopy
EMS	energy management system
EPC	Event-driven process chain
ERP	enterprise resource planning
ESB	enterprise service bus
ET	evapotranspiration
EtherCAT	Ethernet for Control Automation Technology
FBD	Function Block Diagram
FDM	fused deposition modeling
FDTD	finite-difference time-domain
FEG	field emission gun
FEM	finite element method
FFT	fast Fourier transform
FLOPS	floating point operations per second
FOV	field of view
FRET	fluorescence resonance energy transfer
FRFo	fast random forest
FTIR	Fourier transform infrared
GBTA	gradient boosted trees algorithm
GDS	global distribution system
GigE	Gigabit Ethernet
GIS	geographic information system
GMM	Gaussian mixture model
GN	gold nanoparticle
GPGPU	General-Purpose Graphics Processing Unit
GPR	Ground Penetrating Radar
GPRS	General Packet Radio Service
GPS	Global Positioning System
GPU	graphics processing unit
GTO	gate turn off
GUI	graphical user interface
HART	Highway Addressable Remote Transducer
HDMI	High-Definition Multimedia Interface
HMI	human–machine interface
HMM	hidden Markov model
HoQ	House of Quality
HPF	hertzian potential formulation
HSB	hue saturation brightness
HTTP	Hypertext Transfer Protocol
HTTPS	Hypertext Transfer Protocol Secure

HW	hardware
IBAC	Identity-Based Access Control
ICM	Inference Control Module
ICoP	inductively coupled plasma
ICP	inference control policy
ID	identification number
ID3	Iterative Dichotomiser 3
IDE	integrated development environment
IDS	intrusion detection system
IGBT	insulated-gate bipolar transistor
IL	Instruction List
I/O	input/output
IoE	Internet of Energy
IoT	Internet of Things
IP	Internet Protocol
IPC	Industrial PC
IPM	interior permanent magnet
IQ	index of quality
IR	infrared
IRR	internal rate of return
IRT	infrared thermography
IT	information technology
JSON	JavaScript Object Notation
KB	knowledge base
KG	knowledge gain
KNIME	Konstanz Information Miner
KNN	k-nearest neighbors
KNX	Konnex
KPI	Key Performance Indicator
KPM	Kelvin probe microscopy
LabVIEW	Laboratory Virtual Instrument Engineering Workbench
LAN	local area network
LCL	lower center line
LCS	learning classifier system
LED	light-emitting diode
LL	ladder logic
LoRa	long range
LPBFAMP	laser powder bed fusion additive manufacturing process
LSTM	long short-term memory
LS-SVM	least squares support vector machine
LVDT	linear variable differential transformer
M2M	machine-to-machine
MAC	mandatory access control
MAE	mean absolute error
MAS	multi-agent system
MC	microprocessor
MEG	multipole expansion of Green's function

MEMS	microelectromechanical system
MES	manufacturing execution system
Mini PC	mini personal computer
ML	machine learning
MLP	multilayered perceptron
MOSFET	metal-oxide-semiconductor field-effect transistor
MPTS	mercaptopropyltrimethoxysilane
MQTT	Message Queuing Telemetry Transport
MR	measuring range
MRI	magnetic resonance imaging
MRP	material requirements planning
MSB	Manufacturing Service Bus
MSC	mean shift clustering
MSE	mean square error
MWPECVD	microwave plasma enhanced chemical vapor deposition
NA	numerical aperture
ND	nanodiamond
NDT	nondestructive testing
NEMS	nanoelectromechanical system
NFC	near field communication
NIR	near infrared
NM	nanocomposite material
NMOS	N-type metal oxide semiconductor
NN	neural network
NoSQL	not only SQL
NP	nanoparticle
NPV	net present value
N-V	nitrogen-vacancy
NVH	noise vibration harshness
OEE	overall equipment effectiveness
OLE	object linking and embedding
OMA	optical multichannel analyzer
OPC-UA	Open Platform Communications' Unified Architecture
Or-BAC	organization-based access control
OS	operating system
PAC	programmable automation controller
PAN	personal area network
PC	photonic crystal
PCA	principal component analysis
PCB	printed circuit board
PCD	polycrystalline diamond
PCR	polymerase chain reaction
PDCA	Plan-Do-Check-Act
PDE	partial differential equation
PDMS	polydimethylsiloxane
PET	positron emission tomography
PID	proportional integral derivative

PLC	programmable logic controller
PLM	product lifecycle management
PM	project manager
P-MInD	Process Mining Integrating Decisions
PMMA	poly(methyl methacrylate)
PMOS	P-type metal oxide semiconductor
PNN	probabilistic neural network
PnP	press-n-peel
POC	point of care
P&P	plug and play
Profinet	process field net
PSD	position sensitive detector
PTH	pin through hole
PVC	polyvinyl chloride
PWM	pulse width modulation
Q	quality factor
QD	quantum dot
QoS	quality of service
RAM	random access memory
R-BAC	role-based access control
R&D	research and development
RDBMS	relational database management system
RE	reverse engineering
REFM	reverse engineering for manufacturing
RESTful	representational state transfer
RF	radio frequency
RFo	random forest
RFID	radio frequency identification
RGB	red, green, and blue
RM	resource manager
RMEMS	ring MEMS
RMS	root mean square
RMSE	root mean square error
RNN	recurrent neural network
ROC	receiver operating characteristic
ROI	region of interest
ROVER	remotely operated video enhanced receiver
RPM	revolutions per minute
Rprop	resilient back propagation
RPS	radians per second
RT	real time
RTCP	RTP Control Protocol
RTOS	real-time operating system
RTP	Real-time Transport Protocol
RTU	remote terminal unit
Rx	receiver
SCADA	supervisory control and data acquisition

SCARA	selective compliance articulated robot arm
SCM	scanning capacitance microscopy
SCR	silicon controlled rectifier
SD	standard deviation
SDB	signature database
SDK	software development kit
SEM	scanning electron microscopy
SFC	Sequential Function Chart
SIG	semantic inference graph
SIM	semantic inference module
Si-V	siliscon-vacancy
SLM	selective laser melting
SM	smart manufacturing
SMA	shape memory alloy
SMD	surface mount device
SMP	symmetric multiprocessing
SMT	surface mount technology
SMvA	subminiature version A
S/N	signal/noise
SNA	scalar network analyzer
SNR	signal-to-noise ratio
SOA	service-oriented architecture
SOAP	Simple Object Access Protocol
SOC	system on chip
SPECT	photon emission computed tomography
SPM	surface-mounted permanent magnet
SQL	Structured Query Language
SSL	secure sockets layer
ST	Structured Text
STRD	simultaneous transverse resonance diffraction
SVD	single value deposition
SVM	support vector machine
SW	software
SWIR	short wave infrared
SWOT	strengths, weaknesses, opportunities, and threats
T-BAC	task-based access control
TCP	Transmission Control Protocol
TE	transverse electric
TEOS	tetraethylorthosilicate
THT	through hole technology
TI	thermal infrared
TIRF	total internal reflection fluorescence
TLS	transport layer security
T-MAC	task-mandatory access control
TPB	threads per block
TRL	technology readiness level
Tx	transmitter

UAV	unmanned aerial vehicle
UCL	upper center line
UDP	User Datagram Protocol
UML	Unified Modeling Language
UR	Universal Robots
UV	ultraviolet
UWB	ulta-wideband
V-BAC	view-based access control
VD	violation detection
VNA	vector network analyzer
VPN	virtual private network
VSI	voltage-source inverter
WAN	wide area network
WBS	work breakdown structure
WEM	wave expansion method
WLAN	wireless local area network
WP	work package
WPAN	wireless personal area network
WSDL	Web Service Definition Language
WWW	World Wide Web
XGBoost	extreme gradient boosting
XML	eXtensible Markup Language
XMPP	Extensible Messaging and Presence Protocol
XPS	X-ray photoelectron spectroscopy
μPL	Micro-photoluminescence

Index

s

t

u

v

w

x

y